Mobile Computing: Multimedia Technology

Mobile Computing: Multimedia Technology

Edited by **Adam Houle**

Willford Press

New York

Published by Willford Press,
118-35 Queens Blvd., Suite 400,
Forest Hills, NY 11375, USA
www.willfordpress.com

Mobile Computing: Multimedia Technology
Edited by Adam Houle

International Standard Book Number: 978-1-68285-175-3 (Hardback)

Printed in China.

Contents

Preface

Mobile computing has rapidly emerged in the last few decades. It has enabled the world to transmit and share data over large distances. This book is a valuable compilation of topics, ranging from the basic to the most complex advancements such as mobile computing techniques, human-computer interaction, intelligent systems and services, multimedia techniques, etc. It includes state-of-the-art researches and developments in the field of mobile computing. Through this book, we attempt to further acquaint the readers about the emerging concepts in this field.

All of the data presented henceforth, was collaborated in the wake of recent advancements in the field. The aim of this book is to present the diversified developments from across the globe in a comprehensible manner. The opinions expressed in each chapter belong solely to the contributing authors. Their interpretations of the topics are the integral part of this book, which I have carefully compiled for a better understanding of the readers.

At the end, I would like to thank all those who dedicated their time and efforts for the successful completion of this book. I also wish to convey my gratitude towards my friends and family who supported me at every step.

Editor

1

Annotation and Visualization in Android: An Application for Education and Real Time Information

Renato Barahona Neri[1], Gustavo Millán García[1], Holman D. Bolibar Barón[2], Rubén González Crespo[1]
[1] *Universidad Pontificia de Salamanca, Madrid, Spain*
[2] *Universidad Católica de Colombia, Bogotá, Colombia*

Abstract — **By using Augmented Reality applications, users can get more information while interacting with real objects. The popularity of the Smartphones and the ubiquity of an Internet connection within modern devices, offer the best combination for these kind of applications, which can pull content from heterogeneous sources. The goal with this work is to show the architecture and a basic implementation of a prototype for an AR application that displays information (opinions) about physical places as comments overlaid to the place left there by other users, but that also encourage in-situ content creation for collaboration. These applications can also be used in order to improve the interaction between students and physical places, getting facts, or associating quizzes to a specific location; tourism guides, promotions of products, just to mention a few.**

Keywords — **Augmented Reality, Android, Interaction, Annotation, Visualization, Smartphones, Education**

I. Introduction

AUGMENTED Reality (AR) applications let the user get more information about a real object by overlaying computer generated content. This is different from a Virtual Reality approach in which both the world, and thus the objects are completely computer generated [2].

AR applications are not new and since the 1960's there have been some investigations like the ones conducted by Ivan Sutherland [27], Jun Rekimoto [22], Hirokazu Kato and Mark Billinghurst [7] which began to lay the basis of what can possibly be achieved with the available hardware [3], [10].

Emmanuel Dubois proposed taxonomy for AR (Fig. 1) in which there can be two types of systems: Task Focus and Nature of Augmentation. The former refers to the object (real or virtual) involved in the user's task, while the latter to the amount of actions can be done with the object (execution or evaluation).

Almost all AR systems available today have a main focus on annotating real objects, which is the main premise upon which applications like Layar [13], and this prototype are built. Users can get more information, but they not modify the object.

Nature of Augmentation / Task Focus	Execution	Evaluation
Virtual Object	Execution is aumented by defining new modalities for the real object.	Realistic fraphics which mimic real world feedback.
Real Object	Enable the user to perform tasks which are impossible without the computer.	Superimposes visual information on real objects.

Fig. 1. Four types of AR Systems [5].

There are also applications like *Expedition Schatzsuche* [26] that displays virtual content over the real world and let the user interact with these objects, these are the basis for AR digital publications, like the one released by the German newspaper Süddeutsche Zeitung [12].

These image overlay has been used in museums like the Louvre [16] or in Wien [26] highlighting the role of education for AR applications. On the other hand, games like *ARQuake* [30] or *Time Warp* [6] highlight the use of the technology not within a room or a building, but in a city in which the user must use special hardware to "see" the augmented information available.

Nowadays both Smartphones and mobile Internet connections are mainstream; these two factors open the door to an improvement in the quality and quantity of the information available to the user depending on his/her location and which he/she is also capable of sharing, instantly, with others.

Modern lifestyle is related to ubiquity, expectation and the will to consume information immediately (Fig. 2); this is the reason all these mobile devices offer more ways to be on-line and let us, via a social web, share, consume and interact with other people.

Based on unknown surroundings, how can a person gather information of the place? How can this content be viewed?

How can the person can interact and/or generate this content?

Since "technology becomes an important means of both constructing and revealing the reality of a contemporary city" there are "substantial potential of media architecture and interactive spaces to extend human perception" [28].

Context awareness let the user pre-fetch some information based on the physical location of a user, which is a great benefit for both outdoor and indoor AR Applications. Thanks to the information gathered from the sensors (WiFi, GPS) information can be presented to the user based on his/her surroundings.

Although there are context aware applications that let the user "discover" or get information of the surroundings, and also studies have been carried to define methods to gather data for an application to provide a service of interest for the user based on context information [1], the information lives within the scope of the application and it doesn't overlap into the real objects.

One of the first attempts at displaying information to a user in which could be seen as a Mobile Augmented Reality (MAR) application in a city is the *Touring Machine* [8] which offered information to the users via an HUD. Most recently we have applications like *Arbela Layers Uncovered* [24], and, *Astrid's Steps*[18], which rely on historical sites to let people undercover the secrets in the concrete places these applications are designed to work.

There have been advances in the way the information can be organized, maintained and fetched [23], how to get better identification of geotagged images [31][11], how to better navigate based between GPS locations [15][9], and also some example applications of how to use geotagged data in order to leverage the education in Architecture [25] or for virtual tourism [21].

Studies for user expectations on AR Systems and geo localized games [19][29][17] have yielded the importance not only in context awareness and useful information for the user, but also a sense of collaboration both between trusted contact and unknown contacts.

It is interesting to note that in most cases, users expressed the opinion for "creating personalized augmented views of the shopping centre, e.g., by making the environment more cheerful with decorations" or "interested in other users' comments, specially the local people's comments while abroad" [19].

Work done by De Michelis [4] and Ramírez [20] denotes the importance in the use of technology and reality augmentation for collaboration; the fact that the people could access the needed content (perhaps also based on a role, which is out of the scope of this prototype) is important for the collaboration (place annotation, touristic/academic/commercial knowledge base maintenance) to be as smooth as possible.

Applications like the ones mentioned, have specific roles for the people who create content and the ones who consume it, marking a clear line between the place in which the content is created and the actual physical place in which it is consumed.

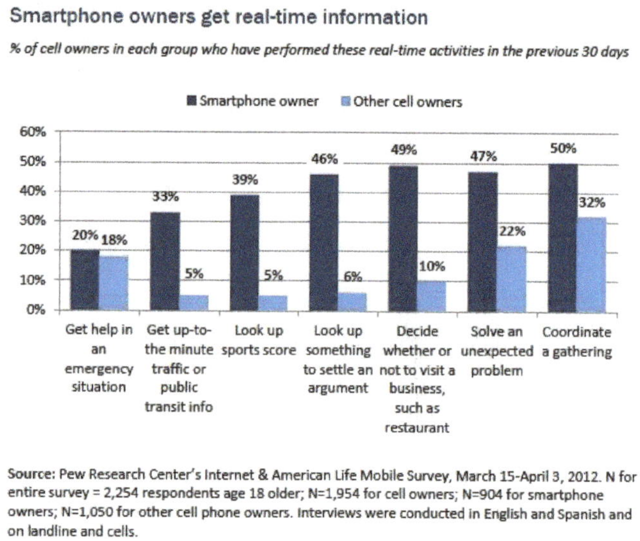

Fig. 2. Real time information in Smartphones [14]

The goal of this work is to demonstrate the basic architecture for an outdoor AR application that encourages the user to annotate in-situ a place based on the GPS coordinates.

The basic annotation for the world will be represented as a yellow square (like a Post-it© note). That way another user can look at these notes and decide on whether the place is worthy of his/her time or not.

In some restaurant is commonplace to let the users mark the walls with comments, however this wouldn't be acceptable in a museum or a store, and that's why AR solves the problem, because all the notes won't be physically attached to a place, but virtually to a location.

These kind of annotations could be viewed both as a mean to gather information, but also as an interactive art installations that "provokes social interactions and their potential to transform existing configurations of public space" [28].

The rest of the article is organized as follows: In Section II we select the development framework. In Section III the architecture for the prototype is described and in Section IV the implementation is showed. Finally Section V has the Conclusions and the future applications that can be based in the prototype

II. Development Framework

There are quite some tools available for AR content; from the SDK for Android, up to the complex applications and ecosystems in which AR Browses live, fetch data and overlay it to the user.

These tool are different and although the concept may seem the same, the way the information is showed to the user, the OS in which the applications can be used and the freedom to create content. For this prototype we evaluated several existing frameworks taking into consideration the following aspects:

Openness. Is the framework open for development without royalties?

Location via GPS. Can the framework use the GPS information to know the location of the user?

Content Generation. Can the user generate his/her own content?

In-situ annotation. Can the user generate the content in the place he intends the content to be shown?

Based in these characteristics we evaluated the following frameworks: *Wikitude*, *TagWhat*, *Metaio/Junaio*, *Layar*, and *KHARMA*. The results can be summarized in the following table (Fig. 3):

Aspect / Framework	Openess	GPS Localization	Freedom of content	In-situ annotation
Wikitude	X	X	X	
TagWhat	X	X		X
Metaio/Junaio	X			
Layar		X		
KHARMA	X	X	X	

Fig. 3. Frameworks

None of the evaluated frameworks offers the whole range of aspects needed to the development of the prototype, and although TagWhat and KHARMA were really close in what they offered, the former is only available in the US and the second is not available for the Android OS.

Since none of the evaluated frameworks could be used the SDK for the Android OS was the development tool chosen for the prototype based on the following:

- Most of the frameworks are closed. Content creations is based on on-line templates and the interaction is limited, because the user doesn't have control over which he/she wishes to annotate.
- Most of the frameworks are capable of more than is needed for the application and, as in a closed system, they are dependent on third party DBs to get all the information.
- Developing a prototype without a framework is an opportunity to delve into the SDK and get to know all the tools available for the user; these knowledge can be used for future projects.
- Getting to know first-hand the way to implement all the requirements for the applications can be translated to solving similar problems in all kinds of applications for the OS.
- This prototype can be the groundwork for a bigger application, which could leverage the information to offer the user touristic information or an education tool for field trips.

III. Architecture

In order to meet the goals of building the prototype of an AR application that let the user "annotate" places and interact with the notes, we must leverage the tools available in the SDK to communicate with the sensors, both location (GPS) and orientation, which are almost in every mobile phone.

The user will be able to, given his/her location, create a place (in case it is not in the DB) and associate a note to the place. These notes will be visible to other users as yellow squares overlaid on the actual physical object. The user can also interact with the notes by reading them.

The proposed Architecture has five components which are grouped within the Model View Controller pattern (Fig. 4).

The Model includes all the relative information of the state of the system in any given moment. Within the Model we can find the *Data Storage* which can be hosted inside or outside the device. Ideally it should be in a server, in order for every user to access it. Here is where the information created by the user is stored.

For the prototype, this component will reside inside the mobile phone; however all the access methods are exposed via an interface, so we are decoupling the implementation, so these module could be implemented in a server without modifying the actual structure.

For the DB we used the SQLite DBMS, although as stated, any RDBMS can be used.

For each of the tables we defined a POJO (Plain Old Java Object) so we can translate the information in the DB into the Objects needed in the application.

The *Location* component has all the elements related to the orientation and location of the device, which are also part of the system's state. When we speak of Location we are talking about the position of the device within Earth's surface. This information is acquired via the GPS satellites. Based on the location, a user can define a new place, which would later be annotated.

As for the Orientation, it refers to the information regarding the way into which the user is pointing the device, in which direction the user is rotating it or holding it. This information is then used to let the notes be displayed as if they were really attached to the place.

Android has built-in interfaces for communication with the sensors. For the GPS we implement both a `LocationListener` and a `GpsStatus`.

Within the Android API there are some guidelines on the usage of these interfaces and also best practices, which let let the developers make the most efficient use of the both the date of the sensors, and the resource systems (i.e. the battery).

For the prototype, since the application should be location aware all the time, we are asking for a change in the position of the phone every moment. This is useful also if, in a future upgrade, a user wishes to record a walking route highlighting and annotating places as he/she walks.

For the orientation sensors the interface `SensorEventListener` let the user interact with the sensors, which in turn, use the accelerometers and the magnetic field information to know how the user is moving/holding the device and, translate this information to the coordinate system of the device.

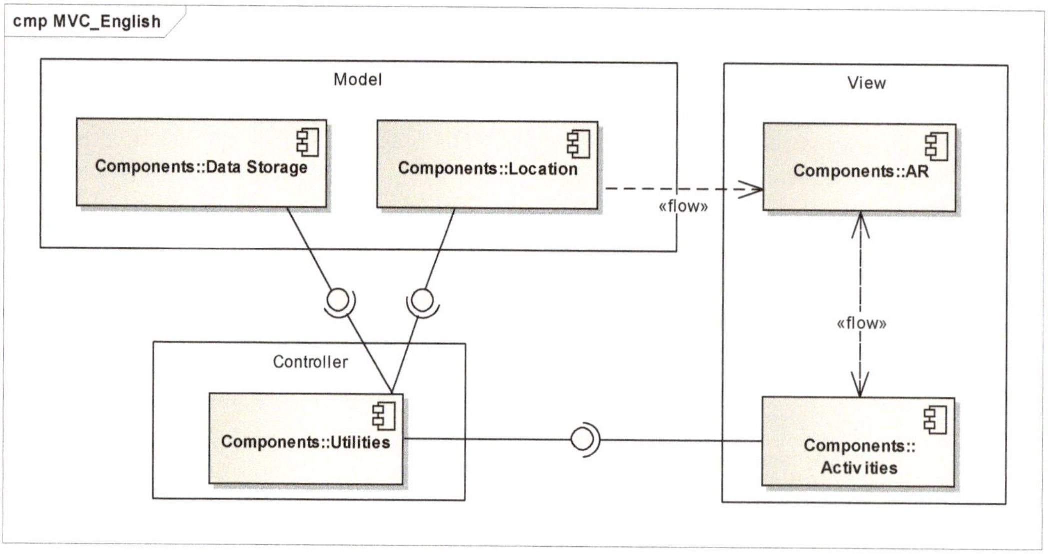

Fig. 4. Proposed Architecture

This is important because it lets the application calculate how the notes should be presented, so the user can really think that they are sticked there. As we progressed with the development we found out that, since the sensor points to the top of the device, so pointing to the North in vertical position and the rotating the phone to a landscape mode, since the "top" of the phone would point to east or west we needed to make some corrections.

The **View** includes the *Augmented Reality* component and this is the one in which all the information is presented to the user and he/she can interact with it.

Based on the GPS information, a place is retrieved and in turn, all the notes are then displayed to the user. As the user pans his device through the place, the notes might be in place thanks to the information of the orientation sensors.

All this is within an `Activity`, base for all the screens in the Android OS. This `Activity` holds all the information of the sensors and starts a View based on the information captured in real time from the camera, via a `SurfaceHolder.Callback interface`.

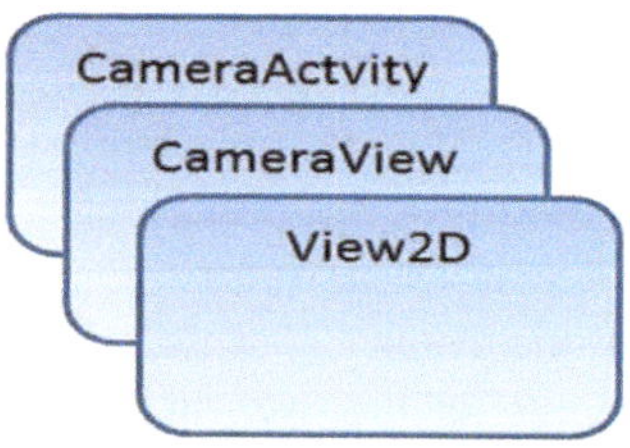

Fig. 5. AR Component overview

Once we can see what the camera sees, we can display our notes, 2D objects, with the possibility of using more complex ways of representing the notes, from images to 3D objects. In order to display all the objects the notes should extend from `View` and on the `onDraw()` method we define how the objects should be rendered according to the position of the device.

The way we use the layers for viewing and showing the context can be summarized in (Fig. 5)

As we can see, we are using two views, one over the other. This way, when the user touches the screen, the upper layer can detect the position of the finger and know if the user is touching a note or not. If he/she is indeed trying to interact with a note (for the prototype clicking on a note is the only way to read it) it triggers the actions defined by the programmers. With these approach we can guarantee that if the user clicks on note "1", then the content for such note will be available and not a random behavior.

Also the *Activities* (which are the name given in Android to all the screens) are enclosed within the View since they are part of the way the user communicates with the system.

Finally in the Controller we define a *Utility* component in which we group the common elements to all the applications and the actions which control the flow of information within the prototype.

Constants. Elements with a fixed value. Interfaces with static variables are used in order to be more efficient. We can find error messages, log messages, etc.

DataSharing. This is the solution to share information between Activities (screens). The most important information shared this way is the location information, that needs to be passed from the sensors to the new note or new place screen and then to the DB.

Listeners. They implement all the behavior for each button in the prototype. For each button an action is triggered and in this listener we can know the button that started the action and then trigger the correct action.

IV. Implementation

Once the prototype was ready, tests were made first to check that all the sensors worked OK and then two phases of tests were made; the first one was to correctly display (both in landscape and portrait mode) dummy information verifying:

- Image display from the camera.
- Image rotation according to the device rotation.
- Note overlay and interaction with the user input.

In these tests we found out that the notes weren't correctly displayed since the information from the sensors wasn't being correctly translated to the device.

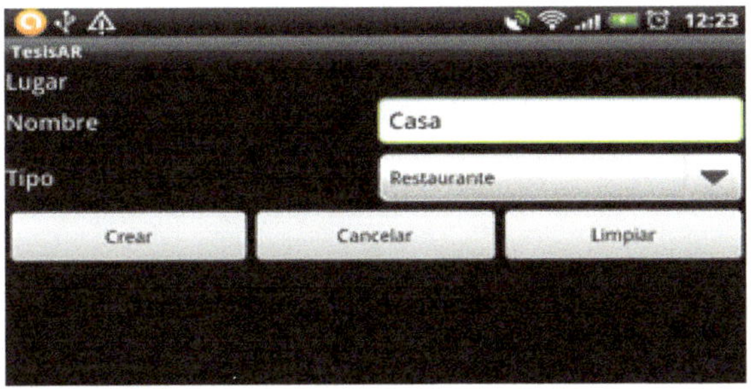

Fig. 6. Place creation

Fig. 7. Note creation

Fig. 8. Notes in Landscape mode

Since they are too fine grained and too sensitive, adjustments were made in order to make them less sensitive so the transition could be smooth enough for the user.

As for the image, corrections had to be made in order to correctly rotate the image, since we found out that we needed to rotate the image to correctly display it when the user switches from the landscape mode to the portrait mode.

For the final test we followed the following steps:

- Creation of a New Place (Fig. 6).
- Creation of Notes for a Place (Fig. 7).
- Overlaying of Notes to the Place (Fig. 8).
- Interaction with the Notes.

All this actions were performed without a flaw. The Place was created, and notes were associated. Once the Notes were retrieved they seemed to be sticked to the walls and the interaction with the user was without a problem. For each note that the user clicked on, the content of that note would appear before him/her.

V. Conclusions

AR applications are not limited to mobile devices, however is in this devices that lays the biggest potential to grow, since new SDKs and the popularity of this devices put them in the hands of avid users and developers.

From medical applications, games, collaboration tools and hardware development pave the way to just search an object and it's properties just by looking at it; imagine getting to a store and realizing which size of clothes is the best for you, buying furniture for your house just placing a virtual object from an on-line catalog interacting with your living-room or in education by letting the students know more about objects that aren't available in a common classroom or training new engineers or doctors on how to give maintenance to an engine, or give attention to a virtual patient.

This prototype uses the characteristics of a Smartphone not just to display information about a place, but encourages the user to generate in-situ such context that will be present to other people, and also, letting the user interact seamlessly with virtual objects in a real environment thanks to the touch screen.

The flexibility of the proposed prototype can be the base for a more complex application, for example a tourist guide, that let the user create content about interesting routes, facts about buildings and recommendations about a nice place to stay, eat or buy souvenirs.

Another application can be educational information about historic places, in which just by pointing the device to the desired location can give the student important information or make in-situ tests, letting them interact with real places leveraging the way they can understand and apply the information given to them.

The schema presented can take advantage of all the research done in the areas of data organization and is open to the use of external and static data sources for the AR content that complement the use of the user created content.

Even though in this case the information is stored locally, there's room for improvement by letting query this knowledge from a server and making this information better and broader, and if it's true that there are some applications that offer similar characteristics, AR can be a novel way to market them.

As applications like Google Glass move from the prototype phase to mass production and consumption, the fact that this pervasive technology is being made available to selected developers can give us a hint about a trend that will be even more popular as the time passes, and as with every technology, the hardware and software evolves, making us wonder what the future will bring to the AR.

Because of this is that, as a final thought, correct privacy and copyright measures must be taken in order to guarantee the safety of all the parties involved. Avoiding security breaches will be as important as the way the information is presented.

REFERENCES

[1] Amato, A., B. D. Martino, et al.; "Semantic Brokering of Multimedia Contents for Smart Delivery of Ubiquitous Services in Pervasive Environments"; International Jorunal of Interactive Multimedia and Artificial Intelligence (2012) pp. 16-25

[2] Azuma Ronald; "A Survey of Augmented Reality"; Presence: Teleoperators and Virtual Environments 6, 4 (Agosto 1997); pp. 355 – 385

[3] Benefon; "Benefon Esc! NT 2002"; http://www.benefon.de/products/esc/ (accesed on 2013/04/30)

[4] De Michelis Gio, De Faoli Flavio, Pluchinotta Constanza y otros; "Weakly Augmented Reality: observing and designing the work-place of creative designers"; Proceedings of DARE 2000 on Designing augmented reality environments (2000); pp. 81-91

[5] Dubois Emmanuel, Nigay Laurence; "Augmented Reality: Which Augmentation for Which Reality?"; Proceeding of DARE 2000, Design Augmented Reality Environments, ACM (2000); pp. 165-166

[6] Herbst Iris, Braun Anne-Kathrin, McCall Rod y otros; "Time Warp: Interactive Time Travel with a Mobile Mixed Reality Game"; MobileHCI (Septiembre 2,5 2008); pp. 234-244

[7] Hirokazu Kato, Mark Billinghurst, "Marker Tracking and HMD Calibration for a Video-Based Augmented Reality Conferencing System"; 2nd IEEE and ACM International Workshop on Augmented Reality 1999; pp.85

[8] S. Feiner, B. MacIntyre, T. Höllerer y otros; "A touring machine: Prototyping 3D mobile augmented reality systems for exploring the urban environment"; Proceedings of First IEEE International Symposium on Wearable Computers (ISWC '97); 1997; pp 74–81

[9] García, C. M., and G. M. Ortega; "Route planning algorithms: Planific@ Project"; International Journal of Artificial Intelligence and Interactive Multimedia (2009); pp. 57-66

[10] Impress Corporation; "Ketai Watch: J-SH04"; http://k-tai.impress.co.jp/cda/article/showcase_top/3913.html (accesed on 2013/04/30)

[11] Jong-Seung Park, Ramesh Jain; "Identification of scene locations from geotagged images"; ACM Transactions on Multimedia Computing, Communications, and Applications (TOMCCAP) TOMCCAP Homepage archive (Volume 9 Issue 1, February 2013); Article No. 5

[12] Junaio/Metaio; "Press Release: SZ Magazin and metaio develop Mobile AR Special"; 2010; http://www.metaio.com/press/press-release/2010/sueddeutsche-zeitung-sz-magazin/ (accesed on 2013/04/30)

[13] Layar; "What is Layar?; 2012; http://layar.com/what-is-layar/ (accesed on 2013/04/30)

[14] Lee Rainie, Susannah Fox. Just-in-time Information through Mobile Connections. Pew Internet & American Life Project, May 7, 2012, http://www.pewinternet.org/Reports/2012/Just-in-time/Main-Report/Findings.aspx (accessed on 2013/01/10)

[15] Magnusson Charlotte, Rassmus-Gröhn Kirsten, et al.; "Navigation by pointing to GPS locations"; Personal and Ubiquitous Computing archive, Volume 16 Issue 8 (December 2012); pp. 959-971

[16] Miyashita T, Meier P, Tachikawa T; "An Augmented Reality Museum Guide"; ISMAR '08 Proceedings of the 7th IEEE/ACM International Symposium on Mixed and Augmented Reality (2008); pp. 103-106

[17] Neustaedter Carman, Tang Anthony, et al., "Creating scalable location-based games: lessons from Geocaching"; Personal and Ubiquitous Computing archive, Volume 17 Issue 2 (February 2013); pp. 335-349

[18] NilssonSusanna, Arvola Mattias, et. al "Exploring place and direction: mobile augmented reality in the Astrid Lindgren landscape"; OzCHI '12 Proceedings of the 24th Australian Computer-Human Interaction Conference (2012); pp. 411-419

[19] Olsson Thomas, Lagerstam Else, et al., "Expected user experience of mobile augmented reality services: a user study in the context of shopping centres"; Personal and Ubiquitous Computing archive, Volume 17 Issue 2 (February 2013); pp. 287-304

[20] Ramirez, J. J.; "Collaborative Geographic Information Systems for Business Intelligence"; International Journal of Artificial Intelligence and Interactive Multimedia (2009); pp. 94-97

[21] Realinho Valentín, Teresa Romão, et al; "Building mobile context-aware applications for leisure and entertainment"; ACE '11 Proceedings of the 8th International Conference on Advances in Computer Entertainment Technology (2011), Article No. 29

[22] Rekimoto Jun, Nagao Katashi; "The World Through the Computer: Computer Augmented Interaction with Real World Environments"; Proceedings of UIST '95 14-17 (Noviembre 1995) pp. 29-36 via [1]

[23] Rodríguez Sánchez, Martínez Romo, et al; "GAT: Platform for automatic context-aware mobile services for m-tourism"; Expert Systems with Applications: An International Journal archive. Volume 40 Issue 10, (August, 2013); pp. 4154-4163

[24] Rozhen Kamal Mohammed-Amin, Richard M. Levy, et al. "Mobile augmented reality for interpretation of archaeological sites"; PATCH '12 Proceedings of the second international ACM workshop on Personalized access to cultural heritage (2012); pp 11-14

[25] Sánchez Albert, Redondo Ernest, et al; "Developing an augmented reality application in the framework of architecture degree"; UXeLATE '12 Proceedings of the 2012 ACM workshop on User experience in e-learning and augmented technologies in education (2012); pp. 37-42

[26] Schmalstieg Dieter, Wagner Daniel; "Experiences with Handheld Augmented Reality"; Proceedings of the 2007 6th IEEE and ACM International Symposium on Mixed and Augmented Reality (2007); pp. 1 – 13

[27] Sutherland E. Ivan; "A head-mounted three dimensional display"; AFIPS '68 Proceedings of the December 9-11, 1968, fall joint computer conference part I (1968); pp. 757-764

[28] Urbanowicz Katarzyna, Nyka Lucyna; "Media architecture: participation through the senses"; MAB '12 Proceedings of the 4th Media Architecture Biennale Conference: Participation (2012), pp. 51-54

[29] Ventä-Olkkonen Leena, Koskenranta Olli, et al., "User expectations of mobile mixed reality service content"; MUM '12 Proceedings of the 11th International Conference on Mobile and Ubiquitous Multimedia (2012); Article No. 52

[30] Wereable Computer Lab; "ARQuake: Interactive Outdoor Augmented Reality Collaboration System"; 2010; http://wearables.unisa.edu.au/projects/arquake/ (accesed on 2013/04/30)

[31] Yen-Ta Huang, Kuan-Ting Chen , et al; "Detecting the directions of viewing landmarks for recommendation by large-scale user-contributed photos"; MM '12 Proceedings of the 20th ACM international conference on Multimedia (2012); pp. 997-1000

2

Reconstruction of High Resolution 3D Objects from Incomplete Images and 3D Information

Alexander Pacheco[1], Holman Bolívar[1], Jordán Pascual Espada[2], Rubén González Crespo[3]

[1] *Universidad Católica de Colombia, Bogotá, Colombia*
[2] *Universidad de Oviedo, Asturias, Spain*
[3] *Universidad Internacional de La Rioja - UNIR, Madrid, Spain*

Abstract **— To this day, digital object reconstruction is a quite complex area that requires many techniques and novel approaches, in which high-resolution 3D objects present one of the biggest challenges. There are mainly two different methods that can be used to reconstruct high resolution objects and images: passive methods and active methods. This methods depend on the type of information available as input for modeling 3D objects. The passive methods use information contained in the images and the active methods make use of controlled light sources, such as lasers. The reconstruction of 3D objects is quite complex and there is no unique solution- The use of specific methodologies for the reconstruction of certain objects it's also very common, such as human faces, molecular structures, etc. This paper proposes a novel hybrid methodology, composed by 10 phases that combine active and passive methods, using images and a laser in order to supplement the missing information and obtain better results in the 3D object reconstruction. Finally, the proposed methodology proved its efficiency in two complex topological complex objects**

Keywords **— Visual hull, octree, ICP (iterative closest point), marching cubes, iso-surface, iso-level.**

I. Introduction

THE object and image reconstruction is a very useful technique that is commonly applied in many areas, such as human faces recognition [1], quality production control [2], aerial images and maps [3], etc. The high-resolution 3D objects are one of the biggest challenges in the field of objects and image reconstruction methods. Overall, the reconstruction of high resolution 3D objects is an expensive and tedious process, since a user usually has to spend so much time on manual work [4]. The objects reconstruction technology has a very important role in archeology and ancient history; these disciplines often require simple, robust and cheap methods for scanning objects, so these can be studied. The poor state of some discovered antiques implies having complex topologies [5].

The 3-D object reconstruction is a procedure for recovering 3-D geometry and color information from an object [6]. Majorly, there are two kinds of methods to make digital reconstructions of 3D objects: the passive methods [7] and the active methods [8]. The passive methods make use of a set of digital images of the object and the active methods make use of data obtained by laser, which is used to scan the object. Each method has its own advantages and disadvantages, each object reconstruction poses its own problems and appropriate solution must be sought, depending on the requirements.

The active methods are quite expensive in comparison to the passive methods, as they make use of controlled light sources, such as a laser scanner. The object data obtained with the active methods is usually very accurate and the reconstruction algorithms are generally computationally faster than those used in passive methods. Generally, these computational processes are responsible for surface constructions and data registration. In some cases, the configuration of these methods is not easy because the laser scanners require proper calibrations to obtain high quality data. Another aspect of these methods is that they are often not able to capture the actual texture of the physical object; it would need to be combined with the use of cameras. The brightness of objects can also be a problem when trying to scan the object with the laser, the light reflected from the object does not allow laser triangulation and therefore data occlusion is generated.

Passive methods can be used with less technical resources and different types of digital cameras. Recovering the 3D object's geometry by using passive methods involves solving problems related to the camera's calibration, occlusion, correspondence and data fusion. Nowadays, passive methods are widely used in reconstruction and recognition systems, and in many cases they can reconstruct objects with a high level of quality. Generally, these systems involve high computational costs [9] , as they obtain many data sets from digital images and these must be processed through several algorithms. As in active methods, passive methods can also have problems obtaining data shape of bright objects because the bright parts of the object generate texture degradation. Passive methods are very appropriate when the goal is to obtain a real object's reconstruction and is necessary to obtain its texture.

There are several methodologies based on passive or active methods; these methodologies often pursue different objectives, such as: improving the quality of the object reconstructions, improving the processing speed, reducing the computational cost of the process, reconstructions of objects in real time, etc. As noted above, both passive and active methods have advantages and disadvantages. This research aims to achieve a specific methodology in order to get high resolution reconstructions of topologically complex objects with great detail and with a relatively low computational cost, or at least lower than the cost involved in passive methods. The proposed method must be able to obtain the object texture in order to achieve a more realistic reconstruction. For the rebuilding of an object of great quality we will combine passive and active methods, the objective is to obtain the best data for the reconstruction of the object. The combination of two methods presents challenges, especially if we take into consideration the objectives of the proposed methodology. The need of applying two different methodologies increases the computational costs, which is a problem to be solved. The combination of active and passive methods in a unique methodology raises the number of challenges, is quite possible that the computational cost due to the application of the two methods is very high. This is a problem that the proposed methodology has to solve, combining the appropriate tasks and algorithms for data processing and reusing the object information and tool calibration.

The proposed method is composed by 10 phases that combine the use of high resolution images and laser scanner in order to supplement the missing information, the object's information obtained is analyzed with image reconstruction algorithms to obtain the best result for the 3D high resolution reconstruction of the object. In order to evaluate the effectiveness and quality of the proposed methodology we have applied it to the reconstruction of topologically complex, detailed objects. We intend to take advantage of the two types of reconstruction methods in order to achieve successful results. Although there is a proposed methodology which is very similar to the one being developed [10], the proposed method uses the NextEngine laser scanner and digital cameras with 13 Mpixels.

This paper is divided into the following sections: Section 2 describes related work. Section 3 describes the proposed hybrid methodology for the reconstruction of 3D objects. In Section 4 we use the proposed methodology to reconstruct two topologically complex 3D objects. Then, in Section 5 we present the results of the application of the proposed methodology. Finally, Section 6 presents conclusions and future work.

II. Related Work

Within the area of reconstruction of high resolution objects from data obtained from images (passive methods [7]) a wide range of solutions have been proposed, depending on the physical property we want to retrieve, such as Space Sculpting Algorithm, Set Level Algorithm, Graph Cut Algorithm, Visual Hull Algorithm [11], Stereo Fitting Algorithm [12, 13, 14, 15]. Several proposals have been combining some of these classic algorithms to improve some aspects of the image reconstruction systems, the main goal of this has been to improve the quality of the reconstruction and the computational costs [8]. Although they are widely used, most of these algorithms do not obtain perfect reconstructions of 3D objects [16, 17]. However, in some cases several solutions achieved high quality reconstructions of particular types of 3D objects, such as human faces [1], industrial parts, simple topological objects, etc. Either way, there is no general solution based on passive methods that is capable of making perfect reconstructions of any type of 3D object, especially in the case of topologically complex objects. Another very common drawback of these solutions is that, in order to slightly increase the quality of the object reconstruction, a great increase of the computation costs becomes inevitable [15]. Additionally, these methods which use digital cameras generate the need to control parameters such as light, color or movement.

There are several types of algorithms that are used commonly in passive methods for the reconstruction and recognition of objects. Some of the most relevant algorithms are: space sculpting, set level, graph cut, Visual hull and stereo fitting.

The space sculpting algorithm consists in initially creating a volume surrounding the surface of interest in order to be able to make iterations on it and reconstruct the shape of the object. In most cases, the refinement is performed under the photo-consistency conditions and starting from an initial volume represented by voxels. Therefore it's an algorithm that commonly uses the color information method and takes volumetric orientation. An example of such solution can be found in [1,18]

The set level algorithm represents the surface as an initial set, which varies in time according to the volumetric density function defined. These algorithms are numerically stable and handle topology changes automatically. Several authors have worked on this type of solution, the most representative being Jean Pons, R.Keriven, and O. Faugeras [19], M. Habbecke and L.Kobbelt in [13].

The graph cut algorithm consists in making a representation of the object's volume in a graph structure with weight on the edges. The problem is finding the most efficient way to cut in a graph with the lowest cost. The weight indicates the discrepancies between images, in other words, it integrates the function of photo-consistency. Several authors have proposed solutions on this methodology, Kolmogorov and R. Zabih in their work of scene reconstruction via graph cuts [20], G. Vogiatzis P. Torr and R. Cipolla in their work of multi-view stereo via volumetric graph-cut [11], A. Hornung and L. Kobbelt in their work of hierarchical volumetric multi-view stereo reconstruction of manifold surfaces based on dual graph embedding [21].

Visual hull algorithm is defined as the intersection of all the possible cones containing the object. It was proposed by Laurentini [22] and several other authors which made use of this algorithm as an initial step to make refinements on it. [23, 24] Visual hull is the maximal object shape that is consistent with the silhouettes of the object. Supposedly, it can be obtained by the intersection of back-projected visual cones of all the silhouettes [24].

The stereo fitting algorithm is designed to find the correspondence of objects between two images. With the use of this information we can find the fundamental matrix that relates the image space with the object space, and then reconstruct the object through triangulation. The following authors have worked with this type of solution: Y. Furukawa and J. Ponce [13]; and Derek Bradley, T. Boubekeur, and W. Heidrich in [25]. Depending on the physical property we want to analyze and characterize in the object, the restrictions and properties described in a method may be applied. There are three types of methods shown in figure 1: the pattern through shading, pattern through shape and the use of the color information.

Besides, the type of algorithm can be applied to other important factors that determine the quality of the reconstruction of the object. Pattern through the shading requires light calibration in order to assume the surface is Lambertian and make use of the properties this assumption implies. Pattern through the shape requires space calibration to allow the use of segmentation and use geometry properties to obtain a representation of the object's pattern. The color information allows photo-consistency verifications between images. The orientation of the algorithm depends on how it is generating the structure of the object. It may be by volumetry (voxel), or through representation by means of the construction of surfaces.

Within the area of reconstruction of high resolution objects from data obtained by scanners (active methods [8]), reconstruction algorithms are based on parametric surface orientation or implicit functions. [26] is an example of algorithms based on parametric surfaces (Delaunay triangulation) and [27] shows an example of algorithms based on implicit functions (signed functions followed by iso-surface extraction [28]).

Active methods have proven their effectiveness and quality in many solutions, but due to the technical cost involved by its use they are not suitable for all scenarios, not always have the resources or can be used to that extent. There are several solutions using hybrid methods, based on a combination of passive and active methods. [29] Combines the active and passive vision and determines viewpoints in a similar way to human intuition, but in the words of the authors the recovered surface is not very accurate, so the solution is not very effective for reconstructing topologically complex objects.

The proposed solution aims to define a novel methodology to reconstruct objects with high quality, including texture mapping. To obtain the best possible quality data, this methodology merges data from active and passive methods. This methodology is based on the solutions that use volumetric techniques to merge laser scan data and digital images for the reconstruction of the surface. This approach will reduce the characteristic reconstruction errors of active and passive methods by merging data, which can be useful when reconstructing topologically complex objects with better quality. To achieve this, the proposed methodology has to solve the problem of computational cost derived from the combination of two methods, this can be done by combining the appropriate tasks and algorithms for data processing and reusing the object's information. Also, if the information is not reused properly, the calibrations in the cameras and the scanner can also mean a great loss of efficiency. For this purposes, the proposal introduces a new approach to reduce the complexity presented by the scanner's laser calibrations, it uses the box enclosing the 3D object convex hull to register the models. Finally, the geometric 3D object is calibrated through the high resolution shapes, obtained from photography, to make texture mapping as realistic as possible.

III. Hybrid Methodology for the Reconstruction of 3D Objects

As mentioned above, the proposed methodology uses both the data obtained from digital images and the data acquired via laser scanner to retrieve the geometric information of the object. This methodology is divided in two stages. The first one would be to use the digital image information alone to generate a reconstruction of the initial model, and the second one is to register both models: the one obtained in the first stage and the object reconstructed by the laser scanner, in order to make a refinement and find the final volume that best represents the 3D object we want to rebuild. The first phase, reconstruction from images, uses space sculpting algorithm, pattern methods through shape and volumetric orientation. Figure 1 shows in detail the methodology and each of its steps. The first stage consists of phases 1 to 7, and the second stage comprises the phases of 8 to 10.

We will describe the tasks involved in each phase of the proposed methodology below.

A. STAGE 1

1) Phase 1: Data Acquisition

The objective of this phase is to collect information concerning the object, mainly its shape, but also the calibration pattern for each object view. There are some physical factors that can affect this data collection, one of them is the brightness of the object. Bright objects, like shiny metal objects are difficult to reconstruct with either the active or the passive method. In the active method, the light reflected from the object does not allow laser triangulation and, therefore, data occlusion is generated; in the passive method, the bright parts of the object generate texture degradation.

In this first stage we must minimize the effect of light in

order to obtain acceptable images to be processed in the following phases. We have developed a photo-montage with the following elements: Light box (own production), professional lights, a semi-professional digital camera, a revolving board and a computer. Figure 2 shows the montage developed for this phase of data acquisition.

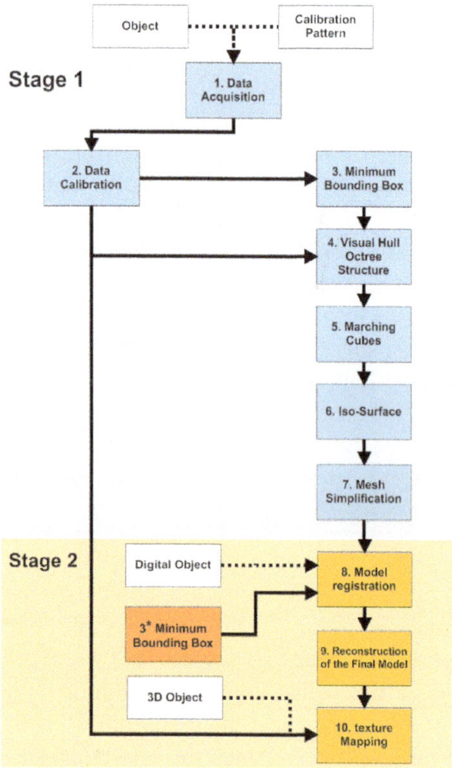

Fig 1. The flow of phases of the methodology for 3D objects reconstruction.

The revolving board allows us to obtain up to 36 views of the object. The lights are calibrated with the light box and the digital camera to minimize the reflection of light on the object as much as possible. The pictures have been taken from the computer in order to obtain more stability and consistency between images.

The background has been arranged as uniformly as possible (white color). However, shadows and dark lines that may affect the process of thresholding of the images can still appear

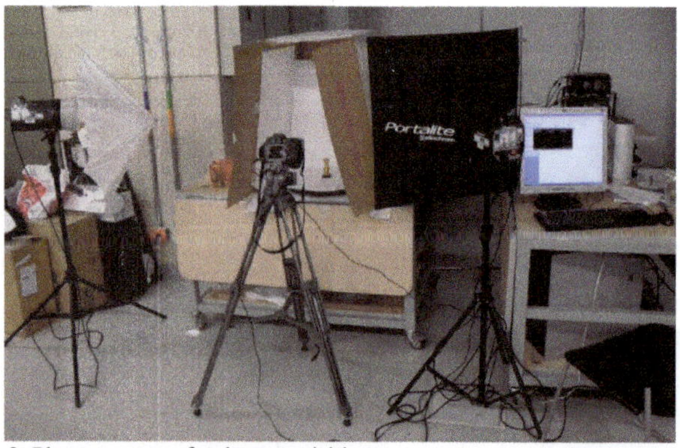

Fig. 2. Photo-montage for data acquisition.

As noted, this phase has the object and the calibration pattern as inputs. The object is photographed in different views and afterwards a thresholding pre-processing is performed to obtain the shape of the object. In practice, many image-editing software can be used to obtain an accurate shape of the object. The calibration pattern corresponds to a table with black and white squares, the result of the pictures taken in each view of the object.

As shown, this first phase has two types of data sets as output. The first data set corresponds to images containing shapes of the object and the second data set corresponds to images with the calibration pattern for the object in each view.

2) Phase 2: Data Calibration

In order to calibrate data we can use a specific open source computer vision software or commercial software. In this case we used Matlab Toolbox [30] developed by Jean-Yves Bouquet. Matlab toolbox allows us to find both the intrinsic and extrinsic parameters of the camera quickly and accurately. In addition, intermediate results can be seen graphically and projections can be created on the calibration to minimize the errors. The output of this phase is the projection matrix M3x4 for each view of the object. The projection matrix is:

$$(1)\ M_{3X4} = K[R|T]$$

Where K is the matrix, containing the intrinsic parameters of the camera, R is the rotation matrix and T is the translation vector of the object in relation to the camera.

3) Phase 3: Minimum Box Surround

In order to speed up the computational processing of data we must know the minimum work space. From **[18]**, "*the 2D projections of N boxes define a 3D convex hull formed by 4N planes*". The minimum 3D box that encloses the 3D convex hull can be calculated if posed as an optimization problem where the objective function becomes one of the 6 variables that define the 3D box we want to recover under the constraints of each 2D plane of the box. The restrictions imposed by each 2D plane of the box are:

$$(2)\ a_i x + b_i y + c_i z + w \leq 0, \quad i = \{1,\ldots,4n\}$$

We can find the six variables that define the 3D box by using the simplex method. In practice, it is necessary to calculate the 2D convex hull of the object in each of the images to find the box enclosing the 2D object, and then project the sides of the box (lines) to find the equation of the 3D plane.

4) Phase 4: Visual Hull

There are several ways to construct the visual hull of the 3D object depending on several factors: the algorithm, the method and approach used, etc. In this method we will use volume-oriented algorithms. In order to do so we have decided to represent the object with octree structure.

The structure of the octree is described in **[31]**. The advantage of using this structure is that it uses little memory as it is described by a string of '1's, '0's, '('s that represent the position of each cube (inside, outside and above respectively). Moreover, the order of the chain is due to the pre-order path in the tree. In order to generate the octree we first take a cube, we project it in all images and we check the iso-level function's value to determine the position of the cube. If the position is 'inside', the cube is drawn, if it is 'outside' it is not drawn, and if it is 'above' it must be subdivided in eight cubes.

This process continues recursively projecting and determining positions to smaller and smaller cubes until we reach the degree of resolution that is wanted. In this case, the digital images have a resolution of 2008x3040 pixels; the recommended depth level of the tree will be 8 or 9.

To define the iso-level function the images that match the shapes must be encoded so the background has a value (e.g. -0.5) and the figure has a value with opposite sign (e.g. 0.5). This way, the boundary of the given object will have an intermediate value, 0. Using these values, the iso-level function would be:

$$(3)\quad f_i(c) = Min V_i(P_i c), \quad i = \{1,\ldots,n\}$$

Where $V_i(P_i c)$ is the value taken by the projection of the 3D point in the 2D image, depending on whether it is inside or outside the silhouette or not.

In order to determine the status of the cube we must compare the iso-level values from the eight points of the cube. If all of them were negative, it would mean that the cube is outside the valid volume range for the model. If all of them were positive, the cube would be within the valid range, but if we'd find both positive and negative values, we wouldn't be able to validate it with certainty, and we'd must therefore continue subdividing.

In practice, we should project points that are not only at the vertex but on the edges of the cube, in order to avoid sampling problems and to be able to determine the position of the cube correctly.

5) Phase 5: Marching cubes.

Once we have obtained the volumetric representation of the object, phases 5 *Marching cubes*, 6 *Iso-surface* and 7 *Simplification* are applied to obtain the mesh that best represents the surface of the object. The marching cubes algorithm takes the iso-level values containing the vertexes of the cube and determines the points cutting the iso-surface to generate the corresponding mesh configuration. As they are cubes, there are 256 meshing possibilities that can be reduced to 15. Once the mesh in each cube has been created, they are merged in order to create the surface that represents the object.

6) Phase 6: Iso-surface & Phase 7: Simplification

In the Phase 6, the iso-surface for the proposed algorithm corresponds with values equal to zero. Then we proceed to apply the phase 7: Simplification. Once the mesh has been obtained, the next step is to go through an optimization process by applying mesh simplification algorithms. These are algorithms of decimation and our intention is to maintain the shape of the surface for a smaller quantity of sampled data.

B. STAGE 2

After the first stage, based on a passive method, the second stage began. At this stage the methodology focuses on data obtained using laser scanner. In order to reduce the computational cost during this stage we use data obtained from the preceding stage.

1) Phase 7: Minimum bounding box

This phase consists in the calculation of the scanned object's workspace. As the object is already a 3D object and is also aligned with the coordinate system so it doesn't present inclination along any axis, the minimum containing box is determined by finding the six variables defining the cube through simple minimum and maximum functions.

2) Phase 8: Model Registration

At this stage we aim to register the two geometric models obtained through the two methods we have used so far: passive and active. As the object that has been reconstructed by passive methods is being calibrated, we want to transport the model obtained by the laser scanner to the coordinate system of the model obtained from the digital image data.

The common way to register models is to calibrate the laser, using methods like the one described in **[32]**, and then apply ICP algorithms (iterative closest point) and find the rotation and translation matrix that relates both models. However, due to the way the acquisition of data has been performed; this paper suggests that both models are brought to register through the minimum box enclosing the object.

The registration process from the bounding boxes can be a manual process in practice, but significantly reduces the complexity of the problem, bearing in mind that the models have been reconstructed by different means and have considerable variations.

3) Phase 9: Reconstruction of the final model

Once we have brought the two objects to register we proceed to generate the final surface of the object detailed and free of gaps.

The idea of this algorithm is to take the visual hull represented by the octree and redefine the states of the cube depending on the intersection of the line defined by the laser spot and the source point of the camera. This algorithm makes sculpting in 3D space on the volume defined by the octree guided by 3D points obtained using the laser scanner. In order to generate the model correct, several cases must be considered, such as the one described in **[10]**.

Each octree cube which is 'inside' or 'on' changes to 'off' if the ray defined by the 3D point and the origin of the camera intersects the surface. Finally, the marching cubes algorithm is applied in order to reconstruct the final, gap-free volume of the object.

4) Phase 10: Texture mapping

To offer realistic 3D digital objects, these should include textures. The textures we used were collected from digital images. As the data is calibrated, texture mapping can be done with any algorithm that may be developed for the purpose. In this instance we are proposing a mapping based on particles such as in **[33,34]**. After phase 10 the reconstruction process of the 3D object was successfully completed. As an output from the application of the methodology, a high definition 3D reconstruction of the object is obtained. This reconstruction has a high degree of quality and texture mapping.

IV. Application of the Proposed Metodology in a Real Scenario

To validate the objectives set at the beginning of this research and illustrate the use of the proposed methodology we used it to reconstruct topologically complex 3D objects. The objects involved in the reconstruction process are two replicas of archaeological objects from the Gold Museum in Bogota, Colombia. We choose these objects because of its morphological properties, great detail and brightness management (Phase 1 & 2). Figures 3a and 3b shows the images of the objects we want to rebuild.

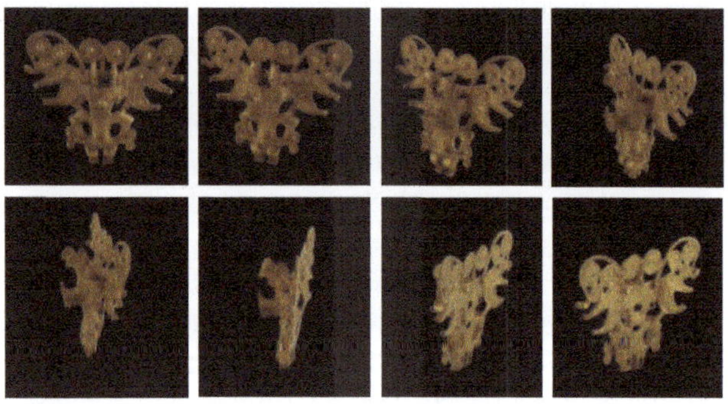

Fig. 3a. Replica from the Gold Museum, Objet 1.

Fig. 3b. Replica from the Gold Museum, Objet 2.

Figure 4 shows the graphic results of the minimum bounding box for the objects (Phase 3). The figure's vertexes are the source point of the cameras and the lines represent the projection of the 2D box surrounding the object. The minimum 3D box is defined in the interior space of intersection of the projections.

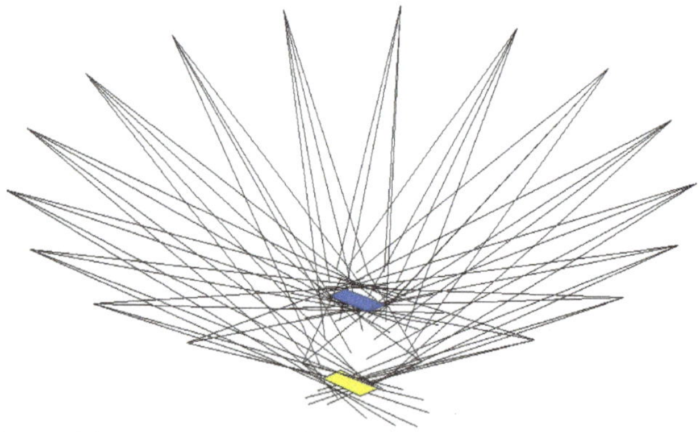

Fig. 4. Minimum Bounding Box (lower and top cover).

For the visual hull's phase we define the structure of the octree given in **[31]**. The right part of Figure 5 shows the simulation results for the octree represented on the left side of the figure.

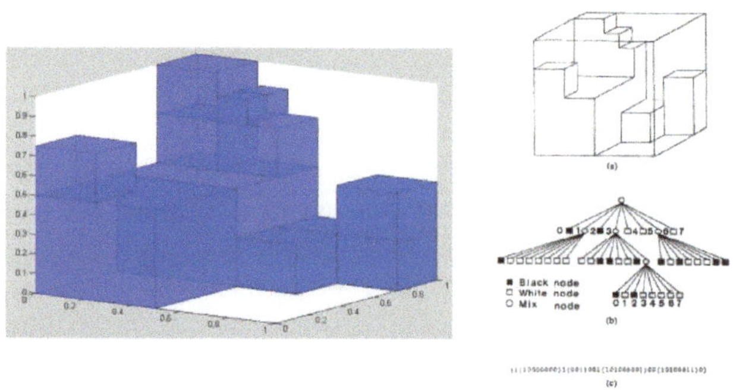

Fig. 5. Representation of the octree.

The results of the octree for the replica from the gold museum are shown in Figure 6. Figure 7 shows the results of phases: 5 Machine Cubes, 6 Iso-surface and 7 Mesh Simplification.

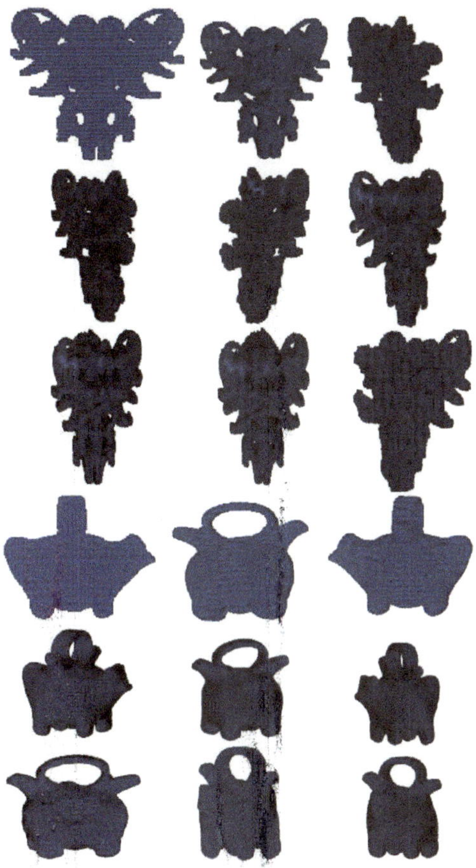

Fig. 6. Octree of objects reconstructed from different views.

Fig. 7. Results of phases: 5 Marching Cubes, 6 Iso-surface and 7 Mesh Simplification.

For the second stage we used the NextEngine 3D desktop scanner. The data obtained by the laser scanner is shown in Figure 8. As shown, the objects have some deficiencies in the reconstruction.

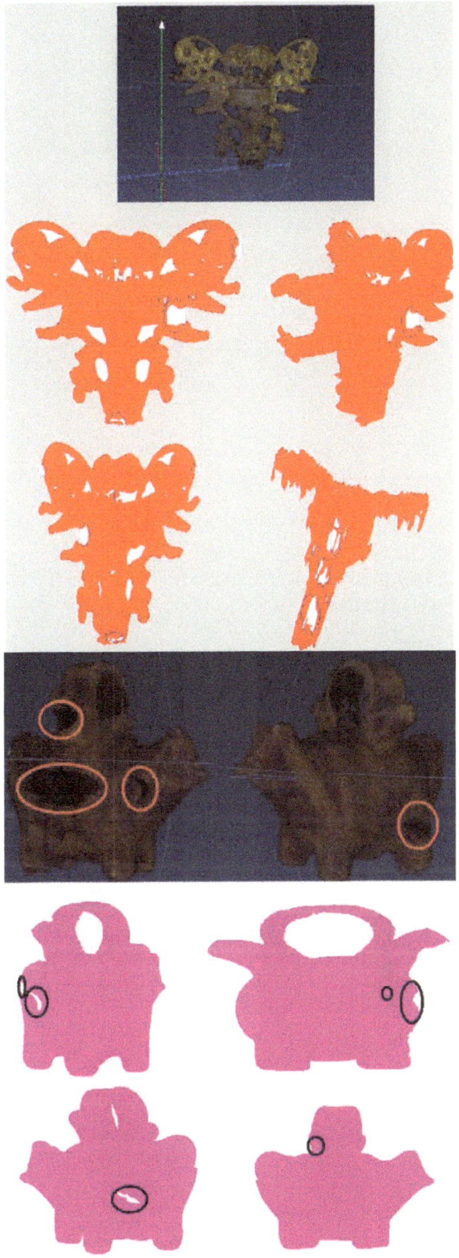

Fig. 8. Data obtained by the laser scanner. Highlighting shows where the failures of reconstructed objects are.

Then we proceed to implement Phase 8 of the methodology: Model Registration, bringing the two models into register, with the help of the minimum encasing boxes that contain the object, as shown in figure 9. In practice, the boxes do not have the same dimensions and the objects must be brought into register manually. However, they are close enough to apply registration algorithms such as ICP (iterative closest point) so that the object is eventually registered.

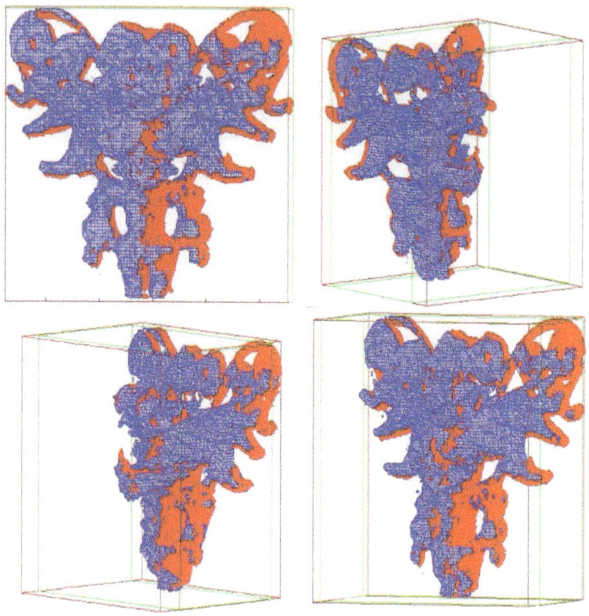

Fig. 9a. Model Registration reconstructed with different techniques. Reconstructed object from images (blue), object reconstructed from laser scanner (red).

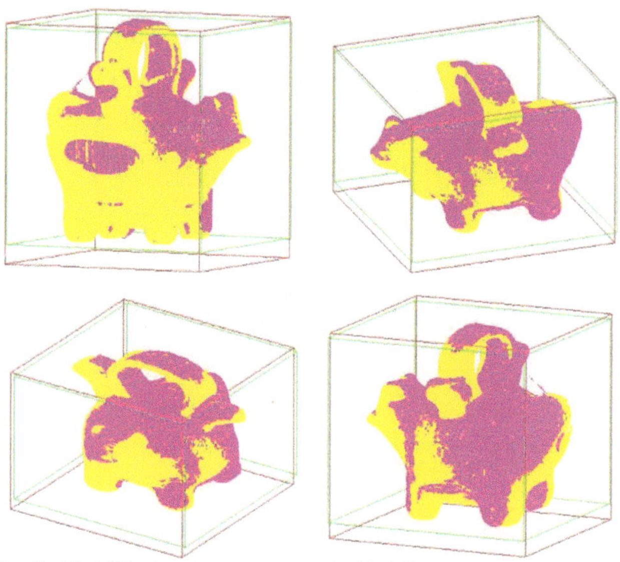

Fig. 9b. Model Registration reconstructed with different techniques. Reconstructed object from images (purple), object reconstructed from laser scanner (yellow).

Finally, in the Phase 10: Texture Mapping we must obtain the merged geometric model and make a photo-realistic texture mapping. An approximation of the final geometry of the 3D objects is shown in Figure 10. After applying the proposed methodology we have successfully managed to build two morphologically complex 3D objects, obtaining a high quality reconstruction of the object, with a detail level significantly higher than those obtained by using passive methods.

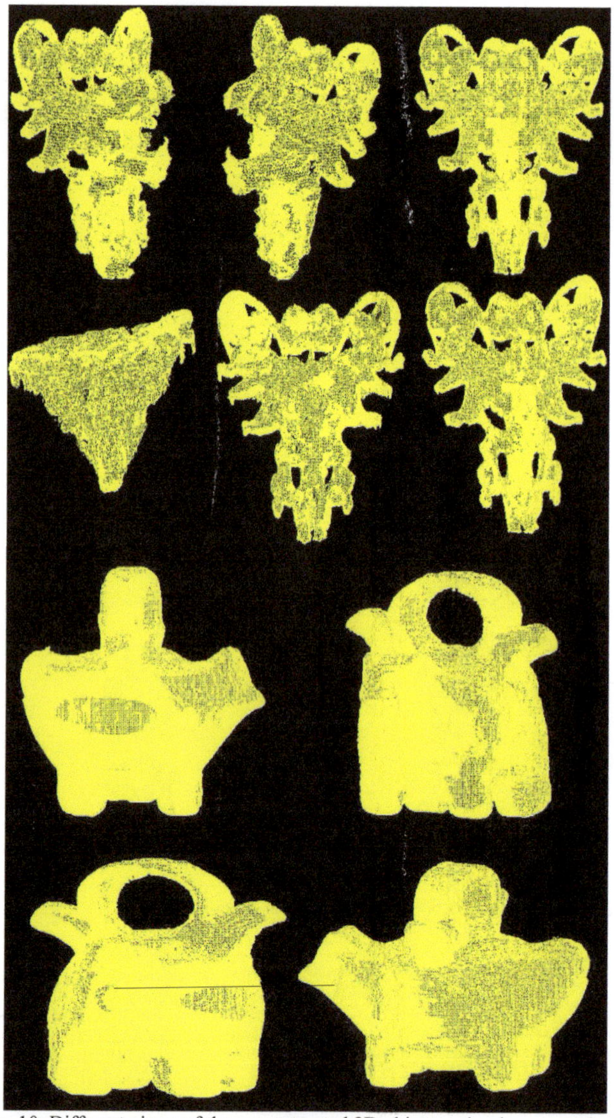

Fig. 10. Different views of the reconstructed 3D objects using images and laser scanner information.

V. Proposal Evaluation

The proposed methodology has been previously used in the reconstruction of two morphologically complex objects. This evaluation examines the relevant aspects related to the objectives set at the beginning of this research.

One of the objectives of the methodology was to obtain quality reconstructions from objects that are difficult to reconstruct, like topologically complex objects with shiny surfaces. To verify that the quality of the reconstructions has increased using the proposed methodology we will compare it with a passive method. A parameter that is often used to validate the quality of an object's reconstruction is the number of errors in the topology generated during the reconstruction. As an object is topologically more complex needs a larger number of voxels to be reconstructed, when more voxels are used the potential for reconstruction errors increases. The error relationship between the two methods is shown in Figure 11, observing results it is possible to detect a function with an associated trend to the occurrence of errors depending on to the number of voxels of the object to be reconstructed.

The combination of two methods would increase the computational cost. The other major objective of this research was to achieve an adequate computational cost, which did not exceed the computational cost of a passive methodology. To achieve this requirement the proposed methodology reuses the object data captured and the calibrations calculated at different phases of the process. The trend of error occurrence in the calibration as a function of the number of voxels is potential with a coefficient of determination of 0.94. Second, calculate the relationship between the time spent on the reconstruction process of the 3D object and the object's voxels. In the graph of Figure 11b the comparison of the proposed methodology with a passive one is shown

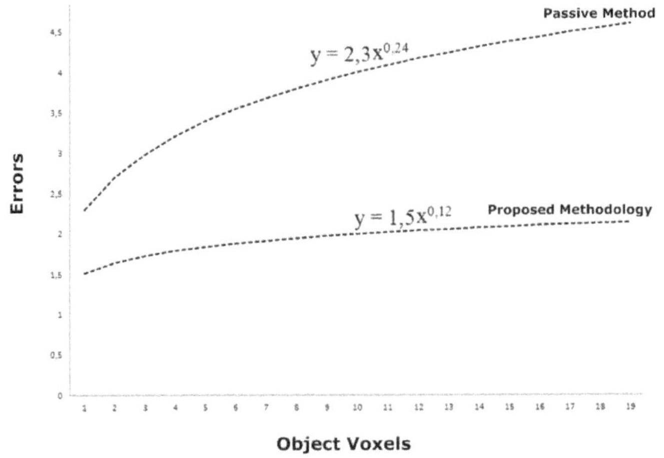

Fig. 11a. Graphical representation of the relationship between the errors and the voxels of the object in the proposed methodology and passive methodology.

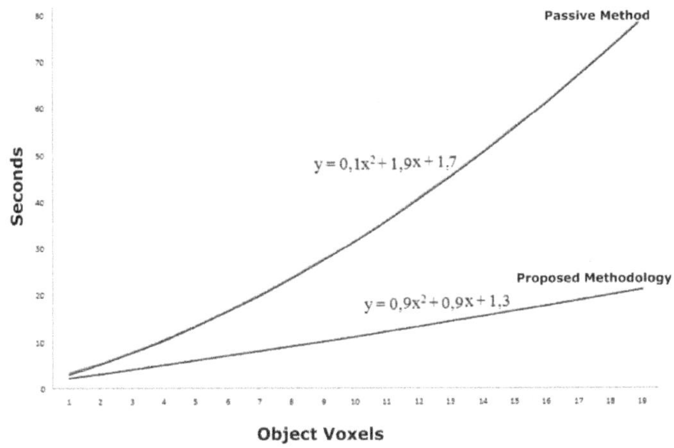

Fig. 11b. Relationship between processing time and the number of voxels of the reconstructed object.

The tendency of the processing time as a function of the number of voxels is polynomial and has a coefficient of determination of 0.99. As a result of this analysis it can be concluded that the proposed methodology improves in several aspects respect passive methods for reconstructing 3D objects, being able to reduce significantly the number of errors and also reducing the processing time used. The improvement of these features is complemented by a more optimal use of the technical resources that enable active methodologies, such as laser scanners.

VI. Conclusions and Future Work

In this paper we propose a novel hybrid methodology for reconstruction of high resolution 3D objects. These objects are also topologically complex and can have bright parts. The methodology is composed by 10 phases that combine active and passive methods, using data obtained from digital images and a laser scanner in order to supplement the missing information and obtain better results in the 3D object reconstruction. The proposed methodology takes advantage of the passive and active methods of object reconstruction, using the level of detail obtained with the laser scanner and the topology recovered by the digital images method. As it has been demonstrated in the Section 5 with the combination method, the proposed methodology has improved the quality of the reconstruction by significantly reducing the number of errors depending on the object's voxels. The proposed methodology solved the problems derived from the need to handle large volumes of object data from multiple methods and the need for different tool calibrations. In the different phases of the proposed methodology, several specific techniques have been applied to reuse object data, calibrations and optimize the computation cost.

Also, we observed that small scanner laser errors generate large errors in the reconstruction of the object. For this purposes the methodology introduces a new approach to reduce the complexity presented by the scanner's laser calibrations, it uses the box enclosing the 3D object convex hull to register the models. The use of digital images in this methodology allows us to obtain the texture-mapping of the object, making it well suited to achieve realistic object reconstructions. The geometric 3D object is calibrated with the high resolution shapes, obtained from photography, to make texture mapping as realistic as possible.

There are many directions for further research. The methodology can be modified at some phase to guide their tasks and algorithms towards achieving better quality and efficiency in the reconstruction of a specific type of topological complex objects, such as industrial material. A modification of the methodology it's being evaluated in order to optimize it for use it in real-time objects recognition under unfavorable circumstances for active recognition.

References

[1] Leo MJ, Manimegalai D, editors. 3D modeling of human faces- A survey. Trendz in Information Sciences and Computing (TISC), 2011 3rd International Conference on; 2011 8-9 Dec. 2011.

[2] Supak M, editor. Mechatronics in a household - sound recording. MECHATRONIKA, 2010 13th International Symposium; 2010 2-4 June 2010.

[3] Suveg I, Vosselman G. Reconstruction of 3D building models from aerial images and maps. ISPRS Journal of Photogrammetry and Remote Sensing. 2004;58(3–4):202-224.

[4] Park J-S. Interactive 3D reconstruction from multiple images: A primitive-based approach. Pattern Recognition Letters. 2005;26(16):2558-2571.

[5] Ducke B, Score D, Reeves J. Multiview 3D reconstruction of the archaeological site at Weymouth from image series. Computers & Graphics. 2011;35(2):375-382.

[6] Grum M, Bors AG Image Processing, 2008. ICIP 2008. 15th IEEE International Conference on; 12-15 Oct. 2008; p. 209-212

[7] Lowe DG. Distinctive Image Features from Scale-Invariant Keypoints. Int J Comput Vision. 2004;60(2):91-110.

[8] Fischer A, Manor A, Barhak Y, editors. Adaptive parameterization for reconstruction of 3D freeform objects from laser-scanned data. Computer Graphics and Applications, 1999 Proceedings Seventh Pacific Conference on; 1999

[9] Modi RA, Goyal SR, editors. An efficient algorithm for 3-D object reconstruction from 2-D images. Image Information Processing (ICIIP), 2011 International Conference on; 2011 3-5 Nov. 2011.

[10] Yemez, Y. and Wetherilt, CJ. 2007. A volumetric fusion technique for surface reconstruction from silhouettes and range data. *Comput. Vis. Image Underst.* 105, 1 (Jan. 2007), 30-41.

[11] Chen G, Shen B, editors. A Visual Hull Algorithm of 3D Reconstruction Based on Interframe Coherence. Virtual Reality and Visualization (ICVRV), 2011 International Conference on; 2011 4-5 Nov. 2011.

[12] Vogiatzis, G., Torr, P. and Cipolla, R. Multi-view stereo via volumetric graph-cuts. CVPR 2005, pp. 391-398.

[13] Furukawa, Y. and Ponce, J. High-fidelity image-based modeling. Technical Report 2006-02, UIUC, 2006.

[14] Habbecke, M. and Kobbelt, L. A surface-growing approach to multi-view stereo reconstruction. CVPR 2007.

[15] Seitz, S., Curless, B., Diebel, J., Scharstein, D. and Szeliski, R. A Comparison and Evaluation of Multi-View Stereo Reconstruction Algorithms. CVPR 2006, vol. 1, pages 519-526.

[16] Huei-Yung L, Jing-Ren W, editors. 3D reconstruction by combining shape from silhouette with stereo. Pattern Recognition, 2008 ICPR 2008 19th International Conference on; 2008 8-11 Dec. 2008

[17] Jang W-S, Ho Y-S. 3-D object reconstruction from multiple 2-D images. Journal 3D Research. 2010;1(2):1-5.

[18] Hernandez, C. and Schmitt, F. Silhouette and stereo fusion for 3D object modeling. CVIU, 96(3):367-392, 2004.

[19] Pons, JP. Keriven, R. and Faugeras, O. Modelling dynamic scenes by registering multi-view image sequences. CVPR 2005, vol. II, pp. 822-827.

[20] Kolmogorov, V. and Zabih, R. Multi-camera scene reconstruction via graph cuts. ECCV 2002, vol. III, pp. 82-96.

[21] Hornung A. and Kobbelt, L. Hierarchical volumetric multi-view stereo reconstruction of manifold surfaces based on dual graph embedding. CVPR 2006.

[22] Laurentini, A. The visual hull concept for silhouette based image understanding, IEEE. Trans. on PAMI 16 (2) (1994) 150.162.

[23] Goesele, M., Curless, B. and Seitz, S. Multi-view stereo revisited. CVPR 2006.

[24] Chari V, Agrawal A, Taguchi Y, Ramalingam S, editors. Convex bricks: A new primitive for visual hull modeling and reconstruction. Robotics and Automation (ICRA), 2012 IEEE International Conference on; 2012 14-18 May 2012.

[25] Bradley, D., Boubekeur, T. and Heidrich, W. Accurate multi-view reconstruction using robust binocular stereo and surface meshing. CVPR 2008.

[26] Boissonnat, J. 1984. Geometric structures for three-dimensional shape representation. *ACM Trans. Graph.* 3, 4 (Oct. 1984), 266-286.

[27] Hoppe, H., DeRose, T., Duchamp, T., McDonald, J. and W. Stuetzle. Surface reconstruction from unorganized points. In *Computer Graphics (SIGGRAPH '92 Proceedings)*, volume 26, pages 71–78, July 1992.

[28] Na L, Bao-de F, editors. An efficient ISO-surface extraction method supported by topology correct. Image and Signal Processing (CISP), 2010 3rd International Congress on; 2010 16-18 Oct. 2010.

[29] Wei F, Bingwei H, editors. Automatic view planning for 3D reconstruction and occlusion handling based on the integration of active and passive vision. Industrial Electronics (ISIE), 2012 IEEE International Symposium on; 2012 28-31 May 2012.

[30] Bouquet, J. *Camera Calibration Toolbox for Matlab. http://www.vision.caltech.edu/bouguetj/calib_doc/.*,2008

[31] Noborio, H., Fukuda, S., Arimoto, S., "Construction of the octree approximating a three-dimensional object by using multiple views," *Pattern Analysis and Machine Intelligence, IEEE Transactions on* , vol.10, no.6, pp.769-782, Nov 1988.

[32] Zhang, Q., Pless, R., "Extrinsic calibration of a camera and laser range finder (improves camera calibration)," *Intelligent Robots and Systems, 2004. (IROS 2004). Proceedings. 2004 IEEE/RSJ International Conference on* , vol.3, no., pp. 2301-2306 vol.3, 28 Sept.-2 Oct. 2004.

[33] Schmitt, F., Yemez, Y., "3D color object reconstruction from 2D image sequences," *Image Processing, 1999. ICIP 99. Proceedings. 1999 International Conference on* , vol.3, no., pp.65-69 vol.3, 1999

[34] Bolivar-Baron H, Pachecho A, Gonzalez-Crespo R. Semantics of immersive web through its architectural structure and graphic primitives. International Journal of Artificial Intelligence and Interactive Multimedia. 2010; Vol. 1, 47-52. http://dx.doi.org/10.9781/ijimai.2010.138.

New Challenges on Crossplatform Digital Contents

Jesús Iglesias Feijoo, Guillermo Amat Gomariz

Zed World Wide

Abstract — **When we speak about devices and platforms, generally we think about those of general use which are currently available (mainly smartphones and tablets). Surely, we would forget all those which are on the way (watches, glasses, cars) and those which are coming. The Internet of Things will transform the technological world in which we are into an amalgamation of devices and interfaces. This paper analyses the challenge for the coming years of getting all these new devices to communicate between them, regardless of their technology and the platforms they use, and it is based on the works done under the Visio Project, funded by the Spanish Ministry of Industry, Energy and Tourism. Finally, a truly universal platform to avoid market fragmentation and provide access to information and services is proposed.**

Keywords — **Apps, cloud, crossplatform, internet of things**

I. Introduction

OVER the past 20 years we have been concerned about the compatibility of classic multimedia content (images, music, and video). It has never been as easy as it is now to play any media format on most popular devices (computers, smartphones, tablets...). This fact has been reached due to a slow process of technical improvements, format standardization processes, and also the adoption of these advanced and standardized formats by manufacturers and developers. This has led to a multi-device and multi-platform media scenario. The achievement has been to ensure the user is able to play media on any device, without needing to have the same content in multiple formats to display on different devices, and thus, avoiding the case of having to pay for every required format.

As an example of the opposite, in the late 90s, although HTML was a widely adopted standard, the emergence of various web browsers led to a real browser war where the most affected were web page developers, as they had to create an almost tailored version of their work for each of these browsers. Even worse, the end users themselves were affected as they lost some freedom regarding what browser to use, because some of them were unable to view certain web pages with certain browsers.

One more case to mention is the appearance of the Apple Store in July 2008 [1]. The iPhone application market revolutionized the market of content for mobile devices. It was the beginning of a revolution but also the beginning of the same old mistake. The rise of this concept and the spread of smartphones have achieved something particularly interesting, since nobody but the developers seem to be concerned about the total lack of cross-platform support for these new multimedia contents: applications.

This document is structured as follows: Section II defines the current problem of content incompatibility, section III explains the implications of security issues in the current scenario while section IV introduces the relation with cloud computing. In section V some existent partial solutions and approaches are explained. Finally, section VI proposes a number of possible solutions to achieve a cross platform of digital content environment.

II. Current Problem

The next section shows that regardless of the type of Smart Object, the problem is common for all of them. It is observed that the strategy designed for smartphones has propagated the same problem to the new technological sectors, as they have all copied the same paradigm of proprietary, closed and completely isolated systems.

A. Smartphones

As these new "application markets" evolve, the various players in terms of operating systems are making their application platforms available for developers, thus consolidating two major companies (Apple and Google) over the rest of competitors:

1. iOS
2. Android
3. Windows Phone
4. Blackberry
5. Others

From the user's point of view, this behavior has created a barrier against freedom of choice of the user device. This occurs, for example, when users buy applications for their smartphones. If, after a few months, they decide to switch to another device with a competing platform, e.g. iOS to Android, then what happens is that previously purchased applications must be acquired again. When a user has to pay again for all his applications, the decision to change one platform for another is a deterrent like no other, as the cost after several years of use, paying for several applications, can be very high and unaffordable.

As a consequence, currently we do not buy applications but the right to use one in certain platform. A similar fact occurred years ago with the attempt to encode media formats using DRM technologies [2], trying to restrict music and videos to certain players. This, completely absurd from the point of view of the user, is however a big deal for the owners of the application stores.

Therefore, a problem for the development of multiplatform content arises. This is what usually is mentioned to explain the lack of such content. But on the other hand, this deficiency makes it difficult to find useful multiplatform content to end users for their consumption. However, the latter, as explained above, is not a serious problem for big players, as it is yielding great benefits for them.

B. Smart TVs

As for Smart TVs, they are undergoing a similar process. Many manufacturers started to launch their new devices based on Android OS but now major TV brands are introducing their own operative systems, aiming at differentiating them from the rest of their competitors:

- Samsung has recently announced Tizen OS for Smart TVs [3].
- Panasonic is going to be using Firefox OS [4]
- LG is already using Webos [5].

Currently, Sony is the only one which is still faithful to Android. The rest of manufacturers have realized that, by doing that, they are increasing Google's profits.

The main conclusion is that we are witnessing a big fragmentation of operating systems for future TVs. This is not necessarily a negative thing; on the contrary, it increases competition and thus, the potential benefits for users. However, this fragmentation means, again, a lack of multiplatform content to allow a total availability of applications. In other words, each manufacturer will have to worry about increasing their applications catalogue on their own.

Smart TV content developers are in the same situation that occurs for the kind of devices that were previously discussed. Anyone who wants to make an application for Smart TVs will be forced to repeat the development as many times as platforms they wish to reach.

C. Connected Cars

Connected cars are one of the markets that most excitement and growth will experiment in the coming years. Traditionally car interfaces have been completely created by each car manufacturer. There have not been major horizontal suppliers of user interfaces. As the vehicles were implementing dashboard displays and infotainment systems, each manufacturer was developing their own UI to suit their needs. Therefore, there is no relationship between one car manufacturer and the rest. Now it is time for these interfaces to evolve providing connectivity, allowing the use of applications and so forth. The conclusion is that this sector has arrived to the same point we explained before: each manufacturer has their own proprietary and closed system and there is no way to get some multiplatform content.

In recent times, this situation has become increasingly complex due to new products being launched by two of the main mobile communication companies: Apple for Car Play [6] and Google for Android Auto [7]. MirrorLink should also be added to this list, as it is the solution proposed by most automotive-related manufacturers subscribed to the Car Connectivity Consortium, except for Apple. Regarding its functioning, these three systems show certain similarities, as they all project the Smartphone screen onto the infotainment system's and allow users to use certain applications, although the full catalogue of installed applications in such devices would not be accessible. All these applications would be adapted to be used in vehicles in accordance with the design, safety and usability standards of the car manufacturing sector.

Therefore, it must be stressed that these systems are not native to the actual vehicles. In fact, despite the fact that only a few commercial proposals including them have been pitched, the vehicles that have them will also count on the classic native system as devised by the manufacturer and also on a button that will grant the user access to these solutions. However, in principle, unless manufacturers design their own customization layers for these systems, users will not be able to control aspects such as the vehicle's air conditioning system or the radio, or to access the vehicle's setup panel. To sum up, users will always have two different systems in their vehicle, both with different interfaces.

It is widely thought that these systems may be the ultimate and universal solution concerning the use of applications in vehicles, but this might still be an unrealistic idea if we take into account what these products can actually achieve and the international safety standards to which they are subjected. Besides, it would also be necessary to bear in mind what the most common consequences to the use of these proprietary solutions might be. What would happen if a user decided to purchase a new telephone which they would not be able to use in their vehicle? Or if different members of the same family used the same vehicle but all of them had devices operating on different platforms? If, in a best case scenario, it were possible to install several systems in the same vehicle, what would the total cost be? The whole picture is becoming considerably more complex than it might have seemed in the first place.

D. Smart Objects

The Internet of Things and its Smart Objects, small devices that carry out specific functions within a wider network, are the ultimate development in the technological world. In order to understand their usability and future expansion, we can list a number of examples. For instance, there could be Small Objects that are able to measure the temperature or the relative humidity index in a certain environment, but there could also be other more elaborate ones that can detect a presence or, to a higher degree of sophistication, help control or prevent diseases. All these devices are connected within wider networks which receive the information gathered by the Small Objects and allow us not only to monitor their activity, but also to carry out actions on the devices so that they respond in real time to the measured conditions.

According to a study by Cisco[9], in 2015 there will be twenty billion connected devices, and this figure could be doubled by 2020. Therefore, Smart Objects are one of the subjects of interest within our industry, since they will allow nearly any everyday device to connect to the Internet and to be incorporated into the communication structures in our home or in our business. These objects would be useful both in an industrial or corporate environment and in a domestic one. In fact, DIY (do it yourself) objects will become one of the main factors contributing to the expansion of IoT networks, along with a decrease in the price of electronic components and the standardization of 3D printers.

Even though Smart Objects could still be considered a brand new element in the grand scheme of things, there is a series of mistakes that are already being made. There are not any standards in place regarding communication and interconnection protocols. There are not any free platforms that allow object networks to be developed universally in a controlled and safe environment. The market is currently monopolized by private solutions that do not interact among themselves, without the existence of open protocols or APIs that make it possible to develop a more sophisticated product.

Let us take the example of a simple "smart bulb" to help us illustrate this situation. The supplier would provide us with an application for our Smartphone from which we would be able to set up a timer for the bulb to go on and off. Thus, when we got home our Smartphone would detect the presence of our lightbulb and, depending on the time, the application would establish whether the bulb needs to be on or off. This simple case is perfectly valid, although it would be far more convenient if we could have a bulb that went on or off according to a much more realistic factor such as the degree of luminosity rather than the time of day. In this case, we should somehow communicate with the bulb and with a sensor. Complications could arise, such as the lack of a common communication protocol between the bulb and any other device, like the sensor. To sum up, nowadays it is difficult to develop Smart Objects networks by different manufacturers and which have a common platform that allows communication amongst them all, carrying out data analyses and making decisions.

We have, once again, come to the same conclusion: the new sector is making the same mistakes when they decide to hinder standardization and the use of open protocols.

III. Security and Privacy

Another important subject that seems to be overlooked for the abovementioned parties is that of security and privacy. Manufacturers will launch devices with the sole objective of making our lives easier but, at the same time, these devices gather highly sensitive information concerning users and their habits, which could potentially be dangerous if a third party accessed them with malicious purposes.

The exploitation of vulnerabilities also affects new devices. Let us give as an example the registered case of hacked televisions [10], by which a hacker would gather information by means of the built-in webcam that these televisions incorporate. This will soon extend to connected cars, in which everything is controlled by means of sensors and servos, resulting in hackers being able to access the braking or the accelerating system, among others [11]. Recently, a BMW executive warned us about the interest that many companies are showing lately on the enormous amount of data issued by modern vehicles [12].

A good example of such behavior on the part of manufacturers is that of traditional routers connecting households and small and medium-sized businesses to broadband networks. A decrease in their price resulted in less investment on the development of their firmware, and therefore on their security and privacy mechanisms, which resulted in our current scenario, with millions of routers sold in the last few years presenting great vulnerabilities, since sales were more important than security [13].

To sum up, given the insufficient quality of security nowadays, it is a matter of time before Smart Objects get hacked. This situation is due to manufacturers having ignored any engineering processes concerning the design of security for their devices. This result in a reduction of the costs incurred in during the development of the product and enables a higher number of products to be launched in a much shorter period of time.

IV. Cloud Computing

The Cloud is an essential component of any application and, according to analysts, the Cloud services market will grow exponentially in the next few years, along with the Internet of Things industry. However, the use of the Cloud has been limited to simply acting as a data storage application, granting said information an "apparent ubiquity", whereas in actual terms it does not provide any advantages for the end user. The Cloud, along with other broadband data lines, smartphones and tablets, has changed the way we create and use content. Along with its flexibility and pay-per-use model, one of the greatest advantages of the Cloud is its actual ubiquitous access to applications and services on the Cloud. However, nowadays this ubiquity is very much limited to its ability to access applications and services adapted to multiple devices, as these solutions can be:

- Expensive, for they require complex and specific developments for each platform

- Incomplete, because, although there is some degree of adaptation of the graphic interface to each device, it is not a dynamic adaptation and, more often than not, a change of device will imply a change in the model of interaction. The majority of services in the Cloud are useful in terms of their server functionality without user interaction, as most of them focus on infrastructure.

If we looked back a decade ago, we would be able to clearly identify the major software developers worldwide. However, nowadays young people's skills at monetizing their innovation and creativity go mostly unnoticed. With technologies like the Cloud, html5 and other capacities, the application economy has changed dramatically. Amateur programmers can now make use of their creativity and innovative spirit to develop new applications or services and offer them through app stores. Unlike in past decades, today it is not necessary to count on a major capital investment in order to start writing code and start selling new applications or services. Costs have decreased drastically as a result of cloud computing and Infrastructure as Service (IasS). The Cloud has changed the way modern-day software is created.

This ubiquity becomes apparent because, in effect, when sharing data among several applications in different devices and from any location, it becomes clear that we have universal access to that information. The context of the knowledge of applications resides on those data in the Cloud and not on the actual application that has been physically installed on a device. But this "universal" availability is a first step that gets taken for granted by any user, it is the bare minimum, but does not go any further than that. The Cloud at a user level is clearly underused, since it does not allow users to go one step beyond, it cannot be used somehow to fill the niche of applications that are independent of platforms or devices. However, we are sure that there would be many more possibilities of use. Later in this article we will present a number of proposals that may contribute to the abovementioned ubiquity not to refer to data only, but also to the actual user interfaces.

V. CROSSPLATFORM DEVELOPMENT

The current model for application development is not resalable, for a developer will have to create as many versions of their application as platforms exist in the market. This not only means smartphones, as we have seen, but any other type of electronic device that is connected.

The need to develop different versions of applications and to adapt them to different devices entails a high cost. Adaptability is essential in order to create a sustainable application ecosystem.

Currently, the different operative systems are at war, which will result in the need to create more versions of the same application if the ultimate goal is to make said applications available to each and every user. This, however, would not be viable in the current scenario. Fragmentation thus becomes unsustainable.

In such a scenario, developers end up having to sacrifice a number of versions for the sake of the two or three versions of their application that would reach the highest number of potential users, thus maximizing monetization. The remaining options become automatically discarded.

The user interface management system that traditional operative systems use (Windows, Mac, Android, etc.) gives programmers more freedom through the use of its APIs, so much so that it enables them to communicate with the user by setting up interactive objects such as buttons or menus with windows and dialogues that are transferred to the user in order to show them the new information. Once the user has filled in the information required by the application, the window or user process dialogue moves on to validate the received information.

This paradigm for the development of user interfaces is not sustainable in a complete cross-platform environment, since the interaction device followed by the user at any given time does not guarantee the existence of windows or buttons (let us imagine, for instance, that the device were only able to recognize and utter speech).

We cannot state, however, that there are not any solutions in the market that advocate for a cross-platform development environment which would allow the coveted approach of "develop once, deploy many times". These, though, are still limited to specific platforms and services. For instance, there are a number of development environments that enable us to automatically export an application into iOS, Android or Windows Phone, to name a few. There are some systems that may include even more versions. However, there are not any universal solutions that are not focused on smartphones only and that apply to all kinds of devices, as we have mentioned: TVs, cars or any kinds of Smart Objects.

If we look beyond the actual devices, we can find even more problems when it comes to dealing with different interaction systems. An application that has been designed to be used in smartphones will not be easily adapted to a TV and the user experience would not be suitable, either. Moreover, if we wanted to use the same application in a small Smart Object, we would be faced with more frustration and problems related to subjects ranging from the actual human-machine interaction (haptic, voice, gestural) to the visualization of results of said interaction. It cannot be expected to use an application with a user interface if we do not have a screen to see it on.

VI. SOLUTIONS

If we take into account both conceptual and lexical design when proposing universal access to the knowledge base of an application, we can state that, even when accurately planned, said universality cannot be guaranteed, as there will be situations where the interface has not been designed for its general use. This is why it will be necessary, in most situations, to develop several versions of said interfaces which can be adapted to the needs of both devices and human groups.

In order to avoid this, said conceptual and lexical levels have to be dynamically generated according to the contextual and interaction requirements of that moment, creating a user interface management system that is intelligent and able to design and construct interactive dialogues in real time that can be especially adapted to the cognitive, perceptive and motor characteristics of an active user, as well as to the technical characteristics of the device that the user is using at that very moment, granting a great variety of users and devices full access to the interface, including those users who suffer from certain disabilities. Thus, the functionality of the application would be completely independent from the interface, which would hugely increase the system's efficiency in terms of usability and user experience. The computing capacity of the Cloud would be a solution to potential load issues in small devices, whose only concern would be the user interface.

As it is impossible to know about the device's interaction characteristics when designing and compiling, a change of paradigm becomes necessary, where the programmer would determine their needs in terms of UI (types of data required to launch a process, in/out parameters, response, etc.) and an independent service would decide in real time how to communicate with the user and their interaction device in order to obtain the information required. We can now introduce a new concept: the concept of Cloud UI, user interfaces in the Cloud.

Nowadays, we are able to maximize the Cloud's calculus power and the ubiquity that it provides us with in order to use them for the greater benefit of users and applications alike. The goal should be that applications become an essential actor in the user interface, rather than just a final solution regarding data treatment and visualization. This idea is not new, however, and there exist solutions and R+D projects based on these ideas.

Active Video[14], with its platform CloudTV, proposes a solution of this kind for TVs and set top boxes, and it is based on transforming the user interface into a video stream. This would enable the user to visualize the user interface from their device and interact with it as if it were a native application. The difference lies in the fact that the data issued throughout the interaction are sent to the Cloud, processed and sent back to the user as a response in the shape of a new video stream.

With the new interface model in the Cloud, we achieve the coveted paradigm of cross-platform applications. Regardless of the device's operative system, brand or model, we only need a small connector that would enable us to automatically use all the applications that are designed to work on such a platform. It would not be necessary to replicate and adapt applications to different systems; it would only have to be done once.

Another element to discuss would be the fact that user interfaces, instead of being tailored to needs, should become description systems. The VISIO project, developed jointly by the University of Oviedo and Zed Worldwide, is based on this concept. Communication with the user is based on a set of minimum requirements in terms of the application's functionality, which are as follows:

- The information the user must have
- How they must have it
- How it should be communicated to the system

Bearing these in mind, the actual user interface gets relegated to the background and what becomes important is the fact that any person would be able to use a specific application regardless of the device or the kind of visualization and interaction interface they are using. Through the use of mechanisms that describe an application's interface and its behavior towards the user, it would be plausible for the same application to be used in completely different devices:

- Smartphones, Smart TVs, connected vehicles and even small Smart Objects,
- Devices with a screen with, for example, small LED displays or even devices without a screen where an interface reads and describes the application or emits sound effects when the task has been processed successfully.
- Devices with various interaction systems: haptic, gestural, voice, etc.

To automatically generate the applications' interfaces in real time would enable us not only to make a general use of them, as we have just seen, but also it would improve the quality of the user experience by gradually adapting the adaptations and/or alterations of the interface to the user context so that they do not occur suddenly. For instance, through the use of logic engines, a rule defines the size of a button as BIG if the degree of driving precision is LOW and the degree of visual precision is HIGH. The engine defines the probability of an interactive object to be LONG as a combination based on the probabilities of the engine precision and the visual precision. Interaction with touchscreens in work or home environments will be completely different to the interaction that takes place in a moving vehicle (such as a car, metro or bus). These situations could be detected (based on the sensors of the actual device) and we would be able to apply various methods of interaction to each of them. Regarding interaction in movement, the degree of precision for the user suffers significant degradation and this would be detected by the system. When we apply logic rules, when we increase the probability of the driving precision to be low, the chances of generating a bigger size button increase. This is exactly what would be expected in the context we have described, since the use of longer buttons increases the level of precision of the interaction in a moving environment. Once the user has left said environment and the driving precision increases once again, the size of the buttons decreases again.

This simple example that appears to be obvious is not viable to be used with the current user interface design paradigms. All the elements are static and would appear in the same format as they have been previously defined, regardless of the conditions of the user context. Again, there is a seeming ubiquity, but it is completely de-contextualized.

User interfaces tend to be designed to satisfy the needs of the "typical user". The classic design of interfaces does not consider human individuality and diversity. As a consequence, the resulting systems present serious flaws regarding user experience, since those who do not fall into the category of "typical user" will face problems and frustration on using these systems. The use of systems based on the definition of interfaces would be a solution for this problem: the adaptation for users with disabilities. Applications would be possible to adapt to visually impaired users or to users suffering from reduced mobility. The actual system can be adapted, for it has not been designed to be used in a specific way, but thinking of its utilization, of the information that it must know and of what the data exchange with said system should be like.

However, we must not claim that this would be a perfect, problem-free solution. There is a major problem that is quite obvious: the total dependence of the system on connectivity, without which it would be impossible to render the user interface. Despite the advantages that the use of local caches might have, as well as the possibility to use the offline mode, it becomes quite clear that this system might still be proposing a limited scenario. In any case, we must also take into consideration that the vast majority of applications we use today, despite unfolding interfaces locally, depend almost entirely on a connection to be able to operate. The apparent ubiquity we mentioned earlier on this paper and that stems from the synchronization of data from the Cloud would still be hindering the optimal use of the application, which, even when it is clearly a weakness, does not necessarily mean that native applications are entirely free from it.

Along the same lines, another issue could be a delay in the network that could cause great frustration to the user if there is not an almost immediate response between call and return. Once again, this is a situation that might well happen today, as our dependence on Cloud services is almost total.

As per the advantages of this system, besides the ones that we mentioned in a previous section of this paper, i.e. the fact that they can be cross-platform, used in multiple devices and for general purposes for every type of user, regardless of their condition, we must mention other additional benefits:

- Scalability. It is scalable not only at a computational level, but also at a functional level. Given the fact that all the processes reside in the Cloud, it is relatively simple to add new models of interaction. This scalability applies equally to the number of devices that can be connected to our system.
- A much more efficient user experience. Improvements in the system can be universally applied to all users and devices. With a numerous user contingent we can undertake mass usability tests, significantly increasing the processes of interaction and thus generating functional improvements.

VII. Conclusion

Every strategy designed to reduce the human and technological limitations and achieve a significant increase in the potential use of any technological tool should include the creation of a wide range of communication channels through which humans and machines can exchange information. According to the type of user or device that makes use of the service, we can adapt not only the communication channel to be used but also the way in which information is perceived by the user, i.e. the interface, using even the knowledge that the system might possess about the user context.

The human-machine interaction must be considered as something more than the simple use of a touchscreen and comprehensively and universally include any type of device, interaction and human condition, so that technology becomes a means to unite, not to separate, for those who use different platforms, for those who suffer from certain disabilities or for those who, in certain situations, cannot use certain types of interaction.

Even though it is down to users to demand these technological advances, it is also the responsibility of device and operative systems manufacturers to approach technology with a long-term vision, through the use of open standards and protocols, thus facilitating benefits for all the parties involved. It might be that today's business models, based on pay-per-download, or even in-app sales and subscriptions that are inaccessible for providers in app stores are not ideal and should be looked into. It may also be convenient to test new models that, in accordance with the ubiquity requirements that future electronic devices might present, continue to allow providers to have satisfactory account balances.

An understanding approach to the user context in the upcoming years will become one of the major axis of online sales channels (particularly from mobile devices), offering the user what can be inferred that they are going to need and/or want according to who they are, where they are and their previous activity with information systems. If these means can be used to "recommend" the purchase of specific goods and services, why should they not be used to improve the interface user experience through the use of contextual ubiquity?

Total ubiquity is advantageous not only for users themselves, but also for any sphere of their everyday lives. It is not a technological matter, but a social and ethical one. The Internet's neutrality is a widely known concept and this neutrality should be made extensive to applications and services, since these represent the digital profile of every individual. The owner should thus be free to export said profile to any platform that they deem appropriate.

Acknowledgements

This paper is being supported by the VISIO Project (http://www.bitmonlab.com/rdi/Visio/Index.html), funded by the "Acción Estratégica de Economía y Sociedad Digital" programme of the Ministry of Industry, Energy and Tourism of Spain. The VISIO project is collaboration between the University of Oviedo, Spain and Zed Worldwide, Spain.

References

[1] Apple App Store press release, July 14, 2008.

[2] http://www.apple.com/pr/library/2008/07/14iPhone-App-Store-Downloads-Top-10-Million-in-First-Weekend.html
[3] Removal of restrictions can decrease music piracy
[4] http://business.rice.edu/jgsb2011_2column.aspx?id=4294967941
[5] SAMSUNG Revolutionizes the Viewing Experience With Innovative New SUHD TV, January 05, 2015. http://www.samsung.com/us/news/24393
[6] Panasonic and Mozilla to Collaborate on New Firefox OS Open Platform for Next Generation Smart TVs, January 6, 2014. http://www2.panasonic.com/webapp/wcs/stores/servlet/prModelDetail?storeId=11301&itemId=705511
[7] LG Electronics Acquires webOS from HP to Enhance Smart TV, February 2, 2013. http://www.lg.com/us/press-release/webos-release
[8] Apple Car Play. https://www.apple.com/ios/carplay/
[9] Android Auto. http://www.android.com/auto/
[10] Mirrorlink. http://www.mirrorlink.com/
[11] The Internet of Things. How the Next Evolution of the Internet Is Changing Everything. Dave Evans, Cisco, Cisco Internet Business Solutions Group (IBSG), April 2011. http://www.cisco.com/web/about/ac79/docs/innov/IoT_IBSG_0411FINAL.pdf
[12] Samsung Smart TV a spy in the living room as webcam hack revealed, August 2, 2013. http://www.slashgear.com/samsung-smart-tv-a-spy-in-the-living-room-as-webcam-hack-revealed-02292655/
[13] Hacker Says Attacks On 'Insecure' Progressive Insurance Dongle In 2 Million US Cars Could Spawn Road Carnage. Forbes, January 1, 2015. http://www.forbes.com/sites/thomasbrewster/2015/01/15/researcher-says-progressive-insurance-dongle-totally-insecure/
[14] BMW sounds alarm over tech companies seeking car data. Fiancial Times. http://www.ft.com/intl/cms/s/0/685fe610-9ba6-11e4-950f-00144feabdc0.html#axzz3PG1kItGq
[15] Lizard Stresser Runs on Hacked Home Routers. Krebs on Security, January 2015. http://krebsonsecurity.com/2015/01/lizard-stresser-runs-on-hacked-home-routers/
[16] Active Video. http://www.activevideo.com/

4

Emerging Technologies Landscape on Education. A review

Luis de-la-Fuente-Valentín, Aurora Carrasco, Kinga Konya, and Daniel Burgos
UNIR Research, International University of La Rioja, Paseo de la Castellana 163, Madrid, Spain

***Abstract* — This paper presents a desk research that analysed available recent studies in the field of Technology Enhanced Learning. The desk research is focused on work produced in the frame of FP6 and FP7 European programs, in the area of Information and Communication Technologies. It concentrates in technologies that support existing forms of learning, and also in technologies that enhance new learning paradigms. This approach includes already adopted and successfully piloted technologies. The elaboration of the desk research had three main parts: firstly, the collection of documents from CORDIS and other institutions related to TEL research; secondly, the identification of relevant terms appearing in those documents and the elaboration of a thesaurus; and thirdly, a quantitative analysis of each term occurrences. Many of the identified technologies belong to the fields of interactive multimedia, Human-computer Interaction and-or related to recommendation and learning analytics. This study becomes a thorough review of the current state of these fields through the actual development of R&D European projects. This research, will be used as a basis to better understand the evolution of the sector, and to focus future research efforts on these sectors and their application to education.**

***Keywords—* Interactivity, Multimedia, R&D European projects, ICT and learning, emerging technologies**

I. Introduction

Trends and hot research topics in a specific field are reflected on the research areas promoted by public and private funding programs. This is the case of the Framework Programs launched by the European Commission that "*bundles all research-related EU initiatives together under a common roof playing a crucial role in reaching the goals of growth, competitiveness and employment*" [35]. Therefore, the identification of the most repeated research topics in projects funded by FP6 and FP7 leads to the identification of the most relevant research topics in Europe since 2002. The work presented in this document focuses on the educational field, and aims to identify the most relevant technologies that have impacted and will impact education in the near future. It is part of HoTEL, a European project that will provide a methodological framework to increase the impact of technological innovations in the educational field.

The desk research is focused on work produced in the frame of FP6 and FP7 programs (projects and Networks of Excellence), in the area of Information and Communication Technologies (ICT). The identified trends have been complemented by those mentioned in related works such as the reports published by the Institute for Prospective Technological Studies (IPTS), and the Horizon Reports produced by the New Media Consortium (NMC). The result is a landscape of the currently most relevant research topics in Technology Enhanced Learning (TEL), and the areas of learning in which these technologies are being applied.

Being one of the goals of this paper the identification of already adopted technologies for learning, we have reviewed the existing literature in order to identify those aspects of TEL that are considered effectively adopted or planned to be adopted in the near future. In particular, we have focused on review studies from the IPTS, who gathered their information from experts workshops, field consultations and desk research. Thus, they provide a complete view of current state of TEL in the European countries.

Another aim of this paper is to identify those technologies that will impact future education. For this purpose, we have focused the analysis of the different editions of the Horizon Report (i.e. Higher education, K-12 and museums) produced by the NMC.

This document is structured as follows, Section II describes the methodology used to perform the desk research and Section III presents the thesaurus used as reference. Sections IV, 0 and VI analyse the information from different perspectives: research trends, research areas and application learning areas, respectively. Section VI.H elaborates on the analysed Networks of Excellence (NoE). Next, Section depicts the future applications of the presented study and,finally, Section VIII presents the conclusions of the study.

II. Methodology

A. Data Sources

This desk research is aimed at identifying technologies enhancing new forms of learning, specially focusing on the research produced in the frame of EU funded programmes. However, for the elaboration of this document we have also considered other information sources.

In particular, we have analysed the EU research projects and NoEs funded by the FP6 and FP7 programs, and studies and reports from the IPTs and the NMC.

1) FP6 and FP7 projects

As stated in [35], the CORDIS repository stores information about research projects funded by the European Union which are carried out by European, research institutions such as commercial organizations, research institutes and universities.

The CORDIS repository stores the project information once the agreement has been signed, and, if there is a change in the contract, updates the stored information upon specific request. The records remain in the repository after the project has ended. Not all projects contain the same set of information fields, however, there is an overall homogeneity that allows bulk retrieval and classification of projects information.

2) FP6 and FP7 Networks of Excellence (NoE)

Information about NoEs is stored and structured in CORDIS in the same way as in the case of the projects. However, a NoE project should not be strictly considered as a research project. Therefore, NoEs have been analysed separately from projects, and they are described in order to complement the conclusions extracted from FP6 and FP7 projects.

3) Institute for Prospective Technological Studies (IPTS)

The IPTS is one of the seven scientific institutes of the European Commission's Joint Research Centre (JRC). The studies carried out by the IPTS lead to the publication of reports, which are focused on specific scientific and technical issues. These reports (around 700) are publicly available for consultation, being some of them quite aligned with the goal of this desk research. Therefore, they have been considered as one of the sources for our analysis.

4) The New Media Consortium (NMC)

For the sake of this desk research, the most popular report offered by NMC: the Horizon Report. Annually published, it is "*a decade-long research project designed to identify and describe emerging technologies likely to have an impact on learning, teaching, and creative inquiry in higher education*".

Since the objective of the Horizon Reports is quite aligned with the goal of this desk research, they have been considered as one of the sources of our analysis.

B. Selection criteria

1) FP6 and FP7 projects and NoEs

From all the FP6 and FP7 projects available for consultation at CORDIS[1], we have selected those particularly interesting in the field of Technology Enhanced Learning. At the CORDIS page, we used the simple search interface as follows:

- Search terms: education, TEL, learn, learning, teach, teaching, train, training and school
- Framework programme: FP6 and FP7 checked.

We reviewed the description of the found projects and discarded those that did not fit with the desk research goals. For example, the *3D VIVANT* project was in the results set, but it was discarded because it is not directly related to education. As a result of this manual review, 102 projects were selected.

The collection of 102 projects used for this desk research contained 7 Networks of Excellence. In a second review round, we examined the project descriptions in more detail and assigned several keywords to each project. More details about the keywords assignment are given in subsection C. As we read the project descriptions more carefully in this second round, more projects were discarded, containing the resulting set a total of 86 projects.

2) Institute for Prospective Technological Studies

From all the reports available at IPTS, we have focused on those regarding ICT and TEL. Therefore, we have filtered the number of documents by applying the criteria described as follows:

- Using the search engine provided at the web page [25], we have restricted the reports to those matching the keyword "ICT".
- Then, we have manually reviewed the project descriptions, and we have selected those that explicitly mention education.
- Finally, we have reviewed the reports given the summary of e-Learning related projects [26] and added them to our selection.

3) The New Media Consortium

In this desk research, we have used the NMC Horizon Reports to analyse trends and predictions on ICT and education. We considered the reports produced since 2009, because older reports are no longer relevant in the identification of trends.

C. Analysis

1) FP6 and FP7 projects

It was the authors' opinion that the *Subjects* field (a classification of projects on thematic areas provided by CORDIS) does not offer sufficient information of the project's research topics. For example, the ELEKTRA project aims at the development of an educational game, which is hardly outlined by its subjects (education, training - information processing, information systems-telecommunications).

Thus, the first task was to obtain a more descriptive set of keywords to summarize each project topics. This was manually done by the authors: firstly, reviewing the project description; secondly, writing keywords that matched with the description. A combination of the identified terms with the terms extracted from the IPTS and NMC reports resulted in the elaboration of a thesaurus that guided the classification of the research topics and the areas of learning they support, to be presented in Section III.

The identified NoEs have been separately discussed, and they are described in Section VII in order to complement the analysis of the projects.

2) NMC and IPTS reports

All NMC Horizon Reports follow the same structure: they choose six different technologies and classify them by the expected time for adoption. Three classifications are available:

[1] http://cordis.europa.eu/fp7/ict/projects/home_en.html

"one year or less", "two to three years" and "four to five years". For each identified technology, the report includes an executive description, a discussion of its relationship with education, and a collection of examples that use technology in the educational field. For the purpose of this desk research, we have compiled all the identified technologies in the reviewed reports, also considering its field of application (given by the report in which they appear) and the time period on which the prediction was made.

The IPTS reports do not follow such a fixed structure and the way the information is presented depends on the specific objective of the report and the project that frames the document. Thus, it is not possible to extract the information from these documents with the methodological approach adopted in the case of NMC. Due to this, and considering the nature of IPTS reports, they have been used to take note of the technologies they explicitly mention. In other words, to determine which technologies were considered relevant for their studies.

Information collected from both NMC and IPTS reports were used in Section IV to analyse research trends in the TEL field, and to perform a time based analysis of the TEL predictions.

III. ICT Thesaurus Used As Reference

During the analysis of the documentation, a number of technologies have been identified. In some cases, the same technology was mentioned using different terms, in other cases, different technologies are related among themselves so that they can be clustered in research areas. With all the identified terms and technologies, we have elaborated a thesaurus that classifies them by research topics and the areas of learning they support, finds synonyms and allows for a more systematic analysis. The thesaurus has been created as follows:

Firstly, the descriptions of all the selected FP6 and FP7 projects were manually reviewed, assigning them keywords. The keywords identified the main and secondary technologies used in the project, as well as the area of learning where the project applies. The reviewers were asked not to use a pre-defined set of keywords, but to be spontaneous. The technologies that appeared at IPTS and NMC reports were also considered, and this time the names were collected as they appeared in the documents. Note that the manual nature of the keywords selection could have introduced an involuntary bias on the study.

Then, synonyms were identified from all the keywords collected at the previous steps, reducing the number of available terms, and translating the spontaneous keywords assigned in the first step, into a collection of more formal terms. Finally, the related technologies and areas of learning were grouped into clusters. The creation of the clusters was guided by the research areas identified in all the reviewed documents and the expertise of the authors of the help desk.

The thesaurus, shown at appendix A, has been used for the classification of FP6 and FP7 research projects. It is divided in two main categories (technologies and areas of learning), which are divided into subcategories (also referred as clusters). The technologies related to Artificial Intelligence and Interactive Multimedia play a transversal role in the thesaurus. That is, these technologies are represented in most of the clusters, but there is no cluster specifically dedicated to them.

IV. Trends In Technology Enhanced Learning

This section presents the review of the NMC and IPTS reports and the conclusions extracted from them. The reviewed documents can be classified in three types:

a) IPTS review reports: this type of documents usually review the existing examples and are focused on past research.

b) IPTS reports identifying trends: documents such as [29], [31] or [33] have the goal of identifying trends that will build the future of TEL.

c) NMC reports: the Horizon Report charts the landscape of emerging technologies and provides a complete collection of the technologies that are expected to impact education in future years.

The analysis resulted in Table I, which summarizes the occurrences of relevant keywords in the reports. In the table, each column represents one of the above mentioned document types, and each row refers to one of the clusters identified in the created thesaurus. Each cell's content determines whether or not a given cluster is mentioned in the corresponding type of documents. For example, the reference to [24] in the first row, second column, means that this document mentions one of the keywords belonging to the UC cluster. Mentions per document, are ordered in decreasing occurrence order per cluster. Giving an idea of the number of reports in which they appear, and their relevance in them.

Note that cells in the table may contain two or more references to the same document. This is because we placed one reference to the document for each different mentioned keyword. For example, document [4] mentions 'digital preservation' and 'electronic publishing', both classified into the CRT cluster, therefore, two occurrences of the reference to this document are in the corresponding cell.

If we consider a technology to be more relevant the more occurrences it has in the reports and the more number of documents where it is mentioned, three different behaviours appear on the table: (i) technologies relevant in past research, but not so relevant in the predictions; (ii) technologies that are as relevant in the past research as they are in the predictions; and (iii) technologies more relevant in the predictions than in past research.

The only cluster that might fall in (i) category is PA, which is mentioned in IPTS predictions but not in the Horizon Reports. In group (ii) we find WEB and SAS clusters. The relevance of web based systems in past TEL research cannot be denied, and the semantic web has counted with the support of the W3C. Finally, most of the clusters fall in (iii) category.

TABLE I
MENTIONS TO THE IDENTIFIED CLUSTERS IN IPTS AND NMC REPORTS

Clusters	IPTS Review Reports (past)	IPTS Trend Reports (future)	NMC Horizon Reports (future)
Ubiquitous computing (**UC**)	[30] [30] [24]	[31] [31] [31] [33] [33] [29]	[4] [4] [4] [5] [5] [5] [9] [9] [9] [6] [6] [7] [7] [8] [8] [3] [3] [10] [11] [12] [1] [2]
Web 2.0 based tools and systems (**WEB**)	[24] [24] [24] [13] [30]	[29] [29] [29] [29] [33] [33] [33]	[7] [7] [7] [2] [2] [5] [5] [3] [6] [8] [12]
Games and Virtual Worlds (**GVW**)	[30] [24]	[31] [31] [31] [33] [33] [29] [29]	[2] [6] [9] [10] [11] [12]
Augmented reality (**AR**)		[29] [31] [33]	[1] [3] [4] [6] [8] [10] [12]
Other technologies (**OT**)	[21] [19]	[33] [33] [23]	[11] [11] [1] [10] [12]
Human computer interaction (**HCI**)		[31] [33]	[1] [3] [6] [8] [9] [10]
Access-to-content related technologies (**CRT**)		[29] [31] [33]	[4] [4] [1] [2] [3]
Learning Analytics (**LA**)		[33] [33]	[1] [2] [9] [10] [11]
Environments and technologies for collaboration (**CSCL**)	[30]	[29] [33]	[7] [7] [12] [2]
Personalized, adaptive technologies (**PA**)	[30]	[31] [31] [31] [33] [29]	
Semantic-aware systems (**SAS**)	[30] [30]	[33]	[5] [8]

It is worth to mention the case of LA, AR and HCI. Such technologies have recently appeared in the TEL world, so they receive no mention in past TEL research. These clusters are highly related to artificial intelligence and interactive multimedia, which emphasizes the relevance of those technologies in the learning world. Surprisingly, CRT cluster has no presence at all in past research reviews. MOOCs and OpenCourseWare will definitively impact TEL research in the future years.

Artificial intelligence technologies play an important role in clusters such as UC, LA and SAS. This emphasizes their relevance in current research and in future trends. Also, interactive multimedia

A. Time analysis of NMC predictions

In this subsection we have visually represented the predictions made by NMC reports in a time line, in order to determine how they evolved over the years. We provide three different figures: one for higher education, another one f or K-12 education (the US equivalent to the 'schools' area of learning), and the last one for museums.

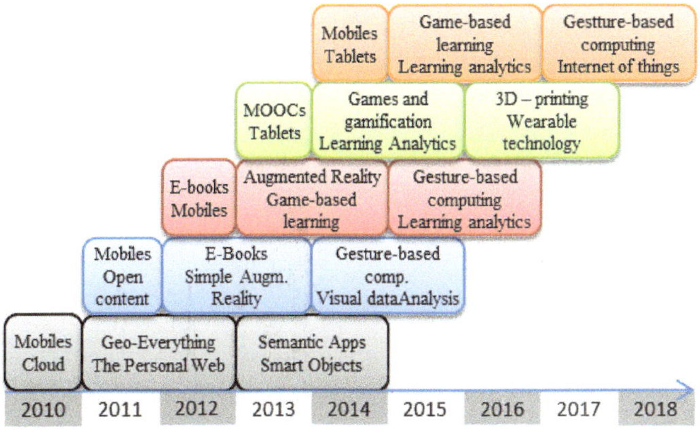

Fig. 1: NMC predictions for higher education

Fig. 1 shows the predictions for higher education. The following observations can be made:

- Since 2009, mobiles adoption is expected to occur in "one year or less". The prediction is repeated year by year but the goal does not seem to be achieved. In the last two editions of the report, 'mobiles' are being replaced by 'tablets'.
- Learning analytics appeared in the 2011 report and, since then, they are expected to have an impact on higher education sooner or later. The same can be said about games.
- Massive Open Online Courses suddenly appeared in the 2013 report, and are expected to have a clear and immediate impact in higher education.
- The 2013 report introduces new and interesting concepts, such as 3D-printing and wearable technology.

From Fig. 2, which shows the predictions for K-12, the following observations can be made:

- Mobiles are expected to be adopted, but the prediction is less "aggressive" than in higher education. However, it seems to be an always live promise.
- Game based learning is expected to play a relevant role in K-12 education. This is coherent with Section VI.A.1).

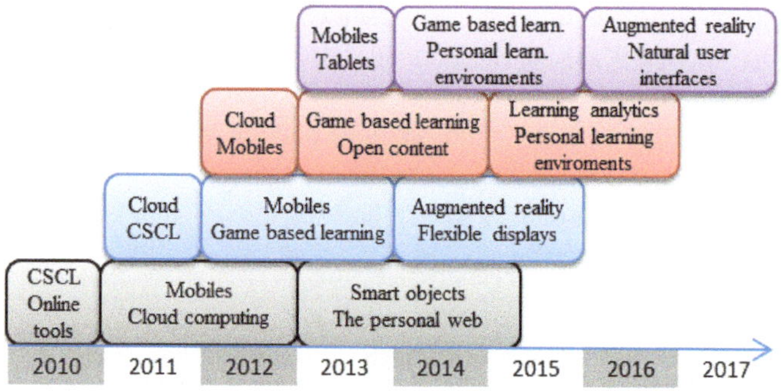

Fig. 2: NMC predictions for K-12

From Fig. 3, which shows the predictions for museums, the following observations can be made:

- Innovative methods to interact with the information are subject of research. Thus, augmented reality, gesture-based computing and natural user interfaces are expected to be adopted. This is consistent with Section VI.I.
- As it happened in higher education and K-12, mobiles are expected to be adopted in the next year (since 2010), but the goal does not seem to get accomplished.

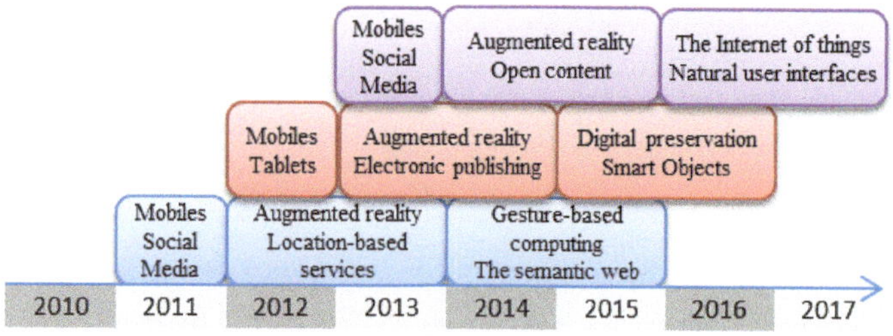

Fig. 3: NMC predictions for museums

The above analysis emphasizes how interactive multimedia is expected to impact the learning world, mainly by means of educational games and augmented reality. Also, artificial intelligence is essential part of learning analytics and is therefore expected to be widely used in higher education and other areas of learning.

V. ICT Research At FP6 And FP7

In this section, we first provide a quick overview of the topics researched in the projects funded by the European Commission within the FP6 and FP7 programs and, after that, we present a cluster-by-cluster analysis based on the thesaurus of Section III.

A first approach to determine the most frequent research topics in the analysed projects was to study the most repeated terms in the project summaries provided by CORDIS. However, the analysis revealed that the most relevant words were 'learning, project', 'knowledge, research', 'support' and 'technology', being therefore difficult to get an overview of the research topics with this method.

The second approach was to analyse the most repeated terms in the manually assigned keywords, which were selected from the thesaurus. The resulting set contains much more technology related terms, and it is therefore more feasible to obtain a quick overview of the most relevant technologies. The terms that appear the most are shown in

TABLE II. It can be seen that technologies related to learning games, content management, interfaces, collaborative environments or semantic analysis (just to mention a few) are frequent topics in FP6 and FP7 research projects, and therefore play a relevant role in recent research on Technology Enhanced Learning.

TABLE II
MOST REPEATED KEYWORDS IN REVIEWED PROJECTS

Term	#	Term	#	Term	#
semantic-aware technologies	11	human computer interaction	5	learning standards	4
Games	10	context-aware systems	5	grid computing	4
learning management systems	9	personalized content presentation	5	content repositories	4
collaborative environments	7	web 2.0	4	augmented reality	4
social networks	5	personal learning environments	4	mobile devices	4

Table III shows the most repeated keywords within the different identified clusters. A project may be related to different clusters, and thus can contain keywords associated to several of them. As a consequence, the table allows the identification of how clusters are related among themselves. Next subsections give an overview of each cluster and depict their mutual relationships. For presentation purposes, some of the keyword lists in Table III have been truncated.

A. Web 2.0 based tools and systems (WEB)

The traditional web, based on static HTML pages and with no chances for interaction, experienced a big change with the widespread use of technologies that allowed dynamic generation of web content. With such approach, the users became able to interact with the web, easily writing and publishing their own content. Then, the traditional web shifted to the web 2.0. Such change also impacted educational systems, so that learners were able to interact with the content and with peers. The "web 2.0 based tools and systems" cluster refers to those learning technologies whose typical interface is web based, designed or not for educational purposes.

According to Table III, the keywords in this cluster show that projects related to WEB are also related to SAS (7 occurrences of 'semantic-aware technologies'), UC (4 occurrences of 'mobile devices', 3 of 'context-aware systems'). LA and PA are also related to the WEB cluster.

B. Ubiquitous computing (UC)

Ubiquitous (or pervasive) computing refers to a type of human-computer interaction where the information processing is integrated into everyday objects and activities. The term ubiquitous computing was coined by Mark Weiser in 1988, and the topic has received much attention since then.

Ubiquitous computing is related to distributed computing, mobile computing, sensor networks, human-computer interaction, and artificial intelligence. There are many ways in which information can interact with people. For example, from a non-intrusive perspective, ambient displays may react to events by changing their decoration; another example could be the TV volume automatically turned down when a visit enters the living room.

Ubiquitous learning refers to the support of anytime-anywhere learning, that is, when the specific goal of the ubiquitous technology is to support the learning process. Intuitively, this technology fits with informal learning approaches, but also applies to formal learning strategies such as shifting the classroom, or activities involving different spatial locations.

Table III shows how the UC cluster is quite related to PA (2 occurrences of 'personalized content presentation') and WEB (2 occurrences of 'social networks').

C. Augmented reality (AR)

Our lives are increasingly surrounded by technology, and it is difficult to think of a daily task that is not supported by any device or gadget. We use computers for work and leisure, and we communicate with others by using smartphones which integrate a lot of functionalities. In this scenario we can use the displays of our devices to view the real world, but another interesting usage is to augment our view of the real word by mixing such view with elements drawn by the computer. Such mixture of the real world with virtual elements, only visible through the corresponding display, is what we call augmented reality.

Such technology has been exploited in games, e-commerce and also learning. The information that we capture with our senses can be enhanced with augmented reality techniques, with the corresponding impact in the learning process.

Projects in the AR cluster are quite related to HCI, with occurrences of 'human-computer-interaction', 'haptic devices' and 'gesture-computing'. Keywords such as 'immersive virtual worlds' or 'simulations' also reveal a kind of overlapping with the GVW cluster.

D. Access-to-content related technologies (CRT)

Before the digital revolution that came with the Internet, the production of educational content was hardly affordable by teachers, who used textbooks created and sold by third parties. However, digital content is easier to create and, more important, easier to reuse. Reusability of learning content increases the efficiency of the authoring process, so the definition of methods for an efficient creation, distribution and use of learning content has been on hot topic in educational research.

Content repositories store large amounts of learning objects, which can be used to compose new and richer learning objects. Interoperability among repositories enables a more agile distribution of such learning objects. Such agile distribution of content required new copyright strategies, compatible with the surrounding technology. Thus, in 2002 the OpenCouseWare initiative (course lessons created at universities and freely published via the Internet) was launched at the MIT and was soon reinforced by the launch of similar projects at different universities. Also, the Creative Commons license defined the copyleft (as opposed to copyright), that provided legal mechanisms to authors to allow free access to their content. The last step towards such open culture is the MOOC movement (Massive Open Online Courses), consisting on online courses (not only content) freely offered via the web. MOOC is one of the current buzzwords in educational research.

Personalization and adaptation, possibly enabled by semantic analysis, are promoted by the large availability of learning content. Thus, the CRT cluster is mainly related to SAS, PA and WEB.

E. Human computer interaction (HCI)

Human–computer Interaction (HCI) involves the study, planning, and design of the interaction between people (users) and computers. It is often regarded as the intersection of computer science, behavioural sciences, design and several other fields of study. Strictly speaking, HCI research brings on innovative computer interfaces, including new way of presenting the information and new ways to introduce human's input.

The HCI cluster is quite related to AR, with 2 occurrences of 'augmented reality'.

F. Learning Analytics (LA)

While we interact with computers, we produce large amounts of information that can be stored and processed. The analysis of such information has been used in fields such as e-commerce to determine the customers' personal habits, and vendors use this technique to offer them personal recommendations, usually aimed at increasing the benefits.

As computers enter in the educational world, learners' activities can be traced and stored and this fact enables the use of the abovementioned analysis techniques. Thus, learning analytics is defined as the measurement, collection, analysis and reporting of data about learners and their contexts, for purposes of understanding and optimizing learning and the environments in which it occurs.

According to Table III, the LA cluster is mainly related to WEB and GVW. However, LA is so recent in the TEL field that not so many projects have been funded in the topic. As a result, there is no meaningful overlap with other clusters.

G. Games and Virtual Worlds (GVW)

The term virtual world is largely used as a synonym of 3-D virtual environment. In such environments, the user plays the role of an avatar and "lives" inside this world. Virtual worlds have evolved from simple text-based interfaces to rich 3-D environments that allow for an immersive user experience, where a real person interacts with objects in computer simulated environments.

Virtual worlds provide means to augment or enhance the way we receive information and have received the interest of educational researchers. In them, interaction with peers is an important part of the environment, so they enable the collaborative and cooperative learning strategies. User engagement in the virtual world can be achieved by gamification techniques, thus enabling serious games to be developed for educational purposes.

Projects aimed at the research of educational game are also usually complemented by very different technologies. As a result, GVW cluster is related to almost all other clusters in the proposed thesaurus.

TABLE III
KEYWORD OCCURRENCES GROUPED BY CLUSTER

LA		GVW		CSCL		WEB		CRT		PA	
information visualization	3	Games	9	collaborative environments	7	learning management systems	8	content repositories	4	learning management systems (WEB)	5
learning analytics	3	immersive virtual worlds	3	computer supported collaborative learning	2	semantic-aware technologies (SAS)	7	open source content	3	personalized content presentation	5
social networks (WEB)	1	simulations	2	online communication tools	2	social networks	5	digital preservation	3	semantic-aware technologies (SAS)	4
social media (WEB)	1	Mobiles (UC)	1	tactile interfaces (HCI)	1	mobile devices (UC)	4	semantic-aware technologies (SAS)	2	personal learning environments	4
online applications for teaching (WEB)	1	information visualization (LA)	1	semantic-aware technologies (SAS)	1	learning standards	4	personalized content presentation (PA)	2	collaborative environments (CSCL)	3
learning management systems (WEB)	1	content repositories (CRT)	1	information visualization (LA)	1	web 2.0	4	grid computing (WEB)	2	electronic tutors	3
Games (GVW)	1	adaptive learning system (PA)	1	electronic tutors (PA)	1	grid computing	4	web interfaces (WEB)	1	web 2.0 (WEB)	2
data mining	1	virtual environments	1	attention management (PA)	1	context-aware systems (UC)	3	super rich online repositories	1	Games (GVW)	2
virtual environments (GVW)	1	tactile interfaces (HCI)	1	artificial intelligence (PA)	1	learning analytics (LA)	3	serious games (GVW)	1	adaptive content	2
student monitoring	1	student monitoring (LA)	1	ubiquitous computing (UC)	1	personal learning environments (PA)	3	semantic interoperability (SAS)	1	emotion-aware systems	2
semantic-aware technologies (SAS)	1	social networks (WEB)	1	Telepresence	1	Interoperability	3	learning standards (WEB)	1	context-aware systems (UC)	2
multimedia content (WEB)	1	Interfaces (HCI)	1	social networks (WEB)	1	collaborative environments (CSCL)	3	content creation	1		
mobile devices (UC)	1	gesture based computing (HCI)	1	personal learning environments (PA)	1						
Interfaces (HCI)	1	emotion-aware systems (PA)	1	immersive virtual worlds (GVW)	1						
e-portfolio (WEB)	1	collaborative environments (CSCL)	1								
computer supported collaborative learning (CSCL)	1	augmented reality (AR)	1								

UC		AR		OT		HCI		SAS	
context-aware systems	5	augmented reality	4	smart-tv	1	human computer interaction	5	semantic-aware technologies	11
mobile devices	4	human computer interaction (HCI)	2	semantic web (SAS)	1	gesture based computing	3	learning management systems (WEB)	4
ubiquitous computing	3	immersive virtual worlds (GVW)	1	context-aware systems (UC)	1	haptic interfaces	2	translation technologies	2
personalized content presentation (PA)	2	haptic interfaces (HCI)	1	wearable technology	1	augmented reality (AR)	2	personal learning environments (PA)	2
smart spaces	2	gesture based computing (HCI)	1	Interfaces (HCI)	1	Interfaces	2	content repositories (CRT)	2
social networks (WEB)	2	Simulations (GVW)	1	interactive whiteboards	1				
Mobiles	2	merge virtual and physical worlds	1	future classroom (UC)	1				
pervasive computing	2	enriched interfaces (HCI)	1	ambient intelligence (UC)	1				
pervasive technologies	2								

H. Environments and technologies for collaboration (CSCL)

The ability to carry out an effective collaboration is a transversal skill that, thanks to constructivist learning approaches, receives lots of attention in the educational research world. Collaboration involves interaction between humans which is, by nature, quite unstructured. Such interaction can be supported by computers. When collaborative techniques are used to foster learning with the support of computers, we call it CSCL (Computer Supported Collaborative Learning).

The case of the CSCL cluster is similar to GVW. That is, projects in this cluster are also usually complemented by very different technologies, resulting in relationships with many other clusters. Among them, PA seems to present a stronger overlapping.

I. Semantic-aware systems (SAS)

In computer science, a semantic reasoner is a piece of software able to infer logical consequences from a set of asserted facts. These facts (called semantic metadata) are usually represented by a formal language such as RDF and refer to textual or multimedia objects stored in a repository, and to users' information.

In the educational field, semantic analysis research is oriented towards the understanding of natural human language (as opposite to computer languages) in order to provide better search mechanisms on learning content, and also to provide better interaction with computers.

The SAS cluster is related to WEB (4 occurrences of 'learning management systems'), PA and CRT.

J. Personalized, adaptive technologies (PA)

According to the learning styles theories, different students achieve learning with different methods, and prefer different content types. However, computer systems usually make no difference among users and present the same content to every student, structured in the same sequence, and offering the same interface.

The personalization and/or adaptation of the learning process may cover the different needs of the different students, so they can achieve learning in a more efficient manner.

The PA cluster is quite related to WEB (5 occurrences of 'learning management systems'), SAS (4 occurrences of 'semantic-aware technologies') and CSCL.

K. Other technologies (OT)

This cluster includes all those technologies that are somehow related to the analysed projects and documents, but have not been included in the defined clusters.

Not so many projects in this cluster have been funded by the FP6 and FP7 programs, however, it is interesting because it presents not so frequent emerging technologies, that can be promising for TEL research. For example, no project related to 3D printing have been found, while the last NMC report mentions such technology in its predictions.

Other interesting technologies are wearable interfaces (also related with UC cluster), or e-books as an innovative way to deliver learning material.

In this section, we depict the use of ICT in Education, in which technologies related to interactivity, multimedia, and artificial intelligence, become a mainstream in the application context. In Table III, we highlight the topics of highest relevance, in which semantics, collaboration, social networks, HCI, context-aware, web 2.0, and personalisation, combined, score 46 out of 85, meaning a 54% of the total keywords. This combination of key concepts related to interactive multimedia and artificial intelligence become a major guideline for current and future research and development projects.

VI. Areas Of Learning

This section analyses the areas of learning to which the technology is being applied. In some projects the proposed system is aimed to a very specific field. For instance, the ELEKTRA project was focused on the development of a game to help teaching grade 8-level optics. On the contrary, the technology researched by other projects, such as E-LEGI, claims to be pedagogically neutral. That is, it could be used in any learning situation.

According to the thesaurus presented in Section III, six different areas of learning were identified:

- Formal education
- Non-formal learning
- Workplace learning
- Museums
- Unspecified field
- Others

The analysis presented in this section is based on

Table IV, which represents the cluster occurrences grouped by areas of learning, an asterisk at column C, row R, means that there is a project that research technological cluster C, and applies to area of learning R.

A. Formal education

Formal education corresponds to a systematic, organized education model, structured and administered according to a given set of laws and norms, presenting a rather rigid curriculum as regards objectives, content and methodology [36].

For the convenience of this study, we have divided the formal education learning area in three main parts: schools (primary education), secondary education, and higher education.

1) Schools

Table IV shows the list of projects that research technologies for schools. The most repeated cluster is GVW, related to games and virtual worlds, which seems logical, since games are used to increase learners' motivation and children are especially prone to like games. Therefore, games appear to be a promising technique to improve learning at schools. Another relevant technology for schools is human computer interaction

(HCI). That is, new interfaces offer more intuitive interactions with computers, and allow children (not yet used to other interfaces) to learn more intuitively.

2) Secondary education

The most relevant technologies for secondary education are the Web and learning analytics. On the other hand, promising techniques such as augmented reality and human computer interfaces are not even mentioned, and there are only two occurrences of ubiquitous computing techniques. It could be argued that innovation in secondary education is limited by the classroom scenario, where mobile phones are in most cases not even allowed.

3) Higher education

In the case of higher education, the Web is the most researched technology, with personalization techniques playing a relevant role. Access-to-content technologies are also relevant in the field. This is logical since initiatives such as OpenCourseWare and MOOCs initially started at universities.

B. Non-formal learning

Educative processes endowed with flexible curricula and methodology, capable of adapting to the needs and interests of students, for which time is not a pre-established factor but is contingent upon the student's work pace, certainly do not correspond to those comprised by formal education, but fit into the so-called non-formal education [36].

Non-formal and informal education terms are frequently used indistinctly. Without entering in a debate of the appropriateness of each term, in this study we have identified that projects that focuses on informal learning are in fact related to what we have defined as non-formal learning. Thus, these two terms have been used indistinctly.

In the field of non-formal learning, the Web is the most researched technological cluster, and research on access to content technologies also plays a relevant role. It sounds logical since non-formal learners are usually intrinsically motivated to learn, so what they need is a good collection of learning materials (provided by the CRT cluster) and a method to interact with such material at their own pace (provided by the WEB).

C. Workplace learning

By 'workplace learning' we refer to those strategies applied by companies in order to allow their employees

TABLE IV
CLUSTER OCCURRENCES IN AREAS OF LEARNING

	WEB	UC	AR	CRT	HCI	LA	GVW	CSCL	SAS	PA	OT	TOTAL
Schools	**	*	*	**	***	*	*****	**	**	***	*	23
Sec. Ed.	****	**		*		****	***	*	**		*	18
High. Ed.	******	*		***		*		**	***	****		20
Non-formal	******	**	*	***			*	*	*	*		16
Workplace	***	*					**	****	**	*****		17
Museums		*	*		**	*	**					7
Not specified	******	***	*	*	***	*	**	**	**	*****	*	27
Not exp. for learning	*******	********	*	*****	***	*		*	****		*	31
Other	*****	*	*		***		***		**	***		18

D. Workplace learning

By 'workplace learning' we refer to those strategies applied by companies in order to allow their employees acquire new competences related to their position.

Environments for collaborative learning are the most researched technologies in this area of learning. Furthermore, this is the area of learning that gives more priority to collaboration technologies (CSCL). In other words, collaborative and cooperative skills are highly demanded at the workplace. On the other hand, personalization techniques are also relevant in this area.

E. Museums

Museums could be considered as a vehicle for non-formal or informal learning. However, the development of new methods to engage visitors by mixing leisure and information delivery has unique characteristics in museums, and several projects are aimed to this specific field. Thus, we have considered museums as an area of learning different to non-formal learning.

Few projects are explicitly devoted to research technology for museums. Among them, human computer interaction techniques and Virtual Worlds are the most relevant technological clusters. In museums, it is important to engage visitors, which requires innovative methods to interact with the exhibits, therefore, research is oriented toward this goal.

F. Unspecific field

The area of learning to which a project applies was determined by extracting such information from the project description, as published at CORDIS. However, there are some project descriptions in which such information is not explicitly stated, so we were not able to classify to which area of learning they belong. This group of 'unspecific field' projects, is divided in two subgroups: firstly, projects whose goal is explicitly oriented towards educational purposes, but do not state to which area of learning they belong. We have classified these projects as 'not explicit area of learning'. And secondly, projects that matched with the selection criteria, but are not explicitly aimed at educational purposes. We have classified these projects as 'not explicitly for learning.'

1) Not explicit area of learning.

This category is composed by projects regarding educational research, but that do not explicitly specifying to which area of learning they apply.

The Web is the most relevant researched technology in this type of projects. As it also happens in the higher education cluster, personalization techniques are researched as a way to improve the educational face of the web. It can be argued that those projects that do not explicitly state their area of application are generically designed, and can be adapted to different educational settings. It is therefore important to achieve a high degree of personalization and adaptation.

2) Not explicitly for learning

Some projects are related to technologies to might be applied to educational settings, but it is not explicitly stated on the project description. In those cases, we classified these projects as *not explicitly for learning.*

Ubiquitous computing (UC) is the hot research topic in this type of projects. Surprisingly, this topic has no major relevance in the rest of the areas of learning. It could be argued that ubiquitous computing is a promising research field, but it is not still clear how to apply it in educational settings.

G. Workplace learning

By 'workplace learning' we refer to those strategies applied by companies in order to allow their employees

H. Workplace learning

By 'workplace learning' we refer to those strategies applied by companies in order to allow their employees acquire new competences related to their position.

Environments for collaborative learning are the most researched technologies in this area of learning. Furthermore, this is the area of learning that gives more priority to collaboration technologies (CSCL). In other words, collaborative and cooperative skills are highly demanded at the workplace. On the other hand, personalization techniques are also relevant in this area.

I. Museums

Museums could be considered as a vehicle for non-formal or informal learning. However, the development of new methods to engage visitors by mixing leisure and information delivery has unique characteristics in museums, and several projects are aimed to this specific field. Thus, we have considered museums as an area of learning different to non-formal learning.

Few projects are explicitly devoted to research technology for museums. Among them, human computer interaction techniques and Virtual Worlds are the most relevant technological clusters. In museums, it is important to engage visitors, which requires innovative methods to interact with the exhibits, therefore, research is oriented toward this goal.

J. Unspecific field

The area of learning to which a project applies was determined by extracting such information from the project description, as published at CORDIS. However, there are some project descriptions in which such information is not explicitly stated, so we were not able to classify to which area of learning they belong. This group of 'unspecific field' projects, is divided in two subgroups: firstly, projects whose goal is explicitly oriented towards educational purposes, but do not state to which area of learning they belong. We have classified these projects as 'not explicit area of learning'. And secondly, projects that matched with the selection criteria, but are not explicitly aimed at educational purposes. We have classified these projects as 'not explicitly for learning.'

1) Not explicit area of learning.

This category is composed by projects regarding educational research, but that do not explicitly specifying to which area of learning they apply.

The Web is the most relevant researched technology in this type of projects. As it also happens in the higher education cluster, personalization techniques are researched as a way to improve the educational face of the web. It can be argued that those projects that do not explicitly state their area of application are generically designed, and can be adapted to different educational settings. It is therefore important to achieve a high degree of personalization and adaptation.

2) Not explicitly for learning

Some projects are related to technologies to might be applied to educational settings, but it is not explicitly stated on the project description. In those cases, we classified these projects as *not explicitly for learning.*

Ubiquitous computing (UC) is the hot research topic in this type of projects. Surprisingly, this topic has no major relevance in the rest of the areas of learning. It could be argued that ubiquitous computing is a promising research field, but it is not still clear how to apply it in educational settings.

K. Other

Finally, there are some cases where the project description states an educational purpose, but the application area does not fit to any of the previous criteria. Topics covered in this 'others' area of learning are digital libraries, social inclusion, conflict resolution, learners with specific needs, music institutions and educational curriculum.

It can be seen that music education puts emphasis on HCI technologies to improve learning. Another observation is that games are used as the vehicle that drives motivation in projects toward social inclusion.

VII. Networks Of Excellence

- AIM@SHAPE: by recognizing the relevance of multimedia objects (called shapes) in different application fields such as scientific simulations or edutainment, this NoE promoted semantic-based shape representations and semantic-oriented tools to acquire, build, transmit, and process shapes with their associated knowledge
- PROLEARN: this NoE worked in personalized adaptive learning and interactive media, with learning resources connected to real-world settings and reusable in different contexts. It investigated issues especially relevant for professional training in SME's and larger companies.
- MUSCLE: aims at harnessing the potential of machine learning for the automatic semantic annotation of multimedia content, creation of interfaces for the exploration of complex content, improvement of interoperability and exchangeability. This NoE also aims at the distribution of such technology to the stakeholders, such as educational institutions.
- KNOWLEDGE WEB: Supporting the transition process of Ontology technology from Academia to Industry is the main and major goal of this NoE. This is achieved by providing support to incorporate this technology, developing high-class education in the area of semantic web, and researching new uses of semantic web and web services.
- GaLA: This NoE that acknowledges the potential of serious games in the support of the learning process. It aims at the integration and harmonization of research activities across European institutions, as well as at the organization of joint research activities and the dissemination of serious games to support learning.
- IRIS: Interactive Storytelling is a major endeavour to develop new media which could offer a radically new user experience, with a potential to revolutionise digital entertainment. European research in Interactive Storytelling has played a leading role in the development of the field, and this creates a unique opportunity to strengthen its position even further by structuring some of its best actors within a Network of Excellence. IRIS (Integrating Research in Interactive Storytelling) aims at creating a virtual centre of excellence that will be able to achieve breakthroughs in the understanding of Interactive Storytelling and the development of corresponding technologies.
- STELLAR: represents the effort of the leading institutions and projects in European TEL to unify our diverse TEL community. This Network of Excellence is motivated by the need for European research on Technology-Enhanced Learning (TEL) to build upon, synergize and extend the valuable work we have started by significantly building capacity in TEL research within Europe, which is required to allow the European Union to achieve its goals via the Bologna Agreement and the execution of the Lisbon Agenda. The European TEL agenda has been set for the last 4 years by the Kaleidoscope network - with a huge strength in pedagogy and scientific excellence, and the Prolearn network with a complimentary strength in technical and professional excellence. We see integrating this excellence and moving on to the higher strategic formation of policy based in leading research is the challenge for the next three years. STELLAR will move beyond the earlier networks by setting a new and critical foresight agenda for TEL via an annually reviewed Grand Challenge programme.

VIII. Future works and Applications

The research presented in this paper is part of the HoTEL European project, a Support Action that will propose a methodological framework to support technological innovations in the educational field, speeding up the innovation cycle. The presented work is an exploratory research, the output of which will serve as the basis for an online consultation, and a number of interviews with acknowledged experts in the field. The overall goal is to better understand how innovation takes place in the learning world, and to elaborate a map of the most relevant technologies that will affect education in the near future. This research is focused on technologies. Next steps include the identification of innovative pedagogical approaches, and the creation of a grassroots innovators network that will benefit from the HoTEL innovation cycle.

IX. Conclusion

This desk research has presented an analysis of the information from a total of 86 research projects and 7 Networks of Excellence funded by the European Union as part of the FP6 and FP7 programs. Also, we have considered the information from reports offered by the IPTS on the educational field, and the predictive Horizon Reports, produced by the NMC.

A detailed analysis of the compiled information has allowed to identify the most relevant technologies, and to produce clusters that group them by similarity. The following 11 different clusters were identified, many of them of the highest relevance to interactive multimedia and artificial intelligence:

1. Web 2.0 based tools and systems
2. Ubiquitous computing
3. Augmented reality
4. Access-to-content related technologies
5. Human computer interaction
6. Learning Analytics
7. Games and Virtual Worlds
8. Environments and technologies for collaboration
9. Semantic-aware systems
10. Personalized, adaptive technologies
11. Technologies that could not be grouped in any other

cluster.

The analysis showed that web based technologies are the most funded topic within EU programs. Furthermore, they are usually researched in combination with other technologies such as semantic technologies or personalized learning content. In fact, in some cases it is not obvious at all to make a clear distinction among the different clusters. The analysis also shows the upcoming interest of ubiquitous computing on learning, especially via mobile and tablet computing. Another interesting conclusion is the irruption of learning analytics, an approach that has recently appeared and is gaining momentum rapidly. The analysis also revealed which technology combinations are more frequent. For example, the new interfaces suggested in the field of Human Computer Information are usually researched in combination with augmented reality techniques.

Another focus of the analysis was the identification of learning areas in which the researched technologies can be applied. The following 4 areas were identified:

- Formal education (including primary, secondary and higher education)
- Non-formal learning
- Workplace learning
- Museums

We also considered projects that do not apply to any explicit area of learning, and also projects that do not even explicitly apply to learning, but whose researched technologies are aligned with TEL research.

The analysis shows the relationship among areas of learning and technology clusters, for example, games are mainly researched for their application in schools, while computer supported collaboration is the preferred topic for workplace learning. Furthermore, a number of projects not explicitly for learning research ubiquitous computing technologies, so it is expected that such technologies will impact in the near future of TEL research.

The IPTS and NMC reports (a total of 34 reviewed documents) reveal predictions on what technologies will be used in the near future of education. It is worth to note how mobile computing is always expected to be immediately adopted, but never seems to be mature enough for an authentic usage. The most recent predictions introduce new and interesting concepts, such as 3D-printing, the Internet of Things, learning analytics, massive online open courses, augmented reality and wearable technology.

ACKNOLEDGMENT

This paper is being supported by the HOTEL Project (grant agreement 318530), funded by the Seventh Framework Programme of the European Commission, under the FP7 Programme-ICT-Call 8; and UNIR Research , private research initiative which supports the Research Group on eLearning and Social Networks TELSOCK at the university.

REFERENCES

[1] L. Johnson, A. Levine, R. Smith, and S. Stone, "The 2010 Horizon Report," New Media Consortium, Austin, Texas, 2010.
[2] L. Johnson, S. Adams, and K. Haywood, "The 2011 Horizon Report K-12 edition," The New Media Consortium, 2011.
[3] L. Johnson, S. Adams Becker, H. Witchey, M. Cummins, V. Estrada, A. Freeman, and H. Ludgate, "The NMC Horizon Report: 2012 Museum Edition.," Austin, Texas, 2012.
[4] L. Johnson, S. Adams, and H. Witchey, "The NMC Horizon Report: 2011 Museum Edition," Austin, Texas, 2011.
[5] L. Johnson, A. Levine, and R. Smith, "The Horizon Report, 2009 Edition," New Media Consortium, Austin, Texas, 2009.
[6] L. Johnson, S. Adams, and M. Cummins, "NMC Horizon Report: 2012 K-12 Edition," Austin, Texas, 2012.
[7] L. Johnson, A. Levine, R. Smith, and T. Smythe, "The Horizon Report: 2009 K-12 Edition," The New Media Consortium, 2009.
[8] L. Johnson, H. Witchey, R. Smith, A. Levine, and K. Haywood, "The 2010 Horizon Report: Museum Edition," Austin, Texas, 2010.
[9] L. Johnson, S. Adams, and M. Cummins, "The NMC horizon report: 2012 higher education edition," Austin, Texas, 2012.
[10] L. Johnson, R. Smith, H. Willis, A. Levine, and K. Haywood, "The 2011 Horizon Report," Austin, Texas, 2011.
[11] L. Johnson, S. Adams Becker, M. Cummins, V. Estrada, A. Freeman, and H. Ludgate, "NMC Horizon Report: 2013 Higher Education Edition," Austin, Texas, 2013.
[12] L. Johnson, R. Smith, A. Levine, and K. Haywood, "2010 Horizon Report: K-12 Edition.," Austin, Texas, 2010.
[13] Ala-Mutka, K. (2009). Review of Learning in ICT-enabled Networks and Communities.
[14] Ala-Mutka, K. (2010a). Discussions on Learning in Online Networks and Communities.
[15] Ala-Mutka, K. (2010b). Learning in Informal Online Networks and Communities.
[16] Ala-Mutka, K., Punie, Y., & Redecker, C. (2008a). ICT for Learning, Innovation and Creativity.
[17] Ala-Mutka, K., Punie, Y., & Redecker, C. (2008b). Digital Competence for Lifelong Learning.
[18] Ala-Mutka, Kirsti, Redecker, C., Punie, Y., Ferrari, A., Cachia, R., & Centeno, C. (2010). The Future of Learning: European Teachers' Visions. Report on a foresight consultation at the 2010 eTwinning Conference. Seville.
[19] Bocconi, S., Kampylis, P., & Punie, Y. (2012). Innovating Learning: Key Elements for Developing Creative Classrooms in Europe.
[20] Bogdanowicz, M., Burgelman, J. C., Gourova, E., & Herrmann, C. (2002). Information and Communication Technologies.
[21] Cachia, R., Ferrari, A., Ala-Mutka, K., & Punie, Y. (2010). Creative Learning and Innovative Teaching: Final Report on the Study on Creativity and Innovation in Education in EU Member States.
[22] Driessen, M., Emmerik, J. van, Fuhri, K., Nygren-Junkin, L., & Spotti, M. (2011). CT Use in L2 Education for Adult Migrants - A qualitative study in the Netherlands and Sweden.
[23] Forge, S., & Blackman, C. (2009). OLEDs AND E-PAPER - Disruptive potential for the European display industry.
[24] Huijboom, N., Broek, T. van den, Frissen, V., Kool, L., Kotterink, B., Nielsen, M. M., & Miljard, J. (2009). Public Services 2.0: The Impact of Social Computing on Public Services.
[25] IPTS. (2013a). Scientific and Technical reports. Retrieved February 12, 2013, from http://ipts.jrc.ec.europa.eu/publications/index.cfm
[26] IPTS. (2013b). Summary of e-Learning related projects. Retrieved February 12, 2013, from http://is.jrc.ec.europa.eu/pages/EAP/eLearning.html
[27] Prato, G. De, & Nepelski, D. (2013). Defining European ICT Poles of Excellence: A Literature Review.
[28] Prato, G. de, Nepelski, D., Szewczyk, W., & Turlea, G. (2011). Performance of ICT R&D.
[29] Punie, Y., Cabrera, M., Bogdanowicz, M., & Zinnbauer, D. (2006). The Future of ICT and Learning in the Knowledge Society.
[30] Punie, Y., Zinnbauer, D., & Cabrera, M. (2008). A Review of the Impact of ICT on Learning.

[31] Rader, M., Boehle, K., Parodi, O., Dachs, B., Weber, M., Friedewald, M., Cremer, C., et al. (2008). European Perspectives on the Information Society: Annual Monitoring Synthesis and Emerging Trend Updates.
[32] Redecker, Christine, Haché, A., & Centeno, C. (2010). Using Information and Communication Technologies to Promote Education and Employment Opportunities for Immigrants and Ethnic Minorities.
[33] Redecker, Christine, Leis, M., Leendertse, M., Punie, Y., Gijsbers, G., Kirschner, P., Stoyanov, S., et al. (2011). The Future of Learning: Preparing for Change.
[34] Redecker, Chrstine. (2012). The Use of ICT for the Assessment of Key Competences.
[35] Commision, E. (2013). CORDIS. Retrieved February 12, 2013, from http://cordis.europa.eu/guidance/about-projects_en.html
[36] Dib, C. Z. (1988). Formal, non-formal and informal education: concepts/applicability. AIP Conference Proceedings (Vol. 173, pp. 300–315). AIP. doi:10.1063/1.37526

Dr. Luis de-la-Fuente-Valentín works as post-doctoral Senior Researcher UNIR. He is focused on the field of Technology-enhanced Learning, and a strong background in learning standards and specifications, learning analytics, information visualization, student centered learning systems, recomendations, mobile learning and gamification techniques. He works in European projects Intuitel and Hotel. He holds a PhD in Telematics Engineering

Aurora Carrasco works as Senior Project Manager at UNIR. She is Responsible for the R&D Projects Office. She has deep experience in European and International Programme and Projects Management (e.g. EC EuropeAid, World Bank, FP7, LLP), and extensive knowledge of the EC's Project Cycle Management (PCM), and in the preparation of international project proposals, and offers. She holds degrees in Biology (BSc and MSc with major in Environment and Education)

Kinga Kónya works as Media Officer of Research & Technology Projects at UNIR. She is responsible for the dissemination and communication of project objectives, research and development activities and results. In parallel, she works at the Postgraduate Department of UNIR. She holds a Bachelor's degree in French Studies (minor in Spanish Iberian-American Studies) and a Master's degree in International Economic Relations. She speaks Spanish, English, French and Hungarian.

Prof. Dr. Daniel Burgos works as Vice-chancellor for Research & Technology, and Director of Engineering at the International University of La Rioja. He is or has been involved in a number of R&D projects like, i.e. Intelleo, Hotel, Edumotion, Inspiring Science Education, EU-USR, Stellar, Gala, IntelLEO, Go-MyLife, Grapple, Unfold, ProLearn, TenCompetence, EU4ALL, NiHao, Kaleidoscope, Sister, Comeln, Veritas, et cetera. He holds degrees in Communication (PhD), Computer Science (Dr. Ing), Education (PhD), and Business Administration.

APPENDIX A: THESAURUS USED AS REFERENCE

Categories	Subcategories	Keywords			
Technologies	Web 2.0 based tolos and systems (WEB)	web 2.0 online communication tools blogs podcasting eportfolio accessible web content e-portfolio	multimedia content web interfaces widgets integration distributed open infrastructure linked data service-oriented	online applications for teaching learning management systems learning standards interoperability cloud computing grid computing	social networks social media foaf online communities social network analysis social software
	Ubiquitous computing (UC)	ambient intelligence internet of things smart objects location based services ubiquitous computing	intelligent contexts pervasive technologies context-aware systems pervasive computing smart spaces	mobiles tablet computing mobile apps mobile devices	mobile learning geo-everything creative classrooms future classroom
	Augmented reality (AR)	augmented reality	simple augmented reality	merge virtual and physical worlds	
	Access-to-content related technologies (CRT)	open source content massively open online courses alternative licensing	super rich online repositories electronic publishing digital preservation	creative commons opencourseware	content repositories content creation
	Human computer interaction (HCI)	human computer interaction gesture based computing natural user interfaces	brain-machine interfaces 3-d holographic displays adaptable interfaces	haptic interfaces enriched interfaces interfaces	tactile interfaces voice interfaces
	Learning Analytics (LA)	data mining big data and learning analytics	learning analytics information visualization	student monitoring	visual data analysis
	Games and Virtual Worlds (GVW)	immersive virtual worlds computer generated simulations simulation of physical, chemical models	simulations for training virtual environments simulations	games game-based learning	serious games 3d virtual worlds
	Environments and technologies for collaboration (CSCL)	collaborative environments computer supported collaborative learning	telepresence	online communication tools	sincronous learning
	Semantic-aware systems (SAS)	semantic web metadata generation	natural language processing semantic-aware technologies	semantic interoperability	translation technologies
	Personalized, adaptive technologies (PA)	virtual mentors electronic tutors realtime assessment monitors artificial intelligence	attention management adaptive content smart learning content adaptive learning system	personalized content presentation recommender systems the personal web	personal learning environments roleswitching emotion-aware systems
	Other technologies (OT)	flexible displays smart-tv	e-books 3d printing	interactive whiteboards	wearable technology
Areas of learning	Formal education	higher education	secondary education	Schools	
	Non-formal learning	domestic learners	broad public	informal learning	lifelong learning
	Workplace learning	SME	industry	workplace	training
	Museums	museum			
	Unspecific field	not explicitly for learning	not specified		
	Other	music institutions digital library	learner with special needs educational curriculum	conflict resolution	social inclusion

APPENDIX B: LIST OF REVIEWED PROJECTS

Project reference	Acronym	Title
507740	CALIMERA	Cultural applications: Local institutions mediating electronic resources access
506811	AMI	Augmented Multi-party Interaction
507835	UNFOLD	Understanding Networks of Learning Design
1765	ACEMEDIA	Integrating knowledge, semantic and content for user-centred intelligent media services
507457	BRICKS	Building resources for Integrated Cultural Knowledge Services
508013	AGAMENON	Pictures from the past: A wireless network of magic digital cameras and palmtops for archaeological travels through the time
510166	COHERENT	Collaborative holographic environments for networked tasks
507826	LEACTIVEMATH	Language-enhanced, user adaptive, interactive eLearning for mathematics
507128	TELCERT	Technology enhanced learning certification - European requirements and testing
507844	CONNECT	Designing the classroom of Tomorrow by using advanced technologies to connect formal and informal environments
2205	E-LEGI	European learning GRID infrastructure
4337	CALIBRE	Co-ordination action for libre software engineering for open development platforms for software and services
4293	AKOGRIMO	Access to knowledge through the grid in a mobile world
511592	MICOLE	Multimodal collaboration environment for inclusion of visually impaired children
4778	ENABLED	Enhanced network accessibility for the blind and visually impaired
516895	PAPERWORKS	PaperWorks : Interweaving Paper and Digital Documents
28025	CALIBRATE	Calibrating eLearning in schools
28051	EMAPPS.COM	Motivating active participation of primary schoolschildren in digital online technologies for creative opportunities through multimedia
27168	ICAMP	intercultural learning campus
26883	I-MAESTRO	Interactive Multimedia Environment for Technology Enhanced Music Education and Creative Collaborative Composition and Performance
27607	MGBL	Mobile games based learning
27952	VEMUS	Virtual European Music schools
27087	TENCOMPETENCE	Tencompetence: Building the European network for lifelong competence development
27728	ARGUNAUT	An intelligent guide to support productive online dialogue
27529	ATGENTIVE	Attentive agents for collaborative learners
27073	COOPER	Collaborative open environment for project-centred learning
28027	LEAD	Technology-enhanced learning and problem-solving discussions: Networked learning environments in the classroom
27391	LT4EL	Language technology for eLearning
26751	RE.MATH	Representing mathematics with digital media
27611	AETHER	Self-adaptive embedded technologies for pervasive computing architectures
27020	ACCESS-EGOV	Access to e-Government services employing semantic technologies
27039	ARISE	Augmented reality in schools environments
27866	ELU	Enhanced Learning Unlimited
28038	PALETTE	Pedagogically sustained adaptive Learning through the exploitation of tacit and explicit knowledge
27451	LOGOS	Knowledge-on-demand for ubiquitous learning
27023	APOSDLE	Advanced process-oriented self-directed learning environment
27656	ECIRCUS	Education through characters with emotional-intelligence and role-playing capabilities that understand social interaction
27986	ELEKTRA	Enhanced learning experience and knowledge transfer
27149	LUISA	Learning content management system using innovative semantic web services architecture
33572	CASPAR	Cultural, artistic and scientific knowledge preservation, for access and retrieval
34545	A-WARE	An easy way to access GRID resources

34567	GRID4ALL	Self-* Grid: Dynamic Virtual Organizations for schools, families, and all
34549	AGENT-DYSL	Accommodative intelligent educational environments for Dyslexic learners
34778	EU4ALL	European unified approach for accessible lifelong learning
33860	QALL-ME	Question Answering Learning technologies in a multilingual and multimodal Environment
216837	ATRACO	Adaptive and trusted ambient ecologies
215064	IMPACT	Improving access to text
215893	REFLECT	Responsive flexible collaborating ambient
215434	GRAPPLE	Generic responsive adaptive personalized learning environment
216267	LIWA	Living web archives
217141	SOCIALNETS	Social networking for pervasive adaptation
212578	LTFLL	Language technology for lifelong learning
215918	80DAYS	Around an inspiring virtual learning world in eighty days
216199	IDSPACE	Tooling of and training for collaborative, distributed product innovation
216356	MATURE	Continuous social learning in knowledge networks
215098	PERSIST	Personal self-improving smart spaces
224047	REPLAY	Gaming technology platform for social reintegration of marginalised youth
224044	INCLUSO	Social software for inclusion of (marginalized) young people
222107	NIW	Natural interactive walking
231551	DL.ORG	DL.org: coordination action on digital library interoperability, best practices, and modelling foundations
231717	TARGET	Transformative, adaptive, responsive and engaging Environment
231266	COSPATIAL	Communication and social participation: collaborative technologies for interaction and learning
231590	INTELLEO	Intelligent learning extended organisation
225938	OPPORTUNITY	Activity and context recognition with opportunistic sensor configurations
231396	ROLE	Responsive open learning environments
257639	ALICE	Adaptive Learning via Intuitive/Interactive, Collaborative and Emotional systems
257886	ARISTOTELE	Personalised Learning & Collaborative Working Environments Fostering Social Creativity and Innovations Inside the Organisations
257872	METAFORA	Learning to learn together: A visual language for social orchestration of educational activities
257617	MIRROR	MIRROR - reflectionive Learning at Work
257566	HEATTRONICS	Innovative Technologies for an Engaging Classroom
258114	NEXT-TELL	Next Generation Teaching, Education and Learning for Life
258453	SIREN	Social games for conflIct REsolution based on natural iNteraction
257831	IMREAL	inmersive reflectionive Experience-based Adaptive Learning
257493	SOCIETIES	Self Orchestrating CommunIty ambiEnT IntelligEnce Spaces
257410	TERENCE	An Adaptive Learning System for Reasoning about Stories with Poor Comprehenders and their Educators
270001	DECIPHER	Digital Environment for Cultural Interfaces; Promoting Heritage, Education and Research
271578	ETOILECASCADESIDEAS	Enhanced Technology for Open Intelligent Learning Environments
288578	TARDIS	Training young Adult's Regulation of emotions and Development of social Interaction Skills
288587	MASELTOV	Mobile Assistance for Social Inclusion and Empowerment of Immigrants with Persuasive Learning Technologies and Social Network Services
288596	LITERACY	ONLINE PORTAL FOR E-LEARNING AND SUPPORTING SOCIAL INCLUSION OF PEOPLE WITH DYSLEXIA
296229	EUCLID	Educational curriculum for the usage of Linked Data
318803	ILEARNRW	Integrated Intelligent Learning Environment for Reading and Writing
318496	INTUITEL	Intelligent Tutoring Interface for Technology Enhanced Learning
317964	JUXTALEARN	Juxtapositioned reflectionive performance enabling science and technology learning
318499	WESPOT	Working Environment with Social and Personal Open Tools for inquiry based learning.
318299	WE.LEARN.IT	European schools enhancing creativity, exploration and science

Gade4all: Developing Multi-platform Videogames based on Domain Specific Languages and Model Driven Engineering

Edward Rolando Nuñez-Valdez[1], Oscar Sanjuan[2], Begoña Cristina Pelayo Garcia-Bustelo[1], Juan Manuel Cueva-Lovelle[1], Guillermo Infante Hernandez[1]

[1] *University of Oviedo, Computer Science Department, Sciences Building, Oviedo, Spain*

[2] *University Carlos III of Madrid, Computer Science and Engineering Department, Madrid, Spain.*

***Abstract* — The development of applications for mobile devices is a constantly growing market which and more and more enterprises support the development of applications for this kind of devices. In that sense, videogames for mobile devices have become very popular worldwide and are now part of highly profitable and competitive industry. Due to the diversity of platforms and mobile devices and the complexity of this kind of applications, the development time and the number of errors within that development process have increased. The productivity of the developers has also decreased due to the necessity of using many programming languages in the development process. One of the most popular strategies is to employ specialized people to perform the development tasks more efficiently, but this involves an increase of the costs, which makes some applications economically unviable. In this article we present the Gade4all Project, consisting in a new platform that aims to facilitate the development of videogames and entertainment software through the use of Domain Specific Languages and Model Driven Engineering. This tool makes possible for users without previous knowledge in the field of software development to create 2D videogames for multiplatform mobile devices in a simple and innovative way.**

***Keywords* — DSL, Gade4all , MDE, VideoGame**

I. Introduction

The use of new technologies (like smartphones, tablets, computers and the Web) as tools for the use of videogames and entertainment applications is quite extended nowadays. This expansion has lead to the appearance of a great diversity of platforms and mobile devices that allow to execute a wide range of applications.

The existance of different types of platforms and mobile operative systems and the constant evolution and appearance of new devices has brought with it an increase of the development time and the appearance of errors during that process. Also, the productivity of the developers of mobile applications has decreased, generally due to the necessity of developing the same application using many programming languages in order to publish it in other platforms and reach a higher number of users in the market.

Also, the development of games is generally aimed to people with extensive knowledge in the development of this kind of applications. Normally, an important part of these applications are developed for mobile devices and the Web [1].

All this suggests that the development of videogames can be much more profitable and optimized if we can abstract the main characteristics of many videogame typologies. This abstraction serves a base for the reuse of code and the generation of multiplatform videogames in an automatic way.

With the objective of minimize the aforementioned inconveniences, improve the efficiency and solve the typical problems of software development [2][3], this project proposes the use of Model Driven Engineering (MDE) [4] to develop a tool that allows to create multiplatform videogames in a quick and simple way. This tool can be used by anyone, even if that person lacks programming knowledge.

By using typologies templates and the most advanced technology for automated generation of code based on Domain Specific Languages and Model Driven Engineering, this tool allows the development of 2D multiplatform games. These applications are automatically generated in native code for platforms like Windows Phone, Android, iOs and HTML5, managing to optimize the development time, raise the efficiency and reduce the number of errors.

The reminder of this paper is structured as follows: in Section 2 we present a summary of the current problems that affect the development of videogames. Section 3 presents the background of the videogames, some videogame development tools and the general concepts of Model Driven Engineering and Domain Specific Language. Section 4 shows the gade4all case of study, Section 5 contains the conclusion and future research directions of this work and Section 6 shows the Acknowledgement.

II. Problems

In their beginning, videogames were developed by very small groups of people and they could only be executed by a small amount of devices, some of which had a very high price. As the videogame industry evolved, so did the size of the development teams and the variety of devices able to execute those videogames. This is because the growing popularity of mobile phones has encouraged software development enterprises to produce a greater number of videogame applications.

Roughly speakning, we could conclude that the development process of multiplatform videogames is facing the following problems:

A. Complexity within videogame development

Due to the existing multiple mobile platforms for the development and deployment of videogames, the development of this kind of applications has become harder. This is because it's necessary of developers to have an extensive knowledge of the videogame's domain, as well as experience in the development of multi-device software.

In that sense, the development of videogames requires not only experience in the development of software, but also a great knowledge of the different programming languages that are used to create multiplatform videogames. This increases the complexity of videogame development and also the necessity of having a highly qualified person that can handle the problem.

B. Huge variety of platforms and technologies

Due to the existence of a high diversity of languages and platforms like Java, Objetive C, Microsoft .NET, HTML5 and others that allow to develop applications for their deployment on devices based on Android, iOS, Window or the different web browsers. This often causes the development of applications for specific technologies to be poorly feasible. This is because each technology fills a different market niche, which reduces the number of potential user for the developed product. Fort the enterprises focused in videogame development, this implies having development teams for each platform. This way, they can make sure that their products will be well positioned in the different market niches and make them more feasable.

C. Increase of development time and implementation costs

This is a problema that arises from the complexity of the development of multiplatform and multi-device software. Due to the great diversity of platforms and devices, it's often necessary to develop the same application for different devices and operative systems. This causes a reduction of the speed and effectivity of the software development process, specially when developing multiplatform products. Currently, if we want to generate native applications for the different existing platforms, we have to develop the same application with different programming languages and in different environments, which considerably raises the amount of time and money necessary for the development of the application.

III. Background

Videogames are applications developed for personal entertainment that are based on the interaction of one or more persons with the electronic device in which its being executed. In most cases, videogames represent virtual environments in which the player can control one or more characters with the objective of fulfilling objectives according to the rules of the game.

Since the 70s, videogames have become popular worldwide and are now a highly profitable and competitive industry. The "trend" of videogames has been kept afloat thanks to the great diversity they offer, as well as the evolution of their technology and creativity. Originally, videogames were created and designed to be executed in computational systems, but in the middle 70s the first home video game consoles appeared and became quite popular in a short time, hitting 13 millions of units only in the United States [5].

The devices and the way of playing have also experienced a big evolution, from the buttons and sticks of the old games to the motions sensors and tactile interfaces of today. This is possible thanks to the technological advances and the inventiveness of the creators of the games, which take advantage of the available resources to expand the possibilities. The revolution of the Web and the mobile devices represents a new era for videogames, as there are millions of them in the Web prepared to be executed in a web browser and many mobile devices and platforms, like Android, iOS, Windows Phone and others.

To facilitate the creation of videogames for all these platforms and devices, several graphic tools have been developed to help videogame creators and users in general to generate videogames in an easy way. This section describes some of the videogame editors that help to perform this task. Also, it shows the general concepts about Model Driven Engineering and Domain Specific Language as software development approaches, which served as a base for the development of the Gade4all platform.

A. Videogame Editors

Nowadays there are many videogame editors that make easier to develop this kind of applications. Some of these tools allow to export videogames to different platforms like Windows, iOS, Android, Flash and HTML5. The main differences between these tools are the supported platforms, the typologies of games that can be generated, the degree of customization of the characteristics and behaviours of the elements present in the games and, most importantly, their simplicity and accessibility. Below are some examples of videogame editors that can be found in the market:

1) GameSalad

It's a tool that allows to develop videogames and export them for its distribution in different devices and platforms: Macs, PCs, iOS smart devices, Android, and HTML5 [6]. The development of games with GameSalad Creator is performed through scenes and actors, whose attributes can be defined to stablish a behavior, physical appearance, sounds and rules. However, while it has many advantages that make easy to develop them, the videogames created with GameSalad can only be edited from this tool, which doesn't make possible for other developers to expand them using the native development environments.

2) GameMaker Studio

GameMaker Studio is a tool for the creation of videogames, aimed both to novice users and experimented professionals of videogame development. It's quite intuitive, because it was originally designed for inexperienced users to learn the basic concepts of programming, understand game's architecture and create games in a simple way [7]. In [8] it's shown how some children designed educative videogames that demonstrated their understanding of the GameMaker software. This tool is based in its own interpreted language called GML (Game Maker Language), which is based in scripts aimed to expert users. The program is regulated by the Drag and Drop philoshopy, so supposedly is not necessary to write a single line of code. It allows the generation of videogames for HTML5 and the platforms iOS, Android, Windows.

3) Stencyl

It's a tool for the creation of videogames developed by StencylWorks [9]. This product offers videogame designers a graphical editor that allows them to create games and export them to iOS, Mac, Windows and Flash formats. Since version 3.0 (version beta) Stencyl allows to export applications to Android and HTML5. The creation of games in Stencyl is based in the definition of scenes and actors. Each scene corresponds to a level in the game, and the different actors are added to those scenes. An actor is any element that is shown in the level, from the graphical elements without added logic or functionality to the characters of the game, with their own rules and behaviors. The development environment of Stencyl is intuitive enough for people with experience in design program, and though it has several default templates, it demands programming and algorithmic logic knowledge, as many of the game's modifications require the creation of algorithm flowcharts.

B. Model Driven Engineering

MDE, or "Model Driven Engineering" is an approach that suggests the use of models as the main axis of the lifecycle of a software and saves time and effort in the development process [10]. Those models are abstractions that are focused in the development of specific applications of a problem.

The implementation of Model Driven Engineering allows the raise of productivity through the reutilization of standardized models, thus simplifying the development process through recurring templates of an specific domain, and, at the same time, creating an standardization of the terminology and the best practices used in the application of the domain.

Model Driven Engineering possesses a wide spotlight architecture and automation with high levels of abstraction in software development. This abstraction promotes the design of simple models with a wide perspective for the resolution of specific problems, which combined with the adequate semantic, raise the total rate of possible automation. The OMG (Object Management Group) has developed a set of standards referred to as MDA (Model Driven Architecture).

According to [10] a model is the description of one or more elements of the domain or real world. There is also the concept of metamodel, which presents the tools that allow the creation a model through its description. Metamodels also describe themselves, generating an extremely high abstraction in which all the models match each other.

The use of Model Driven Engineering is key for the development of this project, as the concepts about the definition of models with high levels of abstraction efficiently adjusts to the developed tool. By applying the principles of this engineering we manage to abstract the main characteristics of the supported videogames, making easier to create a Domain Specific Language that allows the generation of multiplatform games in a simple and easy way.

C. Domain Specific Language

DSL, or "Domain Specific Language" is a usually declarative language that focus on the resolution of problems of particular domain. In many cases, these solutions can be translated to library calls or common subroutines, in which the DSL can be seen as a way of occulting the details of that library [11]

As [12] describe, models can be clearly expressed by using a DSL whose language is specifically designed to meet the specific needs of the application's domain. In that sense, the DSL contains specific concepts of the domain that allow a better understanding of it and ignoring elements that moves away from the reality of the domain.

As Fig.8 shows, one of the main characteristics of the DSL is that it allows the implementation of a definition that is independent from the platform and, once finished, will go through a transformation process, offering the construction of a proyect in a specific platform as a final result. The DSL defined for the Gade4all project makes easier to generate complete applications for the iPhone, Android, Windows Mobile and HTML5 platforms.

Fig.8. Transformation of platform independent models

DSL is crucial for Model Driven Engineering and so for the project. The DSL used for the definition of the videogames contains a high level of abstraction that allows to define all the elements that are necessary for the generation of a videogame, as well as the stablishment of the required values to assign a behavior to the actors. That task is performed through a videogame editor, so the final user doesn't need to know its details.

IV. Case Study

In the last years several operative systems for mobile devices have appeared, among which we can mention: iOS, developed by Apple, Windows Phone, developed by Microsoft and Android, developed by Google. The birth of the Smartphone has unleashed a war in the development of applications for mobile devices based on the aforementioned operative systems.

The necessity of developing applications that can be executed in different platforms has brought with it the appearance of problems that were described in Section II. In order to solve those problems, it's necessary to create a tool that allows to develop applications for different platforms in an agile, efficient and easy way.

In the context of videogames, a tool that allows to reduce the complexity of the developing of multiplatform videogames will also allow to minimize the costs and development time, as well as reducing the appearance of errors in this kind of application.

In this case study we show the Gade4all videogame platform, which has been developed with the purpose of making possible for a wide range of inexperienced users to create multiplatform videogames in an easy way. Because many videogames share a series of characteristics and behavior patterns, Gade4all has defined a high level language that allows the creation of a great variety of videogames in an easy and effortless way, all through the abstraction of the most relevant characteristics of the main videogame typologies.

A. Gade4all

Gade4all is an innovative project that aims to simplify the development of multiplatform videogames for mobile devices and the Web. Through the use of a graphical tool, it allows inexperienced users to develop software and create innovative videogames in a quick and simple way. The Gade4all tool, which has been developed using Model Driven Engineering as a base, facilitates the development of videogames in a dynamic way by using a Domain Specific Language (DSL). This platform allows the definition of the elements, characteristics and rules of videogames in an easy way. The use of this editor as a tool for the development of videogames makes easier to get in touch with the most demanded technologies and devices nowadays, like: Android, iPhone, Windows Phone and HTML5.

This tool allows an agile development of 2D videogames that are automatically generated in native code for multiple platforms, allowing the improvement of the development time, the reduction of the number of errors and the raising of productivity.

The generation of applications in native code offers different advantages, among which we can highlight the real use of the functionalities of each platform and the possible incorporation of improvements by developers with knowledge of the environment. In that sense, the generated games can be used as prototypes for more complex applications.

Besides, the platform is oriented to create videogames and entertainment software of any kind, quickly enough to create them as support applications to advertise recent events, reducing costs and development time.

On its first stage of development, this platform supports a series of videogame typologies like: touch, trivial, strategy, puzzle and platform. These make easier to create videogames for devices like Android, iPhone, Windows Phone y web navigators that support HTML5.

B. Gade4all platform Archytecture

As shown in Figure 2, through the use of a graphical editor, the Gade4all platform allows the creation of videogames easily and intuitively. This allows us to create all the elements, actors and rules of the game and offers all the necessary mechanisms for the configuration and use of the DSL that describes the model of videogames automatically and dynamically. Once the application has been designed with the editor, an integrated transformation engine uses the DSL and the default templates to generate a complete application in native code, which will be ready to be executed in the selected platform. Below we describe the different elements that the Gade4all platform's architecture is composed of:

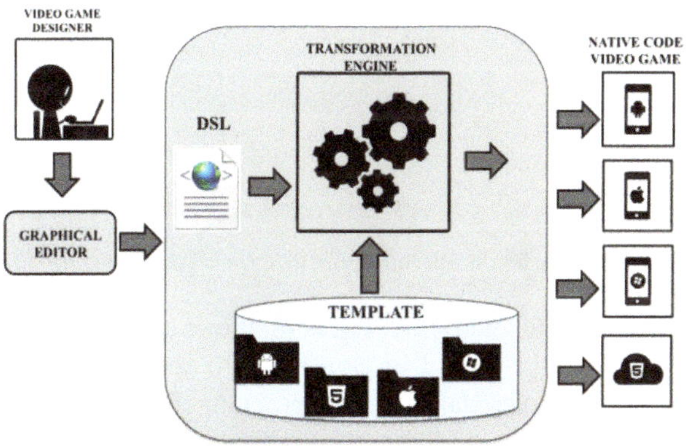

Fig. 9. Gade4all Platform Arquitecture

1) Graphical editor

Many videogames share several patterns of behavior and we intend to take some of the most representative to allow the definition of videogames in a quick and simple way, just by modifying some of the properties of those behaviors.

In order to do this, it's necessary to define the variable elements that make a videogame (graphic aspects, sounds, characters…) and specially their behavior of those elements, so we can stablish the rules of the game.

The graphical editor allows both videogame designers and inexperienced users to define the characteristics and rules of the game through a graphical interface that makes the development of videogames more intuitive.

If a user wants to define a videogame using this tool, he will only have to follow some interactive steps, through user-friendly screens that make the process easier and quicker. This sequence starts with the selection of a typology and ends with the selection of the platform in which we want to deploy the videogame. **Fig. 10** and figure**Fig. 11** show the graphical interfaces of the videogame typology selection screen and the deployment platform selection screen, respectively.

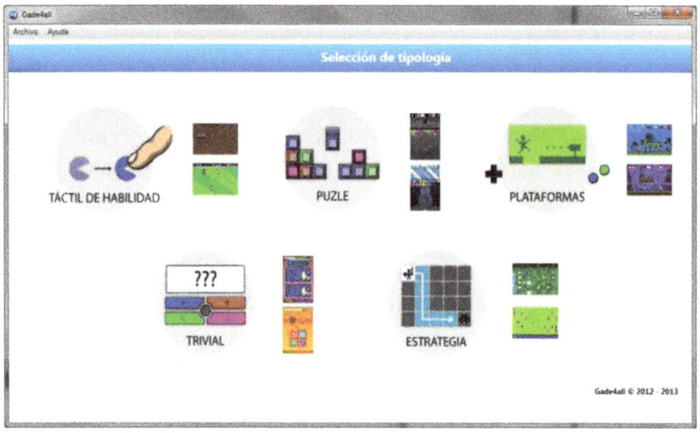

Fig. 10. Interface for the selection of the videogame typology

Fig. 11. Interface for the selection of the deployment platform

The steps to follow can differ from one another, as some of the characteristics and behaviors of the games are different. For instance, a trivial game has different characteristics than a platform game.

Fig. 12 shows the steps that a user must follow to define a platform game, each of one its composed by user-friendly screens that make easier to define the characteristics and behaviors of the videogame even if the user does not possess technical knowledge at all.

Some of these screens represent the pre-configured main elements, which can be used with their default values or can be modified by the use according to the characteristics he wants to include in its videogame.

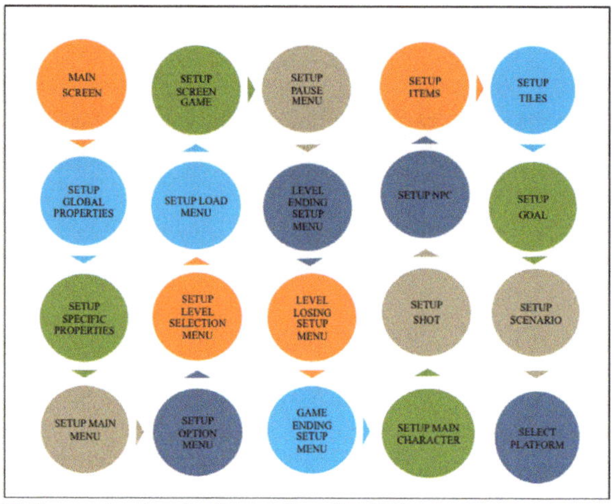

Fig. 12. Screen structure for the creation of a platform game

Fig. 13 shows a graphical interface that allows the user to create a library of characters in a platform videogame. Here is where the main characteristics of the characters are defined: size of the character, number of lives, sounds, animations, etc. The characters defined in this screen are the ones that will be used in the defined scenes defined for the levels of the videogame.

The rest of the screens defined in the editor allow the user to configure the different components that will appear in the videogame: menus (main menu, options menu, pause menu, game, etc), elements (main character, NPCs, obstacles, items, tiles,etc) and scenes (the levels of the videogame). These screens have been defined following the same structure, so the users can interact with them intuitively.

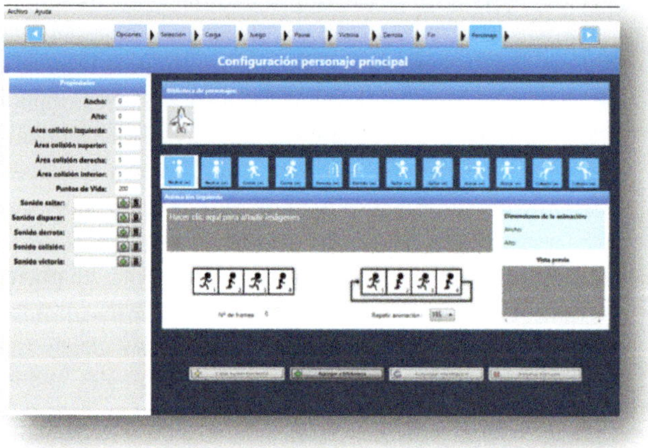

Fig. 13. Interface for the creation of the main character in a platform game

2) Domain Specific Language

On of the main pillars of this platform is the domain specific language that defines the main characteristics of the different videogame typologies supported by the platform. Thanks to the graphical editor, this high level language defined in XML allows videogame designers to define the rules of the game they are developing in a simple way. This language makes easier for inexperienced users to develop videogames in a very intuitive way.

Fig. 14 shows a fragment of DSL corresponding to the "Properties" section of the screen shown in Figure 6. These values are modified by the user from the graphical editor. Generally, the creation process will result in the obtention of a XML file containing all the defined elements corresponding to the DSL of the developed videogame.

```
<character>
    ...
    <character_width>90</character_width>
    <character_height>90</character_height>
    <character_collision_area_left>5</character_collision_area_left>
    <character_collision_area_top>5</character_collision_area_top>
    <character_collision_area_right>5</character_collision_area_right>
    <character_collision_area_down>5</character_collision_area_down>
    <character_health>200</character_health>
    <character_sound_on_jump_source>sound1.mp3</character_sound_on_jump_source>
    <character_sound_on_fire_source>sound2.mp3</character_sound_on_fire_source>
    <character_sound_die_source>sound3.mp3</character_sound_die_source>
    <character_sound_collision_bad_item_source>sound4.mp3</character_sound_collision_bad_item_source>
    <character_sound_collision_good_item_source>sound5.mp3</character_sound_collision_good_item_source>
    ...
    ...
</character>
```

Fig. 14. Fragment of DSL from a platform game

3) Template

To generate native code, we employ a series of templates that abstract a set of common characteristics that can be used as a generic model. This templates are developed as a base for the different platforms that Gade4all supports: iPhone, Android, Windows Phone and HTML5. These projects (Template) abstract a set of attributes and generic methods that make easier to convert and import the characteristics tha result from the implementation of a request from the DSL. **Fig. 15** shows a section of the template where the "Start" button of a videogame is configured. The values of the variables wil be replaced by the transformation engine when the game is generated for the selected platform. This mechanism for the replacement of variables is commonly used on those elements that don't change their behavior too much from game to game, like option menus, stage select menus, etc.

Fig. 16 shows a section of the template where the generation and load processes of the levels of the videogame are performed. This is done with the transformation engine, using SAX (Simple API for XML) technology and Reflection.

This dynamic loading mechanism is used when the behavior of the videogame's elements changes considerably depending on the levels and the typologies. For instance, the positioning and behaviors of the characters within the scene).

Also, because the scene does not know how many elements or characters will appear in every level, we need a a mechanism like this, which offers flexibility to load the elements required for the definition and configuration of a videogame.

```
public int imageButtonStart = %main_screen_start_button_image_source%;
public int heightImageButtonStart = %main_screen_start_button_image_height%;
public int widthtImageButtonStart = %main_screen_start_button_image_width%;
public int posXButtonStartMenu = %main_screen_start_button_x_position%;
public int posYButtonStartMenu = %main_screen_start_button_y_position%;
```

Fig. 15. Template design for the creation of a button

```
tiles[levelLoaded.character_x_position / Tile.Width][ levelLoaded.character_y_position / Tile.Height] =
        LoadStartTile
        (
            levelLoaded.character_width,
            levelLoaded.character_height,
            levelLoaded.character_collision_area_left,
            levelLoaded.character_collision_area_top,
            levelLoaded.character_collision_area_right,
            levelLoaded.character_collision_area_down,
            levelLoaded.character_health,
            levelLoaded.character_sound_on_jump_source,
            levelLoaded.character_sound_on_fire_source,
            levelLoaded.character_sound_die_source,
            levelLoaded.character_sound_collision_bad_item_source,
            levelLoaded.character_sound_collision_good_item_source
            ...
            ...
        );
```

Fig. 16. Template that enables the automatic loading of levels

4) Transformation Engine

The videogames' specifications performed with the DSL are processed by a transformation engine. This engine automatically generates applications in native code, which can be executed in the platforms selected by the videogame designer during the development process. The transformation engine applies a set of algorithms on the specification of the game to convert this "abstract" code into specific code (Java, C#, Objective C, HTML5) that can be executed in a mobile platform or the Web.

This transformation engine allows us to perform the following tasks in the videogame's generation process:

- Copy image and sound resources used in the editor.
- Copy of the files of each level, obtained from the editor, in order to simplify the transformation process.
- Replacement of the values of the DSL's request in the selected templates.
- Automatic loading of the different XML levels configured in the developed videogame.

Fig. 17 shows the process of automatic replacement of values in the variables defined in the template where the main characteristics of a button are abstracted.

```
public int imageButtonStart = %main_screen_start_button_image_source%;
public int heightImageButtonStart = %main_screen_start_button_image_height%;
public int widthtImageButtonStart = %main_screen_start_button_image_width%;
public int posXButtonStartMenu = %main_screen_start_button_x_position%;
public int posYButtonStartMenu = %main_screen_start_button_y_position%;
```

```
public int imageButtonStart = R.drawable.u1;
public int heightImageButtonStart = 216;
public int widthtImageButtonStart = 39;
public int posXButtonStartMenu = 118;
public int posYButtonStartMenu = 120;
```

Fig. 17. Transformation process of a button's characteristics

Finally, once the transformation engine has performed all the processes for the generation of native code, the resulting projects can be imported, modified and compiled in the development environment of the selected platform: Eclipse, Xcode, Visual Studio .Net, etc.

Fig. 18 shows an example of the transformation and compilation process of a videogame developed for the

Android platform, which will generate an application that can be executed in any device that supports this type of application.

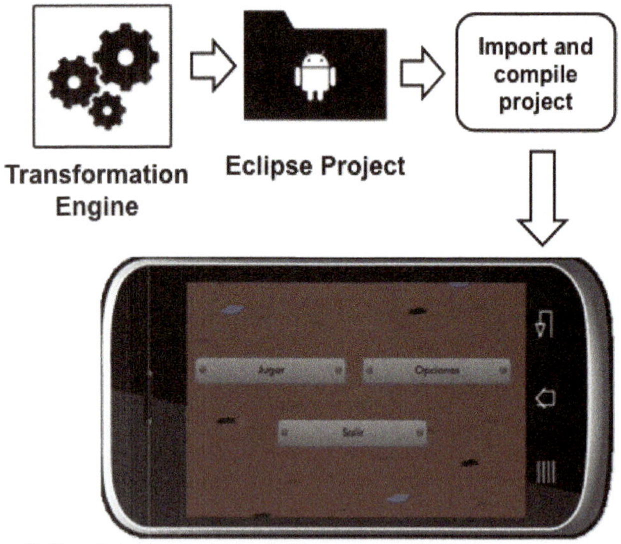

Fig. 18. Transformation and compilation process in Android

C. Supported typologies

Gade4all makes easier to develop videogames for different typologies in a flexible, user-friendly way. These typologies allow users to develop any kind of 2D videogame that matches the supported typologies. To simplify the development process and group the different types of games within a specific classification, the tool allows the development of videogames belonging to the following typologies:

a) Touch ability games

In this kind of typology we move the main character by touching the surface of the mobile phone's screen. As it can be seen in Figure 12 the mechanic of the typology allows users to create videogames based in dodging enemies and gathering items. The objectives of the level can be different: advance a certain amount of distance without crashing into enemies or obstacles, gather a special item, obtain a certain amount of points, etc.

Fig. 19. Ability Touch game developed with the Gade4all tool

b) Trivial Typology

With this typology users can create many types of videogames based on questions that can contain different types of resources: texts, images, sounds, videos, etc. Each question can have one or more valid answers and even a order for the correct answers can be stablished. This typology offers lots of possibilities regarding the kind of questions: memory questions, logic questions, trivial-like questions, questions based on observation, etc.

With the support of this typology, we can generate multiplatform educative videogames to help and motivate students and teachers to take advantage of the new technologies in order to improve the process of learning.

Fig. 20 shows some examples of the type of questions that can be created with this typology. These can contribute to interest teachers in this tool and help them creatings educative videogames as a learning resource.

These resources can include tutorials, math exercises, trivials or mental agility tests to help the students to improve their learning through videogames and entertainment applications.

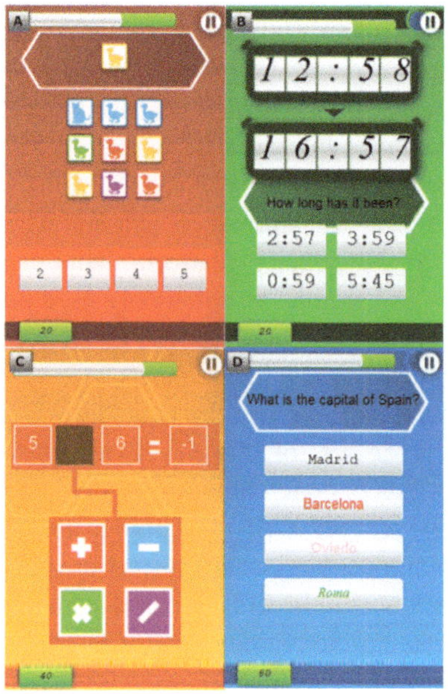

Fig. 20. Trivial game developed with the Gade4all tool

c) Puzzle Typology

As shown in Figure 14, with this typology we can design games in which we can connect pieces or throwing them to others. The mechanics of this typology allow users to create games in which they control the pieces that appear on the screen in order to make them collide with others. There are many possible winning conditions for each level: eliminate all the pieces of the same color, destroy all the pieces of the same color, hit a certain piece, etc.

Gade4all offers a flexibility that makes possible to develop any type of videogames based on this typology, which allows the creation of a wide range of puzzle videogames, in a simple and effortless way.

Fig. 21. Puzzle videogame developed with the Gade4all tool

d) Plataform Typology

In this typology we move the main character by using the controls that appear on the screen: horizontal movement, jump and shoot. As Figure 15 shows, the main character advances through a stage composed of blocks of different types, defeating enemies and dodging their shoots. The character can gather items in order to obtain points and the objective o each level is to reach a certain point of the stage or collecting one special item.

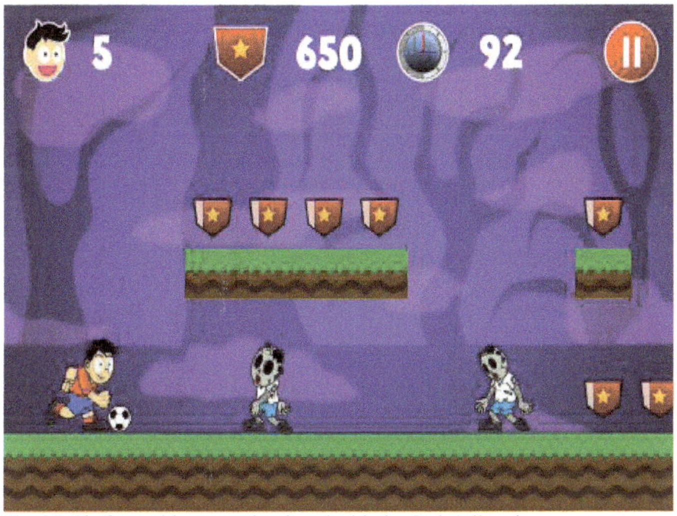

Fig. 22. Platform videogame developed with the Gade4all tool

e) Strategy Videogame

This typology allows user to design classic turn-based strategy videogames. As it can be seen in **Fig. 23**, in this type of videogame we define the aspect of the troops from two armies: the player's army and the computer's. The game can be composed by several levels, with different thematic, terrains and obstacles. The gameplay is turn-based, alternating between the movements of the player and the computer's. There are many possible winning conditions: defeating the leader of the other army, eliminating the whole enemy's army, capturing a flag, etc.

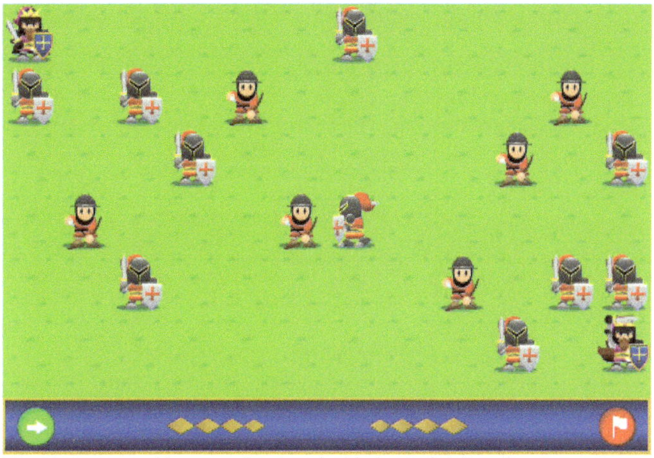

Fig. 23. Strategy videogame developed with the Gade4all tool

D. Comparison of Videogame Editors

In this section we analyze some of the characteristics of the different videogame creation tools that were previously described in this proposal. **¡Error! No se encuentra el origen de la referencia.** shows a general vision of the main similarities and differences between Gade4all and the rest of tools.

It should be noted that, as it can be seen in Table 1, the Gade4all platform offers some significant advantages in comparison to the other tools that were analyzed, among which we can highlight the fact that only Gade4all offers the possibility of developing videogames without requiring knowledge on software development. This makes possible for the tool to be used almost by more people. Also, it is the only tool that allows the complete generation of a videogame in native code for the different platforms, which makes possible for developers to modify or expand previously generated projects and create more complex applications.

1) Main similarities between Gade4all and the rest of videogame editors

After analyzing a set of tools and performing a comparison of these with the Gade4all platform, we can highlight some of the similarities that have been found:

- Possibility of creating actors and characters.
- Definition of sounds for events.
- Graphical editor for the creation and edition of levels.
- Generation of games for different platforms.
- Storage of the characteristics of a videogame for its later modification.

TABLE 1: COMPARISON OF THE MAIN TOOLS OF THE MAIN VIDEOGAME EDITOR

	GameSalad	GameMaker Studio	App Inventor	Stencyl	Gade4all
Previous knowledge on software development	Yes	Yes	Yes	Yes	No
Graphical editor of menus	No	No	Yes	No	Yes
Graphical editor of characters and levels	Yes	Yes	Yes	Yes	Yes
Definition of sounds for events	Yes	Yes	Yes	Yes	Yes
Generation of the complete game in native code, aside from the executable file	No	No	No	No	Yes
Possibility of saving the state of the game	Yes	Yes	Yes	Yes	Yes
Possibility of loading an existing game	Yes	Yes	Yes	Yes	Yes
Editing solutions outside the tool	No	No	No	No	Yes
Modifying levels outside the tool	No	No	No	No	Yes
Checking functioning of a generated level	Yes	Yes	Yes	Yes	Coming soon

S

2) Main differences between Gade4all and the rest of videogame editors.

As it has been specified in previous sections, even though the existing videogame editors share some similarities with the Gade4all tool, there are some important differences that should be mentioned:

- With Gade4all, we can configure the different menus of the game, while with the others (except App Inventor) we can only configure the game screen and its elements.
- The final result of a game created with any of the mentioned editors can only be edited with the same tool and there is no way of using the development's native environments. Gade4all, on the other hand, offers the source code of the games, making possible for them to be modified directlty from the original development environment.
- Aside from having a graphical editor, Gade4all allows the access to the XML file that describes the complete request of the DSL.
- Games developed with Gade4all have a level-loading software module based on XML, which makes possible to add and edit levels without the editor.
- Gade4all doesn't require to type a single line of code, all the behaviors and rules of the game are already defined, which means that anyone can create games without problems, even people with no knowledge on software development. If the code of a level has to be modified, we can go directly to the generated project, but this will require knowledge on the development environment. However, other tools not only require previous programming knowledge, but it's also necessary to learn its own language. For instance, GML to develop with GameMaker.

E. Statistics of videogames developed with Gade4all

With the alpha version of this platform we have developed some videogames that have been deployed in different platforms like Google Play for Android and App Store for iOS applications. Table 1 shows the statistics of the games developed for the different platforms and in different languages. These are just some samples of the videogames that the "University of Oviedo" has published as an associate of the Gade4all project. The other associates of this project have also published their own games, with their own statistics.

TABLE 1: STATISTICS OF VIDEOGAMES DEVELOPED USING THE GADE4ALL TOOL

Product	Download	Platform
Tiro Skeet	2,113	Apple
Tiro Skeet	999	Android
Ace Athlete	1,475	Apple
Ace Athlete	6756	Android
Tennis Adventure	874	Android
Touch Tank Game	220	Android
Platforms Game	94	Android
Strategy game	68	Android
Trivia about Spain	8	Android
Pepins Adventures	9	Android
Puzzle Game	3	Android
Warzone	52	Android

V. CONCLUSIONS AND FUTURE WORK

With the use of the platform presented in this proposal we are trying to fill an important gap in the design and creation of videogames, seeking to simplify the development of this kind of applications for every user, even the ones without programming knowledge.

By following the approach of Model Driven Engineering, we have managed to implement a generic Domain Specific Language which has helped to abstract the main characteristics of the main videogame typologies. This DSL has served to automate all the process of creating videogames for multiple platforms. This automation is reached through the graphical editor, allowing users to develop different types of 2D videogames in a simple and effortless way.

The Gade4all platform helps to increase the competitiveness levels in the development of videogames for mobile devices and the Web, as it makes easier to develop videogames for platforms like Windows Phone, Android, iOs and HTML5 without having to write a single line of code. With automatic generation of code, Gade4all improves the efficiency in the videogame development processes, reducing the development time and the number of errors and increasing productivity.

Currently, this tool can only generate videogames for the typologies shown in this proposal, but the development team is working to implement new ones, as well as improvements of some existing characteristcs, like the inclusion of a videogame library with some pre-designed videogames to make the development process even easier.

On its first stage of development, Gade4all only allows users to generate 2D videogames, but one of its objectives for the future is to support the creation of videogames in 3D.

VI. ACKNOWLEDGMENT

This work was performed by the University of Oviedo under Contract No. MITC-11-TSI-090302-2011-11 of the research project Gade4all. Project co-financed by the Ministry of Industry, Tourism and Commerce under its National Plan for Scientific Research, Development and Technological Innovation.

REFERENCES

[1] M. D. Walker, M. Nilsson, T. Jebb, and R. Turnbull, "Mobile video-streaming," BT Technology Journal, vol. 21, no. 3, pp. 192–202, 2003.

[2] B. Selic, "MDA manifestations," The European Journal for the Informatics Professional. June, 2008.

[3] E. González, H. Fernández, and V. Díaz, "General purpose MDE tools," IJIMAI, vol. 1, 2008.

[4] S. Kent, "Model Driven Engineering," COMPUTER-IEEE COMPUTER SOCIETY-, vol. 2335, no. 2, pp. 286–298, 2002.

[5] T. Donovan and R. Garriott, Replay: The history of video games. Yellow Ant, 2010.

[6] M. Duggan, Making a GameSalad for Teens. Course Technology Ptr, 2013.

[7] J. Elliott, HTML5 Game Development with GameMaker. Packt Publishing, Limited, 2013.

[8] A. Baytak and S. M. Land, "A case study of educational game design by kids and for kids," Procedia - Social and Behavioral Sciences, vol. 2, no. 2, pp. 5242–5246, Jan. 2010.

[9] L. Stencyl, "Stencyl: Design Once, Play Anywhere," 2013. [Online]. Available: http://www.stencyl.com/.

[10] C. E. MONTENEGRO MARÍN, P. A. GARCÍA, J. M. CUEVA LOVELLE, and O. S. MARTÍNEZ, "Aplicación de ingeniería dirigida por modelos (MDA), para la construcción de una herramienta de modelado de dominio específico (DSM) y la creación de módulos en sistemas de gestión de aprendizaje (LMS) independientes de la plataforma.," DYNA, vol. 78, no. 169, pp. 45–52, 2011.

[11] A. van Deursen, "Domain-Specific Languages versus Object-Oriented Frameworks: A Financial Engineering Case Study," in Proceedings of Smalltalk and Java in Industry and Academia (STJA'97), 1997, pp. 35–39.

[12] E. Miotto and T. Vardanega, "On the integration of domain-specific and scientific bodies of knowledge in model driven engineering," Procs. of STANDRTS, vol. 9, 2009.

6

BPLOM: BPM Level-Oriented Methodology for Incremental Business Process Modeling and Code Generation on Mobile Platforms

Jaime Solís-Martínez, Natalia García-Menéndez, B. Cristina Pelayo G-Bustelo and Juan Manuel Cueva Lovelle
Department of Computer Science, University of Oviedo, Spain

Abstract **— The requirements engineering phase is the departure point for the development process of any kind of computer application, it determines the functionality needed in the working scenario of the program. Although this is a crucial point in application development, as incorrect requirement definition leads to costly error appearance in later stages of the development process, application domain experts' implication remains minor. In order to correct this scenario, business process modeling notations were introduced to favor business expert implication in this phase, but notation complexity prevents this participation to reach its ideal state. Hence, we promote the definition of a level oriented business process methodology, which encourages the adaptation of the modeling notation to the modeling and technical knowledge shown by the expert. This approach reduces the complexity found by domain experts and enables them to model their processes completely with a level of technical detail directly proportional to their knowledge.**

Keywords **— Business Process Modeling, BPMN, process transformation, code generation.**

I. Introduction

Software development has seen some great changes in some of its methodologies and techniques, but some of its problems have remained unchanged since their appearance in the mid and late twentieth century. One of these lasting problems is related with requirements engineering, aspect of the software development process that has seen very little evolution. Sommerville and Kotonya stated in their study [1] that there are several problems related with this initial activity of the software development process. The first problem mentioned is that the requirements engineer is not an expert in the application domain being addressed. Another of the difficulties present in this phase of the software development cycle which is also mentioned by these two authors is the fact that natural language is ambiguous, which has also been confirmed by the work of Laue and Gadatsch [11].

This work has been partially funded by the Ministry of Industry, Energy and Tourism of Spain, by the Avanza2 plan and by the European Regional Development Fund through the project "GADE4ALL: Plataforma genérica para facilitar el desarrollo de videojuegos y software de entretenimiento multiplataforma". The code for the project is MITC-11-TSI-090302-2011-11.

The Business Process Management Initiative (BPMI) [24], a group inside the OMG [23], proposed a solution for this problematic situation: Business Process Modeling (BPM), a discipline promoting the implication of the domain experts in the requirements engineering process through the use of a modeling notation that lies between the domain experts' language and the computer experts' knowledge. In its proposal, the BPMI introduced Business Process Modeling Notation (BPMN) [25] as the standard notation for BPM. Although their intention was different, this standardization made the notation grow in size and complexity to the point where non-technical domain experts must undergo a training process in order to understand it and use it properly [5].

Due to this circumstance, some simplifications of BPMN have been conceived, with different success degrees; one of these simplifications is Simple BPMN (SBPMN) [6], defined during previous work at the University of Oviedo with the following objectives: reduce the number of symbols needed to model a process and raise the user's level of abstraction. SBPMN was used to generate applications in a similar scenario to the one presented in this paper [20], although in that case several adaptions of the model were necessary for the code to be generated. The results obtained by SBPMN in the tests carried out were satisfactory but we consider the number of symbols it offers to be too large for non-technical domain experts. In order to solve this situation our first objective is the definition of a BPM level oriented methodology, a system that enables the adaptation of a graphical modeling notation to the skill level presented by the business expert.

In order for our approach to be used in real scenarios we need to achieve two other objectives: encourage business experts to model their processes and enable quick generation of the applications supporting the models created by the experts. We intend to complete these two goals with the definition of two separate but closely related tools: BPLevel Modeler and BPLevel Generator. BPLevel Modeler is a business process modeling tool which supports the level oriented methodology and promotes the involvement of business experts in the requirements engineering process through a simple and intuitive user interface. On the other hand, BPLevel Generator is capable of analyzing the models created with BPLevel Modeler and generate a specific and custom application for each model; in this case the tool is aimed at the computer experts in charge of developing the applications in each scenario.

A summary of the structure of our proposal is shown in Fig. 1, where the whole modeling and code generation process can be overviewed; as it is seen, business experts will be able to create their models through the use of our modeling software and the models generated will be handed to the corresponding IT technicians who will generate the custom applications using our code generation tool. These computer experts will also be involved in the configuration of the XML file containing the graphical details of the custom application, which at this moment needs to be done manually.

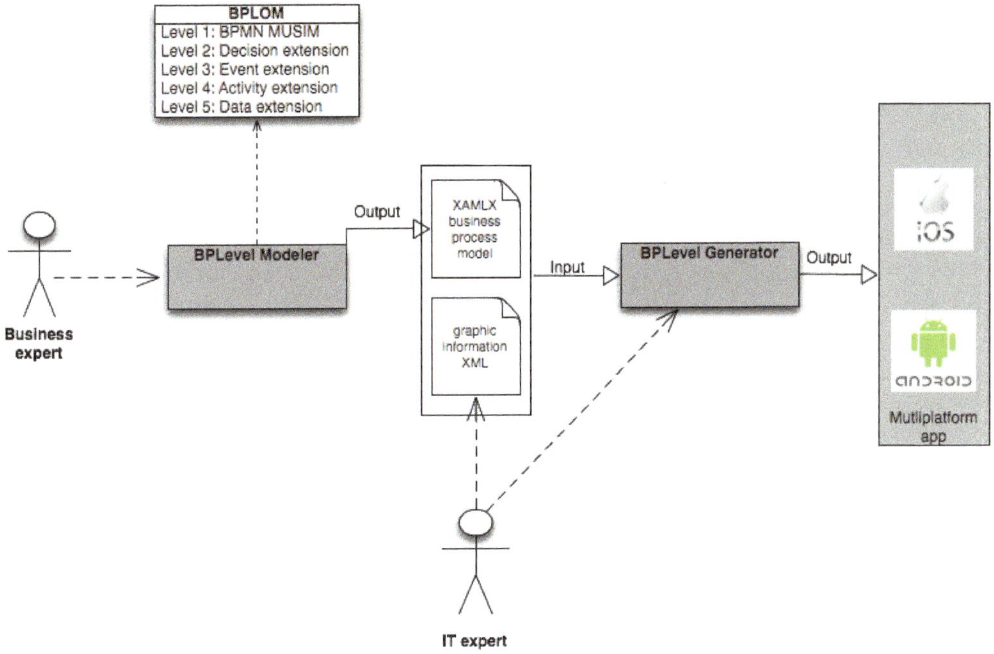

Fig. 1. Graphical overview of BPLOM use scenario

The rest of this paper will be structured as follows: section II will describe some of the existing BPM notations and establish the difficulties that non-technical domain experts undergo when using them. Section III will introduce our BPM level-oriented methodology and the results obtained by its initial level in a real scenario at a Spanish enterprise, where business experts used the initial level of our methodology for modeling their processes. In section IV we will present our tools, starting with BPLevel Modeler and continuing with BPLevel Generator and its intended use. Lastly, in section V, we will identify our conclusions and section VI will establish the future work we intend to carry out in order to improve not only our tools but also our methodology.

II. BPM and Business Process Modeling Notations

BPM is an initiative that encourages domain experts to define their business processes through modeling notations as a way of reducing the difficulties found in the requirements engineering process. This approach is based on the use of notations that are half way between the domain experts' language and the computer experts' knowledge.

There are, mainly, two types of business process modeling notations: graphical notations and textual notations. Bearing in mind our current application scenario, we have studied what we consider to be a representative set of notations that meet all of the following factors:

- High degree of diffusion, including in our revision those notations with higher diffusion degree.
- Wide range of complexity, from the most complex example (BPMN) to the simplest one (SBPMN).
- Domain expert implication, factor which rules out textual notations as a suitable modeling alternative. As stated by Lu and Sadiq [10], graphical notations allow users to represent business processes through simpler semantics and more abstract syntax, circumstance that lowers the complexity domain experts experience during the modeling and verification of the processes.

A. BPMN: the standard notation

BPMN is the standard notation for business process modeling. It was proposed by the BPMI in 2004 and since then it has undergone periodical revisions, being at the time in its 2.0 version. It is the most widely use notation of this type, with more than 70 implementations nowadays.

The two objectives that where stated as principal goals for BPMN in the introductory document written by Stephen White [2] where the following: provide domain experts an understandable and usable notation and reduce the number of existing notations and modeling tools.

1) BPMN's features

BPMN's root element is the Business Process Diagram (BPD), which is composed by a set of activities that represent the actions present in the business process and a group of flow controls entities that establish the order in which these activities are done. The models built with BPMN can be enriched with other elements such as events, choreographies, messages, lanes and pools. BPMN's elements can be divided into the following categories: activities, gateways, conversations, choreographies, events, swimlanes and data.

The number of elements present in the BPMN specification is large, circumstance explained by the standard category of the notation. Version 1.1 of BPMN consisted of 52 different elements whilst the actual 2.0 version has seen an increase in this feature, as it can be seen in the BPMN 1.1 and 2.0 posters linked in the BPMN web page [25].

2) BPMN's disadvantages

The main issue regarding the use of BPMN when dealing with non-technical domain experts for process modeling is its complexity.

Wahl and Sindre have confirmed this fact in a study [5] where they state that a non-technical expert will need training in order to be able to use BPMN in a correct way. This complexity can also be seen in the investigation carried out by Recker in 2007 [7], where through the answers of 590 BPMN users he has been able to establish that there are several entities in BPMN that receive very little use. This fact can be seen in Fig. 2, a graphic classification of BPMN entities based on the use they receive obtained from the results presented in the referred work. The entities in BPMN are classified into three different categories: important, sometimes used and unused.

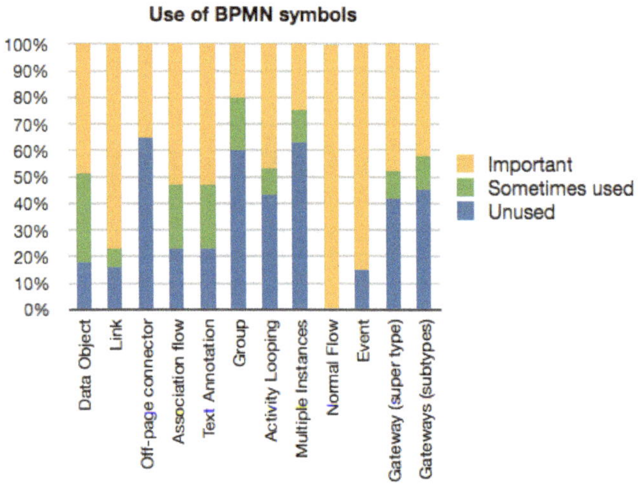

Fig. 2. Graphical representation of BPMN symbol importance obtained in Wahl's and Sindre's work

Stephen White and Miers also mention BPMN's complexity in their BPMN reference guide [8]. In one of the sections of the book they point out that it is not likely for a business analyst or end user, who can also be referred to as a application domain expert, to need all the symbols included in BPMN.

Another issue concerning the usage of BPMN by non-technical domain experts is the great number of BPMN supporting tools that exist actually. Although one of the objectives stated in the presentation of BPMN [2] was to reduce the number of tools with support for BPMN, the situation has gone the other way; this circumstance can be confirmed by a study carried out in 2010 [13] where it is stated that the popularity of BPMN has encouraged the appearance of a greater number of modeling tools with support for the standard. The magnitude of this problem seems to increase when the differences between the tools are notable due to the interpretation of the standard made by the authors and the different approaches that lead to including BPMN entities to the tool or not.

B. UML Activity Diagrams

UML Activity Diagrams, which can also be referred to as UML AD, are one of the types of diagram included in the Universal Modeling Language (UML) [26] specification.

The Activity entity, which represents the different actions that have to be carried out during the execution of the process, is the base of this type of diagrams. The activity entity is accompanied in these diagrams by other artifacts like decisions and parallel activity execution syntax. Although UML AD include entities present in the other business process modeling notations, a study [4] referenced in our analysis establishes that UML AD are less expressive than BPMN.

Even though UML AD do not include such a great amount of entities as BPMN, fact which enables a reduction of the complexity found when modeling business processes, there are some issues regarding the graphical representation chosen for them. This circumstance is triggered by the fact that some of the entities included in the specification share the same graphical representation. For example, the decision entity is represented in the same way as the merge entity, situation that can lead to problems for the understanding of the model by the non-technical domain experts. This circumstance can be seen in Fig. 3, which shows a sample UML Activity Diagram process.

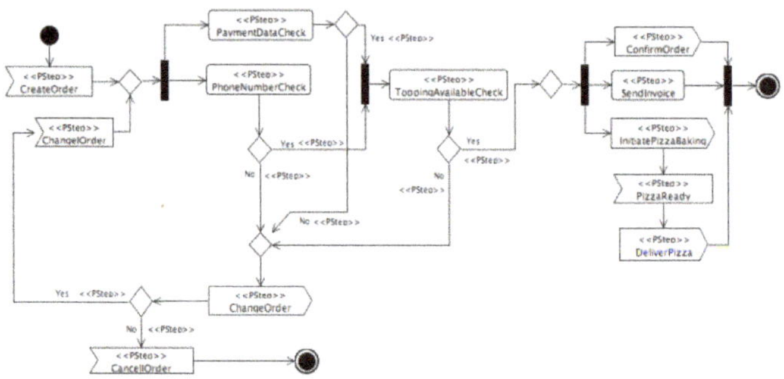

Fig. 3. UML AD sample process diagram

The other aspect that encourages us to avoid using UML AD as a valid notation for process modeling by non-technical domain experts is its abstraction level. As UML is a general-purpose language, UML AD offers the user a very low abstraction level. This aspect goes against one of the main goals in this investigation: usability experienced by non-technical domain experts.

C. jPDL, process modeling language in Java

jPDL, which stands for Java Process Definition Language, is a graphical language for business process definition included in jBoss jBPM [27], a BPM suite written in Java for applying BPM under this platform.

This language appeared as a simplification of BPMN via two different approaches: reducing the number of entities available for the definition of the process and modifying some of the graphical representation of the entities through a color code. As a result of the appliance of these approaches, jPDL has managed to offer a lower complexity level than BPMN and has also achieved an increase in the user's level of abstraction. Fig. 4 shows a sample process diagram built using jPDL, where the color code can be appreciated as the main difference between this diagram and the one generated with UML AD.

An example where jBPM is used to increase the level of automation of the business processes can be found in the work done by Castaño [21]. Having correctly identified the suite's capability for adapting to nearly all business process, Castaño introduces a prototype using jBPM that enables further automation of business processes through the use of data mining. Via this solution, the dependency that some processes may have with human interaction can be reduced.

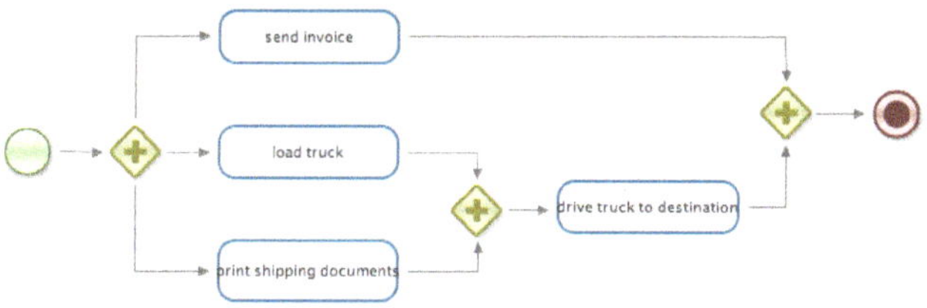

Fig. 4 jPDL sample process diagram

Although the inclusion of this color code favors the comprehension of the models built with jPDL and despite the fact that this language has managed to reduce the number of entities offered by BPMN, we consider that jPDL is not suitable for our investigation. Our main concern regarding jPDL is the user's level of abstraction, as there are two aspects related with the activity entity that make it low: on one hand, the fact that there are various types of activities which have high technical content (the script task, for example) and, on the other hand, the need to edit under some circumstances the XML code behind the graphical representation of the business process in order to configure some of its details.

D. Petri nets, another option for process modeling

Petri Nets [14], defined by Carl Adam Petri in the mid twentieth century, are another option to be considered when dealing with process modeling. There are several features that make Petri Nets suitable for this task, as stated by van der Aalst in his study [15]:

- Formal semantics, which enable the precise and clear definition of process models.
- Graphical nature, which promotes process definition through the use of nodes and transitions.
- Expressiveness, as they support all the primitives used when defining processes.
- The existence of many analysis techniques that can be applied to them. These techniques can be used to evaluate properties and also to calculate performance rates.
- Platform independency, as Petri Nets are not based on any proprietary software.

Despite these properties Petri Nets have and although they can be defined graphically, there is great concern when introducing them to non-technical domain experts: their complexity and low level of abstraction. These two inconveniences are against our usability and business expert focuses so we have decided to avoid using Petri Nets in our scenario, although we think they are suitable for completing modeling tasks under other use circumstances.

E. SBPMN: Simple BPMN

Simple BPMN, also referenced to as SBPMN [6][20], is a reduction of BPMN attempted as a previous investigation of components in our group at the University of Oviedo [28].

SBPMN arose as a possible solution to the problems found with the abstraction and complexity levels of the notations previously presented in this paper. One of the objectives of SBPMN was to reduce the technical knowledge level needed to complete the modeling of a business process by the domain expert. It also tries to avoid the arbitrary use domain experts give to some of the symbols present in BPMN.

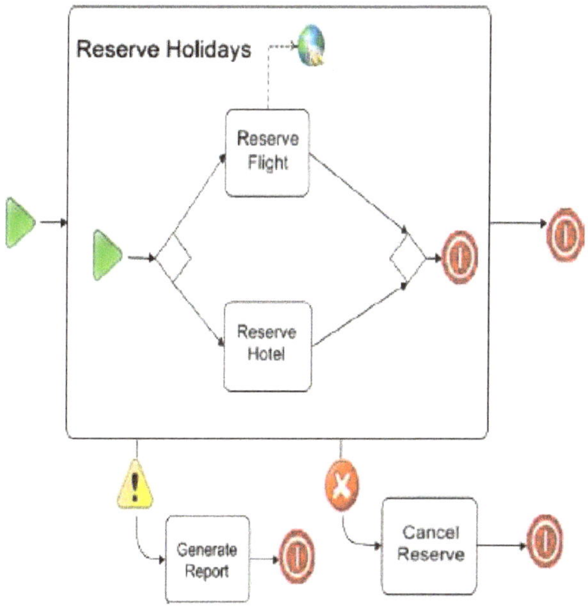

Fig. 5. SBPMN sample process diagram

Fig. 5 shows a sample SBPMN diagram representing the process used to make a trip reservation. As it can be seen, SBPMN offers different graphical representation of some of the entities it borrows from BPMN, which makes the process more intuitive for the user and simpler to understand.

The graph presented in Fig. 6 shows a graphical summary of the results obtained by SBPMN in the tests that were carried out after its definition. These results show that percentage of errors or failures made by domain experts when using SBPMN was less than when using BPMN, showing a reduction rate greater than 20%. This error reduction implies that the skill level shown by the users raised considerably when using SBPMN (more than 80% of skills shown) compared to BPMN (less than 50% of skills shown).

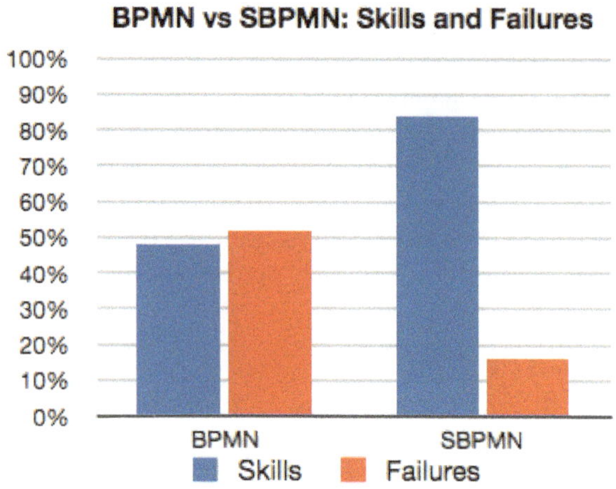

Fig. 6. Comparison of skills and failures between BPMN and SBPMN

The results shown establish that SBPMN is simpler than BPMN for non-technical domain experts when modeling their business processes. The problem concerning this simplification, despite the great effort shown in managing to simplify process modeling to non-technical domain experts, is the fact that the number of entities offered is still big and sometimes can be too complex for non-technical experts; for example, SBPMN proposes the use of various types of activities (human task, simple task and automatic task) and includes some entities with technical background like XML schema data object and data source object.

III. BPLOM: BPM Level Oriented Methodology

Until this moment, we have focused on analyzing several notations based on their degree of diffusion and their complexity. All of the notations presented have managed to model processes correctly but there are several issues which make us discard them as suitable for non-technical domain experts: low abstraction level, big complexity in their use and difficulty in understanding the models at first sight.

A common feature of all of these notations is to present the user a set of entities to use, which cover the basic and the advanced features that can be present in a process model. This circumstance can bring problems to novice non-technical users, as they face the use of complicated and technical artifacts (like events, signals and data objects) that they may not understand fully. This situation can also lead to misuse of some of the entities in the language and, thus, to the specification of a wrong model.

In order to prevent this situation, why not present the user a set of entities that is capable of adapting to his modeling skills and knowledge? This is the basic approach of our proposal: a level-oriented methodology for the application of BPM to any type of business process, promoting the adaptation of the modeling entity set to the expert's knowledge level. This departure point differs from other business process modeling notations like the one introduced by Chinosi and Trombetta in the 2009 edition of the BPM Handbook [18]. The first stage of the referenced methodology is based on reading the documentation related to the process and creating a primary sketched version of the process derived from the interpretation a computer expert has of this documentation. Although this approach could be valid, the ambiguous nature of natural language [1][11] does not recommend it.

Our proposed methodology will be divided into 5 incremental levels. Level 0, also known as BPMN MUSIM [16], will be the initial level and it will offer the minimum number of entities needed for business process modeling. Throughout the following levels we will be introducing more artifacts like events or data objects to the methodology in order for it to gain in expressiveness, trying to reach an expressive power as similar as possible to that in BPMN. The levels will be incremental, so the user will be able to model his process using the entities contained in the current level in addition to those contained in the previous ones.

This gain in expressiveness achieved when going up through the levels is directly related to an increase in the complexity users fin when using the symbols included in the methodology, but as the higher levels are intended for more technical based experts this complexity increase is manageable.

A. Features of the Methodology

The methodology we propose has some key features we would like to point out. These are the following:

- Incremental nature of the levels, which allow the user to model his processes with the entities that adapt to his modeling skills and technical knowledge.
- All processes can be transformed to code and executed, no matter in what level they are in.
- Platform independency. Despite the fact that we have chosen .NET as the target platform in our study, this approach could be used to generate code for any other platform.
- BPM phase schema reduced. As the domain experts are in charge of modeling the processes and these will be used to generate code, there is no need to capture requirements in text form and translate this text into models. This enables the avoidance of common errors in the requirements engineering process [1] and reduces the 5-phase BPM application scenario described by Ryan K. L. Ko [9].

B. Drawbacks of the Methodology

The main disadvantage of our methodology in comparison to BPMN is its expressive power. Although our methodology includes the majority of the artifacts and entities included in the standard, some other components of it have been dismissed due to different reasons (see discarded BPMN entities section of this paper).

Despite this fact, we think the methodology is still capable of modeling any type of business process completely and with no need for any additional symbols, notwithstanding the possibility of further extensions of the entity set available in our approach.

C. Level Description

Once our proposal's main features and limitations have been introduced, the next step is to present each of the five levels that make up our methodology. This presentation will be done by introducing the symbols or entities present in each of the levels with an explanation of their functionality and an overview of their graphical representation.

1) Level 0: BPMN MUSIM

The first level of our methodology is called BPMN MUSIM [16], which stands for very simple BPMN. As the basis of our methodology, this level is intended for novice, non-technical domain experts that want to model their processes through a simple and clear notation. It contains the minimum set of symbols needed for process modeling, 5 entities in total. BPMN MUSIM's main feature as an introductory modeling artifact for novice, non-technical domain experts is the quick and simple learning process domain experts undergo before they start modeling with it.

a) BPMN MUSIM entity selection

As it was stated before, BPMN MUSIM is conceived as the minimum set of symbols needed to define a business process completely. The selection of the entities it includes has been made following the results of several studies [3][7][9]. Particularly interesting are the results of the study carried out by Recker in 2008 [3], which show, through the analysis of several BPMN models, that there is a common subset of entities that are present in the majority of BPMN models produced by experts.

Fig. 7 shows a graphical recreation of the results obtained by Recker in this study in the form of a set diagram. Each of the boxes holds a group of entities of the BPMN symbol set and a number indicating the amount of times these entities appear together in the studied BPMN process models. For example, 116 models have tasks and sequence flow entities in common and 65 processes have these two entities and the start and end events in common; the greater the number inside the box, the greater the chance a business process model has of containing all of the entities inside the box. As the graph shows, the most commonly used symbols in BPMN are: tasks, sequence flow, start event, end event, pools and gateways. Based on this study and bearing in mind pools are mainly used as the representation of the ownership a user has of the business process, we believe 5 entities can compose the minimum set allowing complete modeling of business processes and thus we propose it to be the introductory level for our methodology.

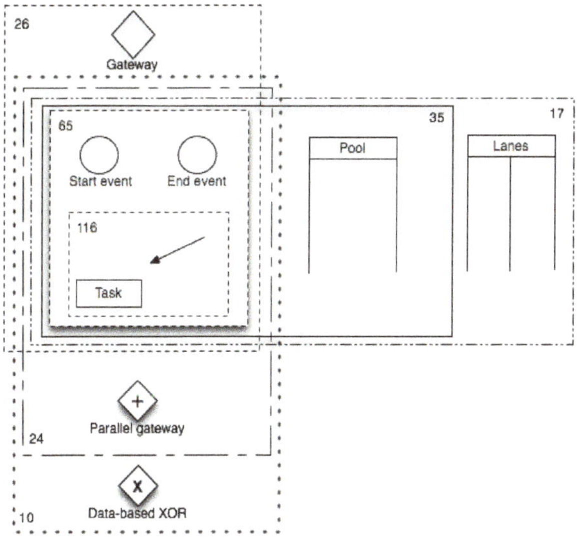

Fig. 7. Results extracted from Recker's 2008 study

b) Level 0 elements

The set of entities included in this level and their graphical representation is the following:

- **Starting point:** All processes modeled with BPMN MUSIM must have a single starting point that will be represented by a green circle.
- **Ending point:** A process defined with BPMN MUSIM can have one or more ending points, which represent the end of the process. The ending point in BPMN MUSIM will be a red circle.
- **Activity:** A rectangular shape with its name inside will represent an activity.
- **Transition:** A transition represents the flow between two elements of the model and will be represented with an arrow. The arrow's head will point the direction of the flow.
- **Decision:** Decisions enable alternative taking in business processes. A decision will be based in an expression to decide upon and two branches: true and false. A diamond shape will represent decisions.

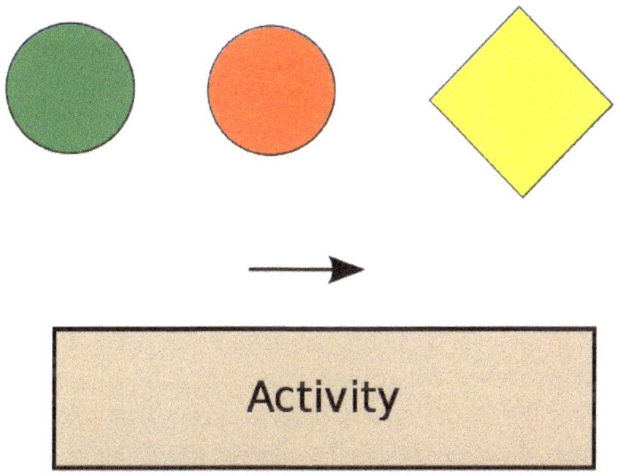

Fig. 8. Graphical representation of BPMN MUSIM entities

The graphical representation of the entities in this level, presented in Fig. X, was chosen due to the following reasons.

- The need to be close to the graphical representation chosen by the OMG for BPMN. As the standard notation, BPMN is widely used not only as it is but also as the basis for other notations and tools (like Microsoft Visio for example). For this reason, if we define a similar graphical representation for our initial level, user's migration to our approach can become a simpler task.
- We have defined first sight differentiable symbols for each of the entities offered in our methodology in order to avoid one of the problems found in UML AD, where two artifacts shared the same shape and this could lead to process experts misunderstanding the models.
- As the coloring approach offered by jPDL seemed to be generating more comprehensible models than those obtained with black and white entity representation, we decided to include color to our shapes. In this way, we continued with the green and red color code for representing starting and ending points and also added colors to the other symbols.
- Recker, Safrudin and Rosemann studied novice user modeling patterns in their work [12] and stated that this type of users understand better models including text and abstract symbols (circles, arrows and rectangles) than those containing concrete figures.

c) BPMN MUSIM's use in real life processes: examples and results

BPMN MUSIM has been used for modeling two real life processes in a Spanish enterprise, trying to demonstrate its suitability for business process modeling.

In order to establish its skills for the modeling of any kind of business process we decided to use BPMN MUSIM to model the following processes: informatics incidence management and recruitment, one of them close to computer science and the other concerning non-technical aspects. Details on the process models and other circumstances regarding the application of BPMN MUSIM to this use case can be found on the two articles [16][17] presented in 2011 at an Iberian congress.

The scenario designed for the application of BPMN MUSIM to these processes started with a brief meeting with the domain experts, where they were introduced to the notation: its entities, their meanings and the first modeling exercises. Once the experts understood the language and were capable of using it, an iterative meeting approach was taken: each of the experts was addressed to come to a meeting with the process models he had done and these would be reviewed with the project's team in order to spot the errors or difficulties; when the review was complete, the expert was set to correct the errors and another meeting was scheduled in the following days in order to undertake further review. This meeting schedule was repeated until the non-technical domain experts marked the process models as definitive.

As the following figures show, BPMN MUSIM has obtained good results in the aspects that were measured after its introduction to the domain experts. In first place, users have stated a better comprehension of the symbols in BPMN MUSIM, as seen in Fig. 9; this circumstance is due to the inclusion of the color code, which enables users to understand models better at first sight. Results also point out, like it is seen in Fig. 10, that the majority of the domain experts didn't spot the need for any additional symbols in BPMN MUSIM in order to be able to model their processes completely.

The interpretation of these results allows us to think that BPMN MUSIM symbols are easier to use than BPMN symbols and that the proposed symbol set is sufficient for non-technical experts to model their processes completely at a basic or early stage modeling level.

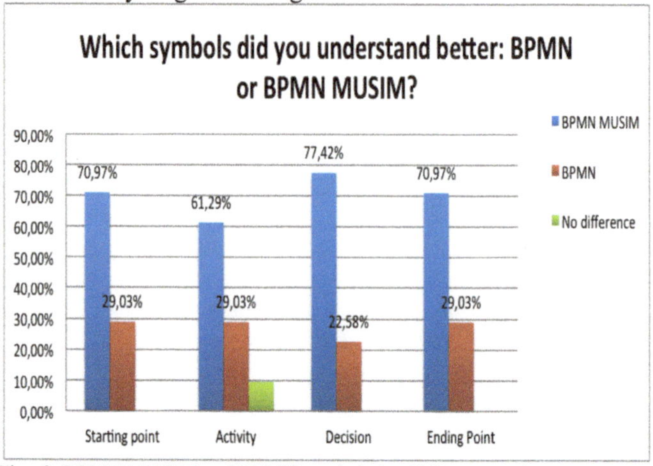

Fig. 9. BPMN MUSIM and BPMN symbol simplicity comparison

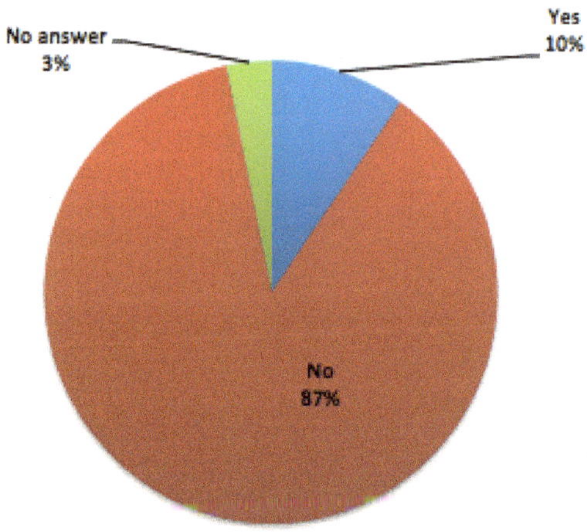

Fig. 10. Pie chart representing need for additional symbols in BPMN MUSIM

2) Level 1: Decision Extension

Once the basic symbols needed for process modeling have been introduced, its time for increasing the expressiveness of the methodology.

Fig. 7, which established the most widely used symbols in BPMN diagrams built by domain experts, showed that the main core of symbols are the ones included in BPMN MUSIM and that these are followed by the use of more complex decision entities. In order to follow the tendency pointed out in this investigation carried out by Muehlen and Recker [3], we have decided that the second level of our methodology is going to be the decision extension.

a) Level 1 elements

The symbol set proposed as a decision extension for the methodology is the following:

- **Parallel decision:** A parallel decision allows the expert to define the execution of two simultaneous paths inside the process. A yellow diamond shape with a cross inside will represent it.
- **Inclusive decision:** Inclusive decisions enable the activation of at least one of their branches, depending on the incoming condition. A yellow diamond with a circle inside will represent an inclusive decision.
- **Join:** The inclusion of these new decision types forces the inclusion of the join entity, the point where the process waits for the completion of the process' paths before moving forward to the next activity. A horizontal line with two incoming transitions and one outgoing transition will represent joins.

Fig. 11. Graphical representation of Level 1 entities

3) Level 2: Event Extension

Once the elements pointed out by Muehlen and Recker as the most used in BPMN have been included in the methodology, it is time to extend our proposal with other type of artifacts provided by BPMN and which are intended for more technical and experienced users.

BPMN gives a lot of importance to the events, entities that reflect the appearance or occurrence of certain actions that alter the normal process flow. The entities of this kind offered by BPMN cover from messages to signals, without forgetting others like time events and errors. With a closer look at the BPMN 2.0 entity set, clearly represented in the BPMN 2.0 poster offered by the OMG [23], there is a main issue regarding the event use in this version of the standard: each of the events proposed in BPMN 2.0 has several graphical representations depending on the place where they appear in the model (beginning, intermediate and end) and if they activate a subprocess or not.

In order to simplify the event model proposed by BPMN for the non-technical domain experts and, at the same time, avoid the appearance of different entities with similar graphical representation, we propose another approach for event modeling under level 2 of our methodology. We call this level the event extension.

a) Level 2 elements

The entities included in the event extension of the proposed business process modeling methodology are the following:

- **Message event:** This type of event allows the user to include message receiving and sending inside a process, enabling communication between different processes and/or users. Messages will have two graphical representations in this methodology: an outgoing message will be represented by an envelope with an arrow pointing up and incoming messages will be represented by an envelope and an arrow pointing down.
- **Time event:** A time event allows the definition of time conditions inside a process, like the fact that a process must wait for a certain activity to end before continuing or also a time lapse. A clock will represent time events.
- **Error event:** Error events define the place where a process stops due to the appearance of an error, ending the process' execution immediately. Error events will be represented by a prohibition signal with the word "Error" inside.
- **Cancelation event:** Cancelation events represent the moment where a process' execution is cancelled. A red colored cross will represent these events.
- **Signal event:** This event allows the user to send a signal to another process in order for it to continue its execution. It is directly related with the time events presented before. A danger signal will represent signal events in this methodology.

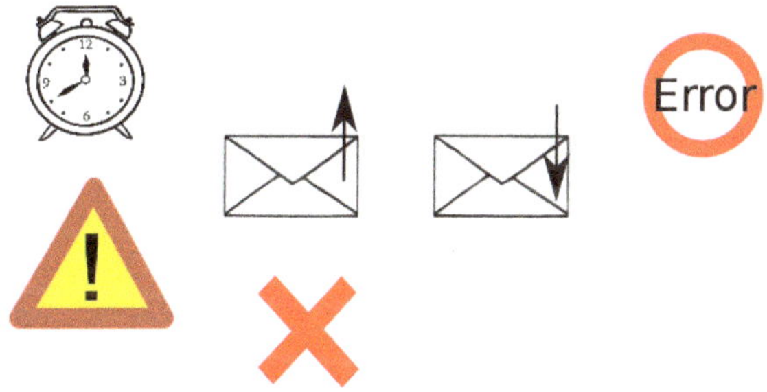

Fig. 12. Graphical representation of event extension entities

b) Using Events in the Models

Once the events have been introduced, there is the need to explain how they can be used for process modeling. All events, except cancelation and error events that need to appear just before an ending point, can appear in any point of the process.

Events are artifacts that make the process stop until their occurrence in order to continue normally. Thus, an event must always appear after the activity that generates it and before the entity whose execution it has to impact.

For example, if we need to model a process where an activity causes the sending of a message and after the delivery takes a decision we would have to place the activity first, then the outgoing message entity and finally the decision. Making the model like this will ensure the process will send the message once the preceding activity has been completed and also it will wait until the message is sent before going on to taking the decision.

4) Level 3: Activity Extension

The fourth level of the proposed methodology is defined as an activity extension for the proposal. Until this moment the methodology only provided the user one type of activity, which represented an automatic or user task.

However, as domain experts move along the levels of the methodology they have more experience with the use of the notation and also their processes need of more rich constructions in order to be modeled precisely. Under these circumstances, the activity extension of the methodology introduces subprocesses and function calling to its artifacts.

a) Level 3 elements

The elements included in the activity extension of the proposed methodology are the following:

- **Call activity:** A call activity references a global task that is repeated throughout several processes. This entity allows the user to call certain a certain task without having to redefine it in each of the processes it appears in; a common call activity could be user identification or login. Call activities are represented by the same entity as the normal activities but including a world globe icon that identifies it as a call activity.
- **Subprocess:** A subprocess is a set of activities that need to be done without any interruption. If any of the activities inside one subprocess produces an error, the process will be terminated. A subprocess is represented by a dashed rectangle that includes all of the activities of the subprocess.
- **Event subprocess:** Event subprocesses are a special type of subprocess, which is triggered after the appearance of an event; it will behave identically as the normal subprocess in case an error takes place. Its representation is the same as the normal subprocess one but the first entity of an event subprocess must be an event (in the following image the event is an incoming message event). The events available for an event subprocess are: incoming and outgoing message events, time events and signal events.

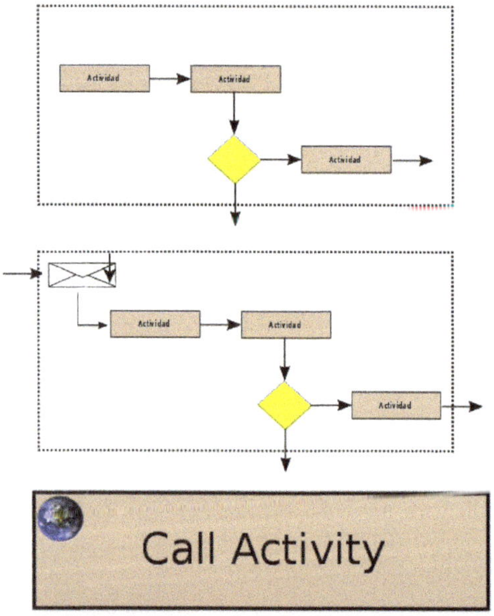

Fig. 13. Graphical representation of the activity extension elements

5) Level 4: Data Extension

The last level of this BPM methodology is defined as a data extension. Until this level, the introduction of technical detail in the models has been kept as small as possible in order to preserve the expert from tying the model to any technical aspect.

However, at this top level of the methodology the experts are considered both experienced modelers and technically prepared, so this status requires the introduction of data elements that enable the definition of information structure and flow through the processes. This detail is directly related with technical application implementation details that are too complex for the novice modelers using the lower levels of the methodology.

a) Level 4 elements

The elements included in the data extension level of the methodology are:

- **Data object:** A data object represents information that flows through the process in different ways (documents, data introduced in a form, etc.). A paper sheet that represents information stored in a computer defines a data object.
- **Object collection:** An object collection represents a set of data objects that flows through the process, like a list of documents for example. A stack of documents, which establish that an object collection is made up of multiple data objects, represents it.
- **Warehouse:** A warehouse represents the moment where a process reads or writes data in a database. This implies that data generated in a process and passed to a warehouse survives the process' instance. The classical hard drive representation with the word "Warehouse" inside will represent this entity.

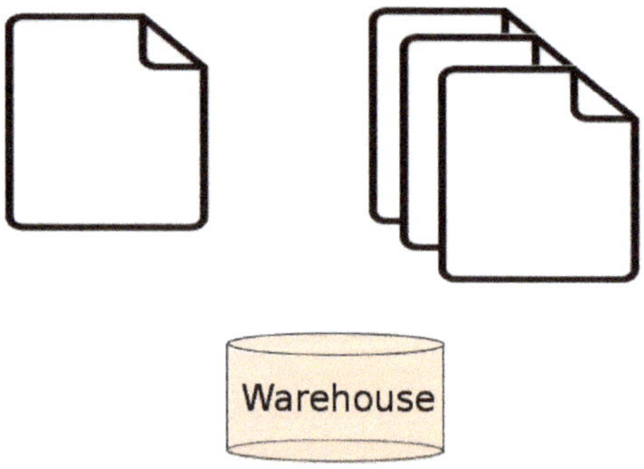

Fig. 14. Graphical representation of Level 4 entities

6) Discarded BPMN Entities

As it has been seen in the presentation of the levels that make up our methodology, there are several artifacts included in BPMN that are not present in our proposal. As the results of the referenced studies show [3][19] several of the modeling entities offered by the standard business process modeling notation experience little use.

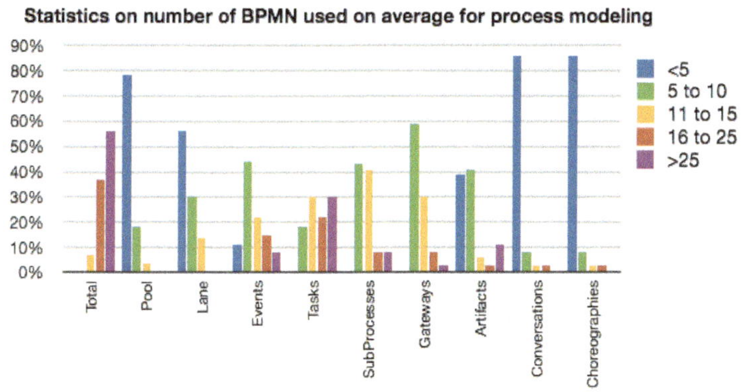

Fig. 15. Results obtained by Chinosi and Trombetta in their work

Fig. 15, a representation of the results from Chinosi's and Trombetta's study published in 2012 [19], shows a bar graph with a measurement for the number of times each of the represented BPMN constructs appear in the models included in the study. As it can be seen, BPMN elements like conversations, choreographies and pools, for example, experience low use by modeling experts. Thus, we have decided to exclude these symbols from this first version of our proposed methodology. Although the results extracted from this study differ in some of its figures from those obtained by Recker [3], the similar low symbol usage trends allow us to rule out some of the least used features found in BPMN.

At this moment we must establish that the exclusion of these symbols is not a definitive decision. As it will be explained in the future work section of this paper, we intend to make our level oriented methodology undergo a thorough testing procedure. One of the main goals of these tests would be to determine the suitability of the selected symbol set for the modeling of all types of business processes. With the results obtained from the tests we will be able to determine the need to include additional symbols to the methodology or discard the inclusion of any other entity to our methodology.

7) Applying BPLOM to a business process

The initial level of BPLOM has been used to model real life business processes at a Spanish enterprise called Isastur. The business processes that were modeled represented to areas of the enterprise with different characteristics: one was the informatics incidence management process and the other was the recruitment process.

This difference in characteristics allowed BPMN MUSIM to be considered suitable for modeling different kinds of processes and at this point we are going to use BPLOM to represent a model from a completely different nature. In this case we are going to use our level oriented approach to illustrate a product catalog application, including the possibility of buying the goods at the en of the process. The different figures in this section will represent the aspect of the process model as it passes through three of the levels in BPLOM: level 0, level 2 and level 4. These levels have been chosen because they correspond to the initial, middle and last stages of the methodology.

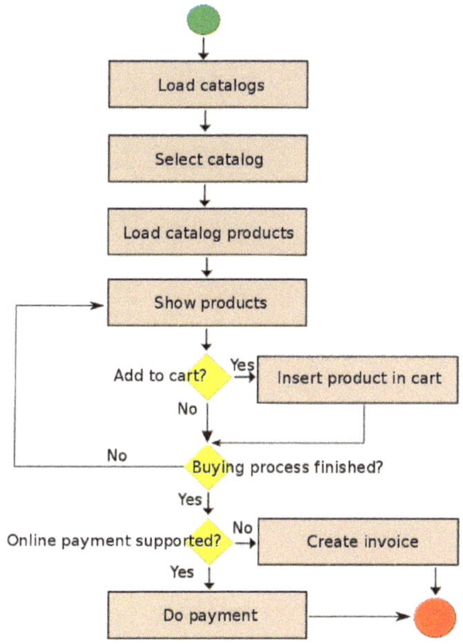

Fig. 16. Level 0 catalog application process model

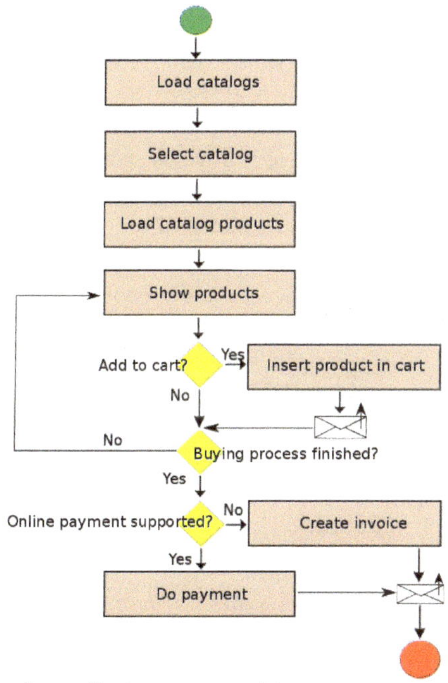

Fig. 17. Level 2 catalog application process model

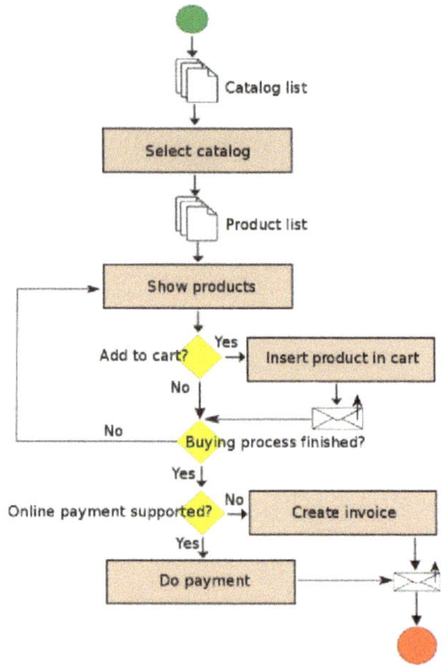

Fig. 18. Level 4 catalog application process model

Looking at the three precedent figures the evolution a process undergoes when passing through the levels of our methodology can be clearly seen. A BPMN MUSIM version of the process is shown in Fig. 16. As it can be seen this process contains no technical detail at all, as this level of our methodology is targeted towards non-technical application domain experts. Once the domain expert has gained some experience with the initial level of the methodology, he would start going up through the following levels, and thus an increase in the number of entities available for modeling and of the notation expressiveness would take place.

A non-technical expert with medium modeling skills would have access to level 2 of the methodology and therefore would design the process seen in Fig. 17. In this case, the inclusion of the outgoing messaging events gives the process additional functionality and shows the richer expressiveness of the notation at this level.

The last process, shown in Fig. 18, represents the use a business process modeling expert with some technical knowledge would give to the methodology. For this particular process the difference between the process models built with level 2 and 4 of the methodology is the inclusion of the object collection entities. This circumstance shows that the business expert that modeled the process has some technical knowledge as he manages to understand the concept of an object collection.

IV. BPLOM Tools: BPLevel Modeler and BPLevel Generator

As it was mentioned in the features section, BPLOM enables the definition and execution of the processes modeled with the proposed entity set. In order to achieve this functionality, BPLOM requires the development of a couple of prototypes that enable the digital definition and transformation of the models. These prototypes have been called BPLevel Modeler and BPLevel Generator.

Despite the platform independent nature of the proposed methodology both prototypes have been built using Windows Workflow Foundation. This is due to the degree of customization that this platform offers for creating a modeling tool like the one that will be introduced. The fact of using this platform for creating our modeling tool has no effect on the platform independency feature shown by the methodology, as it will be explained later in this section.

A. BPLevel Modeler: Graphical Definition of BPLOM Models

BPLevel Modeler is a business process-modeling tool that supports the BPM level methodology proposed in this paper. This tool has been developed under the .NET platform, as Windows Workflow Foundation [22] offers an attractive scenario for developing highly configurable business process modeling tools.

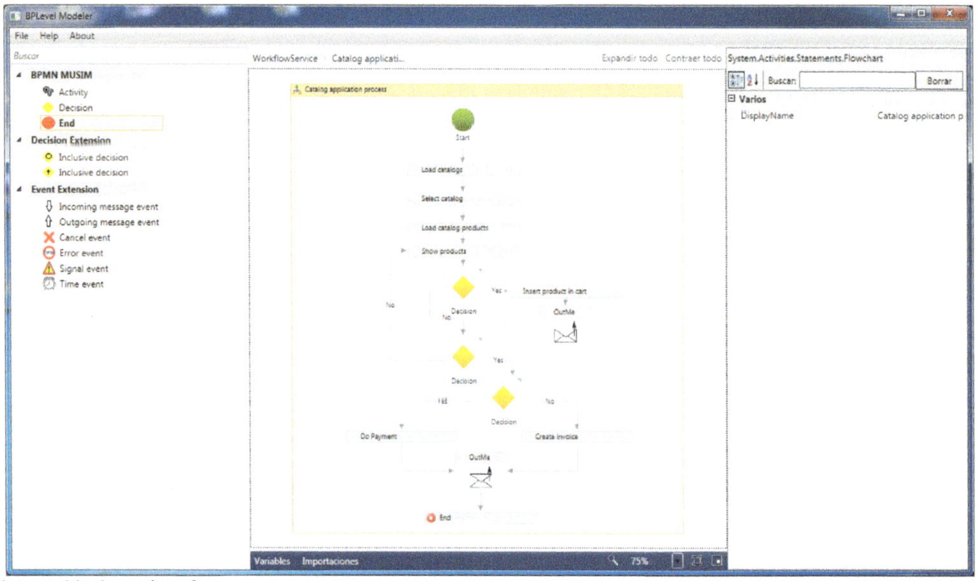

Fig. 19. BPLevel Modeler graphical user interface

The graphical user interface of BPLevel Modeler is shown in Fig. 19. As it can be seen, BPLevel Modeler has been kept as simple as possible in order to adapt to the computer skills shown by non-technical business experts. This tool has only one screen, which was divided into three different areas:

- The entity section, on the left of the screen, contains the different entities that can be used during process modeling. This section has been divided into five different categories, which represent the five levels of the methodology.
- The modeling section takes the center of the interface and is intended for the definition of the business process. It is closely related to the entity section, as the components in it can be dragged into the modeling section to define the desired business process.
- The property section, situated at the right hand side of the tool's screen, contains the property view of the BPLOM elements. The contents of this section depend on the element selected in the modeling section.

As BPLOM is designed as a skill level adaptive methodology, BPLevel Modeler must also adapt to the skill level presented by the user. In order to do so, when the tool is executed it will ask the user his knowledge level and based on the expert's choice the entity section will be adapted to show only the corresponding elements. Fig. 19 shows an entity section corresponding to a Level 3 expert and the business process model for the catalog application for that level.

Although BPLevel Modeler has been developed with Windows Workflow Foundation, which is included inside the .NET platform, the models it generates are considered platform independent. Windows Workflow Foundation stores models in a XML enriched format named XAML, a normal XML file with additional information regarding the position of the elements in the graphical representation of the process. Thus, BPLevel Modeler archives could be transformed using platform independent artifacts (like XSLT stylesheets, for example) and therefore used to generate code for any desired platform. As a matter of fact, in the use case described in this paper BPLevel Modeler files will be transformed using BPLevel Generator to generate mobile device applications for the Android and iOS platforms.

B. BPLevel Generator: Creating custom apps for BPLOM Models

BPLevel Generator is our code generation application. It analyses business process files built with BPLevel Modeler and generates multiplatform applications with specific characteristics for the given BPLOM model. Despite the fact that it is intended for computer experts this code generating tool has also been kept as simple as possible.

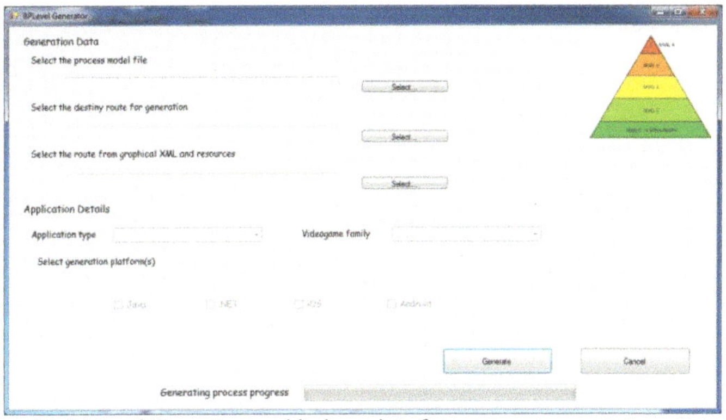

Fig. 20. BPLevel Generator graphical user interface

Fig. 20 shows a screen capture of the BPLevel Generator interface. As it can be seen, this tool requires the user to introduce several pieces of information:

1. The BPLevel Modeler file that represents the process that is going to be used as the basis for the code generation.
2. The path where the application(s) resulting from the generation process will be stored.
3. The path where the XML file with the graphical information of the application and the media resources are stored.
4. The platform(s) that the software is going to be generated for.

It must be pointed out that at this moment not all the generating features of BPLevel Generator are functional, but it is prepared for the inclusion of this functionality as a result of the planned future work.

Once the user introduces this data and starts the generation process, BPLevel Generator begins the analysis of the business model provided by the user and transforms this into the application represented by the details provided by the user. The generation process is divided into the following steps:

- **Preparing the creation structure.** Based on the path provided by the user as destiny for the generation process, BPLevel Modeler prepares the route for receiving the generated app.
- **Creating the custom app.** Code templates of the application(s) for the desired platform(s) are placed inside the path provided as the destiny of the generation process.
- **Configuring the GUI of the application.** BPLevel Generator analyses the XML file containing the graphical details of the application and substitutes the values of these details in the corresponding code class inside the application. BPLevel Generator also copies all of the resources the application needs into the appropriate folder inside the application.
- **Configuring functionality of the application.** Through an analysis of the business process model created with BPLevel Modeler our code generating tool substitutes the needed lines of code inside the application.

```
<main_screen_image>
    <main_screen_image_source>main_background.png</main_screen_image_source>
    <main_screen_image_width>320</main_screen_image_width>
    <main_screen_image_height>480</main_screen_image_height>
    <main_screen_image_x_position>0</main_screen_image_x_position>
    <main_screen_image_y_position>0</main_screen_image_y_position>
</main_screen_image>
```

Substitution

```
#define imageMenu @"main_background"
#define imageMenuHeight 480
#define imageMenuWidth 320
#define imageMenuPosX 0
#define imageMenuPosY 0
```

Creation

```
background = [[UIImageView alloc] initWithImage:[UIImage imageNamed:[NSString
    stringWithFormat:@"%@/%@", inst.imagesDir, imageMenu]]];
[background setFrame:CGRectMake(imageMenuPosX, imageMenuPosY, imageMenuWidth, imageMenuHeight)];
```

Fig. 21. Overview of the phases in the generating process and its result in the application's code for the background image of the main screen

Fig. 21 shows an example of how BPLevel Generator manages to substitute the graphical configuration information inside the application that is being generated. As it can be seen, the XML file contains three types of details for each of the controls or elements inside a given screen of the application: position (x and y coordinates), size (width and height values) and image. For each of the elements these values are stored in a specific class inside the application and then the instances of the elements in each screen are created using these values. The figure shows as an example the graphical configuration of the background image used in the main screen of the iOS application, as the rest of the elements are configured in a very similar way.

V. Conclusions

Requirements engineering is a critical task in the software development process, as it is the first phase done during the construction of any IT application and it establishes the needs and characteristics of the software. Despite this importance, requirements engineering has seen little changes in the last years and difficulties stated by Sommerville and Kotonya in their 1996 work [1] remain valid: the ambiguous character of natural language, also stated by Laue and Gadatsch [11], and the fact that the requirements engineer is not an expert of the application's domain. With these difficulties, another approach for the requirements engineering process arose: business process modeling. This discipline promotes the implication of application domain experts through the use of notations half way between computer knowledge and business domains.

Several business process modeling notations have appeared since BPMN was defined, in some cases trying to reduce the complexity level demonstrated by BPMN. This is the case of SBPMN [6][20], a simple business process modeling notation previously defined at the University of Oviedo. The mentioned complexity rate makes some of the notations difficult to use and understand by application domain experts with little or none modeling and technical knowledge. In addition to the complexity for modeling processes with these notations, several studies on the usage rates of the symbols in BPMN [3][7][19] have demonstrated the low use of some of the symbols included in these notations.

Bearing in mind the figures of these symbol usage studies and the difficulties found by business experts due to the complexity of these notations, we have tried to define a business process modeling methodology with two main goals: adapt the notation's complexity to the modeling and technical knowledge of the business experts and reduce the symbol set available in accordance with the mentioned usage rates. In order to adapt the notation to the expert's conditions, our methodology promotes the definition of five incremental levels, which add symbols to the notation gradually. When using the last level in our methodology, the expert is capable of using 20 symbols for modeling its processes; with this number of symbols, we manage to minimize the loss in expressiveness in comparison to BPMN and, at the same time, make our methodology capable of modeling any type of business process.

Once we defined this level oriented business process methodology, we decided to create two tools to take advantage of the methodology's features: platform independency and adaptive complexity. In first placed we built BPLevel Modeler, our business process modeling tool with support for our incremental level approach. Depending on the level selected by the business expert at the startup of BPLevel Modeler, the tool will automatically adapt its entity section to hold only the symbols available in the selected level; for example, if the user indicates its knowledge level is Level 2, BPLevel Modeler will show in its entity section all the symbols in levels 0, 1 and 2.

The other tool we have created is named BPLevel Generator, a code generating tool intended for IT experts. As business process models created with BPLevel Modeler are stored in an extended XML format and our methodology is platform independent, with BPLevel Generator we are capable of generating custom applications for each business process model constructed. For this generation to take place, BPLevel Generator needs the following information: the business process model file, the XML file with the graphical definition of the user interfaces the application has and the resource files used, the route for the code to be stored in and the deployment platform (Android or iOS). With these pieces of information,

BPLevel Generator is capable of analyzing the business process created with BPLevel Modeler in order to create a custom application with support for that specific process model in the desired platform, taking advantage of the platform independency characteristic of our level oriented methodology.

With this approach we manage to achieve the goals presented at the beginning of this paper. The first objective was to create a modeling notation capable of adapting its complexity to the skill level shown by the user, goal achieved through the definition of the level oriented methodology that fosters the adaptation of the modeling notation to the skills and knowledge the business expert has. We were also keen on involving business experts in the requirements engineering phase; this objective was accomplished with the creation of BPLevel Modeler, a simple, graphical business process modeling tool with support for the level oriented basis of our methodology. Lastly, we intended to use the generated models to generate the code of the custom applications that would give support to this models; with BPLevel Generator we can analyze the models created by the business experts and generate the mobile applications that represent them.

VI. Future Work

This paper includes some usability results that were carried out in order to establish the simplicity of BPLOM Level 0 and the completeness of its symbol set. In order to test the entire methodology we need to carry out some further testing with the rest of the proposed levels, the first point in our future work list. These usability tests will be focused in measuring the following aspects: the entities' graphical representation in comparison with BPMN, as this methodology intends to increase the experts' usability and abstraction levels; and the suitability of each of the levels regarding the entities they include and the order they are presented in, intending to minimize the time needed for the expert to advance through the levels. In this extended testing procedure we will also be able to determine the need to include more symbols present in BPMN to our notation, as it was already said in part six of the third section of this paper.

Another important point in our future work schedule is to change the way in which the XML file with the app's graphical information is created. At this time, the IT expert needs to manually introduce the details included in this document; these details include image files for the backgrounds, positioning of the elements inside the screen and size of these elements. Our main goal is to design and program a graphical user interface that enables the business expert to define the appearance of the different screens that make up the app easily. We are considering a drag and drop approach where the business expert can include backgrounds, buttons and other controls inside a canvas and adapt their look and feel to the style he prefers.

As it could be seen in Fig. 1, the proposed business process methodology is used to generate multiplatform mobile apps. In the use case documented in this paper, the generated software is a catalog app that allows users to scan through product catalogs, add them to their basket and buy them. At this moment the apps are generated for the most widely used platforms for this type of devices: Android and iOS. Although these platforms allow us to reach to the majority of mobile device users, we are interested in offering support not only for other mobile device platforms like Windows Phone but also for desktop and web applications through other platforms like Java and .NET. Moreover, we are also attracted by the possibility of creating other types of applications for the currently supported mobile platforms; for example, BPLOM could be used to model the process behind multimedia mobile apps like interactive books or even videogames, allowing the corresponding business experts to design and generate their own applications.

References

[1] Kotonya, G. and Sommerville, I. "Requirements engineering with viewpoints". 1996. *BCS/IEE. Software Eng. J.* , 11(1): 5-18

[2] Stephen A. White, "Introduction to BPMN". Business Process Management Iniciative (year 2004).

[3] Michael zur Muehlen, Jan Recker. "How Much Language is Enough? Theoretical and Practical Use of the Business Process Modeling Notation". 20th International Conference on Advanced Information Systems Engineering (CAiSE 2008).

[4] Lauri Eloranta, Eero Kallio y Ilkka Terho. "A Notation Evaluation of BPMN and UML Activity Diagrams" (year 2006).

[5] T.Wahl and G. Sindre. "An Analytical Evaluation of BPMN Using a Semiotic Quality Framework". CAiSE'05 Workshops. Volume 1, pages 533–544. FEUP, Porto, Portugal (year 2005).

[6] Fernández, H. F., et al. "SBPMN – An easier business process modeling notation for business users". Computer Standards & Interfaces 32 (1-2):18–28.

[7] Jan C. Recker. "BPMN Modeling – Who, Where, How and Why". BPTrends (year 2008).

[8] Stephen A. White y Derek Miers. "BPMN Modeling and Reference Guide: Understanding and Using BPMN". Future Strategies Inc. (year 2008).

[9] Ryan K. L. Ko. "A computer scientist's introductory guide to business process management (BPM)". Crossroads v.15 n.4, p.11-18 (year 2009).

[10] R. Lu y S. Sadiq. "A Survey on Comparative Modelling Approaches". Proc. BIS'07 (year 2007).

[11] Ralf Laue y Andreas Gadatsch. "Measuring the Understandability of Business Process Models - Are We Asking the Right Questions?". Business Process Management Workshops - BPM 2010 International Workshops and Education Track (year 2010).

[12] Jan Recker, Niz Safrudin y Michael Rosemann. "How Novices Model Business Processes". Business Process Management – 8th International Conference (year 2010).

[13] Zhiqiang Yan, Hajo A. Reijers y Remco M. Dijkman. "An evaluation of BPMN Modeling Tools". Business Process Modeling Notation - Second International Workshop (year 2010).

[14] C.A. Petri. "Kommunikation mit Automaten". PhD thesis, Institut fiir instrumentelle Mathematik, Bonn. (year 1962).

[15] W.M.P van der Aalst et al. "The Application of Petri Nets to Workflow Management". The Journal of Circuits, Systems and Computers, 8(1):21–66. (year 1998).

[16] Jaime Solís Martínez, Vicente García Díaz, Begoña Cristina Pelayo García-Bustelo y Juan Manuel Cueva Lovelle. "BPMN MUSIM: Notación BPMN muy simplificada". 6ª Conferencia Ibérica de Sistemas y Tecnologías de Información (CISTI 2011).

[17] Jaime Solís Martínez, Vicente García Díaz, Begoña Cristina Pelayo García-Bustelo y Juan Manuel Cueva Lovelle. "Isastur Modeler: A tool for BPMN MUSIM". 6ª Conferencia Ibérica de Sistemas y Tecnologías de Información (CISTI 2011).

[18] Michele Chinosi and Alberto Trombetta. "A design methodology for BPMN". Chapter in the 2009 BPM and Workflow Handbook. (year 2009).

[19] Michele Chinosi, and Alberto Trombetta. "BPMN: An introduction to the standard." *Computer Standards & Interfaces* 34.1 (year 2012).

[20] H. Fernandez-Fernandez, E. Palacios-González, V. García-Díaz, B. Cristina Pelayo G-Bustelo, Oscar Sanjuán Martínez and Juan Manuel Cueva Lovelle. "Developing a Business Application with BPM and MDE". International Journal of Artificial Intelligence and Interactive Multimedia (IJIMAI). (year 2009).

[21] A.P. Castaño. "Prototype of assignment intelligent adaptive of service providers inside of ESB with data mining". International Journal of Artificial Intelligence and Interactive Multimedia (IJIMAI). (year 2009).

[22] Windows Workflow Foundation. Last visit on May 5th 2013. http://msdn.microsoft.com/es-es/netframework/aa663328

[23] OMG. Last visit on May 6th 2013. http://www.omg.org/

[24] BPMI. Last visit on May 7th 2013. http://www.omg.org/bpmi

[25] BPMN. Last visit on May 5th 2013. Last update on April 29th 2013. http://www.omg.org/bpmn/

[26] UML. Last visit on May 4th 2013. Last update on April 15th 2013. http://www.uml.org/

[27] jBoss jBPM. Last visit on May 4th 2013. http://www.jboss.org/jbpm

[28] MDE Research Group in the Department of Computer Science, University of Oviedo. Last visit on May 7th 2013. Last update on December 2012. https://sites.google.com/site/mdeootlab/Home

7

Empirical Comparison of Graph-based Recommendation Engines for an Apps Ecosystem

Luis F. Chiroque[1], Héctor Cordobés [1], Antonio Fernández Anta[1], Rafael A. García Leiva[1], Philippe Morere[1], Lorenzo Ornella[3], Fernando Pérez[2], Agustín Santos[1]

[1]IMDEA Networks Institute, Madrid, Spain
[2]U-Tad, Madrid, Spain
[3]Politecnico di Torino, Turin, Italy

Abstract **— Recommendation engines (RE) are becoming highly popular, e.g., in the area of e-commerce. A RE offers new items (products or content) to users based on their profile and historical data. The most popular algorithms used in RE are based on collaborative filtering. This technique makes recommendations based on the past behavior of other users and the similarity between users and items. In this paper we have evaluated the performance of several RE based on the properties of the networks formed by users and items. The RE use in a novel way graph theoretic concepts like edges weights or network flow. The evaluation has been conducted in a real environment (ecosystem) for recommending apps to smartphone users. The analysis of the results allows concluding that the effectiveness of a RE can be improved if the age of the data, and if a global view of the data is considered. It also shows that graph-based RE are effective, but more experiments are required for a more accurate characterization of their properties.**

Keywords **— Recommendation engines, smartphone apps, graph theory, collaborative filtering, flow algorithms.**

I. Introduction

A. Motivation

It is becoming very common in online platforms (shopping websites, online newspapers, online social networks, smartphone apps, etc.) to recommend items to the users that will (hopefully) be of their interest. This trend is becoming so general that Anderson predicted that we are "leaving the age of information and entering the age of recommendation" [4]. The items to recommend are selected by a recommendation engine (RE) that typically leverages the user profile, the context, and historical data. The RE typically has a catalog of items from which to choose its recommendation, and there are spaces in the online platform viewing area in which the recommended product is presented. The context of the user typically includes its past navigation history, including the current viewing context, which may involve a product (e.g., in a shopping website), a piece of news (e.g., in an online newspaper), a user profile (e.g., in an online social network), or the application that is being executed (e.g., in a smartphone).

Recently, the most popular algorithms used in RE are based on collaborative filtering [15]. This technique makes recommendations based on the historical data of all the users and the estimated similarity between them. Typical metrics used for the computation of customers' similarity include Pearson correlation coefficient, adjusted cosine similarity, Spearman's rank correlation coefficient, and mean squared difference.

In parallel with the advances in RE algorithms, we have observed that graph theory and network analysis has been useful in different contexts to extract information from data. This information is not an explicit part of the data, but it is implicitly contained in the underline structure. Examples of this approach are the use of pagerank to identify the most relevant web pages [8], or a recent use we have made of graphs to classify tweets [9]. We believe that graph theory and network concepts can also be useful in the context of recommendation.

B. Contributions

In this paper we present an exploratory work on using graphs to build RE. We have devised, developed, and evaluated RE based on collaborative filtering to promote an ecosystem of smartphone apps. In this ecosystem, the users of the apps get banners advertising other apps that they have not installed (yet). The objective of the RE is maximizing the click-through rate (CTR) of users in these banners, and maximizing the installation of new apps. In addition we have devised one particular RE to promote a specific subset of apps. The proposed RE create models of the ecosystem as networks formed by apps, and use graph theoretic concepts like edges weights or network flow.

The performance of the RE proposed has been evaluated in a real apps ecosystem. Several years' worth of historical data has been used to create the networks that model the ecosystem.

Then, using them, the different RE were put to work with real users for about a week. The analysis of the results obtained has shown big (statistically significant) differences in CTR and installation success of the different RE. The results allow concluding that the effectiveness of a RE can be improved if the age of the data, and if a global view of the data is considered. It also shows that graph-based RE are effective. However, some of the results are puzzling, and hence more experiments are required for a more accurate characterization of the properties of the proposed RE.

C. Structure

The rest of the paper is structured as follows. In Section II we present the problem to be solved. In Section III we describe the RE we have proposed and that will be evaluated in this paper, with the underlying graph they use. In Section IV we present the experiment we have conducted, the results obtained, and some discussion on them. Section V presents previous work related to this paper. Finally, Section VI concludes the paper.

II. PROBLEM STATEMENT

As described, we have a *smartphone app ecosystem.* In this ecosystem, a user that is running an app (called the *publisher*), gets *banners* advertising other apps of the ecosystem it has not installed yet. The objective is to devise a RE that tells the system which app to advertise to a given user at a given time, possibly as a function of the user and the publisher, in order to achieve one (or more) of the following objectives.

- *CTR Maximization:* The objective is to maximize the number of times the user clicks in the banner to get more information about the apps advertised.
- *Installations Maximization:* The objective is to maximize the number of times the user installs the app advertised.
- *Targeted Promotion:* The objective is to maximize the number of times users install a preselected set of apps to be promoted.

An initial hypothesis we will make is that, once a user has clicked in a banner, the probability of installing the corresponding app is roughly same. This has made us concentrate initially in RE for the CTR Maximization and Targeted Promotion objectives. (As will be seen from the results obtained, this initial hypothesis needs some revision.)

III. RECOMMENDATION ENGINES PROPOSED

In this section we describe the recommendation engines we have proposed and evaluated in this paper. In order to describe them, we build graphs from historical user data that convey the essential information that is required by the corresponding RE. Hence, we start describing the graphs we need and use, and then we give the algorithms used by the RE to select an app to advertise.

A. Apps Graphs

All the graphs used in this work will have the set of apps A of the ecosystem as vertices. Moreover, all of them are weighted graphs, and the main difference among them is the weights that are allocated to edges. The graphs used are the following.

Shared Users (SU) Graph. The SU graph is an undirected weighted graph $G_{SU}=(A,E,w)$, where $E=\{\{i,j\}: i,j\in A\}$ and the weight $w(e)$ of an edge $e=\{i,j\}\in E$ is the number of users of the ecosystem that have currently both apps i and j installed.

Aged Shared Users (ASU) Graph. The ASU graph is an undirected weighted graph $G_{ASU}=(A,E,w)$, where $E=\{\{i,j\}: i,j\in A\}$ like in G_{SU}. The difference in this case is that the contribution to the weight $w(e)$ of an edge $e=\{i,j\}\in E$ of a user (that has currently both apps i and i installed) is a function of the time the user has had the apps installed. In particular, let U be the set of all users and $U(a)\subseteq U$ be the set of users that have app $a\in A$ installed. Also, let $age(u,a)$ be the time since user $u\in U(a)$ installed app $a\in A$ (in some suitable units). Then,

$$w(\{i,j\}) = \sum_{u \in U(i)\cap U(j)} \delta^{\min\{age(u,i),age(u,j)\}},$$

where $\delta\leq 1$ is the decay factor. (The intuition is that users that installed an app long time ago are less important for the app.)

CTR Graph. The CTR graph is a *directed* weighted graph $G_{CTR}=(A,E,w)$, where $E = A\times A$ and the weight $w(e)$ of an edge $e=(i,j)\in E$ is the CTR observed when banners with app j are presented to the users with app i as publisher.

B. Recommendation Algorithms

Using the above graphs we can describe now the RE considered in this work.

Shared Users.

Let us consider the SU graph described above. Assuming the publisher app is i, the app recommended j is the one that has the edge with i of largest weight. I.e., $j=argmax_k(w(\{i,k\}):k\in A)$. In this case, this means that j is the app with the largest number of common users with i.

The approach of this algorithm is not new, and it is among the first ideas one may think of when resigning recommendation algorithms.

Collaborative Filtering.

This algorithm also uses the SU graph. Given the user to which the banner will be presented, and the set I of applications the user has already installed, the app j recommended is the one that has a largest aggregate weight with those in set I. I.e., $j=argmax_k(\Sigma_{i\in I}\, w(\{i,k\}):k\in A)$.

Again, this approach is not very novel, since it is common to many collaborative filtering algorithms to use some linear algebra approach that can achieve similar results as this one. For instance, considering the weights of the SU graph as a matrix W, and the applications already installed by the used as a vector v, the algorithm proposed would recommend the app that corresponds to the largest element of the vector $v^T W$.

Aged Shared Users.

This algorithm applies the same process as Shared Users, but in the ASU graph. As far as we know this algorithm is new.

Aged Collaborative Filtering.

This algorithm applies the same process as Collaborative

Filtering, but in the ASU graph. As far as we know this algorithm is also new.

Maxflow.

This algorithm uses the CTR graph with the objective of promoting a preselected subset *P* of apps. The algorithm takes the publisher app *i* and solves a flow maximization problem [11] from *i* to each of the apps in *P*, where the weight of each link is considered its capacity. Then, it recommends the neighbor of *i* whose aggregated flow is the largest. I.e., imagine that the solution of the maximum flow problem from *i* to $a \in P$ sends $f(a,k)$ units of flow across link (i,k). Then, the recommended app is $j = argmax_k(\Sigma_{a \in P} f(a,k) : k \in A)$.

To our knowledge, the Maxflow RE is also new. The intuition behind it is that instead of directly promoting the apps in *P* it is better to promote those that will drive the user to them.

In addition to the 5 RE described, we will consider for reference two trivial algorithms.

Random.

This algorithm recommends an app at random using a uniform distribution over the set of available applications. As we just said, the goal of the random RE is to have a reference with which all the other RE can be compared.

Static Promotion.

This algorithm always recommends one of the applications of the set *P* to be promoted (chosen uniformly at random). It does not depend on the user installed applications, nor the publisher.

IV. Performance Evaluation

In this section we describe the experiment we have conducted in order to evaluate and compare the RE proposed. Then, the results obtained in the experiment are presented and briefly analyzed.

A. Implementation of the Experiment

As mentioned previously, the evaluation of the RE has been done in a real apps ecosystem. This ecosystem is formed by roughly 300 apps with more than 4 million users.

To build the graphs used by the RE and described in the previous section, we have used more than 3 years worth of data. This adds to more than 100 GiB of historical data structure in more than 1.4 billion records. This data has been processed with Big Data technologies (Hadoop, Pig, Hbase [5-7]) in the Amazon Elastic Map Reduce [2] environment. The processing involved cleaning the historic data generated a clean dataset of more than 700 million records of events, containing the user, the publisher, the app advertised (in the banner), the action associated to the event (add, click), and the timestamp.

From the clean dataset just obtained, the above-described graphs were built. The construction of the aged graph ASU used a value of $\delta=0.95$ and the age is measured in units of weeks. It is important to note that the historical data to which we had access did not record explicitly the installation of the apps. The fact that a user had an app installed was extracted from the data because the app appeared as publisher in some event.

Once the graphs were ready, we run an experiment in the real system for about a week (from Jun 2nd, 2014 to Jun 10th, 2014). In this experiment, the different RE recommended the apps shown in banners to the users. In order to avoid cross interference, the same RE generated all the banners for the same user. For the targeted promotions RE proposed (Maxflow and Static Promotion) a manually selected set of 5 apps were chosen to be promoted. At the end of the experiment, banner were shown to more than 300,000 users, and each RE had done more than 130,000 recommendations.

After the experiment, the data of number of banners recommended by each RE, the number of clicks by the user, and the number of apps installed was obtained. It is important to note that the data obtained was cleaned. For instance, multiple clicks associated to the same banner where counted only once. Regarding installations, we assumed that a banner had caused the installation of an app if the app was used (by the user) within 72 hours after the banner was shown.

B. Experiment Results

The basic results obtained in the experiment are presented in Table I. For each RE the table shows the total number of banners that used the RE for recommendation, the number of banners on which the user clicked, and the click through rate, CTR, which is the ratio of the former two values. Additionally, the number of installations from the banner is also shown. Finally, we present two metrics, installation to banners rate (IBR) and installation to clicks rate (IBR), which are the ratio of the number of installations versus the number of banners and the number of clicks, respectively.

TABLE I
RESULTS OF THE EXPERIMENT

RE	Banners	Clicks	CTR (%)	Installs	IBR (%)	ICR (%)
Random	140894	1993	1.41	126	0.09	6.32
Shared Users	133818	2095	1.57	299	0.22	14.27
Aged Shared Users	139417	2258	1.62	390	0.28	17.27
Collaborative Filtering	134790	1966	1.46	329	0.24	16.73
Aged Collaborative Filtering	133623	2204	1.65	375	0.28	17.01
Static Promotion	138922	1929	1.39	215	0.15	11.15
Maxflow	140858	2302	1.63	290	0.21	12.60

As can be seen from the results presented, the CTR observed is different for different RE. Table II presents a comparison of the differences of the CTR achieved by the RE. In each entry of the table it is show (in percentage) the increase in CTR achieved if using the RE of the column instead of the RE of the row. When this number is negative the CTR in fact decreases, and the value is marked in red.

TABLE II
INCREASE OF THE CTR (IN PERCENTAGE) WHEN USING ONE RE (COLUMN) VERSUS ANOTHER (ROW)

	Random	Shared Users	Aged Shared Users	Collab. Filtering	Aged Collab. Filtering	Static Promotion	MaxFlow
Random		10.68	14.50	3.11	16.60	-1.84	15.53
Shared Users	-9.65		3.45	-6.83	5.36	-11.31	4.39
Aged Shared Users	-12.66	-3.34		-9.94	1.84	-14.27	0.91
Collab. Filtering	-3.02	7.34	11.04		13.08	-4.80	12.05
Aged Collab. Filtering	-14.24	-5.08	-1.81	-11.57		-15.82	-0.92
Static Promotion	1.87	12.75	16.64	5.04	18.79		17.70
MaxFlow	-13.45	-4.20	-0.90	-10.75	0.93	-15.04	

Given these differences CTR between the RE used, we want to determine if they are statistically significant. For that, we have computed a z-test [19] to compare the CTR of each pair of RE. For a given pair of RE the null hypothesis is that both populations are extracted from equal distributions (and hence the differences are simply due to statistical noise). The alternative hypothesis is then that the distributions are different. We compute the z value for two RE (numbered 1 and 2) as

$$z = \frac{p_1 - p_2}{\sqrt{p(1-p)}\sqrt{\frac{1}{n_1} + \frac{1}{n_2}}},$$

where n_i is the number of banners of RE i, $p_i = x_i/n_i$ is the ratio between the number of clicks x_i and the number of banners, and p is the pool population, defined as

$$p = \frac{x_1 + x_2}{n_1 + n_2}.$$

The results of all the z-tests for the CTR are shown in Table III. Using an alpha value of 0.05, we reject the null hypothesis if the z value is outside the interval [-1.96,1.96], with a confidence of 0.95. If the null hypothesis is rejected for a pair of RE, it means that with a probability of at least 0.95 the CTR of one is larger than the CTR of the other. In Table III the pairs of RE for which the null hypothesis is rejected have white background, and we use yellow background for the cases in which the null hypothesis cannot be rejected.

As done with the CTR, we have also computed the z-values for the IBR for each pair of RE, with the objective of identifying when the difference in IBR shown in Table I are statistically significant. Again, the null hypothesis is that the differences are only due to randomness. Table IV presents the results, where the white background again means that we reject the null hypothesis with a confidence of 0.95.

TABLE III
Z-VALUES COMPUTED FOR THE CTR OF EACH PAIR OF RE.

	Random	Shared Users	Aged Shared Users	Collab. Filtering	Aged Collab. Filtering	Static Promot.
Shared Users	3.26					
Aged Shared Users	4.44	1.13				
Collab. Filtering	0.97	-2.27	-3.42			
Aged Collab. Filtering	5.01	1.72	0.61	4.00		
Static Promotion	-0.58	-3.83	-5.01	-1.55	-5.57	
MaxFlow	4.76	1.43	0.31	3.73	-0.31	5.32

TABLE IV
Z-VALUES COMPUTED FOR THE IBT OF EACH PAIR OF RE

	Random	Shared Users	Aged Shared Users	Collab. Filtering	Aged Collab. Filtering	Static Promotion
Shared Users	8.93					
Aged Shared Users	11.75	2.93				
Collab. Filtering	10.00	1.11	-1.83			
Aged Collab. Filtering	11.73	2.95	0.04	1.85		
Static Promotion	4.95	-4.13	-7.08	-5.25	-7.07	
MaxFlow	8.05	-0.99	-3.97	-2.12	-3.98	3.18

Finally, we present the results of the different RE as promoters of specific apps. As described above, we have devised a RE, Maxflow, specifically for targeted promotion of apps, and used it for promoting the 5 chosen apps.

In addition, as mentioned, we implemented a trivial RE, Static Promotion, which only recommends the 5 apps to be promoted. In Table V we present the number of installations that these RE achieved for each of the four apps to be promoted, numbered from 1 to 5. For comparison, we also show the numbers of installations achieved with the other RE.

C. Discussion

Table I shows significant differences between the RE used. The first fact to note is that, as expected, both Random and Static Promotion have very low CTR and IBR. All the other algorithms have a CTR that is at least 3% higher than Random and 5% higher than Static Promotion (see Table II). The difference in IBR is even higher, were every RE achieves at least a 133% increase over Random and 40% over Static Promotion.

TABLE V
INSTALLATIONS OF THE 5 APPS PROMOTED

RE	App 1	App 2	App 3	App 4	App 5
Random	2	0	4	0	4
Shared Users	18	1	13	45	31
Aged Shared Users	13	0	15	50	42
Collaborative Filtering	7	0	22	36	39
Aged Collaborative Filtering	13	0	22	47	43
Static Promotion	35	17	38	70	47
Maxflow	0	0	2	11	46

Comparing the CTR of the rest of RE, there is a significant difference between Shared Users versus Aged Shared Users, and Collaborative Filtering versus Aged Collaborative Filtering. This difference leads to conjecture that the preferences of the users change over time. This is the reason why the RE that take that into account this evolution behave well. Somewhat surprising is the high CTR achieved by Maxflow, which has the second largest CTR, since the objective of this RE is not to maximize the CTR.

Table III shows that the differences between the CTR are statistically significant many cases. In particular, in terms of CTR, the z-test divides the RE into two groups. One group has CTR that is statistically smaller than the other. Random, Static Promotion, and Collaborative Filtering form the group of low CTR. The group of high CTR includes Shared Users, Aged Shared Users, Aged Collaborative Filtering, and Maxflow. Observe that a larger alpha value in the z-test would differentiate the RE further.

Looking at the ICR columns in Table I, we can see that the values in the column differ significantly. This disproves our initial hypothesis that, once a user clicks in a banner, she has a similar probability of installing the app. The conclusion is that it is not enough to aim at maximizing CTR if the objective is to get app installations. For instance this causes that Maxflow is the RE with the lowest IBR from those not for reference. Moreover, the values in Table IV show that this difference is statistically significant. From this table we can conclude that all RE are more efficient in terms of installations than Random. Also, that all the "smart" RE are more efficient than Static Promotion (of course, this is natural since the target of this RE is not maximizing installations).

Table IV also shows that the effectiveness in installations of Aged Shared Users and Aged Collaborative Filtering is higher than the other RE. This reinforces the conjecture that the preferences of users change over time and this changes has to be taken into account by the recommendation system.

Finally, regarding targeted promotion, in Table V we can observe that Maxflow achieves a low number of installations for the 5 promoted apps, especially compared with Static Promotion (but even versus all the other RE except Random). This result is disappointing, and it requires further study. Our conjecture is that the experiment conducted was too short to observe the effect of Maxflow, which promotes the apps that lead to other apps. Other lines to explore are the modification of Maxflow in two ways. First, Maxflow must be tested using an IBR graph instead of the CTR graph (since, from a previous discussion CTR is not the critical metric if we want installations). Second, the graph used by Maxflow must consider aging, since as we have observed this is an important aspect of the data. In any case, another conclusion we obtain from Table IV is that using Static Promotion for targeted promotion of apps seems like a valid option.

V. RELATED WORK

The most common approaches to the recommendation problem can be grouped into three types.

- Collaborative filtering [15]: In this approach users are represented by an N-dimensional vector of items, and the recommender looks for users who have similar rating patterns as the target user. Then, it uses the ratings from those like-minded users to make a recommendation for the target user.
- Cluster models: This approach divides the customer base into many segments, and treats the recommendation task as a classification problem. Segments are created using a clustering, or some other unsupervised learning algorithm.
- Search-based methods: In this approach, given the target user's purchased and rated items, the algorithm constructs a search query to find other popular items by the same author, artist, or director, or with similar keywords or topics.

As an example, Amazon uses its own recommendation algorithm, called item-to-item collaborative filtering [17], to personalize the online store for each customer. The algorithm is focused in finding similar items, not similar customers, and hence it scales independently of the number of customers. However, the challenge is to make it scalable with the number of items in the product catalog.

Most of the collaborative filtering algorithms we have found in the literature assume that user preferences remain stable and consistent over time [14]. We believe this is not generally the case, and our conjecture is supported by the fact that in our experiment the RE that considered aging performed very well.

Methodologies for the evaluation of RE have been proposed in [18] and [12]. Other aspects, like advertising effectiveness and Return of Investment (ROI) on social networks, have been a big topic of discussion for advertisers in the past decade [3]. ROI has been typically measured through econometric models that measure the impact of varying levels of advertising (Gross Ratings Points, GRP) on sales, on purchases decision, and choices made. (Finding improved methods of measuring ROI is still an important area of research.) A classical introductory paper is due to Danaher and Rust [10]. Taylor [20] has summarized the current focus of research on advertising.

This is not the first paper that presents approaches based on graphs for recommendation systems. Huang et al. [13] proposed to build a bipartite graph of users and items, where user vertices are connected with item vertices if the user bought or gave a good evaluation to the item. The authors estimate the interest of a given user in a give item by aggregating the weights of short path between the user and the item in the graph. Lien and Phuong [16] extend the users-items bipartite graph with weights representing the evaluation the users gave to the items. Regarding flows, Adomavicius and Kwon [1] used a maximum flow algorithm for maximizing the diversity of the recommendations (instead of improving the recommendation accuracy as we do).

VI. Conclusions

We have presented a collection of graph-based recommendation engines, and have tested them in a real ecosystem of smartphone apps. The results obtained drive us to conjecture that using graphs for recommendation is a promising line of research. However, more experiments are needed in order to verify or disprove this conjecture.

In this work we have built recommendation engines that used graphs of items. We believe that graphs of users could also be very useful for recommendation. However, these graphs tend to be must larger (of several million nodes in our real system, versus a few hundreds of item), and processing them requires using more powerful computational systems and developing scalable algorithms.

References

[1] G. Adomavicius, Y. Kwon, "Maximizing Aggregate Recommendation Diversity: A Graph-Theoretic Approach," in *Workshop on Novelty and Diversity in Recommender Systems,* 2011.

[2] Amazon Web Services, Inc., *Amazon Elastic MapReduce,* 2014. [Online http://aws.amazon.com/elasticmapreduce/; accessed 6-August-2014].

[3] T. Ambler, J. H. Robers, "Assessing marketing performance: don't settle for a silver metric," *Journal of Marketing Management,* 24 (7,8), 733-750, 2008.

[4] C. Anderson, "The Long Tail: Why the Future of Business Is Selling Less of More," *Hyperion,* 2006.

[5] The Apache Software Foundation, *The Apache Hadoop project,* 2014. [Online http://hadoop.apache.org/; accessed 6-August-2014].

[6] The Apache Software Foundation, *Apache HBase,* 2014. [Online http://hbase.apache.org/; accessed 6-August-2014].

[7] The Apache Software Foundation, *Apache Pig,* 2014. [Online http://pig.apache.org/; accessed 6-August-2014].

[8] S. Brin, L. Page, "The Anatomy of a Large-Scale Hypertextual Web Search Engine," *Computer Networks* 30(1-7): 107-117, 1998.

[9] H. Cordobés, A. Fernández Anta, L. F. Chiroque, F. Pérez, T. Redondo, A. Santos, "Graph-based Techniques for Topic Classification of Tweets in Spanish," *IJIMAI* 2(5): 31-37, 2014.

[10] P. J. Danaher, R. T. Rust, "Determining the optimal return on investment for an advertising campaign," *European Journal of Operational Research,* 95:511-521, 1996.

[11] L. R. Ford Jr., D. R. Fulkerson, *Flows in Networks*, Princeton University Press, 1962.

[12] J. L. Herlocker, J. A. Konstan, L. G. Terveen, J. T. Riedl, "Evaluating collaborative filtering recommender systems," *ACM Transactions on Information Systems,* vol 22(1), pp. 5-53, 2004 .

[13] Z. Huang, H. Chen, D. Zeng, "Applying Associative Retrieval Techniques to Alleviate the Sparsity Problem in Collaborative Filtering", *ACM Transactions on Information Systems,* vol. 22(1), pp. 116–142, 2004.

[14] D. Jannach, M. Zanker, A. Felfernig, G. Friedrich, *Recommender Systems: An Introduction,* Cambridge, 2010.

[15] Y. Koren, R. M. Bell, "Advances in collaborative filtering," in *Recommender Systems Handbook,* Springer, pp. 145–186, 2011.

[16] D. T. Lien, N. D. Phuong, "Collaborative filtering with a graph-based similarity measure," in *International Conference on Computing, Management and Telecommunications (ComManTel),* 2014.

[17] G. Linden, B. Smith, J. York, "Amazon.com recommendations: Item-to-item collaborative filtering," *IEEE Internet Computing,* 7(1), 76-80, 2003.

[18] G. Shani, A. Gunawardana, "Evaluating Recommendation Systems," in *Recommender Systems Handbook,* Springer, pp. 257-297, 2011.

[19] R.C. Sprinthall, Basic Statistical Analysis, 9th Edition, *Pearson Education Group,* 2011.

[20] C. R. Taylor, "Hot topics in advertising research," *International Journal of Advertising,* 32(1), 2013.

8

CAIN-21: Automatic adaptation decisions and extensibility in an MPEG-21 adaptation engine

Fernando López[1,2], José M. Martínez[2], Narciso García[3]

[1] VINTEC, Universidad Internacional de la Rioja, Spain, *fernando.lopez@unir.net*
[2] VPULab, EPS - Universidad Autónoma de Madrid, Spain, *{f.lopez, josem.martinez}@uam.es*
[3] GTI, ETSIT - Universidad Politécnica de Madrid, Spain, *narciso@gti.ssr.upm.es*

Abstract **— This paper presents the progress and final state of CAIN-21, an extensible and metadata driven multimedia adaptation in the MPEG-21 framework. CAIN-21 facilitates the integration of pluggable multimedia adaptation tools, automatically chooses the chain of adaptations to perform and manages its execution. To drive the adaptation, it uses the description tools and implied ontology established by MPEG-21. The paper not only describes the evolution and latest version of CAIN-21, but also identifies limitations and ambiguities in the description capabilities of MPEG-21. Therefore, it proposes some extensions to the MPEG-21 description schema for removing these problems. Finally, the pros and cons of CAIN-21 with respect to other multimedia adaptation engines are discussed.**

Keywords **— ontology, multimedia, adaptation, decision, mpeg-7, mpeg-21**

I. Introduction

As time goes by, the variety of multimedia formats and devices has significantly increased, and still does. Multimedia content providers need to distribute their photos, videos and audio to a wide-range of devices and independently of the underlying delivery technology. *User-centric adaptation* [1] places the user in the centre of multimedia services and is also referred as *Universal Multimedia Experiences* (UME) [2].

The MPEG-21 standard [3] addresses the construction of a general multimedia framework that is consistent with the idea of UME. The MPEG-21 description tools enable the representation of a large set of concepts and relationships. MPEG-21 relies on the XML Schema to define the structure of the content and define an *implied ontology* in the text of the standard. Parts of the standards have been extended with description languages with a higher level of expressiveness. Particularly, the *explicit ontology* is represented using semantic description languages such as OWL (Web Ontology Language). The multimedia research com-munity has frequently accepted and used this MPEG-21 (pseudo)-ontology.

This paper compiles the evolution and final state of an adaptation engine named CAIN-21 [4] (Content Adaptation INtegrator in the MPEG-21 framework)[1]. The main purpose of CAIN-21 is to automate interoperability among multimedia formats and systems. Interoperability is implemented by means of an extensibility mechanism. With this mechanism, pluggable software tools are incorporated to progressively address wider ranges of adaptations. CAIN-21 automates interoperability by incorporating a decision mechanism for multimedia adaptation. This mechanism selects the adaptation tools and parameters that have to be executed to adapt multimedia. Furthermore, CAIN-21 exploits multi-step adaptation. Multi-step adaptation enables the combination and execution in several steps of the pluggable adaptation tools. With multi-step adaptations the range of feasible adaptations that can be achieved increases.

CAIN-21 also aims to provide a framework in which multimedia adaptation tools can be integrated and tested. The representation of the multimedia elements has to be formalized in order to make these tests[2] repeatable. To represent the multimedia elements of the tests, a set of MPEG-21 description tools have been selected. Currently, MPEG-21 is the most comprehensive multimedia description standard for the deployment of multimedia applications/systems. However, in practice description standards never cover 100% of the concepts. In the case of CAIN-21, we have encountered some difficulties using the MPEG-21 description elements. These difficulties were solved extending the description tools and implicit ontology that MPEG-21 provides. After presenting CAIN-21 architecture, this paper discusses these issues. We consider helpful to highlight it for people involved in the construction of multimedia adaptation systems, especially if they are determined to provide MPEG-21 interfaces to their users. The clarification of these problems may also be useful for people who intend further interaction with other non-MPEG-21 compliant multimedia systems.

[1] The *CAIN-21 software* together with a *CAIN-21 demo* are publicly available at *http://cain21.sourceforge.net*

[2] This paper demonstrates our proposal with an empirical study. We use the term *test* (instead of *experiment*) to indicate that its execution always yields the same results.

The publication in [4] summarizes the interfaces, the architecture of CAIN-21, and the evolution from Early CAIN to CAIN-21, and provides a preliminary comparison with other adaptation engines. Now that CAIN-21 has reached a stable and mature state, this publication supersedes [4] by providing an extended, comprehensive and updated description of CAIN-21.

In particular, this new publication describes the delivery and adaptation methods used in CAIN-21 as well as the binding modes. The publication also describes and provides usage examples of the *ConversionCapabilities* and *ConversionCapabilities* description tools, incorporates the properties relationships, the KISS principle behind this design, the use of composed properties to address complicated relationships, and proposes new ideas, such as the distinction between implied and explicit ontologies and the advantages of considering MPEG-21 as a simple (pseudo)-ontology. The updated multimedia adaptation engines comparison in Section VI adds *ConversionLink* to the comparison, adds new aspects to the comparison (i.e., multistep, extensibility and semantic adaptation), discusses the reasoning behind the different approaches taken over the years and the pros and cons of the different decision methods. Finally, this publication appends several tests that illustrate the multimedia adaptation method proposed in this paper, and justifies the need for the proposed extensions to the MPEG-21 standard.

In the rest of this paper, Section II reviews the state of the art concerning semantic web description and the description tools that MPEG-21 provides for multimedia adaptation. It also introduces some automatic multimedia adaptation techniques. Section III describes the main features and elements of CAIN-21. Section IV offers innovative description tools that fill the description gaps identified in the standard and justifies their usefulness. Section V provides a set of tests that demonstrate and validate these extensions. Section VI provides a comparative analysis between CAIN-21 and other multimedia adaptation engines. Finally, Section VII gathers the innovations and advantages of the adaptation techniques explained in the paper and it provides some conclusions.

II. STATE OF THE ART

A. Semantic web for multimedia

The Semantic Web [5] aims to represent knowledge in a format that can be automatically processed without human intervention. For this purpose the machine must be capable of understanding the concepts and relationships thereby described. The *Semantic Web Stack* [5] defines a stack of languages in which each layer uses the description capabilities of the layer below it to provide a higher level of expressiveness. In this stack, the technologies up to RDF, OWL and SPARQL have been standardized and accepted. The term *ontology* is used to refer to the concepts (usually defined with a formal *vocabulary*) and relationships in a specific domain. This ontology is frequently represented with OWL creating a *semantic graph*. The technologies in the top of the stack use the semantic graph to infer additional knowledge. Currently, it is not clear how to implement the technologies on the top of the stack. Automatic reasoning has been frequently proposed to infer this additional knowledge. However, the results of these top-level technologies are still limited to achieve the ultimate aim of the Semantic Web: the sharing, processing and understanding of data by automatic systems in the same manner that people can do.

To build a multimedia system that automatically manages and understand multimedia content, it is crucial to define the ontology of its multimedia concepts: Fig. 1 depicts this idea. Bold lines represent better levels of understanding. The figure shows that the user is capable of understanding the meaning of the media, but has more difficulties reading the description of the content. For instance, it is easier for the user to identify a dog in a picture than to interpret its MPEG-7 description [6]. On the other hand, the computer can extract information from metadata more easily than it can analyse the corresponding media resource.

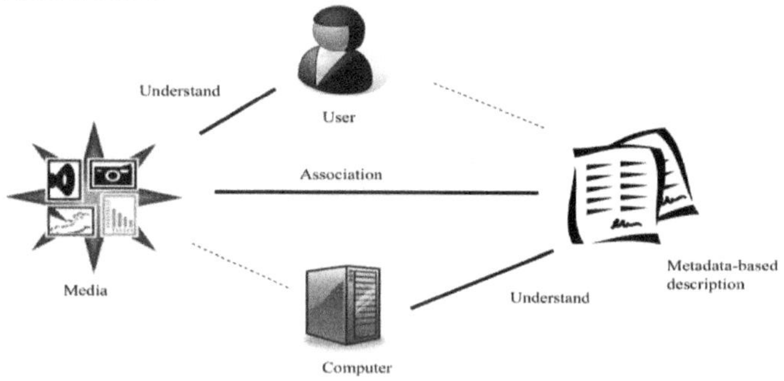

Fig. 1: Semantic description of multimedia content

In the field of multimedia, two widely accepted implied ontologies are the MPEG-7 [6] standard for the media content and MPEG-21 [3] for the whole multimedia system. These standards make use of metadata to achieve a better understanding of multimedia. Particularly, these standards propose several vocabularies to represent a detailed description of the meaning of the multimedia elements. MPEG-7 and MPEG-21 tend to define the external interface of the multimedia systems and leave the algorithms that implement it (e.g. reasoning) to the industry and research community. The W3C Consortium has also initiated a project to represent multimedia ontology called *Multimedia Vocabularies on the Semantic Web* [7]. Even although this standard fully exploits OWL, which has a higher level of expressiveness and ability to represent knowledge, at time of writing, the majority of multimedia research relies on MPEG-7 and MPEG-21.

B. MPEG-21

This section reviews the state of the art for the MPEG-21 description tools to which this paper contributes. A complete description for the MPEG-21 standard can be found in [3].

1) Content description and conditional elements

The notion of *Digital Item* (DI) is a fundamental concept within MPEG-21. A DI is a general representation for any multimedia element. This element can represent both the multimedia content and the multimedia context. MPEG-21 Part 2 [3] standardises the representation of a DI in the case of multimedia content. A DI may contain one or more *Component*[3] elements. Each *Component* includes one *Resource* element and zero or more *Descriptor* elements. The *Resource* element references the media and the *Descriptor* element provides metadata for this media. The MPEG-21 allows optional, alternative and conditional elements. For the purposes of this paper we are only going to describe conditional elements. A *conditional element* is an element of the DI that appears only when certain conditions are true. Certain elements of the DI are configurable, i.e., their content varies depending on the value of *Predicate* elements. A *Predicate* element can take the values *true*, *false* or *undecided*. The *Choice* element enables a “menu”. The options of this menu are provided through *Selection* elements. The *Selection* elements are used to define in runtime the values of the *Predicate* elements. The MPEG-21 standard does not define how the values of the *Predicate* elements are obtained. These values can be asked to the user or automatically decided by the multimedia system. The value of some *Predicates* can be even unknown in runtime, in which case they take the *undecided* value. A *Condition* is a conjunction (and operator) of one or more predicates. The *Condition* elements are used to specify which elements of the DI are valid in runtime. Only the elements for which the *Condition* is true are considered part of the DI. For instance, several *Component* elements may contain a *Condition* element. In runtime, only the *Component* whose *Condition* is true is considered part of the DI.

2) MPEG-21 adaptation tools

MPEG-21 Part 7 [3] has defined a set of *description tools* (or merely *tools*) for multimedia adaptation. These tools do not specify how the adaptation has to be performed; they only gather the information necessary for adapting a DI. These tools are collectively referred as *Digital Item Adaptation* (DIA) *tools*. The instances of these tools are referred as DIA *descriptions* (or merely *descriptions*). This section reviews the *Usage Environment Description* (UED) tools, the *DIA Configuration* tools and the *ConversionLink* tools.

3) UED tools

These tools enable the description of the terminal capabilities, the network constraints, the user’s characteristics, preferences and natural environment. The term *usage environment description* (or merely *usage environment*) refers to an instance of one or more *UED* tools. Further description of the *UED* tools can be found in [3].

[3] MPEG-21 capitalises and italicises XML description tools. This paper adopts this rule.

4) DIA Configuration tools

The DI author can use the *DIA Configuration* tools to recommend how to adapt the content to the usage environment. Specifically, the *DIA Configuration* tools include two tools to drive the adaptation. The first tool allows the DI author to indicate how to obtain the options of the *Choice* conditional mechanism explained above. For this tool, the standard defines only two values: *UserSelection* indicates that the selection has to be done by the user. *BackgroundConfiguration* indicates that the system has to automatically perform this decision.

The second tool is the *SuggestedDIADescription*. The DI author uses this tool to point out which parts of the DI or DIA descriptions have to be used to decide the adaptation. Specifically, XPath [8] expressions are used to provide this information. For instance, the DI author may recommend using the *Format* element of the *VideoCapabilitiesType* in the *UED* to make the adaptation decision. A further description of these tools can be found in [3].

5) ConversionLink tools

The *ConversionLink* tools appear in [9] to complement the *BSDLink* tools. The *ConversionLink* tools are intended to address generic adaptation (e.g. transcoding, transmoding, summarization) whereas the *BSDLink* are intended for scalable bitstream adaptation. The MPEG-21 standard defines a *conversion* as a processor (software or hardware) that changes the characteristics of a *Resource* or of its corresponding *Descriptor* elements. The *ConversionLink* tools include the *ConversionCapabilitiesType* tool. This tool expresses the types of conversions that a terminal is capable of performing. The content of this tool is not standardised, instead, it provides a derivation-by-extension mechanism allowing the inclusion of conversion descriptions. A further description of these tools can also be found in [10].

C. Multimedia adaptation-decision making methods

Typically, multimedia adaptation is performed in two phases, which usually execute in a sequential manner [11][13][14][15]. Firstly, a *decision phase* is used to evaluate which adaptations best suits the constraints of the usage environment. Secondly, in the *execution phase*, these conversions are performed on the media and metadata conveyed in the DI. For the decision phase, two different methods have been widely investigated in the literature:

1) Quality-based methods [11][12][13] (also referred as *optimisation-based methods*) aim at finding the adaptation parameters that maximise the quality (also referred to as *utility*) resulting from the adaptation to the constraints of the usage environment. These methods operate by solving an optimisation problem in the Pareto frontier. Frequently, the MPEG-21 Part-7 DIA tools have been used to point out these relationships between the adaptation parameters and corresponding utilities.

2) Knowledge-based methods [14][15][16] have been used primarily to determine whether a conversion can be executed and which parameters must be supplied to adapt the content. These methods usually consider the concatenation of several conversions in a sequence. They have also been referred as multi-step adaptation.

CAIN-21 (described in Section III) combines both methods in sequence. Firstly, the knowledge-based methods use the media format to decide which conversions have to be carried out in order to adapt the content to the usage environment. This method is further explained in [17]. Secondly, certain "intelligent" conversion tools incorporate the capability to select the parameters that optimise their output. The quality-based methods that CAIN-21 incorporates are demonstrated in [18].

D. Related multimedia adaptation engines

This subsection introduces related multimedia adaptation engines. Section VI compares these adaptation engines with CAIN-21.

Mariam [10] has studied the applicability of a standard *AdaptationQoS* description tool to drive general (scalable and non-scalable) resource adaptation. This investigation concludes developing the ConversionLink[4] adaptation engine togueter with the *ConversionLink* description tool. The *ConversionLink* tool was later standardized in [9]. This tool has already been described in Subsection II.B.

Debargha et al. [11] explained the basis of the *AdaptationQoS* description tool and its usage. Christian et al. [12] builds on this description tool to implement the idea of coded-independent resource adaptation for scalable resources. To this end, they have researched the *BSDLink* tools (introduced in Subsection II.B). In [12] they explains the use of the notion of Pareto optimality and multi-attribute optimisation to identify the scalable layers that best suit the terminal constraints [16].

Jannach et al. [15] developed the koMMa framework in order to demonstrate the use of Artificial Intelligence planning in multistep multimedia adaptation. They exploited Semantic Web Services to address interoperability. They also proposed an extensibility mechanism by means of pluggable Web Services.

Anastasis et al. describe the DCAF adaptation engine in [20]. This research showed how to use heuristic genetic algorithms to identify the parameters of the *AdpatationQoS* description tool. The *UED* and *UCD* description tools are used to represent the context of the adaptation. The notion of Pareto optimality is also introduced to rank the possible decisions.

Davy et al. [21] have built on the aforementioned *AdaptationQoS*, *BSD* and *UED* description tools to develop the NinSuna adaptation engine. This engine provides both coding-format independence and packaging-format independence. The major innovation of this engine is leveraging Semantic Web technologies to accomplish semantic adaptation decisions. The semantics are explicitly represented with RDF tuples and in this way they introduce formal semantics in the exiting MPEG-21 adaptation description tools.

E. Delivery and adaptation methods

From the standpoint of the media client, there are two main media *delivery models* [22]: *download*, where the client starts to play the media content after completely receiving the media from the server, and *streaming* where media content is played while data reception is in progress. *Streaming servers* usually cover two methods to deliver video to the users:

1) Live video. Broadcast of live events in real time. This streaming is useful when the client expects to receive video as soon as it is available. Live events, video conferencing, and surveillance systems are commonly streamed over the Internet as they happen with the assistance of broadcasting software. The video recording software encodes a live source (video or audio) in real time and transfers the resulting media to the streaming server. The streaming server then serves, or "reflects", the live stream to clients. Regardless of when different customers connect to the stream, each sees the same point in the stream at the same time.

2) Video On Demand (VOD). Each customer initiates the reception of the media from the beginning, so no customer ever comes in "late" to the stream. For instance, this mode can be used to distribute movies to users who play those movies at different times.

According to the moment at which the adaptation takes place; media adaptation can be divided into three *adaptation modes*:

1) Offline Adaptation mode (*OffA mode*). The adaptation is performed in the background and before the media is available to the user. This mode is adequate for on demand media delivery. However, this mode is not suitable for live video because the user is expecting to watch the video event as soon as it occurs. This adaptation requires previous knowledge of the feasible terminal capabilities and network bandwidth. The media can be prepared for several terminals of network capacities. The main limitation of the OffA mode is that the user's preferences and natural environment constraints are not taken into account. These parameters are unknown when the media repository is created. While creating a repository of adapted resources for each user's profile is possible, it is unmanageable from a practical point of view when the number of user's profiles increases.

2) On Demand Adaptation mode (*OdA mode*). Adaptation takes place at the same time that the user asks for the resource. In this mode, the client's characteristics, preferences and natural environment can be taken into account. However, if the resource adaptation process is time consuming, the user has to wait until the whole resource is adapted. Therefore, this adaptation results useful for small resources (e.g. images), but can become unacceptable for long resources (e.g. video or speech).

3) Online Adaptation Mode (*OnA mode*). As with the OdA mode, user's characteristics, preferences and natural

[4] Note that the symbol ConversionLink is not italicized to refer to the adaptation engine. However, it is italicized to refer to the description tool.

environment can be taken into account. In this mode also the adaptation begins as soon as the user asks for the resource. However, in contrast to the OdA mode, in the OnA mode the resource begins to be delivered to the user before the whole resource has been adapted. This adaptation is appropriate for long resources (and perhaps also for small resources). The drawback of this approach is that, in general, implementing this solution efficiently is difficult. In OnA mode we need to ensure that media data fragments are delivered to the client in time to maintain playback continuity. The advantage is that once implemented, the OnA mode can be reutilized to simulate the OffA and OdA modes.

III. CAIN-21: SYSTEM DESCRIPTION

This section sequentially describes the CAIN-21 software interfaces, the architecture and the control flow. Section IV builds on this section to specify the description tools that CAIN-21 utilizes and justifies their extension.

A. Software interfaces

CAIN-21 serves adaptation requests through two external software interfaces (see Fig. 2 below): (1) The *media level transcoding interface* performs *blind adaptation* (i.e. semantic-less adaptation) of a media resource. In addition to the media level, this interface can also perform system level adaptation, i.e., videos composed of one or more audio and visual streams. The media level transcoding operations are implemented in the *Tlib* module. This module includes conventional software libraries such as *ffmpeg*, *imagemagick* as well as Java Native Interface (JNI) custom libraries. (2) The *DI level adaptation interface* is in charge of performing system level (semantic or blind) adaptations. In this case metadata is used during the adaptation.

The DI level adaptation interface complies with the MPEG-21 representation schema. The *Content DI* conveys the media resource together with its metadata to be adapted. To drive the adaptation, CAIN-21 uses four *DIA* description tools. Only the *Content DI* and *DIA* description tools follow fully the MPEG-21 recommendations. For the point of view of these interfaces, CAIN-21 is a replaceable black box. Fig. 2 provides a view of CAIN-21 consistent with the idea of an adaptation engine that the MPEG-21 Part-7 framework proposes.

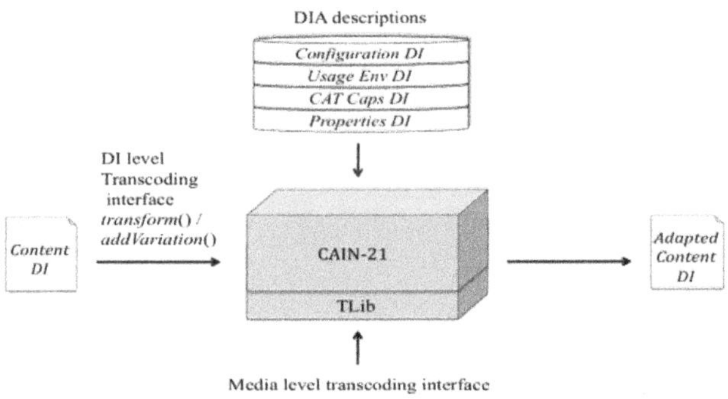

Fig. 2: Software interfaces of CAIN-21

In CAIN-21, metadata-based adaptation [23] is performed through the DI level interface and at the *Component* level. An MPEG-21 *Component* includes a media resource (in the *Resource* element) and its metadata (in the *Descriptor* element). The *Descriptor* elements use MPEG-7 Part 3, Part 4 and Part 5 [6] to describe the multimedia content. The DI level adaptation interface provides two different operations. The first one modifies the existing *Component* and the second operation adds a new *Component* element to the DI. More specifically: (1) the *transform*() operation takes a *Component* from the *Content DI* and modifies its media resource and metadata in order to adapt it to the usage environment; (2) the *addVariation*() operation takes a *Component* from the *Content DI* and creates a new *Component* ready to be consumed in the usage environment. At the end of this adaptation, CAIN-21 adds this adapted *Component* to the *Content DI*.

B. Architecture

This section provides a detailed description of the CAIN-21's modules. Fig. 3 depicts CAIN-21's functional modules and the control flow along the adaptation process. The rest of this subsection explains the modules and description tools in the figure.

1) Adaptation Management Module (AMM)

The AMM is responsible for coordinating the entire DI level adaptation process. Modules below the AMM perform different tasks initiated by the AMM.

2) Adaptation Decision and Execution Modules (ADM and AEM)

Subsection 8.C explained that frequently adaptation engines divide the decision and the execution into two different phases. Firstly, a decision phase is used to decide which adaptation best suits the constraints of the usage environment. Secondly, in the execution phase, these adaptation actions are performed on the media conveyed in the DIs. CAIN-21 also includes this distinction implemented in the *Adaptation Decision Module* (ADM) and the *Adaptation Execution Module* (AEM), respectively.

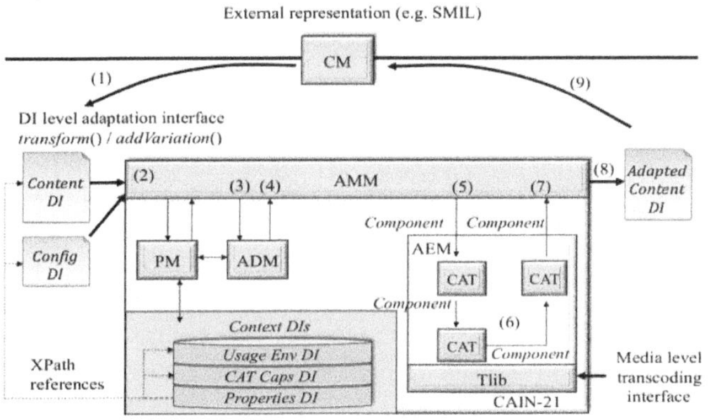

Fig. 3: Modules and control flow within CAIN-21

3) Conversions and Component Adaptation Tools (CATs)

As explained in Subsection II.B, MPEG-21 Part-7 defines a

conversion as the process that changes the characteristics of a resource. In general, a conversion performs the act as defined by the MPEG-21 Part-6 term *adapt*. In CAIN-21, a *Component Adaptation Tools* (CATs) is a pluggable software module that implements one or more conversions. Multi-step adaptation allows for the sequential execution of the conversions implemented in one or more CATs. The ADM uses metadata to determine the sequence of conversions and parameters that should be executed over a *Component* element of the *Content DI*. Subsequently, the AEM executes such sequence of CATs on the original *Component*. When a CAT is executed, both the conversion to execute and the parameters of the conversion have to be provided. If CAIN-21 receives multiple requests to adapt the same content to the same usage environment, a caching mechanism speeds up this process by bypassing the execution of the Planner and Executer several times.

During their execution, CATs have the option of appending information to the *Descriptor* element of the *Component* so that subsequently CATs can use it. We use the name *static decisions* to refer to metadata-based decisions. Static decisions do not depend on the resource content (only the *Descriptor*) and the ADM is responsible for these decisions. On the contrary, we use the term *dynamic decisions* to refer to adaptation decisions that perform operations over the resource content. Dynamic decisions cannot be taken until the resource is available and the CATs take them. These dynamic decisions usually correspond to semantic decisions or quality-based decisions. Frequently semantic decisions assume particular content (e.g. faces, soccer, news items, violent scenes in the movie). For example, in [19] we assume the existence of faces in the images. Quality-based decision methods have been described in Subsection II.C and demonstrated in [18].

4) Context Repository

As further described in Subsection IV.A, CAIN-21 defines a type of DI referred to as *Context DIs*. These *Context DI* elements store *DIA* descriptions with information concerning the context in which the adaptation takes place.

The *Context Repository* in Fig. 3 includes the three *Context DIs*. The *Usage Environment DI* describes the available usage environments using several MPEG-21 *UED* elements (i.e. instances of the *UED* tools). Each *CAT Capabilities DI* describes the different *conversions* that a CAT is able to perform. Each conversion has a set of valid input and output properties along with their corresponding values. The relationships among these elements are described in more detail in Subsection IV.C.

CAIN-21 includes an addressing mechanism in which changes in the metadata descriptors will not imply changes in the underlying source code. This mechanism is described in detail in Subsection IV.E. The mechanism represents all the multimedia information by means of *properties*. Each property has one *key* and one or more *values*. The advantage of this representation is that it suits the decision mechanism that we have developed for CAIN-21 [17]. The *Properties DI* is intended to store a set of keys and corresponding *xpointer*() [24] expressions providing access to the actual values. In Fig. 3, dashed arrows indicate that the *xpointer*() expressions in the *Properties DI* are stored in the other DIs.

5) Configuration DI

The *Configuration DI* is a *DIA* description indicating which description of the terminal, network and user – from the ones available in the *Usage Environment DI* – to use during a adaptation request. Subsection II.B explained that MPEG-21 recommends using the *Choice* descriptor and *DIA Configuration* description tool to specify the adaptation to perform. CAIN-21 does not use this standard mechanism; instead it uses the *Configuration DI* to indicate the parameters of the adaptation to perform. Subsection IV.A justifies this change and explains the advantages that this proposal yields.

6) Parsing Module (PM)

The PM is responsible for resolving the values of the aforementioned properties. Firstly, the PM accesses the *Properties DI* to obtain the set of property keys and corresponding *xpointer*() expressions. Secondly, after resolving these expressions, the values of these properties are generated. During this step, the rest of the metadata is loaded from the *Content DI*, *Configuration DI*, *Usage Environment DI* and *CAT Capabilities DI*. After parsing the different DIs, all the metadata is represented as a set of properties. The value of these properties can be multi-valued (e.g. *bitrate* = [1000..200000], *audio_format* = {*aac*, *mp3*}).

7) Coupling Module (CM)

A wide range of multimedia representation standards exists to represent multimedia content (e.g. HTML, SMIL, NewsML, MPEG-4 BIFS). CAIN-21 can be integrated into heterogeneous multimedia systems that may be using external representation technology (i.e., non-MPEG-21 technology). The CM is the gateway that enables such integration. To this end, this module transforms the external representation of multimedia into an MPEG-21 compliant input *Content DI* that afterwards CAIN-21 processes. In addition, the CM is responsible for transforming the adapted output *Content DI* into its external representation. Instances of the CM are interchangeable modules created to interact with different external representations. In practice, there is a semantic gap during this interaction with the external multimedia description standards, i.e., a direct correspondence between the external descriptors and the MPEG-7/21 descriptors might not exist. To provide these additional meanings, MPEG-7 Part 5 offers a set of open *Classification Schemes* (CSs) [6], which indicates what these external descriptors mean.

C. Control flow

The numbers in Fig. 3 indicate the control flow of the tasks in the adaptation process. (1) When interacting with external systems, the CM transforms the external multimedia representation into a *Content DI* that CAIN-21 can process. (2)

The *Content DI* together with a *Configuration DI* arrives via the DI level interface *transform*() or *addVariation*() operations. (3) The AMM is in charge of coordinating the whole DI level adaptation process. Specifically, the AMM invokes in sequence the ADM and the AEM to (4) decide and (5) execute the corresponding adaptation on the original *Component*. (6) The CATs use the *TLib* services to adapt the media resource. The CATs might also change or append information to the *Descriptor* element of the *Component* so that the subsequent CATs may use it. (7) Once all the conversions of the sequence have been executed, (8) the AMM returns the adapted *Content DI* to the caller. (9) Frequently, the adapted *Content DI* may need to be transformed to an external representation and in this case, the CM performs this transformation.

IV. CAIN-21'S EXTENSIONS TO THE MPEG-21 SCHEMA

CAIN-21 uses the description tools that MPEG-21 standardises. The following subsections identify a set of limitations and ambiguities in the description capabilities of MPEG-21. They then propose some extensions to the MPEG-21 description schema. The additions are justified in order to remove these limitations and ambiguities. The following subsections also discus how these extensions make possible to address a new range of multimedia adaptation problems.

A. Content DI, Context DI and Configuration DI

Subsection II.B explained that in MPEG-21 framework different DIs are used throughout the consumption and delivery chain. The DIs can be classified according to their purpose. One initial approach in the literature has divided the DIs into *Content DIs* and *Context DIs*. The *Content DI* is a DI intended to carry out the multimedia resource and corresponding metadata. The *Context DI* is intended to contain a description of the usage environment. The notions of *Content DI* and *Content DI* have been considered by the MPEG-21 standard (see for instance [25]) although they have not been finally incorporated to the standard. However, some authors have informally used these notions in their systems [26][27].

Particularly, these authors have used the term *Context DI* only to reference the usage environment [25][26][27]. In [28], we proposed to extend the idea of *Context DI* to represent the context information. Particularly, in CAIN-21 there are three types of context elements: the *Usage Environment DI*, the *CAT Capabilities DIs* and the *Properties DI*. Subsection III.B described these elements.

Furthermore, CAIN-21 configures the adaptation using the *DIA Configuration* description tools (described in Subsection IIII.B). After an adaptation request, the *DIA Configuration* tools can be used to specify the target usage environment. Although there are scenarios in which the *DIA Configuration* tools is applicable, we have identified two limitations in the standard *DIA Configuration* mechanism:

1. The standard *Content DIs* uses the *Choice* description element to enclose alternative adaptation options, which depends on the available terminals. This produces a dependency between the *Content DI* (which contains the *Resource* and optionally a *DIA Configuration* description) and the *Usage Environment DI*. This dependency implies changing the *Content DI* whenever the *Usage Environment DI* is modified (e.g. one of the terminal descriptions is changed).

2. *DIA Configuration* assumes that the entire usage environment is known when the *Usage Environment DI* is created.

The idea of using three DIs avoids the first limitation:

1. The *Content DI* with the multimedia resource and corresponding metadata.

2. The *Context DI* that acts as a database where usage environment, adaptation capabilities and metadata properties under consideration are stored.

3. The *Configuration DI* that includes a *DIA Configuration* description.

The *Configuration DI* also solves the second limitation: the *Content DI* and the *Context DI* are created and stored in CAIN-21 during its development or deployment. The *Configuration DI* is dynamically created to provide to CAIN-21 information about the adaptation request to be performed.

Next section describes the *ARC* description tool that the *Configuration DI* conveys. The main aim of our proposal is that the *Content DI* will not be modified when the *Usage Environment DI* changes.

B. The ARC description tool

Section IIII.B described the two *DIA Configuration* description tools that MPEG-21 Part 7 standardises: (1) The *UserSelection*/*BackgroundConfiguration* elements indicate whether the DI *Choice*/*Selection* mechanism must be presented to the user or automatically decided by the system. (2) The DI's author uses the *SuggestedDIADescriptions* to suggest which *DIA Description* elements should be used for the adaptation. Both methods assume the existence of a negotiation mechanism. Authors such as [26][29] have followed this approach incorporating the *DIA Configuration* description in the DI to be consumed. CAIN-21 is not a network agent (as in the *DIA Configuration* usage model developed in [3]) but a middleware providing an API. Previous subsection introduces the problem of selecting zero or one instance of the standard MPEG-21 Part 7 *UED* description tools (i.e., *Terminal*, *Network* and *User*[5] elements) from the *Usage Environment DI*. If we relax the network agent negotiation assumption we can utilise the *DIA Configuration* to specify the particular usage environment. CAIN-21 extends the *DIA Configuration* to provide this information, i.e., it defines a third *DIA Configuration* tool (non-considered in MPEG-21). This extension is called *Adaptation Request Configuration* (*ARC*) tool. Consider, for instance, two terminals in the *Usage Environment DI*, a mobile terminal and a laptop terminal. In this case, an *ARC* description can be used

[5] Currently CAIN-21 does not consider the *NaturalEnvironment* description tool, but its inclusion would be a direct process.

to indicate the target terminal. The *Content DI* and the *Usage Environment DI* can be deployed before starting the adaptation engine. On the contrary, the *ARC* description is only created when an adaptation is going to be executed.

C. CAT Capabilities

The large quantity of multimedia adaptations that could be envisioned makes it unfeasible to implement all of them. Subsection III.B has introduced the notion of pluggable CATs. Their adaptation capabilities are described in *CAT Capabilities DIs* (also introduced in Subsection III.B). One CAT can be used as soon as this CAT and its corresponding *CAT Capabilities DI* are plugged in CAIN-21.

1) CAT Capabilities and Conversion Capabilities

The notion of *CAT Capabilities* was introduced in [28]. The following paragraphs describe the current *CAT Capabilities* description tool of CAIN-21 and compare it with the standard *ConversionLink* [9].

Subsection II.B explained that MPEG-21 Part 7 Amendment 1 defines a *conversion* as an (software or hardware) element capable of performing multimedia adaptation. The original *CAT Capabilities* only allowed describing one conversion. The final *CAT Capabilities* can incorporate several conversion elements. During the development of CAIN-21, we observed the practical fact that conversion capabilities are not always easy to describe with only one conversion. With some types of adaptations, we need to divide the capabilities of an individual *CAT Capabilities* element into several *Conversion Capabilities* elements. Consider, for example, a CAT that is capable of accepting JPEG and PNG images, but PNG images are accepted only in greyscale, whereas JPEG images are accepted in both colour and greyscale. In this case, the *CAT Capabilities DI* must be split into two separate *Conversion Capabilities*. The first *Conversion Capabilities* element states that PNG images are accepted in greyscale. The second *Conversion Capabilities* element states that JPEG images are accepted in both colour and greyscale.

The second major feature implies the description of the values that properties can take. In the CAIN-21 decision process, preconditions, postconditions and parameters can take several possible values (e.g. *format* = {*mpeg*-1, *mpeg*-2, *mpeg*-4}). We have modified the description of the conversions so that each input and output property can take multiple values.

```
<dia:DIA xmlns="urn:vpu:cain21-cat-capabilities"
         xmlns:dia="urn:mpeg:mpeg21:2003:01-DIA-
NS"
         xmlns:mpeg7="urn:mpeg:mpeg7:schema:2001"

xmlns:xsi="http://www.w3.org/2001/XMLSchema-
instance">
  <dia:Description xsi:type="CATCapabilitiesType"
id="video_transcoder_cat">

<CATClassName>es.vpu.cain21.cats.VideoTranscoderCA
T</CATClassName>
  <Platform>
   <ValueSet>
    <Value href="Windows XP">Windows</Value>
    <Value href="Linux">Linux</Value>
    <Value href="Mac OS X">Mac OS X</Value>
   </ValueSet>
   </Platform>
   <!-- Online MPEG conversion using the ffmpeg
library -->
   <ConversionCapability
xsi:type="ConversionCapabilityType"

id="online_mpeg_transcoder">
    <ContentDegradation>0</ContentDegradation>
    <ComputationalCost>1.0</ComputationalCost>
    <Preconditions>
     <URL>
      <AnyValue/>
     </URL>
     <Binding>
      <ValueSet>
       <Value href="urn:mpeg:mpeg21:2007:01-BBL-
NS:handler:HTTP">HTTP</Value>
       <Value href="urn:mpeg:mpeg21:2007:01-BBL-
NS:handler:FILE">FILE</Value>
      </ValueSet>
     </Binding>
     <Content>
      <ValueSet>
       <Value
href="urn:mpeg:mpeg7:cs:ContentCS:2001:2">Audiovis
ual</Value>
       <Value
href="urn:mpeg:mpeg7:cs:ContentCS:2001:4.2">Video<
/Value>
      </ValueSet>
     </Content>
     <FileFormat>
      .............
     </FileFormat>
     <Bitrate>
      <RangeValueSet from="5000" to="1000000"/>
     </Bitrate>
     ..............
    </Preconditions>
    <Postconditions>
     .................
    </Postconditions>
   </ConversionCapability>
   <!-- On Demand MP4 conversion using the ffmpeg
command -->
   <ConversionCapability
xsi:type="ConversionCapabilityType"

id="ondemand_mp4_transcoder">
     ..................
     ..................
  </ConversionCapability>
 </dia:Description>
</dia:DIA>
```

Listing 1: CAT Capabilities DI example

The XML Schema of the *CAT Capabilities* description tool that we propose is available in the file *ccatc.xsd* of the CAIN-21 software. The *CATCapabilitiesType* represents a CAT. The *ConversionCapabilitiesType* represents each conversion that the CAT is capable of performing. Listing 1 shows a fragment of one of the *CAT Capabilities DIs* fully available in the CAIN-21 demo. The CAT comprises two *ConversionCapability* elements named *online_mpeg_transcoder* and *ondemand_mp4_transcoder*. The *Preconditions* and *Postconditions* elements contain information related to the media format that each conversion accepts and produces. These properties are inspired by MPEG-7 Part 5. Note that the properties can be single-valued or multi-valued by means of the *ValueSet* element. The

RangeValueSet element enables the description of ranges. The *AnyValue* element represents a placeholder whenever the value of the parameter must be provided, but every value is acceptable.

2) Comparison with the ConversionLink

Subsection II.B describes that MPEG-21 Part 7 Amendment 1 has standardised the *ConversionLink* description tool. This tool provides a means for linking steering description parameters and conversion capabilities description. *ConversionLink* uses the *ConversionCapability* element to describe the adaptation capabilities.

The *CATCapabilitiesType* of CAIN-21 is defined as a derivation by restriction of the *DIADescriptionType* of MPEG-21. Therefore, the *CATCapabilitiesType* can be seen as a (non-MPEG-21 standardised) *DIA* description tool.

More specifically, the *ConversionCapabilityType* of MPEG-21 is a generic container that used the following type to enable any description:

```
<any                          namespace="##other"
processContents="lax" minOccurs="0"/>
```

The *ConversionCapabilityType* of CAIN-21 is defined as a derivation by extension of this *ConversionCapabilityType*. Therefore, the *ConversionCapabilityType* of CAIN-21 can be seen as an instance of the generic *ConversionCapabilityType* that MPEG-21 provides. In particular, CAIN-21 describes the conversions by means of preconditions and postconditions. This description model suits the automatic decision mechanism of CAIN-21.

Authors such as [10] also use the *ConversionCapability* element[6] together with the *ConversionLink*. In this case, the author makes use of RDF tuples to describe the adaptation capabilities and its semantics. CAIN-21 instead uses preconditions and postconditions that best suit its decision-making mechanism.

D. Binding modes

Subsection II.E explained the Offline/On-demand/Online adaptation modes. CAIN-21 supports all these modes. Subsection II.E has highlighted the difference between adaptation and delivery. Although CAIN-21 is focused on adaptation, delivery is supported to a certain extend. *Binding modes* have been introduced in CAIN-21 to support media delivery. In particular, delivery can be envisioned as a type of adaptation. The binding modes indicate the delivery mechanism that the conversion uses to receive and transmit the media (such as *FILE*, *HTTP* or *RSTP*). This work proposes to use the *mpeg21:Handler* description tool of the *Bitstream Binding Language* (BBL) [30]. The binding modes are used with two purposes: (1) to transfer the media between CATs in a sequence of CATs and (2) to transfer the media from the last CAT in the sequence to the consumption terminal. TABLE shows the binding modes currently available in CAIN-21. The *INPROCESS* binding mode allows efficient transfer of the media resource between CATs.

In CAIN-21 each *ConversionCapabilities* element must provide in its preconditions and postconditions the available *binding modes*. For instance, in Listing 1, the first *ConversionCapabilities* element supports *FILE* and *HTTP* in its preconditions (i.e. in the input of the corresponding conversion). The *Terminal* element of the *Usage Environment DI* must also indicate the delivery modes that it supports to receive media. Listing 3 below shows how the binding modes of a terminal are provided into its terminal description. Listing 1 and Listing 3 show that the binding mode, of both the *online_mpeg_transcoder* and the terminal, can take more that one value. In these examples, both the conversion described in Listing 1 and the terminal described in Listing 3 support the *FILE* and *HTTP* binding modes.

The current release of CAIN-21 includes one CAT (named *HttpVideoStreamingCAT*), which only purpose is to provide HTTP video delivery. If necessary, the decision mechanism automatically adds this CAT to the sequence of CATs. Specifically, this CAT is added at the end of the sequence when the last CAT of the sequence does not provide *HTTP* binding mode in its postconditions (for instance, because the CAT only provides *FILE* binding mode in its postconditions) and the terminal binding mode is defined as *HTTP* only capable.

TABLE I
BINDING MODES PROPOSED BY CAIN-21

Binding mode	Description
urn:mpeg:mpeg21:2007:01-BBL-NS:handler:INPROCESS	In-process technique used to transfer information between CATs. In the case of CAIN-21, objects loaded in memory use the pull model to request data by means of a memory buffer.
urn:mpeg:mpeg21:2007:01-BBL-NS:handler:FILE	Can read/write files provided in the URL. This is an appropriate binding for OdA mode
urn:mpeg:mpeg21:2007:01-BBL-NS:handler:TCP	Can read/write TCP sockets. The IP+port are provided in the URL. This is an appropriate binding for OnA mode.
urn:mpeg:mpeg21:2007:01-BBL-NS:handler:HTTP	Can read/write HTTP protocol. The IP+port are provided in the URL. This is an appropriate binding for OnA mode.
urn:mpeg:mpeg21:2007:01-BBL-NS:handler:RTSP	Can read/write RTSP protocol. The IP+port are provided in the URL. This is an appropriate binding for OnA mode.

E. Properties DI

The *Properties DI* tool gathers all the information required by the multimedia adaptation process following a declarative approach. The main purpose of this tool is that changes in the set of multimedia properties do not imply changes in the

[6] The author uses the name *ConversionDescription* to refer to the notion of *ConversionCapability*.

underlying source code. Particularly, with the *Properties DI* tool all the information is described consistently using so called *multimedia properties*. These multimedia properties include the *Content DI*, the *Usage Environment DI* and the *CAT Capabilities DI*. Each property is represented as a label with an associated XPath [8] expression.

1) Addressing mechanism

Even though the PM is still responsible for parsing the documents and loading them in memory, the ADM does not directly access these properties. In this way, changes in the metadata do not imply changes in the underlying source code. Instead, these changes imply only modifying the *Properties DI*.

```
<dia:DIA xmlns="urn:vpu:cain21-properties-di"
         xmlns:dia="urn:mpeg:mpeg21:2003:01-DIA-
NS"
         xmlns:mpeg7="urn:mpeg:mpeg7:schema:2001"

xmlns:xsi="http://www.w3.org/2001/XMLSchema-
instance">
 <dia:Description xsi:type="PropertiesDIType">
  <DIProperties>
    <Property name="genre" required="false"

xpath="/Item/Descriptor/Statement/Mpeg7/Descriptio
nUnit/Genre/@href"/>
  </DIProperties>
  <ComponentProperties>
   <Property name="id" required="true"
xpath="/@id"/>
   <Property name="url" required="true"
xpath="/Resource/@ref"/>
   <Property name="mime_type" required="false"
xpath="/Resource/@mimeType"/>
    ..................
   <ComposedProperty name="visual_frame"
required="false">
    <Value
xpath="//Mpeg7/Description/MediaInformation/MediaP
rofile

//MediaFormat/VisualCoding/Frame/@width"/>
    <Value
xpath="//Mpeg7/Description/MediaInformation/MediaP
rofile

//MediaFormat/VisualCoding/Frame/@height"/>
   </ComposedProperty>
  </ComponentProperties>
  <CATProperties>
   <Property name="id" required="true"
xpath="/@id"/>
   <Property name="cat_class_name" required="true"
xpath="/CATClassName"/>
    ....................
  </CATProperties>
  <ConversionProperties>
   <Property name="id" required="true"
xpath="/@id"/>
   <Property name="content_degradation"
required="true"
             xpath="/ContentDegradation"/>
   <Property name="computational_cost"
required="true"
             xpath="/ComputationalCost"/>
   <!-- Input properties -->
   <Property name="pre_url" required="true"
xpath="/Preconditions/URL"/>
   <Property name="pre_binding" required="true"
xpath="/Preconditions/Binding"/>
   <Property name="pre_content" required="true"
xpath="/Preconditions/Content"/>
    ..............
   <!-- Output properties -->
   <Property name="post_url" required="true"
xpath="/Postconditions/URL"/>
   <Property name="post_binding" required="true"
xpath="/Postconditions/Binding"/>
   <Property name="post_content" required="true"
xpath="/Postconditions/Content"/>
    .................
  </ConversionProperties>
  <UsageEnvProperties>
   <TerminalProperties>
    <Property name="id" required="true"
xpath="/@id"/>
    <Property name="binding" required="true"

xpath="/TerminalCapability[@type='cde:HandlerCapab
ilitiesType']
          /Handler/@handlerURI"/>
     ..........
   </TerminalProperties>
   <NetworkProperties>
    <Property name="id" required="true"
xpath="/@id"/>
    <Property name="max_capacity" required="false"

xpath="/NetworkCharacteristic/@maxCapacity"/>
    <Property name="min_guaranteed"
required="false"

xpath="/NetworkCharacteristic/@minGuaranteed"/>
   </NetworkProperties>
   <UserProperties>
    <Property name="id" required="true"
xpath="/@id"/>
     ..............
    <Property name="pref_focus_of_attention"
required="false"

xpath="/UserCharacteristic/ROI/@uri"/>
   </UserProperties>
  </UsageEnvProperties>
 </dia:Description>
</dia:DIA>
```

Listing 2: Properties DI example

The expression of each property points out to the part of the DI where its values are located. XPath expressions are relative to the document. Therefore, the *Properties DI* stores only the XPath of the property. The document that contains these properties is determined during the execution of the adaptation. The *Configuration DI* (introduced in Subsection IV.A) is used to identify these documents. Furthermore, properties are only resolved on-demand. In this way, properties that are never used are not extracted from de DIs. Internally, CAIN-21 uses *xpointer*() [24] expressions to reference both the document and the XML element or attribute to be accessed. The standard Xalan processor [31] is used in our work to gather all these properties.

The *Properties DI* schema that we propose is available in the file *cpr.xsd* of the CAIN-21 software. The *PropertiesDIType* is defined as a derivation by restriction of the MPEG-21 standard *DIADescriptionType*. This type includes four important elements that correspond to the five groups of properties: *DIProperties*, *ComponentProperties*, *CATProperties*, *ConversionProperties* and *UsageEnvProperties*. Listing 2 shows the more relevant parts of the current *Properties DI* of CAIN-21 (the whole document is available in the file *pr.xml* of the CAIN-21 demo). For instance, in Listing 2 the *ConversionProperties* element contains the property *pre_url* whose XPath expression is "*/Preconditions/URL*". On resolving this XPath expression in Listing 1, *AnyValue* is obtained indicating that the conversion accepts any value for this property. As another example, on resolving the *pre_binding* property in Listing 2, the *FILE* and *HTTP* binding modes are obtained from Listing 1.

2) Properties and relationships

Subsection III.A introduced the Semantic Web. Semantic Web languages such as OWL allow explicitly representing and storing concepts and their relationships in a semantic graph. Software tools such as Protégé facilitate loading this graph from disk to memory. Frequently, automatic-reasoning techniques use this graph to search for relationships among the values and to infer additional information.

In CAIN-21, the concepts are represented by means of properties and the relationships are limited. Specifically, relationships are just intended to assist the matching algorithm developed in [17]. In this way, CAIN-21 uses a delimited subset of the rich relationships that the Semantic Web provides.

The *Properties DI* complies with the *Keep It Short and Simple* (KISS) design principle. This principle recommends to avoid unnecessary complexity and construct systems as simple as possible, but no simpler. The main purpose of the *Properties DI* is not information inference, but to elude changes in the decision algorithm when the metadata under consideration evolves. In contrast, depending on the reasoning techniques, changes in the relationships of the semantic graph imply changes in the underlying reasoning algorithms.

In a nutshell, the matching mechanism tests whether the input of one CAT accepts the output of the previous CAT in the sequence. The matching mechanism also tests whether the terminal accepts the output of the last CAT. In order to perform these tests, usually the simple properties-based representation mechanism has demonstrated to be enough [17]. These tests demonstrated its suitability and also demonstrated that the matching mechanism operates efficiently. However, during the tests, we encountered that occasionally it is convenient to consider more complicated relationships between properties. For instance, to maintain the ratio in the adapted media the width and height should be considered together. In these cases, we use *composed properties*. Listing 2 shows an example of these composed properties. The *visual_frame* property uses the *ComposedProperty* element to gather the width and height elemental values. In the representation schema of CAIN-21, these elemental values can be represented by means of ranges or as a placeholder accepting any value.

F. Extensions to the UED

In particular, this document has identified the following handicaps in the current UED:

1. The *mpeg21:TerminalType* does not include any reference to the modalities of the content that the terminal consumes (images, video, audiovisual, audio, etc). The *mpeg7:Content* description serves this purpose by the *mpeg7:ContentCS* classification scheme

2. The terminal does not provide any description of the binding modes, i.e., the delivery mechanism (such as *HTTP* or *RTSP*) used to consume content as described in Subsection IV.D.

3. The standard MPEG-21 Part 7 *UED* tools do not specify whether the properties of the terminal are mandatory or optional. For instance, if the terminal is defined using the *mpeg21:AudioCapabilitiesType*, does it mean that the adapted media must include audio? Or does it mean that this audio format could be consumed if present?

These semantic gaps include both properties that can be inferred and properties that cannot be inferred (ambiguities). The first gap semantic can be addressed by inference [32] and the other two gaps by extending the current *mpeg21:TerminalType*. More specifically, the first gap can be addressed by inferring the media content (image, video, audiovisual, audio) from the *mpeg21:TransportCapabilitiesType* (illustrated in Listing 3).

To demonstrate how to address the other two gaps, Listing 3 shows a portion of the description of the terminal with *id="iphone"* from the CAIN-21 demo. The extensions that this subsection discussed are marked in bold. The XML Schema with these changes is publicly available in the file *cde.xsd* of the CAIN-21 software.

```
<Terminal id="iphone" xsi:type="cde:TerminalType">
 <TerminalCapability
xsi:type="cde:HandlerCapabilitiesType">
  <Handler handlerURI="urn:mpeg:mpeg21:2007:01-
BBL-NS:handler:FILE"/>
  <Handler handlerURI="urn:mpeg:mpeg21:2007:01-
BBL-NS:handler:HTTP"/>
 </TerminalCapability>
 <TerminalCapability
xsi:type="cde:CodecCapabilitiesType">
  <cde:Decoding
xsi:type="cde:TransportCapabilitiesType">
   <cde:Format
href="urn:vpu:cs:FileFormatCS:2009:3gpp">
    <mpeg7:Name xml:lang="en">
     3GPP file format
    </mpeg7:Name>
   </cde:Format>
  </cde:Decoding>
  <cde:Decoding
xsi:type="cde:VideoCapabilitiesType">
   <cde:Format
href="urn:vpu:cs:VisualCodingFormatCS:2007:1">
    <mpeg7:Name xml:lang="en">
     H.264 Baseline Profile @ Level 1.1
    </mpeg7:Name>
   </cde:Format>
   <cde:CodecParameter
xsi:type="CodecParameterBitRateType">
    <BitRate >32000</BitRate>
   </cde:CodecParameter>
  </cde:Decoding>
  <cde:Decoding
xsi:type="cde:AudioCapabilitiesType"
optional="true">
   <cde:Format
href="urn:mpeg:mpeg7:cs:AudioCodingFormatCS:2001:4
.3.1">
    <mpeg7:Name xml:lang="en">
     MPEG-2 Audio AAC Low Complexity Profile
    </mpeg7:Name>
```

```
      </cde:Format>
      <cde:CodecParameter
   xsi:type="CodecParameterBitRateType">
        <BitRate>7950</BitRate>
      </cde:CodecParameter>
    </cde:Decoding>
  </TerminalCapability>
    ....................
</Terminal>
```

Listing 3: Extended mpeg21:TerminalType

The proposed extensions to the *mpeg21:TerminalType* are:

1. Representing the binding modes of the terminal in the *cde:HandlerCapabilitiesType* description tool. This element makes reference to the *mpeg21:Handler* description tool. shows how to describe that the iPhone terminal supports the *FILE* and *HTTP* binding modes.

2. Mandatory and optional constrains are instances of the hard and soft constraints model developed in [17]. To provide this description, CAIN-21 extends the *mpeg21:TerminalType* with the *optional* attribute. Listing 3 shows how to signal that the audio stream is optional using the optional attribute in the *cde:AudioCapabilitiesType*. If this attribute is absent, CAIN-21 considers the terminal description as a mandatory constraint.

V. TESTS AND VALIDATION

The CAIN-21 demo, publicly available at *http://cain21.sourceforge.net*, provides several tests demonstrating the multimedia adaptation approach of this paper. This section focuses on demonstrating and validating the extensions to the MPEG-21 standard proposed in Section IV.

Subsection III.A described the DI level adaptation interface. Both operations of this interface – i.e., *transform*() and *addVariation*() – have been used in the tests. In addition, to cover a wide range of multimedia adaptations, both images and videos have been selected for the tests reported in this paper. CAIN-21 can also convert images to video through the *Image2VideoCAT*. Its *image_2_video* conversion has also been covered in the tests.

A. Transforming an image to an small video terminal

Test 1 illustrates how a *Content DI* with an image (named *photo.xml*) is adapted to the *id="iphone"* video terminal (shown in Listing 3). The full description of these elements is available in the CAIN-21 demo. The *transform*() software interface receives a *Configuration DI* (described in Subsection IV.A) to indicate the target terminal. The PM of CAIN-21 (described in Subsection III.B) uses the *Properties DI* to gather the properties of the *Content DI*, *CAT Capabilities DIs* and *Usage Environment DI*. After that, the ADM (introduced in Subsection III.B) produces the following sequence of conversions *initial* → *image_transcoder* → *image_2_video* → *ondemand_video_transcoder* → *goal*. In this sequence, *initial* represents the properties of the original *Content DI*. The *Preconditions* and *Postconditions* of *image_transcoder*, *image_2_video* and *ondemand_video_transcoder* are described in their corresponding *ConversionCapabilities* elements as explained in Subsection IV.C. Lastly, *goal* represents the properties adapted content. The *image_transcoder* conversion transcodes the image format and size to the preconditions of the *image_2_video* conversion (i.e., JPEG image format and 3:4 aspect ratio). The *image_2_video* conversion accepts only JPEG images and produces only MPEG-2 video. *The ondemand_video_transcoder* (whose conversion capabilities appear in Listing 1) transcodes the MPEG-2 video to the constraints of the terminal (3GPP according to Listing 3). In this Test 1, the ADM has selected the *FILE* binding mode the conversions steps. This happened because all the conversions provide this transfer mechanism in their *Preconditions* and *Postconditions* description tools.

If we change the terminal of Test 1 from *"iphone"* to *"http_nokia_n95"*, we have the didactic Test 2. This test fully demonstrates the usefulness of the binding modes. In Test 2, CAIN-21 produces a sequence with four conversions *initial* → *image_transcoder* → *image_2_video* → *ondemand_video_transdoder* → *http_delivering* → *goal*. Specifically, CAIN-21 has added to the end of the sequence the *http_delivering* conversion to change the binding property from *FILE* to *HTTP*. In Test 1 the *"iphone"* terminal supported the *FILE* delivery mechanism (see Listing 3), which corresponds to the binding property at the output of *ondemand_video_transcoder*. Therefore CAIN-21 did not add the *http_deliveding* conversion at the end of the sequence. However, in Test 2, the *"http_nokia_n95"* terminal only supports the *HTTP* binding mode, and therefore CAIN-21 has added the *http_delivering* conversion at the end of the sequence. This conversion has the *FILE* binding mode in its preconditions and the *HTTP* binding mode in its postconditions: this indicates that the purpose of this tool is to transfer the input file using the *HTTP* standard protocol.

B. Summarizing variations of video news items

Test 3 summarizes and adapts a *Content DI* containing a news item to three different terminals [33]. Listing 4 shows the original *Content DI* to be adapted.

```
<DIDL xmlns="urn:mpeg:mpeg21:2002:02-DIDL-NS"
      xmlns:xsi="http://www.w3.org/2001/XMLSchema-
instance"
      xmlns:cdi="urn:vpu:cain21-di"
      >
 <Item xsi:type="cdi:ItemType">

  <!-- Classification -->
  <Descriptor>
   <Statement mimeType="text/xml">
    <Mpeg7 xmlns="urn:mpeg:mpeg7:schema:2004">
     <DescriptionUnit xsi:type="ClassificationType">
      <Genre
href="urn:mpeg:mpeg7:cs:ContentCS:2001:1.1.13">
        <Name xml:lang="en">Natural disasters</Name>
      </Genre>
      <Genre
href="urn:mpeg:mpeg7:cs:ContentCS:2001:1.5.1">
        <Name xml:lang="en">Political</Name>
      </Genre>
     </DescriptionUnit>
    </Mpeg7>
```

```
   </Statement>
  </Descriptor>
  <!-- Original content -->
  <Component xsi:type="cdi:VideoComponentType"
id="original">
   <Descriptor xsi:type="cdi:Mpeg7DescriptorType">
    <!-- MPEG-7 MediaDescriptionType describing the
resource -->
    ................
   </Description>
   <Resource mimeType="video/mpeg"
ref="../mesh/didl/flood2video.mpg"/>
  </Component>
 </Item>
</DIDL>
```

Listing 4: Original DI to be summarized and adapted in Test 3

The MPEG-7 *ClassificationType* description type indicates that the news item contains natural disaster and political content. The video is stored in a *Component* element with *id="original"*. This *Component* contains and MPEG-7 *MediaDescriptionType* description of the *Resource* element. The original video is MPEG-1 video and has a resolution of 720x576. This video is summarized according to the methods explained in [33]. Subsequently, the DI is adapted to three terminals. The terminals for the adaptation are all MPEG-2 terminals and have, respectively, screen sizes of 720x576, 352x288 and 176x144. In Test 2, the *addVariation*() operation is used to create the adapted videos in additional *Component* elements of the *Content DI*. Listing 5 shows the adapted *Content DI* with four *Component* elements: the original video and three summarized and adapted variations. The MPEG-7 *VariationDescriptionType* description type indicates that the original video (with *id="original"*) has three variations in the *Component* elements with IDs *"big-sum"*, *"medium-sum"* and *"small-sum"*.

```
<DIDL xmlns="urn:mpeg:mpeg21:2002:02-DIDL-NS"
      xmlns:xsi="http://www.w3.org/2001/XMLSchema-
instance"
      xmlns:cdi="urn:vpu:cain21-di"
      xmlns:mpeg7="urn:mpeg:mpeg7:schema:2004">
 <Item xsi:type="cdi:ItemType">
  <!-- Classification -->
  .......................
  <!-- Original content -->
  <Component xsi:type="cdi:VideoComponentType"
id="original">
   .........
   <Descriptor xsi:type="VariationDescriptionType">
    <VariationSet>
     <Source xsi:type="AudioVisualType">
      <AudioVisual>
       <MediaLocator>
        <MediaUri>#original</MediaUri>
       </MediaLocator>
      </AudioVisual>
     </Source>
     <Variation priority="1">
      <Content xsi:type="AudioVisualType">
       <AudioVisual>
        <MediaLocator>
         <MediaUri>#big-sum</MediaUri>
        </MediaLocator>
       </AudioVisual>
      </Content>
      <VariationRelationship>
       summarization
      </VariationRelationship>
     </Variation>
     <Variation priority="2">
      <Content xsi:type="AudioVisualType">
       <AudioVisual>
        <MediaLocator>
         <MediaUri>#medium-sum</MediaUri>
        </MediaLocator>
       </AudioVisual>
      </Content>
      <VariationRelationship>
       summarization
      </VariationRelationship>
     </Variation>
     <Variation priority="3">
      <Content xsi:type="AudioVisualType">
       <AudioVisual>
        <MediaLocator>
         <MediaUri>#small-sum</MediaUri>
        </MediaLocator>
       </AudioVisual>
      </Content>
      <VariationRelationship>
       summarization
      </VariationRelationship>
     </Variation>
    </VariationSet>
   </Descriptor>
   ...............
   <Resource mimeType="video/mpeg"
ref="../mesh/didl/flood2video.mpg"/>
  </Component>
  <!-- Big size summarized content -->
  <Component xsi:type="cdi:VideoComponentType"
id="big-sum">
   .................
  </Component>
  <!-- Medium size summarized content -->
  <Component
xsi:type="cdi:VideoComponentType"id="medium-sum">
  </Component>
  <!-- Small size summarized content -->
  <Component xsi:type="cdi:VideoComponentType"
id="small-sum">
   ..........
</Component>
 </Item>
</DIDL>
```

Listing 5: Summarized and adapted DI in Test 3

Test 3 uses three *Configuration DIs*. These *Configuration DIs* use the *ARC* descriptions (described in Subsection IV.B) to request the adaptation to three terminals respectively labelled as *"720x576"*, *"352x288"* and *"176x144"* in the *Usage Environment DI*. For Test 3, we needed to create a CAT named *RawVideoCombinerCAT*. Its *CAT Capabilities* appear in the file *raw_video_combiner_cat.xml* of the CAIN-21 demo. This CAT was necessary to retrieve the summarized video from the summarization module (further explained in [33]) through two TCP sockets: one for raw WAV audio and one for RAW video. To this end, we created an additional binding mode (see Subsection IV.D) labelled as *urn:mpeg:mpeg21:2007:01-BBL-NS:handler:TCP*.The three terminals in Test 3 were defined with the standard *HTTP* binding mode in TABLE . During the adaptation, the ADM produced a sequence with three conversions: *initial* → *raw_video_combiner* → *online_video_transcoder* → *http_delivering* → *goal*.

C. Extensions demonstrated in the tests

To recapitulate, justify and validate the extensions to the

MPEG-21 schema that this paper proposes the following conclusions are offered:

1. The proposes description schema enables the description of multiple terminals respectively labelled in the tests of this section as *"iphone"*, *"http_nokia_n95"*, *"720x576"*, *"352x288"* and *"176x144"*. To indicate the target terminal of the adaptation this information has to be provided. As MPEG-21 does not define a description tool for this purpose, Subsection IV.B has proposed this description tool.

2. To enable automatic adaptation decision, the inputs and outputs of the conversion tools have to be provided. Subsection IV.C proposed the *CAT Capabilities* description tools. The feasible inputs and output properties are defined using the *Preconditions* and *Postconditions* elements.

3. For automatic decision, it is also necessary to know how the media is going to be transferred to both the next CAT and the target terminal. This justifies the introduction of the binding mode in the description schema.

4. The modality of the content appears in the *mpeg7:Content* description tool. However, this information is not provided by the *mpeg21:TerminalType* description type. During the decision process the modality of the content that the terminal accepts has to be determined. The inference rule described in Subsection IV.F can be used in this case. Specifically, from the *mpeg21:TransportCapabilitiesType* description tool of Listing 3 it can be inferred that the content has to be visual or audiovisual. See [32] for a further explanation of this mechanism.

5. In Test 2 the decision process needs to know the binding mode to identify that the *http_delivering* conversion has to be added to the sequence. This information removes ambiguity and validates the first extension in Subsection IV.F.

6. Before adding the *optional* attribute to the *mpeg21:AudioCapabilitiesType* description (extension 2 in Subsection IV.F), CAIN-21 did not encounter a sequence for Test 1 and Test 2. This happened because the output of the *image_2_video* conversion did not contain this information. This problem has been further described in [32]. Labelling the audio as *optional* (see **Listing 3**) allows for ignoring the audio properties during the computation of the sequence.

VI. MULTIMEDIA ADAPTATION ENGINES COMPARISON

This section provides a comparative review of six multimedia adaptation engines, which operate in the MPEG-21 framework: ConversionLink [10], koMMa [15], BSD [12], DCAF [20] NinSuna [21] and CAIN-21. These engines have been introduced in Subsection II.D. TABLE shows the year of publication that this paper is analyzing.

A. Aspects to compare

The comparison based on six aspects, namely:

1. The automatic *decision-making method* that the engine implements.

2. Whether the engine supports *multi-step* adaptation.

3. Whether the engine provides a *complete-solution*, i.e., finds all the solutions.

4. The *extensibility mechanism* (if any).

5. The *multimedia content* that the engine is prepared to adapt.

6. The *semantic adaptations* that the engine considers.

TABLE II
SUMMARY OF COMPARATIVE OF MULTIMEDIA ADAPTATION ENGINES

	ConversionLink	koMMa	BSD	DCAF	NinSuna	CAIN-21
Year	2005	2007	2008	2008	2010	2013
Decision-making method	Ad-hoc	Knowledge-based	Quality-based	Quality-based	Quality-based	Knowledge-based + Quality-based
Multi-step	No	Yes	Yes	No	No	Yes
Complete solutions	Unspecified	No	Ranking	Ranking	Unspecified	Knowledge-based + Ranking
Extensibility mechanism	Yes	Yes	No	No	Yes	Yes
Multimedia content	Images + Video + Audio	Image + Video	Scalable content	General video	Scalable content	Images + Video + Audio
Semantic adaptation	Scene adaptation	OWL description	gBSD + IOPins	gBSD + IOPins	RDF + gBSD + SOIs	ROI

Subsection II.C divided automatic decision-making methods into quality-based methods and knowledge-based methods. koMMa and CAIN-21 rely on knowledge-based methods, whereas BSD, DCAF and NinSuna rely on quality-based methods. ConversionLink is a generic description engine that does not specify the algorithms used to make the adaptation decisions. BSD and DCAF engines use the notion of Pareto optimality. CAIN-21 also uses quality-based decisions during a second step (see, [18] for a further discussion on how CAIN-21 implements these quality-based decisions). Whereas BSD, Ninsuna and CAIN-21 rely on classical multi-attribute optimisation methods, DCAF exploits genetic algorithms to compute this optimization.

Section I introduced the advantages that multi-step adaptation provides. These advantages are frequently studied in knowledge-based methods. The koMMa and CAIN-21 adaptation engines use these methods. BSD is mainly devoted to performing the adaptation of the scalable resource in one step. Nonetheless, the authors have also studied the problem of distributed adaptation, which corresponds to the idea of multistep adaptation in different nodes.

In reference to completeness, quality-based methods usually obtain a complete solution, i.e., all the feasible solutions are obtained and ranked: this is the case of BSD, DCAF and CAIN-21. More specifically, these engines create a ranking among the available solutions. Well-known quality metrics such as PSNR or VQM [34] are used to create this ranking. Regarding the knowledge-based methods, koMMa only extracts one solution. CAIN-21 analyses all of them using both the knowledge-based and quality-based decision methods. NinSuna and ConversionLink do not specify the completeness of their decisions.

The idea of extensibility appears in ConversionLink, koMMa, NinSuna and CAIN-21. Both the *ConversionLink* and the *CATCapabilities* description tools include the standard *ConversionCapabilities* [9] description tool. The differences between these descriptions were discussed in Subsection IV.C. BSD and DCAF do not examine their own extensibility. NinSuna discusses its extensibility regarding its format independence.

In reference to the supported media, BSD and NinSuna are particularly effective dealing with scalable media, while DCAF deals with general video resources. ConversionLink, koMMa and CAIN-21 are intended to deal with a wider range of media resources. Specifically, ConversionLink and CAIN-21 can manage images, audio and video, whereas koMMa provides adaptation tests involving images and video. The scalable content adaptation implemented in BSD and DCAF is one of the adaptations that CAIN-21 incorporates. Moreover, [18] discusses how scalable video adaptation is carried out inside a CAT called the SVCCAT. The scalable content adaptation corresponds to to the idea of resource conversion in the case of ConversionLink.

In reference to semantic adaptations, *ConversionLink* allows scene level adaptation. It addresses the question of semantic adaptation of documents based on temporal, spatial and semantic relationships between the media objects. koMMa relies on Semantic web Services to describe its adaptation capabilities and to identify the sequence of conversions. BSD and DCAF use the *gBSD* [3] and the *AdaptationQoS* with *IOPins* [3] description tools. *IOPins* are linked to semantics annotating the video stream on a semantic level. NinSuna uses RDF to describe semantic relationships. It also uses the *gBSD* description tools to provide semantic adaptation for the selection of *Scenes Of Interest* (SOIs) as well as for frame-rate reduction. CAIN-21 makes use of *Regions of Interest* (ROIs) inside some CATs such as the *Image2VideoCAT*. Tests involving semantic adaptation in CAIN-21 have been reported to [19].

B. Adaptation approaches comparison

This section describes the reasoning behind the different approaches chosen for the adaptations engines described above.

In the design of ConversionLink it can be observed an effort to create a general MPEG-21 description of multimedia adaptation, but without paying attention to the underlying adaptation algorithms. Several adaptations are described, but they are ad-hoc adaptations of a specific media item, that is, the decision and adaptation methods do not generalize to make them reusable for other media contents or formats without modifying the underlying implementation.

The major contribution of koMMa is to demonstrate how Web Services are able to represent and calculate multimedia adaptation sequences. koMMa studies in depth the semantic description and planning of the sequence, but defers the study and exploitation of the signal level features of the media to be adapted.

The reasoning behind quality-based methods (i.e., BSD, DCAF and NinSuma) is to find the parameters that maximize the quality or utility of the adaptation. These parameters exist or are applicable only to specific media formats (e.g., scalable video), and hence these methods do not aim to accomplish the adaptation of the widest possible range of multimedia formats. Conversely, knowledge-based methods (such as koMMa or the first phase of CAIN-21) focus on the reusability of the adaptation algorithms as a mean to archive the widest possible range of adaptations. With this purpose, knowledge-based methods propose the use of pluggable adaptation tools, and elaborate a decision method that, without human intervention, finds the adaptation tools and corresponding parameters to accomplish each adaptation scenario.

CAIN-21 contributes to the previous ideas by proposing the combination of knowledge-based and quality-based methods in two steps. Firstly, a descriptions-based method that finds all the feasible adaptations, secondly the CATs use media features to select the parameters that maximize the quality or utility of the adaptation.

VII. Contributions and Conclusions

The objective of the Semantic Web is to represent knowledge in a format that can be automatically processed without human intervention. This paper contributes to this objective by introducing the idea of implied and explicit ontology, envisioning MPEG-21 as a implied ontology, and demonstrating how this (pseudo)-ontology is enough to accomplish multimedia adaptation decision-making automatically (i.e., without human intervention in the decision process).

This paper has explained CAIN-21, its extensibility mechanism and the infrastructure to perform automatic adaptation decisions. So, assuming that enough CATs are available, CAIN-21 is capable of managing all content that can be represented as a DI.

As said, CAIN-21 embraces MPEG-21 and shows its good level of expressiveness, as most of new descriptions for concepts and relationships can be represented with this standard. However, this paper has identified and discussed several handicaps in the MPEG-21 description capabilities. As discussed in Section 4, extensions were provided to solve these handicaps. In particular, the paper identifies gaps in the MPEG-21 schema to represent the UED, in the binding modes, in the *ConversionCapabilities* to represent preconditions and postconditions, and in the *mpeg7:Content* description tool.

Another important unique aspect of CAIN-21 is that semantic and quality-based adaptations have been put apart from the knowledge-based decision mechanism and transferred to the CATs. Therefore, our proposal for the knowledge-based decision method makes the adaptation engine independent of the semantics in the content to be adapted. Specifically, the independence is achieved by making decisions according to the media format. Subsequently, quality-based and semantic adaptation for particular content (e.g. soccer, news items) can be integrated inside the CATs. As can be seen in Table , CAIN-21 combines these two major decision-making methods and integrates a complete algorithm, i.e., an algorithm that identifies all the feasible adaptations that produce content satisfying the usage environment constraints.

The paper also includes the reasoning behind and comparison of the different multimedia adaptation decision approaches.

REFERENCES

[1] J. Lachner, A. Lorenz, B. Reiterer, A. Zimmermann, H. Hellwagner. "Challenges Toward User-Centric Multimedia". Proceedings of Second International Workshop on Semantic Media Adaptation and Personalization (SMAP 2007), pp 159-164, London, UK, Dec. 2007.

[2] F. Pereira, I. Burnett. "Universal multimedia experiences for tomorrow". IEEE Signal Processing Magazine, 20(2):63-73, Mar. 2003.

[3] I. Burnett, F. Pereira, R. V. de Walle, R. Koenen (editors). "The MPEG-21 Book". John Wiley and Sons, Apr. 2006.

[4] Authors. "CAIN-21: An extensible and metadata-driven multimedia adaptation engine in the MPEG-21 framework", Lectures Notes in Computer Science, Springer Verlag, vol. 5887, pp. 114-125, 4th International Conference on Semantic and Digital Media Technologies (SAMT 2009), Dec. 2009.

[5] N. Shadbolt, T. Berners-Lee, W. Hall, "The Semantic Web Revisited". IEEE Intelligent Systems. 21(3):96-101, May 2006.

[6] B. S. Manjunath, P. Salembier, T. Sikora (eds.). "Introduction to MPEG-7 Multimedia Con-tent Description Interface". Wiley & Sons, Jun. 2002.

[7] WWW Consortium (W3C). "Multimedia Vocabularies on the Semantic Web". http://www.w3.org/2005/Incubator/mmsem/, Jul. 2007 (last verified Jun. 2012).

[8] WWW Consortium (W3C). "XML Path Language (XPath) Version 1.0", Nov. 1999.

[9] ISO/IEC 21000-7:2005. " Information technology - Multimedia framework (MPEG-21) - Part 7: Digital Item Adaptation, AMENDMENT 1: DIA Conversions and Permissions", Jan. 2005.

[10] M. Kimiaei. "Adaptation de Contenu Multimédia avec MPEG-21: Conversion de Ressources et Adaptation Sémantique de Scènes". PhD. Thesis. Ecole Nationale Supérieure des Télé-communications, Jun 2005.

[11] D. Mukherjee, E. Delfosse, J.G. Kim, Y. Wang. "Optimal adaptation decision-taking for teminal and network quality-of-service", IEEE Transactions on Multimedia vol. 7(3), pp. 454-462, Jun. 2005.

[12] C. Timmerer, "Generic Adaptation of Scalable Multimedia Resources". Verlag Dr. Muller, May 2008.

[13] M.C. Angelides, and A.A. Sofokleous, A game approach to optimization of band-width allocation using MPEG-7 and MPEG-21, Multimedia Tools and Applications, pp. 1-23, DOI:10.1007/s11042-011-0981-0. Mar. 2012.

[14] B. Girma, L. Brunie, J.M. Pierson. "Planning-Based Multimedia Adaptation Services Composition for Pervasive Computing". Proceedings of 2nd International Conference on Signal-Image Technology Internet based Systems (SITIS 2006) pp. 132.143, Hammamet, Tunisia, Dec. 2006.

[15] D. Jannach, K. Leopold. "Knowledge-based multimedia adaptation for ubiquitous multimedia consumption", Journal of Network and Computer Applications, 30(3):958-982, Aug. 2007.

[16] Feng X., Tang R., Zhai Y., Feng Y., Hong B., "An MPEG-21-driven multimedia adaptation decision-taking engine based on constraint satisfaction problem", Proc. SPIE 8878, Fifth International Conference on Digital Image Processing (ICDIP), 2013.

[17] Authors, Bounded non-deterministic planning for multimedia adaptation, Applied Intelligence, Springer, vol. 36:1, pp. 29-60 Mar. 2012.

[18] Authors. Improving Scalable Video Adaptation in a Knowledge-Based Framework. Proceedings of the 11th International Workshop on Image Analysis for Multimedia Interactive Services (WIAMIS 2010), ISBN 978-1-4244-7848-4, Desenzano, Italy, Apr. 2010.

[19] Authors. "Automatic adaptation decision making using an image to video adaptation tool in the MPEG-21 framework". Proceedings of the International Workshop on Image Analysis for Multimedia Interactive Services (WIAMIS 2009), pp. 222-225, London, UK, May 2009.

[20] A. A. Sofokleous, M. C. Angelides, "DCAF: An MPEG-21 Dynamic Content Adaptation Framework". Multimedia Tools and Applications, 40(2):151-182, Nov. 2008.

[21] D. Van Deursen, W. Van Lancker, W. De Neve, T. Paridaens, E. Mannens, R. Van de Walle, "NinSuna: a fully integrated platform for format-independent multimedia content adaptation and delivery using Semantic Web technologies". Multimedia Tools and Applications, 46(2):371-398, Jan. 2010.

[22] J. Lee, "Scalable continuous media streaming systems", Ed. Wiley. Jul 2005.

[23] P. van Beek, J.R. Smith, T. Ebrahimi, T. Suzuki, J. Askelof. "Metadata-driven multimedia access". IEEE Signal Processing Magazine, 20(2):40-52, Mar 2003.

[24] WWW Consortium (W3C), "XPointer xpointer() Scheme", http://www.w3.org/TR/xptr-xpointer/, Dec. 2002 (last verified Jun. 2012).

[25] J. Bormans, J. Gelissen, A. Perkis, "MPEG-21: The 21st century multimedia framework". IEEE Signal Processing Magazine, 20(2):53-62, Mar. 2003.

[26] S. De Zutter, R. Van de Walle, "Enhanced Quality of Experience in Heterogeneous Environments from a Content Author's Perspective". Proceedings of Sixth FirW PhD Symposium, Ghent University, Nov. 2005.

[27] [M. Einhoff, .J. Casademont, F. Perdrix, S. Noll. "ELIN: A MPEG Related News Framework". Proceedings of the 47th International Symposium (ELMAR 2005), pp. 139-140, Zadar, Croatia, Jun. 2005.

[28] Authors. Towards a fully MPEG-21 compliant adaptation engine: complementary description tools and architectural models. Lectures Notes in Computer Science, Springer Verlag, vol. 5811, pp. 155-169, 6th International Workshop on Adaptive Multimedia Retrieval (AMR 2008), Jun. 2008.

[29] L. Rong, I. Burnett, "Facilitating Universal Multimedia Adaptation (UMA) in a Heterogeneous Peer-to-Peer Net-work". Second International Conference on Automated Production of Cross Media Content for Multi-Channel Distribution, 2006. AXMEDIS apos;06. pp. 105-109. Dec. 2006.

[30] ISO/IEC 21000-18:2008 FDAM 1"Information technology - Multimedia framework (MPEG-21) - Part 18: Digital Item Streaming", Mar. 2008.

[31] Apache Java Xalan XSLT Processor. Available online at http://xml.apache.org/xalan-j/ (last verified Jun. 2012).

[32] Authors. Automatic adaptation decision making in the MPEG-21 framework: mechanisms and complementary description tools. Proceedings of the 11th Inter-national Workshop on Image Analysis for Multimedia Interactive Services (WIAMIS 2010), ISBN 978-1-4244-7848-4, 12-14 April, Desenzano, Italy, Apr. 2010.

[33] Authors. Instant Customized Summaries Streaming: a service for immediate awareness of new video content. Lectures Notes in Computer Science, Springer Verlag, vol. 6535, pp. 24-34, (ISBN 978-3-642-18449-9), 7th International Workshop on Adaptive Multimedia Retrieval (AMR 2009), Madrid, Spain, Sep. 2009.

[34] C. Timmerer, V.H. Ortega, J.M. González, A. León, "Measuring quality of experience for MPEG-21-based cross-layer multimedia content adaptation", ISBN: 978-1-4244-1967-8, Proceedings of the 2008 IEEE/ACS International Conference on Computer Systems and Applications (AICCSA 2008), pp 969-974, Doha, Qatar, Apr. 2008.

Analysis of Log File Data to Understand Mobile Service Context and Usage Patterns

Bernhard Klein, Ivan Pretel, Sacha Vanhecke, Ana B. Lago and Diego Lopez-de-Ipiña,
DeustoTech –Deusto Institute of Technology, University of Deusto, Bilbao, Spain
Department of Electronics and Information Systems, Ghent University, Belgium

***Abstract* — Several mobile acceptance models exist today that focus on user interface handling and usage frequency evaluation. Since mobile applications reach much deeper into everyday life, it is however important to better consider user behaviour for the service evaluation. In this paper we introduce the Behaviour Assessment Model (BAM), which is designed to gaining insights about how well services enable, enhance and replace human activities. More specifically, the basic columns of the evaluation framework concentrate on (1) service actuation in relation to the current user context, (2) the balance between service usage effort and benefit, and (3) the degree to which community knowledge can be exploited. The evaluation is guided by a process model that specifies individual steps of data capturing, aggregation, and final assessment. The BAM helps to gain stronger insights regarding characteristic usage hotspots, frequent usage patterns, and leveraging of networking effects showing more realistically the strengths and weaknesses of mobile services.**

***Keywords* — Mobile services, technical acceptance model, log-data analysis, human-computer interaction**

I. INTRODUCTION

USER acceptance in field trials is still mostly evaluated through questionnaires and focus interviews. Mobile applications are, however, much stronger related to real mobile behaviour as people carry their devices with them. Because of the dependency of mobile applications' functionalities on the user situation answers to general questions about the application can often not easily be given.

A relative new approach for mobile services is the observation of application usage through data loggers. A data logger records application events or errors jointly with other usage or system related metadata. To support daily activities successfully, mobile applications should not interrupt the activities, provide a reasonable interaction/benefit ratio to the user, and provide community leveraging beyond exploitation of personal experience. Analysing usage hotspots, usage frequency and usage type allow researchers to speculate about potential strengths, weaknesses or even problems for the surveyed service.

In this work, we present a behaviour assessment framework that describes the systematic collection of behaviour data and guides researchers in their log data analysis. With such an analysis approach researchers can gain more insights about first and long term service impressions, acceptances issues correlated with the user experience and the success of subsequent product evolvement steps.

The article is structured as follows. The next section discusses related works. Section 3 describes the method how to capture, aggregate, and represent data. In Section 4 the behaviour assessment model is defined. A preliminary case study is given in Section 5. Section 6 concludes the article.

II. RELATED WORK

In order to perform a study focused on mobile services the first step is to compare, analyse and find the differences and connections between existing data loggers, concepts and conclusions related to the mobile services evaluation field.

Lab-based evaluation frameworks log information in a controlled environment using specific devices and specific users. The main advantages of the lab-based frameworks are the highly controllable environment and the collection of data, which is cheap and easy. However, the context, which is the most influential factor in the mobile services field, is not considered and it can hardly be simulated. Many simulation tools produce highly inaccurate results because of the context. Furthermore, several agents also alter the results of user experiments. The experts who lead the experiment and the tasks performed by the users can not only alter the execution of experiments but also evoke situations that would never happen in real environments. The users may also add biased results during the execution of the experiments [1] because they suffer several problems such as test-anxiety [1]: during the task performance the highly test-anxious person divides his attention between self-relevant and task-relevant variables; due to the self-focussed attention the user of the mobile service may not show real behaviour. Further, in many tasks such as phone calls, it would be subjectively annoying for many users to be in a room with observing researchers.

On the other hand the field-based evaluation frameworks (see Table 1) capture information in real environments. They commonly use added cameras and human observers to capture information from the interactions. Furthermore, this kind of framework tries to bring the lab to the field. For example, the Usertesting platform [2] not only brings methods like the think-aloud verbal protocol but also records the user's feedback with a webcam; finally it reproduces the interaction again enabling the annotations during it. Using this kind of techniques means that although the task is performed in real environment, it is changed and consequently, the interaction altered. Another tool related to Usertesting is the Morae Observer [3] tool. It captures all the interaction data and indexes it to one master timeline for instant retrieval and analysis; it generates graphs of usability metrics. Both tools are focused on the interaction because they are centred on capture of screen interaction and the user's feedback through filming the face or recording comments. Another group of tools such as ContextPhone [4] and RECON [5] are focused on the context capture. They capture the surrounding environment through mobile sensors. This capturing technique retrieves a lot of real data without influencing the interaction but the user's feedback is lost. In order to fill the lack of the user's feedback other tools like MyExperience [6] and SocioXensor [7] use techniques like self-reports, surveys and interviews mixed with the context capture. These tools are quite powerful and flexible because the user has at any time the complete control about when participate in an application acceptance survey. In case, he has been interrupted in the survey he can resume it to a later point of time.

To sum up, to acquire valid interaction data about mobile services, it is essential to capture objective information to solve questions like when, where, how long, etc. users are really interacting with a service. These questions can hardly be determined with a lab-based framework. The field-based evaluation frameworks can provide deeper and more objective information, but the added agents such as cameras and invasive evaluation methods (e.g. think-aloud verbal protocols) have to be removed. In order to do so, the best way to capture interaction data is by registering information through a mobile device using a tiny capture tool. This tool should log the context via the built-in mobile sensors and logging the key interaction events.

III. Mobile Service Assessment through Behavior Analysis

A framework for automatically logging and processing data for evaluation has been developed. In the following we briefly explain the different behaviour capturing and aggregation phases and the architectural requirements.

A. Data Logging and Aggregation Overview

As can be seen in Fig. 1 the framework distinguishes four main phases:

1. Data Capture: A data logger component installed separately on the mobile device records event and error data triggered by the mobile service. Examples for logging data are: service start and stop times, UI events e.g. buttons pressed, screen transitions, any changes in settings and erroneous data entries, exceptions and any unexpected system behaviour. These data are complemented with additional user contexts (e.g. provider and subscriber data), service information (e.g. queries/results, content data, screen stay duration) and device contexts (e.g. location data) for further evaluation.

TABLE I
Properties for the Logging Tools

Tool	Capture Techniques	Data	Report
Usertesting	Screen, webcam and microphone	Interaction, user information and user's feedback	Reproduce the screen interaction
Morae Observer	Screen, webcam and microphone, observer	Interaction, user information and user's feedback	Reproduce the interactions and calculate graphs
ContextPhone	Mobile sensing and interaction event logging	Interaction, device status and environment	Mobility patterns detection
RECON	Interaction event logging and mobile sensing	Interaction, device status, user information, user's feedback, and environment	Trace Data analysis Engine
MyExperience	Wearable hardware sensing, mobile sensing, audio recording and user surveys	Interaction, device status and environment	Performance analysis, SMS usage and mobility analysis
SocioXensor	Interaction event logging, survey, interview	Interaction, user, device status and environment	SQL database

2. Transfer Protocol: Logging data is periodically (e.g. daily) transferred to an analysis component hosted on the Internet. To minimize the influence on mobile service performance the transfer process is only started if the mobile device remains in an idle execution state.

3. Data Aggregation: The analysis component parses the incoming logging data and interprets the raw data log format with a parser. A filter process removes out-of-bound values, spatio-temporal inconsistencies, and entries that do not conform to preset criteria. Following this filtering step the log data are aggregated through clustering analysis.

4. Data Visualization: From the results tables, graphs and diagrams are generated for the researcher. Furthermore, the entire log is automatically annotated so that each entry is

written out for human readability and annotated to get basic derived information such as duration and transitions.

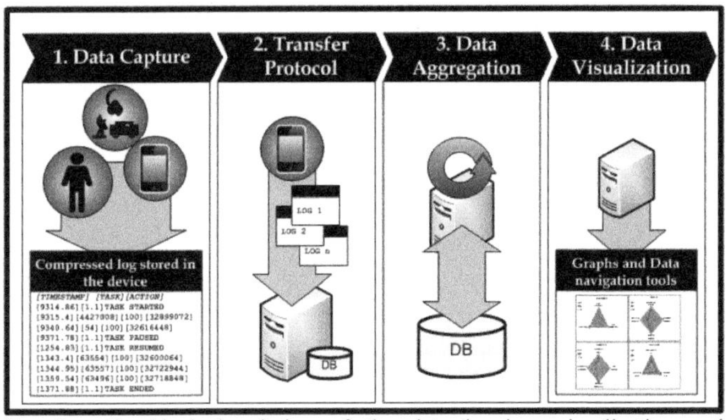

Fig. 1. Graphical description of the data logging, visualization and aggregation process.

B. System Architecture

The Neurona evaluation framework [8] was extended to meet the BAM requirements. This platform shown in Fig. 2 is based on three main components: the mobile device component, the connector component, and the analysis server component.

The Mobile Device component is software installed in the user's mobile and logs user interactions; it is formed by the Logger/App interface, Logger Module and Context Information Module. The Logger/App interface is a tiny software library used to send interaction events to the logger module. The logger module stores the interaction data and shows brief questionnaires about the interaction experience to capture the user's feedback; these questionnaires are shown at the end of the interaction to not disturb the experience. Another element is the context information module, which provides context information acquired from the built-in mobile sensors and the mobile Operative System.

The Analysis Server component is hosted in a web server; this component is formed by the Data Aggregation Module, the Visualization Module, the Applications Manager and the Usergroup Administration. The Data Aggregation Module receives logged data and calculates normalized information to store it in the system database. The expert who wants to check the normalized information can do it using the Visualization Module; which shows advanced graphs. The Applications Manager enables the expert to register into the system, update and remotely configure prototype applications. The Usergroup Administration module registers users and devices, assigns applications and exposes several administration options related to the relations between users, applications and experts.

Finally, the connector between the explained elements transfers the logged information generated by the Mobile Device component to the Analysis Server component. It is divided in two main elements: the mobile interface and the server web service. Basically the mobile interface checks the state of the device and if the user is not interacting with the device it sends logged data to the web service hosted in the Analysis server. To minimize the required transfer bandwidth logging data is encoded in memory saving format and decoded later to a human readable format when the logging data has been received by the Analysis Server.

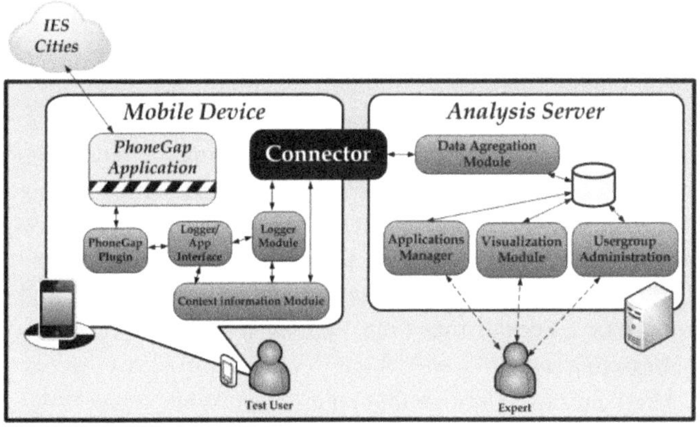

Fig. 2. System architecture of Neurona Tool.

C. Integration with IES Cities through PhoneGAP

For the IES Cities project the Neurona framework has been extended with an additional integration plugin using the PhoneGap tool. The IES Cities mobile applications are developed using the PhoneGap technology. PhoneGap is a framework for building cross-platform mobile apps with standards-based Web technologies. Developers who use PhoneGap may only use HTML, CSS, and JavaScript to implement several applications to remotely access existing backend and cloud processes while the device is connected to the Internet. In particular, thought this framework the developed applications which can access to the IES Cities services.

Among the variables which are relevant to the model presented by this work several sources for completing the model information should be studied as well as the different kind of information sources which can be found inside a mobile device (in particular, in Android OS operating system devices). It is important to mention that this model captures and handles sensitive information (e.g. location of the user). Consequently, applications which use BAM model should ask for several Android permissions: The *INTERNET* permission to open network sockets, check the weather and upload the logged information to the main server. The *ACCESS_FINE_LOCATION* permission to access precise location from location sources such as GPS, cell towers, and Wi-Fi. The *RECORD_AUDIO* permission is used to measure the noise level. With *ACCESS_NETWORK_STATE* our capturer accesses detailed information about networks. Finally, the *ACCESS_WIFI_STATE* permission is used to retrieve information about Wi-Fi networks.

PhoneGap applications cannot access sensors, in order to solve so; a PhoneGap plugin has been developed. PhoneGap Plugins need to be implemented for each platform, to validate the system the data capture in focused on Android platforms. The developed plugin is made up by a JavaScript file which contains functions to call from the IES Cities mobile application and log interaction events, including the current context. There are two main steps to including the developed plugin in IES Cities mobile application: referencing the explained JavaScript file and importing the native code (the extended Neurona evaluation framework library) that will be called through the JavaScript file.

This plugin has only 6 main functions: *log_start_task*, *log_pause_task*, *log_resume_task*, *log_end_task*, *log_interaction* and *log_error*. These functions capture the timestamp of the interaction, the current context and the object with witch user is interacting. They only require three parameters: the event type, which will be explained bellow, the identification of the visual structure for a user interface and last but not least, the object (i.e. buttons, textboxes…) the user is interacting with.

A task can pass through four main states: When a task is not started yet (NOT STARTED), when a task is started and its user is interacting to achieve the goal of the task (STARTED), when a task is started but its user is not interacting to achieve it (PAUSED) and when the task is finally terminated (END). After seeing the main states we will see several events which can be triggered to change the state of a task, these events should be generated by the mobile applications using the plugin functions. During a task performance a user can trigger two main events: START_TASK (at the beginning of the task: *log_start_task*) and END_TASK (at the end of the task: *log_end_task*). Additionally but not compulsory there exist two others: if user leaves the task (e.g. because of an incoming phone call) PAUSE_TASK event (*log_pause_task*) is produced. Where user decides to continue the task RESUME_TASK event (*log_resume_task*) is triggered. When a task is started two events related to the interaction of the user can be triggered. The INTERACTION event (*log_interaction*) means that a user is interacting in the right way. This event should be triggered when a user is achieving little microchallenges inside the goal of the task. The ERROR event (*log_error*) means that a user has made a mistake during the interaction process.

IV. The Behaviour Assessment Model

A proven performance assessment method considering concurrent aspects has been the Balanced Scoreboard (BSC) approach. Aligning each of the dimensions systematically helps get a better impression about different influencing factors.

A. Dimensions of the Behaviour Assessment Model

This leads to following six different dimensions illustrated in Fig. 3:

1. Remote service search: This dimension is based on the categories Planned Execution Scenario and Service Actuation. In order to fulfill end-users need to plan activities ahead of a trip; users require the capability to explore the service offer according to given properties. The retrieval quality depends on the query power e.g. different search concepts and the query success rate. An example is a map based discovery tool, which retrieves services according to locations selected on a map.

2. Nearby service discovery: The dimension founds on the categories Spontaneous Execution and Service Actuation. As mobile services are much stronger correlated with the daily life of end-users an important requirement is to raise their attention to an adequate service offer in a seamingless manner. A successful implementation depends on the reasoning power (that compare the current users' context and the intended service context) and the number of directly consumed services (reasoning success).

3. Service creation/provision: The categories Planned Execution and Service Interaction define this dimension. Complex mobile services require often too much knowledge from the user to execute them easily on the spot. Therefore, services should offer any type of service creation, personalization or reservation functionality so that they can be consumed better in time constrained situations. The editing complexity and the service content quality are important indicator examples to determine this dimension.

4. On-the-spot service consumption: The dimension is constructed through the categories Spontaneous Execution and Service Interaction. Since users on the move often follow other real-world activities it is important that the attention needed to execute the service is kept to an absolute minimum. The navigation complexity (effort) and the quality of the content provided by the service are important indicator examples.

5. General platform activity services: This dimension stems from the categories Planned Execution and Central Provision. All general service aspects influencing the provision quality e.g. power consumption and error handling account for this dimension.

6. Community services: This dimension is founded on the category Spontaneous Execution and Community Networking. Tools that consider community behaviour can help in structuring the knowledge space further and lead to more transparency in the community. Examples are best-of ranking lists, member reputation lists and content recommender systems. For instance car sharing opportunities can be more easily evaluated by users and improve their selection. Suitable example indicators are the lurker ratio (active community participation) and the degree of community transparency achieved with previously mentioned community services.

	Service Actuation	Service Interaction	Service Networking
Planed activity execution	Remote Service Search	Service Creation/ Provision	General Platform Services
Spontaneous activity execution	Nearby Service Discovery	Spontaneous Service Consumption	User Ratings & Community Services

Fig. 3. Dimensions of Behavior Assessment Model.

B. The Balanced Scoreboard Assessment Approach

These six dimensions focus on realistic service usage. This emphasises the valuation of a service by the way how end-users apply services to solve given problems. Such behaviour patterns have the potential to tell us about underlying reasons why specific service fail or become well accepted. Recording such behaviourally relevant data also allow the emulation of service usage in respect to given user's context. Both aspects are important for developers to continuously improve the service. According to the BSC approach, the intention is to find a few aggregated indicators that quantify a given dimension. The indicator must meet the requirements of reasonability and measurability. A general problem of social surveys is to translate the indicators into precise measures. The abstract classes of measurement types, correspond hereby with different event and error logging data types. To achieve comparability between different numerical scales of measurements e.g. an event/error frequency scale, a function has to be defined which maps selected scale areas on specific quality rating values. Since humans perceive the influence of various indicators for a given dimension differently, weight coefficients are used to balance the influence of individual indicators. Both mapping function properties and weight coefficients can be obtained through a profiling questionnaire prior to the field trials.

Finally, the results of an analysis and evaluation are typically held in a spreadsheet for detailed analysis and visualised by a radar chart for a summarised representation (see Fig. 4). For visualization by a radar chart, the six dimensions are equally arranged. The scaling is adapted appropriately according to the distribution of the measurement results with its positive orientation towards the origin. For a better visualisation of the consequences of the results, each scale can be subdivided in fulfilled (positive centre areas), and not fulfilled (negative edge areas).

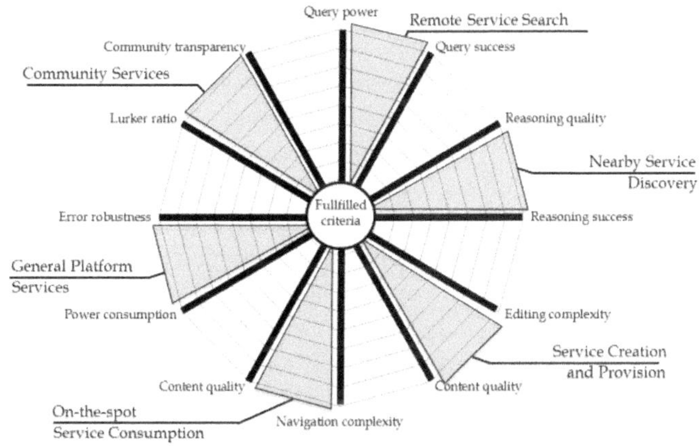

Fig. 4. Visualization of the behavior model with six dimensions (grey color) and example indicators (black color).

V. The IES Cities Project

The IES Cities project aims at providing an open platform so that users are able to use, produce and provide information for internet-based services improving their knowledge about the cities they live. The project represents an extension of other European projects (see m:ciudad, MUGGES, and uService) focusing on service creation, mobile peer-to-peer services, and the integration of open linked data sources. Open linked data refers here to a concept of publishing structured data e.g. pollution data provided by the local government so that it can be interlinked and become more useful. City services, created and provided through the platform enable citizens to wrap this type of data. They adapt to the current context of a citizen through smartphone embedded sensors. End-users may not only to receive information from the city but also generate real-time content which complements, enriches and updates the data available through the open data model associated to the municipality. These services are intelligent as they come with business logic to automate the management of parking spaces, public transport, pollution, health states and many other aspects of urban life, user interactions are eased and automated. In order to facilitate its usage the platform will be provided as mobile and web-based platform. It is assumed that sensor networks are already deployed in the cities and become integrated as part of the project.

In the following the screen interaction model for the *IES-Improve Your Neighbourhood* service is exemplarily explained (see Fig. 6). This screen model includes screen sequences for querying the appropriate IES services (dashed box), creating and consumption of IES services and the service/report rating. In order to create or view *IES-Improve Your Neighbourhood* reports users first have to query for adequate IES service. After selecting the *IES-Improve Your Neighbourhood* service, users can decide a) to create a new report, b) look for and c) rate existing reports. The IES service obtains user data through the screen, retrieve sensor data to personalize IES service and open linked data to provide up to date content.

VI. Case Study of the IES Cities Prototype

A. The Assessment Process

Applying the BAM analysis technique requires specific preparation steps. These include the definition of indicators for each dimension, correlating them with available logging data, appropriate balancing of these measurements with weight factors, the execution of field trials and representing the results. Fig. 5 gives a complete overview about the evaluation process.

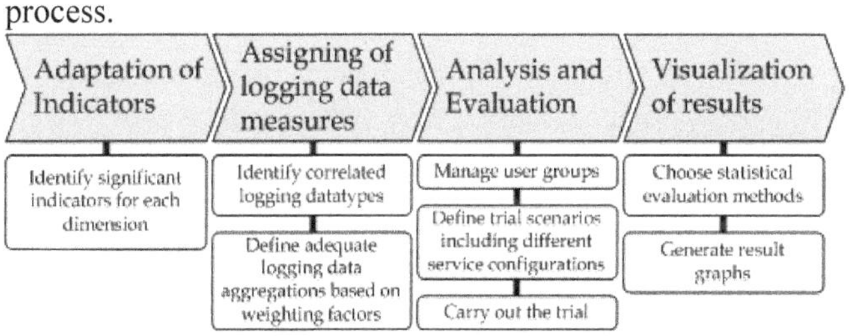

Fig. 5. Different phases of the assessment process and correlated activities.

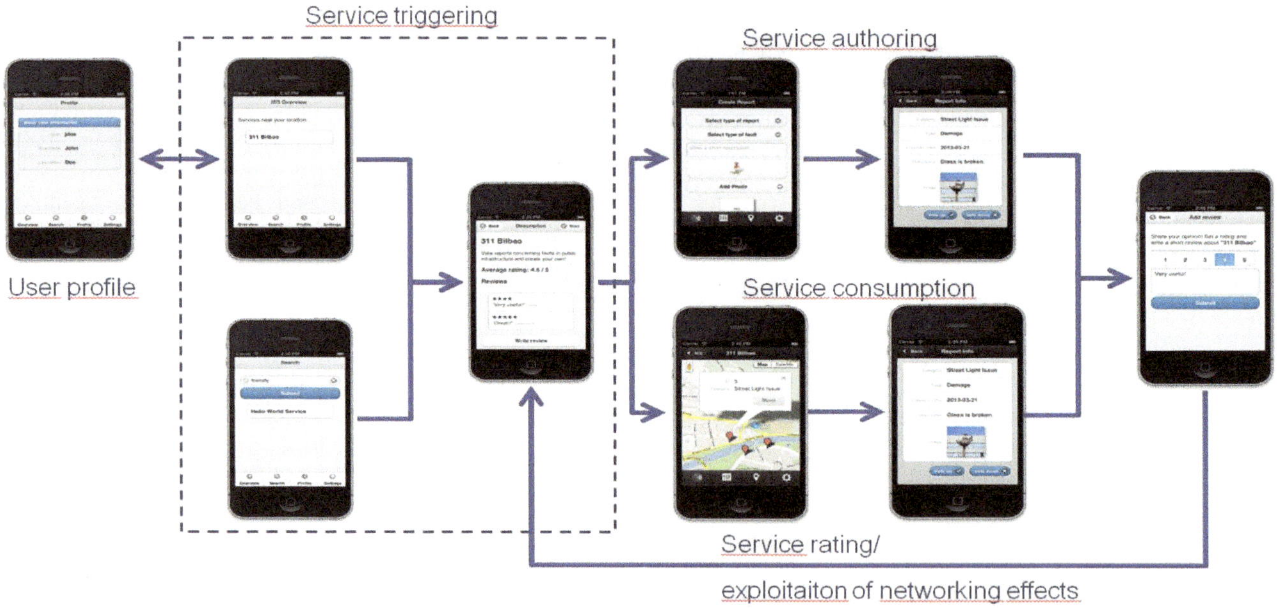

Fig. 6. Screen flow graph of IES – Improve Your Neighbourhood Service

In the following specific modifications to the BAM model in respect to the IES Cities project are presented:

- Adaptation of indicator structure. First, adequate indicators have to be identified which align with the specific dimension of the BAM model. The IES service interaction model considers six different behavior dimensions in accordance with the BAM model. These include activites to trigger city services through seach and discovery functions, activites for the service creation, provision and consumption and finally rating functions to exploit further networking effects on the service and report level.
- Assignment of logging data types. In this phase, logging data types are correlated with the indicators. During the prospective field trial, the Neurona Framework will obtain application and sensor data by plugging into the IES player through the PhoneGAP tools.
- For the dimension service remote search and nearby service discovery, the query success rate and the average result quality are relevant indicators. Therefore the number of queries issued before a service is consumed (query success rate) is counted. The result quality can be measured by the average distance between the current user location and the location to which the report has been assigned. Hereby it is assumed that reports related to a given issue closer to the user a more relevant. The dimension service creation/provision and consumption is determined by the indicators editing/viewing complexity and service quality. The editing/viewing complexity is closely correlated with the time a user spends to create/provide and consume services. Therefore, the logger obtains timestamps when ever an editing and consumption screen flow is started and ended. The average duration of the editing/viewing process can later be correlated the process complexity and even reveal if a given process has been interrupted. The dimension General Platform Services is defined by the indicators power consumption (measured as consumed energy units per day) and error recovery quality (reciprocal number of occurred errors per day). And finally, the dimension community services are defined by the lurker ratio (measured as provider-consumer ratio of a consumed service) and the consumed service quality (measured through average rating of consumed services).
- Analysis and evaluation. Generally an iterative evaluation approach is recommended starting from early prototypes up to the final mobile service. In order to compare the development progress it is important not to vary the measurement criteria. It is assumed that the explanatory power of the BAM model increases with a stronger concretization of the mobile service during the development cycle. All these logging data types have different value ranges. For future evaluation it is important to translate these into an given grading scheme. This is usually achieved through mapping functions which assign chosen value ranges to specific grade value. A series of experiments will therefore be conducted in the laboratory as soon as the IES platform has been completed. Parameterized aggregation services at the Neurona framework will then automatically convert the cleaned logging data to the adequate grading format. Since the previously specified indicators contribute to the successful execution of service with varying degree it is important to specify adequate weighting factors. It is therefore mandatory to learn more about the characteristic service behaviour.
- Preparing the result visualisation. The results of the test group are analysed and evaluated with statistical methods and visualised according to the radar graph approach. First, goals for every single indicator should be determined before the actual start of the field trials, in

order to compare these to the empirical results. Then, initial and long-term service usage should be compared in order to identify entry barriers. Correlated with this analysis, is the grouping of logging data according to the technical technical experience (see technical adoption model described by Rogers [12]). By comparing innovators, early adopters, early majority, late majority and laggards a reasonable priority list of future service modifications can be determined. Finally it is also important to analyse logging data from different trials in order to see to what extend applied service modifications have led to an acceptance improvement.

B. Visual Evaluation of IES Cities

We emulate the advantages of the IES Cities project based on logging data obtained during the field trial executed for the MUGGES project. In this field trial logging data from 30 potential end-users have been collected during a two week period. Each study participant was given a mobile phone with the previously developed MUGGES software installed. The users were given specific tasks e.g. to describe preferred pathways by providing reports for favorite point of interests. The connector component transfers periodically event data to the analysis server for further evaluation.

Applying the BAM approach (with a rating range from 0 – very good till 3 very bad) a service provider can come, for example, to the following simplified conclusions concerning the explained dimensions in Section 4:

- Remote activity discovery: With increasingly more created services users applied more sophisticated search approaches (keyword-based and map-based search) to compensate the small screen size. The discovery function seems to work well for the majority of the trial users (rating 1.5).
- Situation-aware activity recommendation: The overall distance between the location assigned for a selected service and the trial user has been quite far (up to 1 km). Besides the sparse distribution of the provided services another reason has been the bad performance of the location technology. Provider could conclude that the recommendation service is not sufficient (rating 2.5) for the current spontaneous usage scenario.
- Mobile activity preparation: The service creation process took a lot of time, not short enough to create mugglets on-the-go. People compensated this by distributing the creation process in several phases. The mobile activity preparation is not sufficient (rating 3.0) in the current development stage.
- On-the-spot activity support: The services in general have high information quality for the user, as they come with a environment map, text descriptions, comments and photos. Above that, the real-time notification feature helped people to stay up-to-date. Provided services thus have been very useful (rating 1).
- General provider services: The peer-to-peer service sharing approach has lead to an high power consumption and the error rate has been quite high. Service provider may conclude that the execution of shared services in the Internet cloud may be a better option.
- User-created services: Platform usage has been high since users could create their own personal service based on the offered service templates. Especially in later stages during the project service ratings have been found very useful (rating 1) to identify popular services or confirm reports.

The radar graph shows some important weaknesses. Recommender systems, the mugglet creation process and the provider infrastructure still make an everyday usage difficult (see Fig. 7). Comparing these logging data results with the questionnaires conducted after the trial backs these findings. But more importantly, user perceptions were not always clear enough to pinpoint the exact problems with the platform infrastructure. The evaluation with the BAM is more differentiated and considers some critical aspects that influence the acceptance of this mobile service significantly.

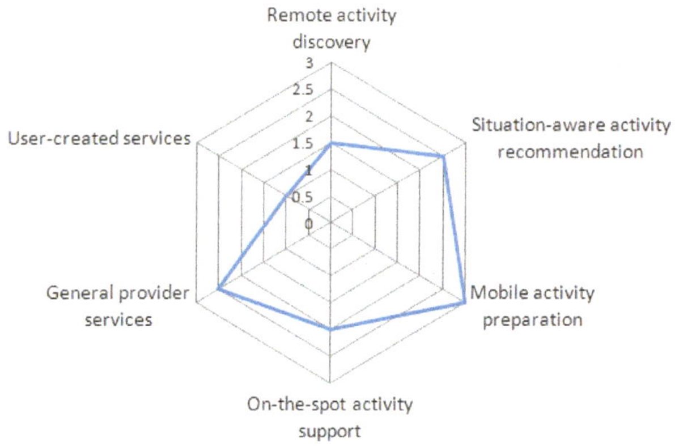

Fig. 7. Evolution of the IES Cities based on MUGGES logging data.

VII. Conclusions and Future Work

We introduced the BAM as an instrument for the analysis and evaluation of the user acceptance for mobile services. The BAM is characterised by a structure that helps to identify systematically a balanced set of important, individually measurable and independent acceptance criteria. The application of the BAM is guided by a process model that supports all phases from the development of acceptance criteria over the measurement of relevant indicators to the evaluation and visualisation of the derived results. Using the BAM reveals several insights:

- *First and permanent usage patterns*. Analyzing the radar graph at the beginning of the trial and later phases of the trial shows can show entry barriers of the mobile service. Results obtained in later stages show how people exploit mobile service strengths but also compensate potential weaknesses of the service.
- *Usage patterns for different technical adoption groups*. According to Rogers technical diffusion model user groups are divided in innovators, early adopters, early

majority, late majority and laggards. Clustering logging data according to these groups may reveal interesting insights how the technical experience influences service usage. These observations are especially valuable to define a priority of feature improvements for the mobile service.

- *Behaviour changes in different product development stages.* As the development of the mobile service evolves comparing results with earlier trials can help to confirm if the applied feature modifications fulfil the intended improvements.

REFERENCES

[1] U.-D. Reips, S. Stieger: Scientific LogAnalyzer: A Web-based tool for analyses of server log files in psychological research. Behavior Research Methods, Instruments, & Computers, 36, pp. 304-311 (2004)

[2] J.C. Cassady, R.E. Johnson: Cognitive Test Anxiety and Academic Performance. In: Contemporary Educational Psychology 27, pp. 270--295 (2002)

[3] UserTesting.com - Low Cost Usability Testing, http://www.usertesting.com

[4] Morae usability testing tools from TechSmith, http://www.techsmith.com

[5] M. Raento, A. Oulasvirta, R. Petit, H. Toivonen: ContextPhone: A Prototyping Platform for Context-Aware Mobile Applications. In: IEEE Pervasive Computing 4, 2, pp. 51-59 (2005)

[6] K.L. Jensen: RECON: Capturing Mobile and Ubiquitous Interaction in Real Contexts. In: Proceedings of MobileHCI 2009, Bonn, Germany. (2009).

[7] J. Froehlich, M.Y. Chen, S. Consolvo, B. Harrison, J.A. Landay: MyExperience: A System for In situ Tracing and Capturing of User Feedback on Mobile Phones. In: MobiSys'07, June 11–14, 2007, pp. 57--70. ACM, San Juan, Puerto Rico, USA (2007)

[8] H. ter Hofte, R. Otte, A. Peddemors, I. Mulder, What's Your Lab Doing in My Pocket? Supporting Mobile Field Studies with SocioXensor. In: CSCW2006 Conference Supplement, Nov. 4-8, Banff, Alberta, Canada (2006)

[9] I. Pretel, A.B. Lago, Capturing Mobile Devices Interactions Minimizing the External Influence. In Proc. of UBICOMM, pp. 200-205 (2011).

[10] Ajzen, Icek (1 December 1991). "The theory of planned behavior". Organizational Behavior and Human Decision Processes 50 (2): 179–211.

[11] M. Colbert, A diary study on rendezvousing: Implications for position-aware computing and communications for the general public, In Proc. of Group Conference (2001)

[12] B. Klein, J. Perez, C. Guggenmos, O. Pihlajamaa, I. Heino, & J. Ser (2012). Social acceptance and usage experiences from a mobile location-aware service environment. In J. Ser et al. (Eds.), Mobile lightweight wireless systems (Vol. 81, p. 186-197). Springer Berlin

[13] E. M. Rogers, Diffusion of Innovations, Glencoe: Free Press (1962).

10

A REST Service for Triangulation of Point Sets Using Oriented Matroids

José Antonio Valero Medina and Ivan Lizarazo Salcedo

Francisco José de Caldas District University, Colombia

Abstract — **This paper describes the implementation of a prototype REST service for triangulation of point sets collected by mobile GPS receivers. The first objective of this paper is to test functionalities of an application, which exploits mobile devices' capabilities to get data associated with their spatial location. A triangulation of a set of points provides a mechanism through which it is possible to produce an accurate representation of spatial data. Such triangulation may be used for representing surfaces by Triangulated Irregular Networks (TINs), and for decomposing complex two-dimensional spatial objects into simpler geometries. The second objective of this paper is to promote the use of oriented matroids for finding alternative solutions to spatial data processing and analysis tasks. This study focused on the particular case of the calculation of triangulations based on oriented matroids. The prototype described in this paper used a wrapper to integrate and expose several tools previously implemented in C++.**

Keywords — **Mobile devices, oriented matroids, spatial data representation, services, triangulation.**

I. Introduction

In the context of spatial data representation, triangulation of a set of points is essential for many applications. A typical example is the generation of triangulated irregular networks (TINs) to represent ground topography. Another example is the representation of spatial objects using *simplicial complexes* which allows breaking complex objects into simpler underlying geometries. In the case of complex objects in the two-dimensional Euclidean space, simple geometries such as triangles provide a close approximation of spatial objects.

Whatever be the purpose of triangulation, a computational solution is necessary to perform an optimal calculation. Traditional algorithms are based on strategies that rely on geometric representation both of the point set and the Euclidean space embedding it. That is the case of greedy algorithms ([1]) which build triangles using an iterative process based on three criteria: (i) selecting the minimum distance between pairs of points, (ii) not revisiting diagonals that have already been visited, and (iii) testing intersection of new diagonals with those previously obtained. This algorithm has $O(n^3)$ time complexity.

A proposal for simple polygons [2] breaks down the input polygon into monotonous chains to subsequently perform triangulation of each sub-polygon, thus obtaining $O(n)$ time complexity. Another proposal initially finds the Voronoi diagram [3] associated with the set of points, and then calculates the Delaunay triangulation. All of the above cited proposals use traditional geometric representations leading to algorithms suffering from a significant computational complexity. However, in recent decades, there has been an interest for using alternative mechanisms to improve computational efficiency for spatial data processing. [4] has studied the particular case of triangulation of a set of points based on oriented matroids, who implemented several tools using the C++ language.

This article describes the implementation of a prototype REST service for triangulation of a set of points obtained by global positioning system (GPS) receivers. This application targets mobile devices such as tablets and cell phones, using Android OS. The prototype wraps components generated by [4] with the purpose of testing oriented matroids as a means to solve problems in which spatial location matters.

The next section of this paper introduces basic concepts on matroids and oriented matroids as well as the proposal of [4] for triangulation of a point set using only purely combinatorial oriented matroids. The third section describes REST services based on entities that are stored in a persistent repository. This section also makes a description of the application development architecture for Android environment, based on the Eclipse IDE.

The fourth section discusses the architecture of the system and its scheme of access and operation. The fifth section makes a presentation of an overview of the used tools. The sixth section presents a prototype test and the last section formulates conclusions and final remarks.

II. From the Geometric to the Combinatorial

This section introduces basic concepts of oriented matroids. For a more comprehensive treatment, interested readers are referred to [5].

A. Matroid

The concept of matroid was introduced by [6] in the article

entitled "On the abstract properties of linear dependence". Whitney describes the approach with respect to a given matrix and its columns, in such a way that any subset of these columns forms a matrix with a particular range. Considering the columns as abstract elements, a matroid with range given by the number of linearly independent columns is formed.

According to [**¡Error! No se encuentra el origen de la referencia.**] from a matrix over any field, it is possible to define a matroid. In particular, the interest from a computational point of view is on finite fields, such as the Galois fields $GF(p^k)$, which give exactly p^k elements when p is a prime number. With $k = 1$ the $GF(p)$ field can be seen as the set $\{0,1,2,\dots,p-1\}$ with operations of addition and multiplication module *p*. The obtained matroid is called a *vector matroid.*

The concept of graphic matroid is introduced in [7] based on the set of edges of a graph and the set of subsets of arcs of the same graph that does not contain all the arcs in any cycle, with the peculiarity that non isomorphic graphs may have isomorphic matroids.

The formulation of affine matroid [8] is based on the concept of affine independence against affine sets (in $\mathbb{R}^2$ they are the empty set, the points themselves, the straight lines and the plane itself); thus, given a set of points in the plane and a set of affine independent subsets of them, there is a matroid called *affine matroid.*

B. Oriented Matroids

According to [8], geometry of matroids is based on what affine or linear sets provide but it lacks a structure of convexity. There is neither notion of duality or intermediation between the points of a straight line, nor existence of hyperplanes separating the space into two half-spaces. According to [8], vector spaces do not have enough structure to support a theory of convexity; therefore, it is not sound to expect matroids to do it. Authors state the need of providing an additional framing (i.e. an orientation) to the matroids from which the convexity may arise. In a nutshell, that is the theory of oriented matroids.

In [9], it is contended that the oriented matroid of a finite set of points draws information from the relative position and orientation from the configuration, which can be provided by a list of signs that encode the orientation of all its bases. Although in the passage from a specific point's configuration to its oriented matroid, metric information is lost, many of the structural properties have their counterpart in the combinatorial level of oriented matroid. That is to say that oriented matroids describe the structure of incidence between the elements of the matroid and their respective hyperplanes, as well as they encode the position of the elements relative to the hyperplanes; i.e. which items fall on the positive side, which ones on the negative side and which ones inside the hyperplane [10].

C. Computational calculation of triangulations

In [4], it is introduced a procedure to perform triangulation from a set of points based only on the combinatorial structures of the associated oriented matroid. Authors state that triangulations are a basic means for decomposing complex objects into simpler ones. According to [4], for a configuration $\mathcal{A}$ with n $d-dimentional$ points, a T subset of subsets of $\mathcal{A}$, each one made of $d+1$ elements, is a triangulation if and only if

$$\cup_{\sigma\in T} conv\ \sigma = conv\ \mathcal{A} \tag{1}$$

and

$$conv\ \sigma \cap conv\ \sigma' = conv(\sigma \cap \sigma')\ \forall\sigma,\sigma' \in T \tag{2}$$

The contribution of [4] is that, instead of using expensive linear programming with exact arithmetic to ensure the two aforementioned conditions, authors use purely combinatorial checks based on the oriented matroid $\mathcal{A}$. To check whether the first condition is met the set of all circuits of $\mathcal{A}$ $is\ used$ and it is verified that for every pair of subsets of $\sigma,\sigma' \in T$ there exists a circuit $(Z^+,Z^-) \in \mathcal{A}, Z^+ \subseteq \sigma\ y\ Z^- \subseteq \sigma'$.

On the other hand, to check purely combinatorial fulfillment of the second condition, the set of all co-circuits of $\mathcal{A}$ is used. There is a co-circuit for each affine hyperplane $(d-1)-dimensional$ extending by subsets of $\mathcal{A}$, which includes all points of $\mathcal{A}$ that are on the positive side (C^+) and all the points of $\mathcal{A}$ that are on the downside (C^-) of the hyperplane. For final checking of the second condition, it is verified that, for each $\sigma \in T$, each cocircuit of $\mathcal{A}$ contained by itself, whose sets C^+ and C^- are both not empty, must have a $\sigma' \in T$ that also contains itself.

In accordance with [4], as long as circuits of $\mathcal{A}$ determine in a purely combinatorial fashion their co-circuits and vice versa, and since the matroid of $\mathcal{A}$ is defined by anyone of such co-circuits, the number of possible triangulations of $\mathcal{A}$ set depends only on its oriented matroid. The passage from the geometry to the combinatorics associated with $\mathcal{A}$ is established through the concept of its chirotopo [11] [12].

$$\chi: \begin{cases} \binom{\mathcal{A}}{d+1} \rightarrow \{+,-,0\} \\ (i_1,i_2,\dots,i_{d+1}) \mapsto sign\left(\det\left(a_{i_1},a_{i_2},\dots,a_{i_{d+1}}\right)\right) \end{cases} \tag{3}$$

A chirotopo assigns its orientation to each ordered base of $\mathcal{A}$. The circuits of $\mathcal{A}$ can be calculated based on all subsets of $d+2$ elements of $\mathcal{A}$ and its cocircuits, using all its subsets of d elements and the calculation of the associated chirotopos. In [4] proposes several algorithmic solutions for triangulation, ranging from obtaining one of the several possible triangulations to obtaining all possible number of triangulations, by using only combinatorial structures associated with the oriented matroid of the points set. These algorithms were implemented in the package TOPCOM (Triangulations of Point Configurations and Oriented

Matroids) looking for a minimum time complexity and a maximum efficiency [4].

III. Technological Development Environment

This section discusses web development concepts useful for using oriented matroids as alternative means for representation and implementation of computational solutions involving spatial location of data.

The main reason for adopting a Web implementation was to provide high availability for the triangulation functionalities proposed by [4]. In a similar way, the choice of the mobile devices environment with Android support leaned on the idea of having a source of sets of points that would be highly available. As it is well known, using a mobile device such as a GPS enabled tablet or a cell phone is something quite common these days.

A. REST webservices

Possibilities for providing online Web capabilities are very different, including solutions based on APIs such as Sockets, XML-RPC or RMI-based components. This article used Web services REST - REpresentational State Transfer [13], as a platform for implementation that provides transparent access to resources possibly including persistent repository storage.

REST was originally introduced as an architectural style for building systems hypermedia distributed on a large scale. The architectural style REST rests on four principles:

--Identification of resources via URIs which outline a set of resources identifying which elements customers interact with.

--Uniform interface which allows user to manipulate the resources using a predefined set of operations: *create* (PUT), *read* (GET), *update* (POST) and *delete* (DELETE).

--Self-descriptive messages, through which resources are uncoupled from their representation to allow content's access in a variety of formats (XML, JSON, etc.).

--Interactions with state through hyperlinks are enabled causing that resource's interactions be stateless, i.e. request messages are self-contained.

The REST services build on well-established W3C protocols (HTTP, XML, etc); thus, their creation should be fairly simple. Effort to building REST services clients is low since testing can be done using any Web browser.

On the other hand, since messages are contained in the URI, a major constraint arises when the associated data set is large enough to exceed the maximum size to be considered a well formed URI.

REST constrains the interface of a resource to its generic uniform interface with predefined operations, and there is very little to choose in terms of available operations. Therefore, a lot of effort should be put on defining which resources need to be exposed. Furthermore, it is necessary to assess whether the four operations are applicable to each resource exposed. In addition, it is also necessary to establish what is the application's semantics associated to each individual resource.

Since REST services are built directly over the HTTP transport protocol, nothing has to be decided about the communication protocol to use. On the other hand, REST has no preset format to adjust the data with. As this issue can be negotiated, it is possible to use different formats such as XML, JSON, or even the SOAP itself.

B. Development environments for Android

Android is undoubtedly one of the most popular platforms for mobile devices in the world such that the manufacturers of this devices use it in a very high proportion. Android is an operating system based on Linux, an open, free and cross-platform operating system, so there are a very nice set of tools for application development. Android provides all the necessary interfaces for developing applications that require access to the basic functionality of mobile devices (i.e. network interface, GPS, etc.) in the Java programming language.

All of the functionalities required in the development of an application for Android are available in its Software Development Kit (SDK). As alternative it is possible to use libraries released by Google as part of its Google Play Services.

The primary IDE environment is the open source tool Eclipse, which also has a plugin, which provides an Android image, with which it is possible to perform application development for Android on any operating system using this emulator for testing [14].

Each Android application uses an own process identified by its ID, and it is the only process accessing user files. The devices have a unique focus, the main application, which is the application visible on the screen, but they can run several applications in the background, each one with its own stack of tasks. The pile of tasks is the sequence of execution of processes in Android. They consist of activities that are stacking as they are invoked, and can only be terminated when the tasks above are completed.

IV. Web Service for Triangulation of Sets of Points

As mentioned above, the prototype service of point set triangulation was planned in order to be a highly available service, so it was arranged as a Web service REST with the access mode shown in **Fig. 1**.

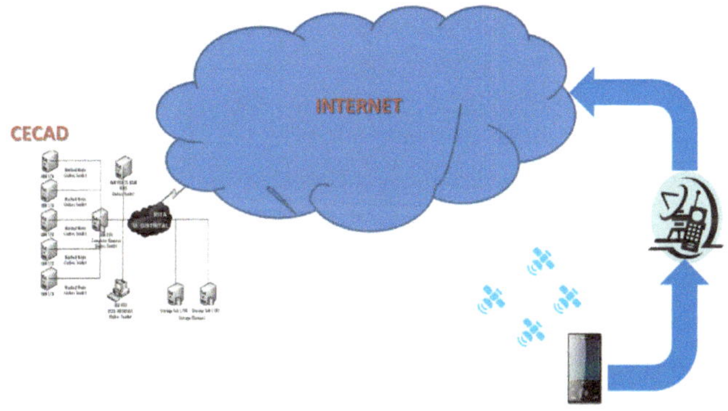

Fig. 1 Access mode of triangulation service based on oriented matroids.

The REST web service was deployed on GlassFish in the Center for High Performance Computing (CECAD) at *Francisco José de Caldas District University*. The service can be accessed from any mobile device connected to the internet via a provider of cellular mobile data, or discovered through a wireless network. The mobile device should have a good coverage of GPS satellites to guarantee a minimum level of accuracy and availability of geographical coordinates.

Fig. 2 shows the components architecture diagram of the system following [15]. For a set of points obtained from the mobile device, it was created a REST server with two REST services from Entity Classes based on the repository defined on the service provided by Java Persistence. *ConjuntodepuntosFacadeREST* is a REST service responsible for generating a group ID of points per session triggered by a connected mobile device with the *read* operation (GET) enabled. This operation returns a list of points formed by a single point with identifier and coordinates dummy but with the new group ID. The reading based on identifier operation (GET {id}) for this REST service was also adapted to allow any connected client to obtain the associated set of points saved to the repository within a given time interval.

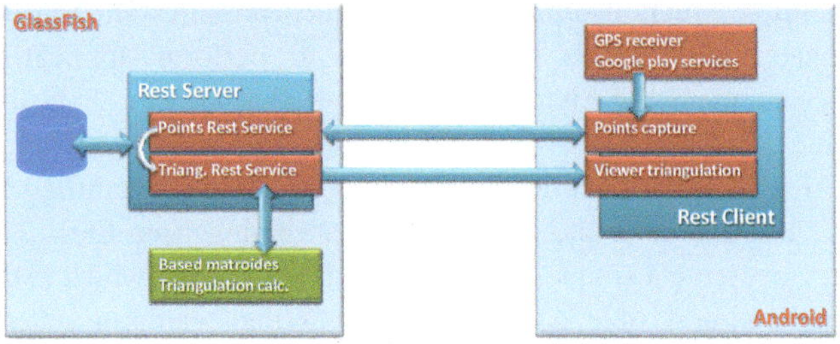

Fig. 2 Triangulation service based on oriented matroids architecture diagram.

TriangulacionesFacadeREST is a REST service created to get the possible triangulations from the current set of points. For this REST service, it was adapted the reading operation based on identifier (GET {id}). Thus, a connected client can request all possible triangulations at any time. For the calculation of the triangles, the REST service was supplemented with a thread responsible for carrying out the wrapper of the *points2triang* component of the software implemented by [4].

The legacy *points2triang* component receives the set of points by its standard input and delivers the calculated triangulations (if it is possible to find any) by its standard output. In order to enable the communication between this component and *TriangulacionesFacadeREST* service, the standard input and output of the component was respectively redirected from and to two auxiliary files. The former is created by the REST service with the coordinate transformation (from the geographic system coordinates to Transversal Mercator) of the points set that the mobile client provide. The latter is generated by the legacy component and used by feed backing the *TriangulacionesFacadeREST*.

Finally, *TriangulacionesFacadeREST* returns back to the mobile client a set of edges for the different possible triangulations associated with the set of points in a session with a client.

The client was implemented based on asynchronous tasks in order to reduce the complexity of the main thread and prevent application crashes.

As mentioned above, the client was developed for the Android system using the API provided by Google Play Services to get coordinates obtained from the receiving device's GPS.

V. TOOLS

This section describes tools used for implementation of the prototype.

A. Building of the web service

The Netbeans IDE was used for the construction of the web service (WS-Rest), following a six-step procedure:

--Creation of tables *ConjuntoDePuntos* and *Triangulaciones*.

--Creation of a new project and addition of connection drivers to the database in Derby.

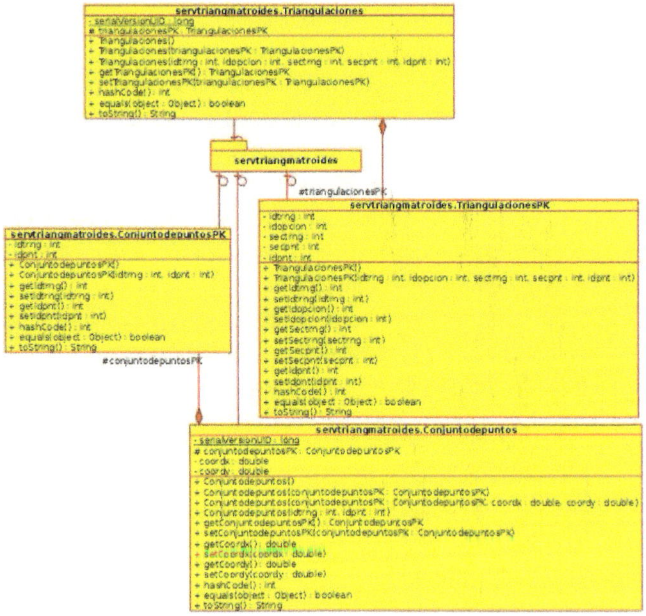

Fig. 3 Entity classes class diagram.

--Creation of a resource (persistent entity) for each table in the database (Fig. 3).

--Creation of *ConjuntodepuntosFacadeREST* and *TriangulacionesFacadeREST* REST Web services for each Entity Class created in the previous step (Fig. 4).

--Creation of methods that implement the REST operations that will be used in each REST service created. The *ConjuntodepuntosFacadeREST* class is adapted using the *find* (GET {id}) and *findAll* (GET) methods. The *TriangulacionesFacadeREST* class is adapted using the *find* (GET {id}) method.

--Creation of the *Trng* class for the Wrapper of the legacy component responsible for carrying out the triangulation based on the oriented matroid associated to the set of points.

The REST services were available at a public URL allocated by the CECAD of the District University.

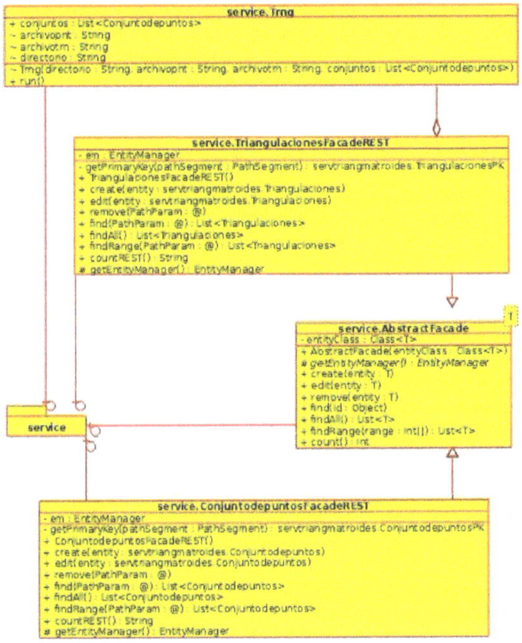

Fig. 4 Services class diagram.

B. Building of the Android (WS-Res) client

For the construction of the Android client providing access to the web service (WS-Rest), following tools were used: Eclipse IDE, the Android Development Tools (ADT) plugin for Eclipse and the Android SDK. Steps for the installation of the mentioned tools were as follows:

--Installing the Eclipse IDE, downloading it from the URL *http://www.eclipse.org/downloads/*. The installation is doing by simple unzip of the downloaded file and run the eclipse program from the eclipse folder.

--Installing the SDK Android available from the URL *http://developer.android.com/sdk/index.html*. After downloading the Android SDK, the installer was run, specifying a path to the JDK. Then, it was needed, through the SDK Manager, to install the "*Android SDK Platform-tools*", "*Android 4.3 (18 API)*" and "*Android 2.2 (8 API)*" platform, and extra "*Android Support Library*" package.

--It was also necessary to create an environment for Android emulation, using the AVD (*Android Virtual Device*) manager of the Android SDK. This emulated environment is useful for test development for Android without using an actual mobile device.

--To complete the tools installation it was necessary to install the *Plugin Android Development Tools* (ADT) for Eclipse using the option "*Install New Software of Eclipse*", providing the URL *https://dl-ssl.google.com/android/eclipse/*, and selecting the two *Developer Tools* and *NDK Plugins* packages.

--Once the software tools were installed, the Android application project in Eclipse and the layout of the GUI client were created.

To support all of the required functionality, it was necessary to make the main activity (*MainActivity*) to extend the *FragmentActivity* class and to implement *ConnectionCallbacks* and *OnConnectionFailedListener* interfaces of the *GooglePlayServicesClient* package, as well as the *LocationListener* interface of the package *com.google.android.gms.location* (Fig. 5).

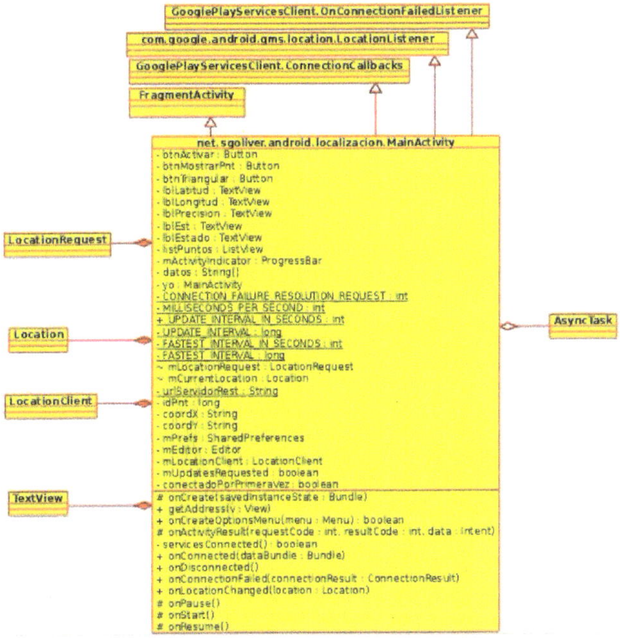

Fig. 5 Client class diagram.

VI. PROTOTYPE TEST

For interaction with the points set triangulation REST service, four asynchronous tasks were defined.

-- A task to obtain the ID of the points set associated with the session initiated by the client; this task starts from the *onCreate* method of the main activity class (Fig. 6).

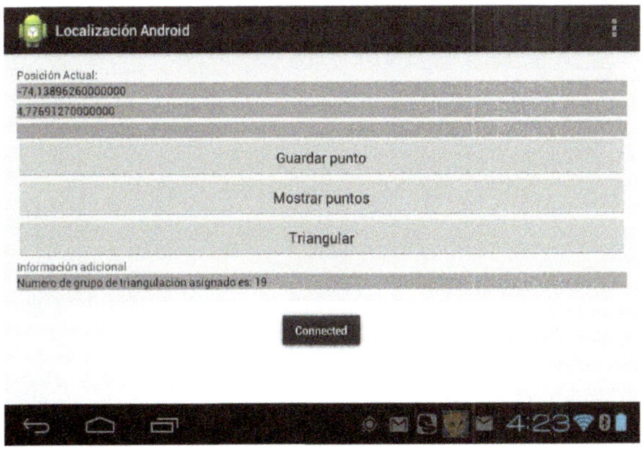

Fig. 6 Triangulation ID assigned on client interface starting.

--When an *onLocationChanged* event is thrown by the Google Play Services software component, a popup message is put on the user interface whit the geographic coordinates provided by the GPS device and the client component put them in the Posición Actual controls zone ().

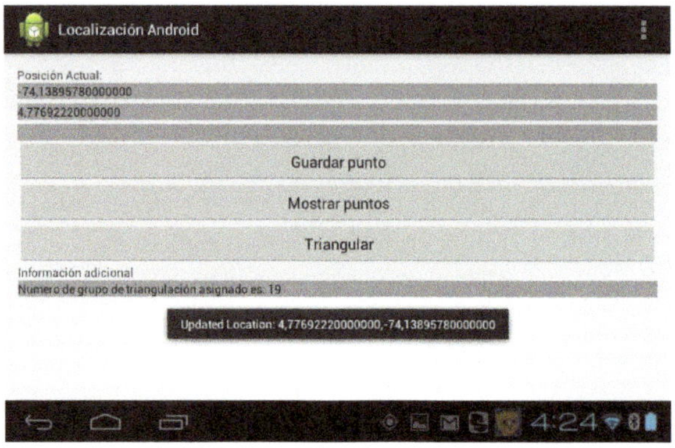

Fig. 7 Detection of Updated Location events on the user interface.

--A second task for the insertion of the point selected by the user (through the action of the "*Guardar punto*" button on the user interface) provided by the GPS receiver and displayed in the user interface in the field "*Posición actual*" (**Fig. 8**).

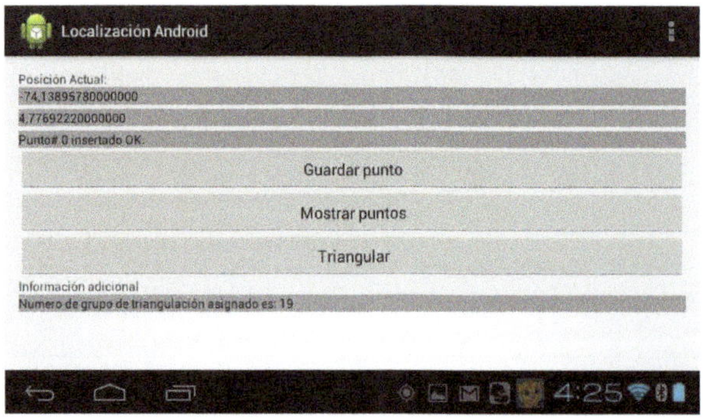

Fig. 8 Storage by the service using the user interface.

--A third task for obtaining the set of points stored so far from the REST service (through the action of the "*Muestre puntos*" button of the user interface) (**Fig. 9**).

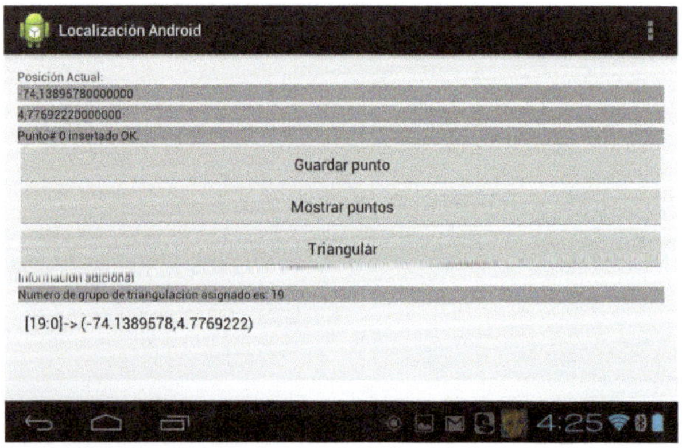

Fig. 9 Stored Points set provided by the service.

--Finally, a task to obtain the edges of the possible triangulations reachable from the saved set of points. Both the list of points and the edges of the possible triangulations are presented in the *ListView* that is shown at the bottom of the user interface.

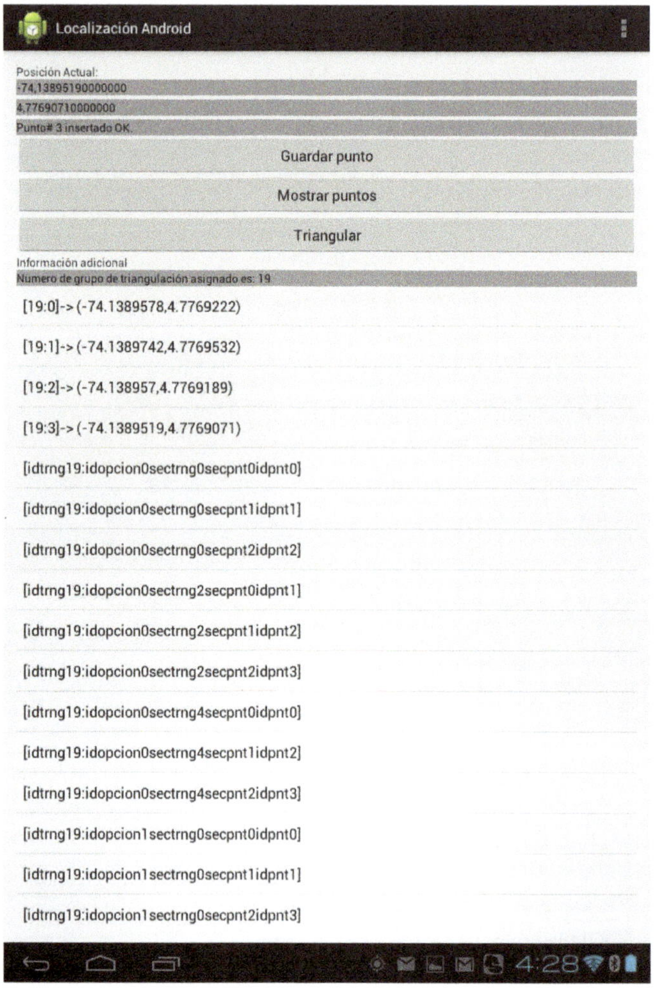

Fig. 10 Stored points set obtained from the service and its possible triangulations.

Fig. 10 shows the user interface with four points provided by the mobile device (a Toshiba Excite 10 inches tablet with a wireless link) and their possible triangulations. For each possible triangulation (identified by *idopcion*), the triangle is shown (identified by *sectrn*). For each triangle the point sequence (indicated by *secpnt*) and the point's id (identified by *idpnt*) is shown. The test used four very near points only a few meters apart and in this case, the particular point's arrangement produced the only two possible triangulations in agreement with the based matroid combinatorial algorithm used.

Coordinate transformation from the original World Geodetic System (WGS-84) spatial reference system into the Transverse Mercator projection was conducted using the minimum latitude and longitude of the set of points as the *latitude of origin* and *central meridian* parameters. This spatial coordinate's projection is needed as GPS receivers provide geographic coordinates.

The test was also conducted using a LG Optimus L7 Cell smartphone. The following explanation focuses on illustrating the REST's server side of this test. **TABLE I** shows the 12 points provided by the smartphone. These points were located at a farther distance than the points in the previous test.

TABLE I
POINT SET LIST ON SERVICE SIDE

Geographic coordinates JSON format World Geodetic System (WGS-84)


```
[{"conjuntodepuntosPK":{"idtrng":23,"idpnt":0},
    "coordx":-74.1465482,"coordy":4.7710368},
{"conjuntodepuntosPK":{"idtrng":23,"idpnt":1},
    "coordx":-74.1429136,"coordy":4.7686211},
{"conjuntodepuntosPK":{"idtrng":23,"idpnt":2},
    "coordx":-74.141702,"coordy":4.7678158},
{"conjuntodepuntosPK":{"idtrng":23,"idpnt":3},
    "coordx":-74.13895,"coordy":4.7768594},
{"conjuntodepuntosPK":{"idtrng":23,"idpnt":4},
    "coordx":-74.1328556,"coordy":4.7716231},
{"conjuntodepuntosPK":{"idtrng":23,"idpnt":5},
    "coordx":-74.1354637,"coordy":4.7729735},
{"conjuntodepuntosPK":{"idtrng":23,"idpnt":6},
    "coordx":-74.1429136,"coordy":4.7686211},
{"conjuntodepuntosPK":{"idtrng":23,"idpnt":7},
    "coordx":-74.1366037,"coordy":4.7763712},
{"conjuntodepuntosPK":{"idtrng":23,"idpnt":8},
    "coordx":-74.1365644,"coordy":4.7764313},
{"conjuntodepuntosPK":{"idtrng":23,"idpnt":9},
    "coordx":-74.1465482,"coordy":4.7710368},
{"conjuntodepuntosPK":{"idtrng":23,"idpnt":10},
    "coordx":-74.140837,"coordy":4.7762281},
{"conjuntodepuntosPK": {"idtrng":23,"idpnt":11},
    "coordx":-74.1409072,"coordy":4.7760984}]
```


The point primary key *conjuntodepuntosPK* is composed of the triangulation id *idtrng* and the point id *idpnt*. *coordx* and *coordy* are the point´s meridian and latitude respectively.

When the client mobile device asks to performing the triangulation, the original coordinates are projected as shown in **TABLE II**, using for this test the values *4.7678158* and *-74.1465482* (the minimums) as the *latitude of origin* and *central meridian* parameters respectively.

TABLE II
PROJECTED POINT SET LIST

Id	X	Y
0	3561	0
1	890	4032
2	0	5376
3	10000	8429
4	4210	15190
5	5703	12296
6	890	4032
7	9460	11031
8	9527	11075
9	3561	0
10	9302	6335
11	9159	6257

The projected coordinates were scaled by a 10 factor.

TABLE III shows the legacy component result returned to the triangulation service REST.

TABLE III
THE LEGACY COMPONENT RESULT


```
[T[0]:=[0->12,3:{{0,1,2},{0,2,3},{2,3,4},{3,4,7},{4,7,8},{3,7,8},{0,3,10}}];]
[T[1]:=[1->12,3:{{0,1,2},{0,2,3},{2,3,4},{3,4,7},{4,7,8},{3,7,8},{3,10,11},
    {0,10,11},{0,3,11}}];]
[T[2]:=[2->12,3:{{0,1,2},{0,2,3},{4,7,8},{3,7,8},{0,3,10},{2,4,7},{2,3,7}}];]
[T[3]:=[3->12,3:{{0,2,3},{2,3,4},{3,4,7},{4,7,8},{3,7,8},{0,3,10},{0,2,6}}];]
[T[4]:=[4->12,3:{{2,3,4},{3,4,7},{4,7,8},{3,7,8},{0,3,10},{1,2,3},{0,1,3}}];]
[T[5]:=[5->12,3:{{0,1,2},{3,4,7},{4,7,8},{3,7,8},{0,3,10},{0,3,4},{0,2,4}}];]
[T[6]:=[6->12,3:{{0,1,2},{0,2,3},{2,3,4},{0,3,10},{3,4,8}}];]
[T[7]:=[7->12,3:{{2,3,4},{3,4,7},{4,7,8},{3,7,8},{1,2,9},{2,3,9},{3,9,10}}];]
[T[8]:=[8->12,3:{{0,1,2},{0,2,3},{3,4,7},{4,7,8},{3,7,8},{0,3,10},{3,4,5},{2,4,5},
    {2,3,5}}];]
[T[9]:=[9->12,3:{{0,1,2},{2,3,4},{3,4,7},{4,7,8},{3,7,8},{2,3,10},{0,2,10}}];]

...

[T[514]:=[514->12,3:{{0,1,2},{0,2,10},{2,4,8},{3,8,10},{2,8,10}}];]
[T[515]:=[515->12,3:{{0,1,2},{2,4,5},{2,3,10},{0,2,10},{2,3,8},{4,5,8},{2,5,8}}];]
[T[516]:=[516->12,3:{{4,7,8},{3,7,8},{2,4,7},{3,9,10},{1,3,9},{1,3,7},{1,2,7}}];]
[T[517]:=[517->12,3:{{3,4,7},{4,7,8},{3,7,8},{3,10,11},{0,10,11},{1,2,3},{3,4,5},
    {2,4,5},{2,3,5},{0,1,11},{1,3,11}}];]
[T[518]:=[518->12,3:{{3,4,7},{4,7,8},{3,7,8},{3,10,11},{1,2,3},{3,4,5},{2,4,5},
    {2,3,5},{1,3,9},{9,10,11},{3,9,11}}];]
[T[519]:=[519->12,3:{{3,4,7},{4,7,8},{3,7,8},{3,10,11},{0,10,11},{0,3,11},
    {0,1,3},{3,4,5},{2,4,5},{1,3,5},{1,2,5}}];]
[T[520]:=[520->12,3:{{3,4,7},{4,7,8},{3,7,8},{3,10,11},{0,10,11},{0,3,11},
    {0,2,6},{3,4,5},{2,4,5},{0,3,5},{0,2,5}}];]
[T[521]:=[521->12,3:{{0,1,2},{3,7,8},{0,3,10},{0,2,4},{3,5,7},{0,3,5},{0,4,5},
    {4,5,8},{5,7,8}}];]
```


The point primary key *conjuntodepuntosPK* is composed of the triangulation id *idtrng* and the point id *idpnt*. *coordx* and *coordy* are the point´s meridian and latitude respectively.

In this case, the particular point's arrangement produced 521 triangulations (all the possible ones). TABLE III shows only the result's head and tail. Each possible triangulation takes no more than the point id's as the process uses only the point set combinatorial structure.

Finally, the four last triangles of the last possible

triangulation of the response to the client mobile device are shown in **TABLE IV** using the JSON format.

TABLE IV
TRIANGULATIONS LIST ON SERVICE SIDE

...

{"triangulacionesPK":{"idtrng":23,"idopcion":521,"sectrng":10,"secpnt":0,"idpnt":0}},
{"triangulacionesPK":{"idtrng":23,"idopcion":521,"sectrng":10,"secpnt":1,"idpnt":3}},
{"triangulacionesPK":{"idtrng":23,"idopcion":521,"sectrng":10,"secpnt":2,"idpnt":5}},
{"triangulacionesPK":{"idtrng":23,"idopcion":521,"sectrng":12,"secpnt":0,"idpnt":0}},
{"triangulacionesPK":{"idtrng":23,"idopcion":521,"sectrng":12,"secpnt":1,"idpnt":4}},
{"triangulacionesPK":{"idtrng":23,"idopcion":521,"sectrng":12,"secpnt":2,"idpnt":5}},
{"triangulacionesPK":{"idtrng":23,"idopcion":521,"sectrng":14,"secpnt":0,"idpnt":4}},
{"triangulacionesPK":{"idtrng":23,"idopcion":521,"sectrng":14,"secpnt":1,"idpnt":5}},
{"triangulacionesPK":{"idtrng":23,"idopcion":521,"sectrng":14,"secpnt":2,"idpnt":8}},
{"triangulacionesPK":{"idtrng":23,"idopcion":521,"sectrng":16,"secpnt":0,"idpnt":5}},
{"triangulacionesPK":{"idtrng":23,"idopcion":521,"sectrng":16,"secpnt":1,"idpnt":7}},
{"triangulacionesPK":{"idtrng":23,"idopcion":521,"sectrng":16,"secpnt":2,"idpnt":8}}]

Each possible triangulation (identified by *idopcion*) shows the triangle (identified by *sectrn*), the point sequence (indicated by *secpnt*), and the point's id (identified by *idpnt*).

VII. Conclusion

The implementation of the prototype presented here allowed the reuse of a legacy tool implemented in C++ for triangulation of point configurations in two-dimensional spaces (for a discussion of performance details see [4]). The inherited tool was integrated in a Rest Web service using the point settings obtained from an Android mobile device.

Android client's interaction for coordinates' collection and use of the triangulation service was greatly simplified by the adoption of REST services. However, adaptation of the semantics of the operations GET and PUT, did not turn out as "natural" as it might be expected.

As definition of the services REST, for both the sets of points and the possible triangulations, was based on the Entity Class associated with the persistence provided by Java Persistence Service, the interchange format output by the triangulation component was not as self-descriptive as the one supplied by the service.

The projected coordinates were scaled by a ten factor in order to have integer values (as required by *points2triang*, the legacy component) but having a decimeter precision. The introduction of this scale factor did not affect in any way the outcome because the procedure used by the legacy component takes in account only the combinatorial structure of the associated oriented matroid.

The REST service asynchronous interaction facilitated the client implementation faults' tolerance, mainly the GPS receiver disconnection one, being necessary only to define the state variables with REST service session as static due that Google Play Services destroys and creates newly its objects after each disconnection.

No statistical validation was done because that was beyond the scope of the work, but this could be the subject of future work.

The matroid-based triangulation service can be reached at the IP address 200.69.103.29 on the http port 22095. The Android Client component (UbicacionGeografica.apk) and the other resources (i.e., IDE projects for Netbeans and eclipse-kepler for the services and client respectively) are available at IDE projects temporary site.

Acknowledgments

We would like to thank the Francisco José de Caldas District University's Center for High Performance (CECAD) for their prompt and ongoing support, and to providing the computational resources required for the implementation of the prototype and the allocation of a public IP address to access the REST services.

References

[1] Worboys, M. and Duckham, M. (2004) GIS: A Computing Perspective, 2Nd Edition, CRC Press, Inc., Boca Raton, FL, USA.

[2] Garey, M. R., Johnson, D. S., Preparata, F. P., and Tarjan, R. E. (1978) Information Processing Letters 7(4), 175 – 179.

[3] deBerg, M., Cheong, O., vanKreveld, M., and Overmars, M. (2008) Computational Geometry, Springer Berlin Heidelberg, third edition.

[4] Pfeifle, J. and Rambau, J. (2003) Computing triangulations using oriented matroids In Michael Joswig and Nobuki Takayama, (ed.), Algebra, Geometry and Software Systems, pp. 49–75 Springer Berlin Heidelberg.

[5] Valero, J. A. (2013) UD y la Geomática 0(7).

[6] Whitney, H. (1935) American Journal of Mathematics 57, 509–533. Oxley, J. (2007).

[8] Stell, J. and Webster, J. (2007) Computers, Environment and Urban Systems 31(4), 379 – 392 Topology and Spatial Databases.

[9] Richter-Gebert, J. and Ziegler, G. M. (2004) In Jacob E. Goodman and Joseph O'Rourke, (ed.), Handbook of Discrete and Computational Geometry, Boca Raton, FL, USA: CRC Press, Inc.. pp. 129–151.

[10] Anderson, L. and Delucchi, E. (2012) Discrete & Computational Geometry 48(4), 807–846.

[11] De Loera, J. A., Rambau, J., and Santos, F. (2010) Triangulations: Structures for Algorithms and Applications, Springer Publishing Company, Incorporated, 1st edition.

[12] Tsukamoto, Y. (2013) Discrete & Computational Geometry 49(2), 287–295.

[13] Pautasso, C., Zimmermann, O., and Leymann, F. (2008) In Proceedings of the 17th International Conference on World Wide Web WWW '08 New York, NY, USA: ACM. pp. 805–814.

[14] Manuel Báez, Álvaro Borrego, J. C., Cruz, L., González, M., Hernández, F., Palomero, D., deLleraand Daniel Sanz, J. R., Saucedo, M., Torralbo, P., and ÁlvaroZapata (2010) Introducción a Android, E.M.E. Editorial ©.

[15] Kovachev, D. and Klamma, Framework for Computation Offloading in Mobile Cloud Computing, R. 12/2012 2012 International Jorunal of Interactive Multimedia and Artificial Intelligence 1(7), 6–15. DOI: 10.9781/ijimai.2012.171

11

Automatic Multimedia Creation Enriched with Dynamic Conceptual Data

Angel Martín, Haritz Iribas, Ion Alberdi and Naiara Aginako
Digital Tv & Multimedia Services, Vicomtech-IK4 GraphicsMedia.net, Donostia-San Sebastián, Spain

***Abstract* — There is a growing gap between the multimedia production and the context centric multimedia services. The main problem is the under-exploitation of the content creation design. The idea is to support dynamic content generation adapted to the user or display profile. Our work is an implementation of a web platform for automatic generation of multimedia presentations based on SMIL (Synchronized Multimedia Integration Language) standard. The system is able to produce rich media with dynamic multimedia content retrieved automatically from different content databases matching the semantic context. For this purpose, we extend the standard interpretation of SMIL tags in order to accomplish a semantic translation of multimedia objects in database queries. This permits services to take benefit of production process to create customized content enhanced with real time information fed from databases. The described system has been successfully deployed to create advanced context centric weather forecasts.**

***Keywords* — Context-aware Multimedia, Dynamic Multimedia Production, Content Driven Database, SMIL**

I. Introduction

In the recent years, motivated by the rapid rise of the Internet, new challenges have arisen from the increasing amount of audiovisual data that are becoming available. Furthermore, the traditional multimedia production has experienced a deep change due to the arrival of new communication paradigms and the evolution and convergence of new technologies. In this context, distributed multimedia presentations are getting more and more popular for web users. SMIL (Synchronized Multimedia Integration Language) [1] is a W3C standard, designed for describing multimedia presentations which combine audio, video, images, text or any other media.

When the creation is concerned, in order to reduce time and cost of multimedia production, both the possibility of working on them collaboratively and reusing existing content are mandatory. For reutilization purpose the use of meta-information has become a necessity [2]. The increase of multimedia content, and the growing difficulty to search, filter and manage such content, require an effective and efficient multimedia search and retrieval system [3]. In addition, the Web [4] may also be employed as a medium to connect a group of authors, which may be geographically distributed, empowering users to be more creative and fostering social interactions enabling cooperative production [5]. One of the main advantages of team working is to gain efficiency in time to solve complex problems, in order to reach a better final result. Currently, the majority of authoring tools goes online enabling real time creative collaboration processes through cooperative frameworks.

Audiovisual producers gather different multimedia contents and graphic objects composing more attractive and enriched contents in order to engage the audience while consumers are willing to embrace new technologies inherited from other application areas to get richer experiences [6]. However, content-aware research is usually placed at the end of the production process, instead of integrating the solution within the production chain where the different multimedia objects involved are still isolated. This fact facilitates multimedia analysis workout through exploiting content knowledge inside the description of the creation process, in terms of timeline and spatial schedule of multiple objects in the scene [7]. A drawback related to the majority of solutions is the interoperability, they operate over private formats that hide the creation process and makes not feasible to exploit this information for other purposes. Another major trend of context-aware multimedia research that could take advantage of multimedia creation descriptors focus on dynamic generation of multimedia presentations.

This paper proposes an automatic dynamic content generator platform based on SMIL standard. Regarding automatic generation of video content in order to address the gap between visualization layer and dynamic data retrieval for databases, we have designed standard-based definitions. The main benefit is the optimization of the costs of the producing process, as well as saving time. In our context, the creation of multimedia presentations, the simplest way for that is the reusability of previously created multimedia videos and templates into new ones. Besides, in this automatic generation of content, we are able to insert new 2D-3D objects as the spatial relation among different objects of the scene is defined in the production process. Our work has been tested in a weather forecast platform.

The rest of the paper is organized as follows. In Section II, we introduce the related work involved in the development, while in Section III we explain the implementation which includes architecture and the main features of our platform. In section IV we state the validation of the platform in the meteorological domain. Finally, we give some conclusion remarks in Section V.

II. RELATED WORK

We can classify the related work section into several broad categories: collaborative SMIL authoring tools; SMIL players; and the metadata exploitation for search and retrieval and Content Management System (CMS).

There are several advantages of using the web standard SMIL instead of tailoring a custom XML. First, the standard compliance eases the interoperability with other solutions or platforms and future developments. Furthermore, it provides access to a huge audience through web browsers. There are several authoring tools for developing multimedia presentations in SMIL. Some of the research solutions like H-SMIL-Net [8], LimSee2 [9] and SMILAuthor2 [10] provide SMIL authoring interface for multimedia objects. They are designed to interpret the temporal behavior of a multimedia presentation, associate hyperlinks with media objects, enable interactivity, and describe the layout of the presentation on a screen. Other authoring tools [11], [12], add to these works the opportunity to work collaboratively to make the SMIL file enriching the final multimedia presentation with different points of view, while reducing the cost and production time. Our approach lives on the previously works aggregating support to additional features described below.

The players can be divided into two categories. On the one hand, the specific SMIL players like GRiNS [13], Ambulant [14] and SmilingWeb [15]. SmilingWeb is cross-platform player for SMIL 3.0 presentations contained in web pages. On the other hand, it should be mentioned that some of the most important media players support SMIL, such as Real Player, Totem, QuickTime and Windows Media Player. However, none of them supports 3D animations and virtual environments rendering that enhance the experience generated by our platform.

The amount of multimedia content that has to be managed has already become unaffordable without a good CMS. This requires an extensive use of multimedia metadata [16]. The SMIL standard provides extra information of the different multimedia objects involved in a content using metadata [17]. Some studies have analyzed how SMIL metadata can apply to the indexing and abstracting of multimedia documentation [18]. Besides, it can be seen how SMIL metadata is used to store information of different templates in databases in some papers [19]. It facilitates the search and retrieval of these templates, making the system more efficient to reuse content. Here, the proposed platform gathers the human knowledge on semantic tags through the multimedia objects contextualization performed in the authoring tool enabling the connection with databases that collect real time information.

The platform described in this paper integrates a SMIL collaborative authoring tool, a multimedia presentations player and a metadata tag aggregator to facilitate the reuse of contents, saving the information of templates in a database.

In addition, we have created a system that using ontologies allows retrieving data from different databases dynamically, so multimedia presentations can be automatically generated using the generated SMIL templates. It should be mentioned that the platform produces videos from these multimedia presentations. Other features provided by our approach include the possibility to add an avatar or virtual character, that makes the created videos more friendly, but also an integrated text to speech converter.

III. IMPLEMENTATION

The architecture of our platform is based on SMIL standard and Gstreamer [20] open source multimedia processing framework. Our approach has extended the semantic meaning of some tags of SMIL in order to perform dynamic content creation while keeping standard compliance. It is achieved by bridging semantic context connections of the different contents involved, such as multimedia elements defined in the outline and scheduled in the layout, with data stored in a database. Once the required data are retrieved from the database according to semantic matching algorithms, Gstreamer is responsible for the creation of the video designed in the SMIL template.

Meanwhile, as we described in the previous section, the SMIL document itself helps the platform to facilitate the work among different people. SMIL describes not only the play out outline but also the production process so it is perfect to trace the activity of each team member in the cooperative creation of multimedia contents. Regarding concurrency and consistency, the SMIL structure centered on parallel or sequence relations between the different elements involved eases the required control. This means, each member can place a multimedia content inside a sequence without timing constraints or in parallel with others establishing just the z-order, from background to foreground, and the position in the layout. The changes are updated in the timeline and the different lines where concurrent contents are represented in order to provide the team current situation awareness. The authoring process driven by a visual representation of the timeline and the different lines available for each content helps to build a consistent SMIL document that depicts the visual designed outline and layout.

Taking into account that the present research in collaborative environments focuses on authoring tools and the required communication and concurrency control on top of XML modules based on Web Services (WS), we have developed a Service Oriented Architecture (SOA) which integrates WS communication and control solutions. However, we will focus on the real novelty around dynamic content generation with data retrieved from databases.

A. Extending SMIL specification

For our approach, SMIL document capabilities potential in terms of content management is twofold depending on the metadata granularity. On the one hand, SMIL provides suitable features in terms of CMS where the metadata, that describe the document content, enable indexing of a large library of generated documents while metadata also provide recommendation solution basis in a similar way that MPEG-7 [2] does but easing web publishing and processing. On the other hand, SMIL document templates also support creation of dynamic multimedia reports and contents through the semantic definition of the different multimedia contents involved. This means, aggregated metadata can be exploited in order to achieve dynamic generation of multimedia contents enabling temporal and spatial visual templates that can be fed with real time retrieved data from databases in order to perform updated and customized videos according to user preferences [21]. This way, our developed platform bridges the gap between data visualization and data storage leveraging data available from databases.

However, no tags for this kind of circumstances are conceived at all in SMIL. So, we use existing tags within a new semantic context in order to keep backwards compatibility with current players. These tags would ease the mapping translation from concepts within the SMIL document, defined like a template, to database (DB) queries.

We use the standard <meta> and <metadata> tags as they are suitable from the point of view of aggregating semantic meaning for the different visual objects scheduled in the document while keep fully standard compliance.

Three different semantic layers have been defined:

1) *First layer:* Metadata related to the general document template suitable for ownership and semantic keyword indexing based CMS.

2) *Second layer:* Semantic keywords defining Primary Key (PK) field name and related fields from DB tables.

3) *Third layer:* Multimedia objects and DB fields mapping.

The different roles of the added attributes are scheduled in the Fig. 1.

The meta-information included in the first layer is stored in a database for more exhaustive searches, and to facilitate reuse of content. In the depicted example the SMIL document represents a "weather report" template to show the "weather figures forecast".

Not just semantic keywords are defined for the general SMIL document, but also for different multimedia objects involved in the second layer. This way advanced CMS indexing is supported easing advanced search of objects inside produced contents by the authoring tool.

The subtemplate structures, defined in the second layer, provide a gathered semantic scheme of required fields from the DB ("attribute") around a PK ("subtemplate_key") where the keywords ("content") provide semantic concepts related to DB table fields. In the example depicted the concept employed as PK is "city" and the required fields to fulfill the necessary data are "city name", "forecast icon" and "wind speed". So, these definitions establish the items in order to infer the SQL database queries.

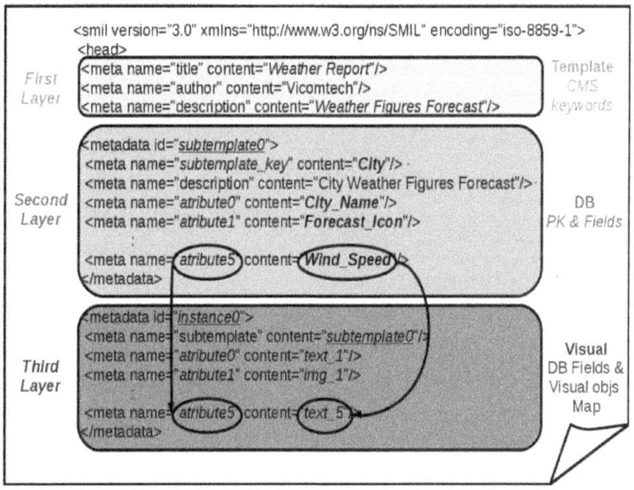

Fig. 1. Semantic data for dynamic performance.

Finally, in the third layer, multimedia object IDs involved in the SMIL document ("content") are mapped to the DB fields ("attribute"). So contents from layer 2 and 3 are linked through the attributes correspondence. Again, according to the Fig. 1, the values of the layer 2 attributes like "wind speed" retrieved from the fields of different DB tables for a specific PK, in this case a "city", would be linked to a layer 3 text ("text_5") respectively. This way the aggregated metadata enable the DB middleware to translate layer 2 structure to DB queries such as SQL

$$select\ f(X)\ from\ T(Y)\ where\ f(Y) = target\_P$$

where:

- f(X) is the result of semantic matching of keyword X with field names from the DB.
- T(Y) is the result of semantic matching of keyword Y with table names from the DB.
- target_PK is the target primary key.

In our example X would be "wind speed", Y "city" and target_PK the city-ID whose forecast parameters will be introduced in the resulting multimedia forecast report.

The definition of subtemplates in the second layer permits the reutilization of subtemplates when creating new multimedia contents in order to reuse a previous design within the same template document. The subtemplate structure has been conceived like a "Class" of object oriented programming languages while instance structure has been designed as an instance of object oriented programming languages. Therefore, the same dataset definition could have different visual instances within the same template depending on the PK value provided. For example, according to the elements defined in the SMIL document depicted in the Fig. 1 if the designer is interested in create a multimedia forecast for several cities, he should just copy the existing instance and add as many as he wants changing the "instance id" and keeping the "subtemplate id" reference.

B. Dynamic multimedia platform architecture

Once we have defined the extension to SMIL standard in order to both CMS features to reuse the content and make a platform that can create videos with dynamic data automatically; in this section, we give an overview of the developed platform based on Gstreamer to render the result.

This platform provides a solution for:

- User edition through the authoring tool to create both multimedia templates and contents.
- Applications and services that request the automatic generation of a certain content according to the context of the user or periodic time scheduled events.

Fig. 2 shows the general architecture of the approach where individual building blocks are described below:

1) *Authoring tool:* Here, users can work collaboratively in the implementation of the multimedia content. Moreover, to facilitate user work, and remove learning curve of the SMIL language, we have developed a multimedia edition tool based on SMIL standard that supports aggregation of metadata tags according to an ontology in order to accomplish DB incrustation.
2) *Context Scheduler:* Third party application or user context aware service that wants to provide user-centric multimedia content.
3) *DB Translator:* Semantic translator from layer 2 metadata and specified PK values, from the authoring tool or a context scheduler, to DB queries. The translator engine is deployed on top of OWL-API and Pellet solutions [23]. The response is the result of perform the query.
4) *Template DB*: SMIL document templates in a database organized around layer 1 metadata.
5) Template Manager: CMS that manages the library of SMIL document templates.
6) *Content DB:* Multimedia DB with contents for specific applications such as weather forecast reports.
7) *Gstreamer Render:* Multimedia processing platform able to mix and orchestrate contents both online and offline contents (Content DB).
8) *SMIL Parser:* Web requests and responses processor to create a content in SMIL format.
9) *Server:* Web service manages two kind of requests:

a) *Available multimedia templates:* Request includes target topics. So, level 1 keyword based multimedia template search is performed by the Template Manager. The response includes all the recommended templates and their layer 2 and 3 data to be translated to DB queries.

The designed workflow starts when the Authoring Tool for human active creation or Context Scheduler for automatic multimedia generation, requests through a WS some available multimedia templates to provide a multimedia content to a specific user. Here context is performed by means of some semantic keywords that define the multimedia target type.

The Template Manager retrieves SMIL documents exploiting information from layer 1 matching with the request keywords. The response includes exclusively the layer 2 and 3 data of the filtered SMIL templates.

b) *Multimedia content creation:* Request includes level 3 with real DB values for each multimedia item. The generation of the video according to the involved SMIL templates and the data provided is performed. The response is the URL with the generated video.

In case there is a request for multimedia content creation from dynamic data, DB Translator takes charge of processing layer 2 data to transform them into DB queries. Afterwards layer 3 elements are filled by Context Scheduler or Authoring Tool with the figures and values resulting from the DB queries. Here it is important to highlight that values from layer 3 linked to a text object will produce text rendering. While if it is related to image or video objects it refers an URL to an existing content. Last but not least, a text value referred to an audio will unleash a Text To Speech (TTS) audio generation.

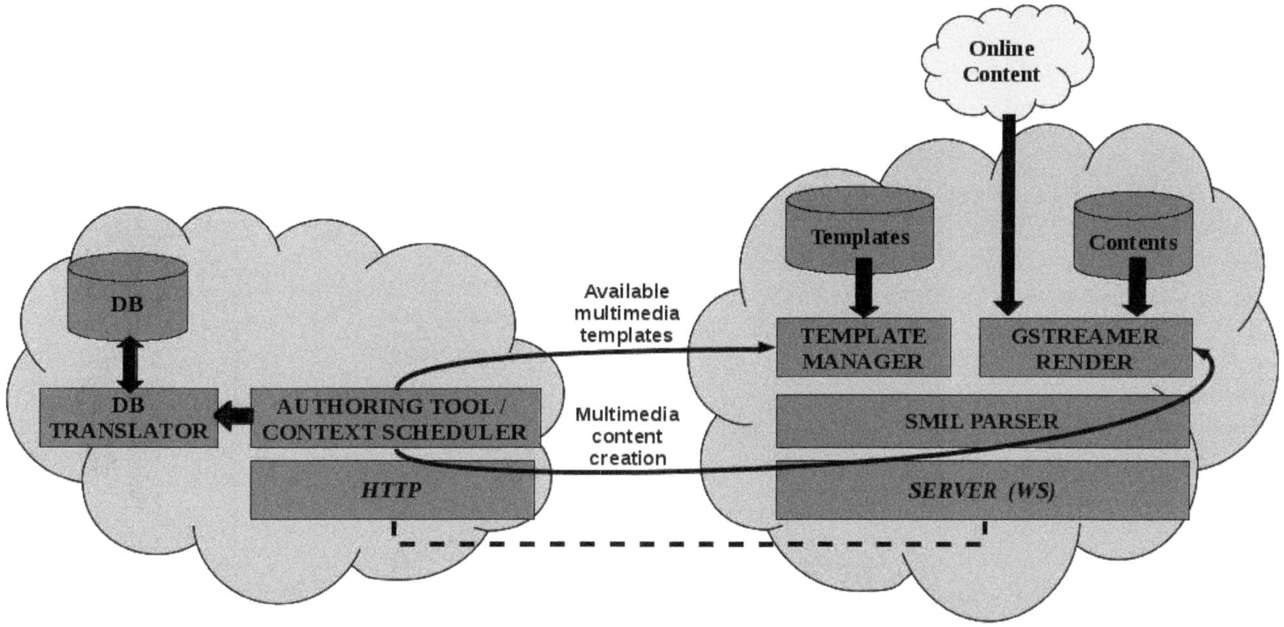

Fig. 2. Architecture for dynamic multimedia creation.

c) Design of ontology: We have designed an ontology in order to define the concept relations between the dynamic content saved in the databases. Therefore, it is possible to manage the information of different databases in order to extract more exact information. The template editor has a limited access to the definition of DB queries so is necessary to implement anarchitecture in order to develop a more powerful tool.

In the meteorological context, we have defined an ontology drawing all the relations between the parameters of weather stations. Consider a real weather station, it contains several meteorological or environmental sensors that measure different forecast parameters. But only measures are relevant, we have to consider more information as sensor model, sensor picture, geolocalization... The definition of the attributes of the concepts is highly relevant.

Ontology permits inserting PK in a total transparent manner from the user point of view. When creating a new template, user inserts a meteorological station ID and the system, using the defined ontology and its relations, shows the possible parameter definitions.

Attributes give the capacity of adequate the presentation of the parameters to the user. Therefore, depending on the nature of each concept, different rules are applied to new objects.

Once all the required data of the desired templates form the final fulfilled SMIL document, it is employed to request the multimedia SMIL driven creation through the WS.

For the process of creating the video, we have developed a SMIL Parser that reads the SMIL document and creates a DOM [22] representation. It extracts the data into a format that we can easily use to create a Gstreamer pipeline. After the platform generates the pipeline with the marks of time and space indicators, GStreamer module executes the pipeline. The GStreamer Render generates the video results and responds with the URL of the video. The Gstreamer core module, gnonlin is in charge of managing the temporal and spatial positions of the different multimedia elements that make up the output video.

To sum up, we have designed a context scheduler that according to time, user preferences and context defines the required PKs in the automatic generation case; or the user himself, in case of on demand authoring, to later perform the renderization request by inserting the retrieved data from the DB inside the corresponding instance of the SMIL document. In order to accomplish our objectives, we have developed all the modules defined in our platform which permits the total control of the automatic content creation.

C. Content-aware production inline

Multimedia applications' next move focuses on 3D-2D fusion and user centric and context-aware that needs of advanced multimedia analysis algorithms to achieve realistic experience and automatic creation of dynamic contents customized to the user context and preferences.

The availability of isolated visual objects within the production chain eases the introduction of 2D-3D objects through the <z-index> SMIL attribute associated to each object. We have exploited the intrinsic information from the production process to remove issues associated to research solutions usually placed at the end of the production chain, such as segmentation, clustering, depth map and more in particular for 3D-2D fusion T-unions, optical flow, partial occlusions, etc. As we keep information of the production process in the SMIL document, we manage to make the later fusion of 3D-2D objects and effects realistically.

During the design process of our content generator platform, we have defined a module to control the z-index of the objects completing a scenario. Therefore, during the creation process of new dynamic content we are able to insert new 2D-3D data to enrich the already defined scenario making the content production affordable.

D. Authoring Tool

The authoring tool to create new objects has been designed as a video editor adapted to user needs. This video editor is divided in two parts. On the one hand, we have the usual editor that permits the mix of audios, videos, texts, images and avatars in the easiest way possible for the multimedia items creators. On the other hand, the collaborative editor permits the reuse of old video templates and videos templates that other collaborators made, and the blend of them for creation of new material.

The whole editor has three main areas structure. On the right there is a preview and options area, on the left there is an elements list and bellow those areas, there is an edition area with the buttons that are necessary for it.

- Elements List

On the top-left side of the interface there is a list of elements with the content library that will be used for the creation of the templates. With the below buttons we can add and remove elements, with the buttons plus and minus respectively, to the list so that the user could use them in the edition.

- Edition Lists

In the video editor there are three elements lists, in the first list are audio elements, in the second list are background elements and the speeches of avatars and in the third list lessons main elements.

The user can manage intuitively all the elements; add, erase and change elements position and duration with the buttons placed at the bottom of the interface. It is so easy to use that e.g. if the user wants to add an element to any list, the system identifies the type of the element and adds to the corresponding list, if it is an audio element it adds in the first list while if it is a video or an image the user just has to define if he wants to add it as a background element or as a main element of the learning object.

- Edition in Collaborative Web Editor

SMIL has different tags for the definition of timing process and the most important for our collaborative system is <par> and <sec> tags.

The elements inside the tag <par> are reproduced all in parallel being grouped to support complementary pedagogical information. The tag <sec> permits the reproduction of elements in sequence enabling sequential outline construction. It is really important because permits to reproduce an element after other without an explicit definition of the start time. The system knows that as soon as one element ends next one must start, maintaining the established order. The user just has to define the order of the elements and when saves it, the system takes the information of the different files and saves in a resultant file.

To sum up, according to the strategy of keeping the developed platform simple, experts can create new objects through defining new multimedia templates or modifying existing ones through a desktop editor application described before. Afterwards the created objects are pushed automatically to the server.

IV. VALIDATION

To test the approach tailored for weather forecast production, validation work has been mainly driven by semantic translation of multimedia objects in database queries by checking correct connection of meteorological concepts, defined by metadata, to real time acquisition databases. SMIL rendering compliance was also performed, where we have created different templates with all kinds of items, including pictures, videos, texts and audios.

Using the authoring tool, we have designed a wide range of meteorological report templates in which different weather maps have been generated as depicted in the Fig. 3. On the one hand, we have produced weather maps where the multimedia elements are not interrelated with semantic information layer. On the other, we defined maps with the items related to weather stations from which data are obtained. In this case tests performed achieve multimedia results automatically adapted to geographical user context and preferences. We also proved the correct implementation of the defined ontologies concepts by retrieving the requested parameters using semantic high-level concepts as weather stations.

As mentioned above, templates are very useful because allow us to create new presentations each day with the same structure, but with new data. These data are acquired from the database, using the weather stations as a PK identifier. In other words, when creating a new video, system translates the metadata attached to a multimedia template item, for example that visualizes the atmospheric pressure, and it takes the value automatically from the database of a specific station according to the user context system.

After testing, we have seen that when introducing elements, the definition of <z-index> in SMIL document has allowed us to use 3D elements, such as avatars, in a direct manner. We resolve the problem of occlusions as the depth of the objects in the scene is already defined.

About the metadata aggregation, the platform makes even easier the use of templates and databases. Access to a database via the web platform makes it more intuitive for inexperienced users. User can provide human understandable meteorological concepts that would be translated to real time acquired atmospheric measures. This way the user just has to provide a geographic context to establish the database PK. After selecting this information, the system automatically creates the queries for the database so that the user does not have to worry about anything. In addition to manually creation of videos, it has proven the periodic creation of videos in an automatic manner.

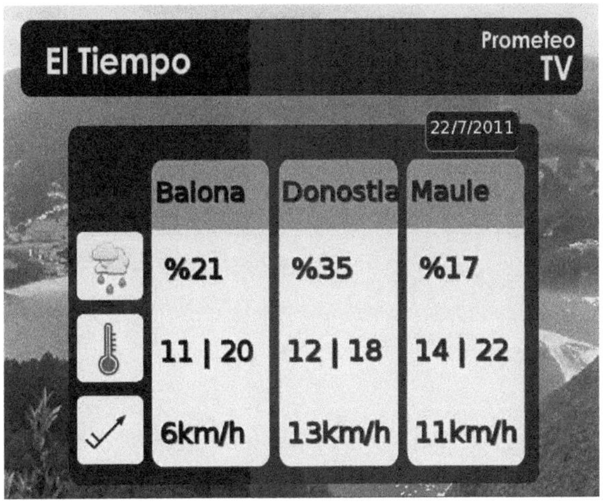

Fig. 3. Platform produced weather report.

This testbed based in the weather information platform does not have real time performance restrictions or high scalability requirements, but it faces dynamic contexts describing a solution extensible from time and localization features to context user-centered multimedia applications. The automation of this process has enormously reduced the cost of producing new contents for weather forecasts and the time needed for the creation of new videos which implies that new content can be updated with higher frequency.

V. CONCLUSION

Creating new multimedia contents and keeping them as a template in order to ease reutilization, decreases dramatically the amount of digital content replicated in edition platforms. Moreover, the capability of creating this content collaboratively permits new envisaged creation possibilities and sharing of content. The SMIL documents considered as templates also ease the reutilization of production work according to reuse and recycle philosophy of green computing minimizing the set of needed resources.

The redefinition of the metadata tag of SMIL standard permits the declaration of objects in SMIL presentations. Giving a semantic meaning to these metadata, reveals the possibility of creating new updated multimedia content, including dynamic data, that automatically minimizes the production costs of lots of contents. This property becomes crucial in repetitive processes which must be updated with new data, for example in weather reports.

The proposed format to describe the content outline makes feasible to exploit this information for other purposes related to content-aware research, such as auto-summaries, 2D-3D fusion or visual near-duplicated research. This one may be relevant for rights holders to provide control of their content enabling copyright policies and content management through automatic processes to identify, claim, and apply policies to partial or entire content.

Future work will include more complex relations in terms of database structures that will also need of enhanced semantic translators from context based keywords to existing databases. We also plan to take benefit of the multimedia object segmentation provided by the SMIL templates in order to boost visual analytics applications improving the search algorithms.

REFERENCES

[1] W3C, SMIL Specification, http://w3.org/AudioVideo
[2] Cornejo, C. M., Ruiz-Rube, I., Dodero, J. M. "Semantic Management of Digital Contents for the Cultural Domain", Recent Trends in Information Reuse and Integration, Springer, pp 211-226, 2012
[3] HongJiang Z., "Multimedia search: Past and current approaches", IEEE 10th Workshop on Multimedia Signal Processing, 2008
[4] Olson, G.M, Zimmerman, A., and Bos, N., Eds., "Scientific Collaboration on the Internet" MIT press, Cambridge, MA, 2008
[5] Sung ,M.Y., Lee D.H., "A Collaborative Multimedia Authoring System", Lecture Notes in Computer Scien-ce, Volume 3033/2004, pp.311-318, 2004
[6] Jansen, J., Cesar, P., Bulterman, D., Stevens, T., Kegel, I., and Issing, J., "Enabling Composition-Based Video-Conferenfing for the Home", IEEE transactions on Multimedia, IEEE, Volume 13, pp. 869-881, 2011
[7] R. Kernchen, S. Meissner, K. Moessner, P. Cesar, I. Vaishnavi, M.u Boussard and C. Hesselman, "Intelli-gent Multimedia Presentation in Ubiquitous Multide-vice", IEEE Computer Society, pp. 52-62, April-June 2010
[8] S. Bouyakoub , A. Belkhir, "H-SMIL-Net: A Hierarchical Petri Net Model for SMIL Documents", Proceedings of the Tenth International Conference on Computer Modeling and Simulation, p.106-111, 2008
[9] Deltour, R., Layaida, N., Weck, D., "LimSee2: A Cross-Platform SMIL2.0 Authoring Tool", The European Research Consortium for Informatics and Mathematics – ERCIM News. n. 62 , 2005
[10] Chun-Chuan, Y., Chen-Kuei, C, Yung-Chi, W., "Extension of Timeline-based Editing for Non-deterministic Temporal Behavior in SMIL2.0 Authoring," Journal of Information Science and Engineering, vol. 24, no. 5, pp. 1377-1395, 2008
[11] Sung, M. Y., Lee, D. H. "A collaborative authoring system for multimedia presentation", Proceedings of the IEEE International Conference on Communications, 1396-1400, 2004
[12] El Saddik, A., Rahman, Md.A., Hossain, M.A., "Authoring Multimedia Objects in Collaborative Ambient Intelligent Virtual Environment", IEEE International Workshop on Haptic Audio Visual Environments and their Applications, 2005
[13] Bulterman, D.C.A., Hardman, L., Jansen, J., Mullender, K.S., Rutledge, L., "GRiNS: A GRaphical INterface for creating and playing SMIL documents" WWW7 Conference, Computer Networks and ISDN Systems, vol. 30, pp. 519–529, 1998
[14] Bulterman, D.C.A., Jansen, J., Kleanthous, K., Blom, K., Benden, D., "Ambulant: a fast, multi-platform open source SMIL player", Proceedings of the 12th annual ACM international conference on Multimedia, 2004
[15] O. Gaggi, L. Danese. "A SMIL player for any web browser", Proceedings of International Conference on Distributed Multimedia Systems, p 114-119, 2011
[16] Garcia, R., Celma, O. "Semantic Integration and Retrieval of Multimedia Metadata", 5th Int.Workshop on Knowledge Markup and Semantic Annotation, 2005
[17] Bulterman, D., Rutledge,L., SMIL 3.0 Flexible Multi-media for Web, Mobile Devices and Daisy Talking Books
[18] Gil, B., Pérez, J. R., Sánchez, R., "Gestión de documentación multimedia con Synchronized Multimedia Integration Language (SMIL)." // Scire: representación y organización del conocimiento. 10:2, p 87-97, 2004
[19] Jiang, D., Liu, Z., Zhong, L., "A smil-based multimedia system for mobile education", 2nd IEEE International Conference on Broadband Network & Multimedia Technology, IC-BNMT '09, p 724-728, 2009
[20] Gstreamer web page, http://gstreamer.freedesktop.org/
[21] Steele, R., Lubonski, M., Ventsov, Y., and Lawrence, E., "Accessing SMIL-based Dynamically Adaptable Multimedia Presentations from Mobile Devices", Proc. Of Int. Conf. on Inf. Technology: Coding and Compu-ting, ITCC04, 2004
[22] W3C, DOM Specification, http://www.w3.org/DOM/

12

Semantic Brokering of Multimedia Contents for Smart Delivery of Ubiquitous Services in Pervasive Environments

Alba Amato, Beniamino Di Martino, Salvatore Venticinque
Department of Information Engineering,
Second University of Naples, Aversa, Italy

Abstract — **With the proliferation of modern mobile devices having the capability to interact each other and with the environment in a transparent manner, there is an increase in the development of those applications that are specifically designed for pervasive and ubiquitous environments. Those applications are able to provide a service of interest for the user that depends on context information, such as the user's position, his preferences, the capability of the device and its available resources. Services have to respond in a rational way in many different situations choosing the actions with the best expected result by the user, so making environment not only more connected and efficient, but smarter. Here we present a semantic framework that provides the technology for the development of intelligent, context aware services and their delivery in pervasive and ubiquitous environments.**

Keywords — **Intelligent Agents, Context Awareness, Mobile Augmented Reality.**

I. Introduction

WITH the proliferation of modern mobile devices having the capability to interact each other and with the environment in a transparent manner, there is an increase in the development of applications specifically designed for pervasive and ubiquitous environments[14]. These environments are characterized by the presence of devices such as sensors, actuators, RFID tags, etc... which make them vastly interconnected, with distributed intelligence. Thus, context-aware applications are able to provide a service of interest for the user that depends on context information, such as the user's position, his preferences, the ability of the device and its available resources (memory, cpu, peripherals). Moreover these services have to be able to respond in a rational way in many different situations choosing the actions with the best expected result, so making environment not just more connected and efficient, but smarter. This leads, as a result, the improvement of the quality of life.

Location awareness is relevant in various fields, in order to provide more efficient services to the user. Popular examples provide support to the users on a visit of a museum or of an archaeological site, for shopping at the supermarket, on the go within a city, to a nurse for the administration of drugs, etc...

Here we present a semantic framework that provides the technology for the development of intelligent, context aware services and their delivery in pervasive and ubiquitous environments. Besides semantic-based techniques have been used for dynamically and efficiently data processing, retrieval and delivery. In order to provide a pervasive environment with advanced smart capabilities services, we integrated those technologies that enable objects of the real world to become part of the *Internet of Things*.

The paper is organized as follows: Section II introduces related work. The problem statement is described in Section III. In Section IV the framework is described and in Section V two case studies are presented. In the end conclusion and future work are discussed.

II. Related Work

Many research contributions focus on the development of software/hardware architecture and frameworks for mobile context-aware applications, also based on agent technology and semantic techniques.

Context Broker Architecture (CoBrA) project from Chen et al. [1] is an agent based architecture for supporting context-aware systems in smart spaces (e.g., intelligent meeting rooms, smart homes, and smart vehicles). Central to this architecture is an intelligent agent called context broker, which maintains a shared model of context on the behalf of a community of agents, services, and devices in the space, and provides privacy protections for the users in the space, by enforcing the policy rules that they define. However their approaches did not address various characteristics of context information such as classification and dependency. The Service-Oriented Context-Aware Middleware (SOCAM) project introduces by Gu et al. [2] is an architecture for the building and rapid prototyping of context-aware services. It provides efficient support for acquiring, discovering, interpreting and accessing various contexts to build context-aware services. The main contribution of this work is the use of an ontology that allows for the description of the context in a semantic way that is independent of programming language. However the implementation can provide and process only contextual information that is associated the meeting domain.

In particular, several frameworks for development and delivery mobile context-aware applications are used in the domain of cultural heritage and recommendation systems. CIMAD [3] (Common Infrastructure/Context Influenced Mobile Acquisition and Delivery of cultural heritage data) is a framework supporting the development of cultural heritage "services" designed and implemented within the EU Network of Excellence EPOCH (Excellence in Processing Open Cultural Heritage). CIMAD aims to address a wide range of context-aware and multi-channel services in the cultural heritage domain [4], i.e. data acquisition, content delivery, monitoring, and management. The innovative aspect of CIMAD is that it helps the development team in mastering the complexity of building, customizing and integrating site specific cultural heritage services, thus making the different implementation activities more productive and cost-effective. This framework has many similarities with our work, but it does not focus on the backend support for experts in the domain of the Cultural Heritage in order to augment the archaeological site as an editor of that site or a guide for the tourist (i.e. map editing, semantic annotation and localization of multimedia contents).

The EasiShop framework [5] assists consumers with their shopping experience. The Shopping Agent is a mobile agent that is responsible for managing the shopping list and finding the best suited products that match consumers preferences, interacting with the retailers Retailer Agents that is hosted by the wireless device in the shops. Shopping Agent is able to cross-compare the products and their prices so making informed decisions and deriving recommendation that could be passed to the consumer. In our approach semantic is used for knowledge representation and management, but the ontology is designed by experts of the application domain. It is used both as a common vocabulary to support interoperability among heterogeneous remote services, local applications and users, and for intelligent discovery of media contents. Besides we also provide tools for experts in the domain of the application in order to augment the market (i.e. map editor, semantic annotator).

III. Problem Model

The formal objective of the framework is to define a model for delivering contents and services that provide the greatest utility for the user according to the knowledge that can be inferred using pervasive devices. The autonomic decision process is driven both by an updating process of the knowledge itself and by a set of rules that are defined by the expert of the application domain. The problem is modeled as an event driven system described by the following set of parameters:

-- A set of Events $E = \{(e_1, t_1), \cdots, (e_{ne}, t_{ne})\}$ represent perceptions about the environment, detected by the device at a certain time;

-- A set of Believes $B = \{b_1, \cdots, b_{nb}\}$, to be used for describing the knowledge in the application context is described by a domain ontology. The ontology has to be defined for the specific use case, by an expert of the application domain;

-- The user knowledge at the time t, about the environment and about his own preferences, is described by a set of concepts and individuals belonging to the ontology B.

For each user i we have a profile $P_i \subset B$;

-- A function $F : (E,P) \rightarrow P'$ updates the user's profile when new events have been notified to the system;

-- A set of goals: $G = \{g_1, \cdots, g_{ng}\}$, whose achievement improves the user's utility;

-- A set of actions: $A = \{a_1, \cdots, a_{na}\}$ can be performed by the system to achieve the goals, so increasing the user's utility;

-- A set of triggers, one for each goal:

$$\{\overrightarrow{w_1}(P), \overrightarrow{w_2}(P), \ldots, \overrightarrow{w_{nw}}(P)\}, \overrightarrow{w_i} : P \rightarrow B^{na.}$$

W is a $[nw, na]$ matrix of boolean values, that must be defined to start a number of actions, when some believes of the user's profile have been updated. A trigger is computed if the correspondent goal needs to be achieved. $A'^T = A^T W^T W I$, where I represents a $[na, nw]$ diagonal matrix of boolean values and A'^T is the resulting subset of action to be started.

For a chosen case study we have to define the set of perceptions, the ontology, the available actions, and the set of triggers. Let us define a multi-objective metric for the user utility as $U(G)$, that is the measure of the achievement of each goal. Objective of the framework is the maximization of such utility executing the minimum subset of actions that help to achieve the active goals.

The set of triggers must satisfy the following condition: $U(G(A)) > U(G(A \cup a_j))\ \forall j: a_j \subset A - A'$.

In fact if an action cannot improve the user's utility, it can bother the user and at least it affects the performance of the system. In a general scenario the function F can be implemented by an expert system, but if $E \in B$ it simply updates the user profile by adding, removing or replacing perception in the user's profile P.

In the following we introduce a technological framework that allows for implementing the proposed model and two case studies.

IV. The Framework

In Fig. 1 the architectural solution of the framework is shown. The framework is composed of different tools and applications.

On the left side of Fig. 1, the user is using his device that hosts a light agent that is able to perceive information from the field by pervasive sensors. The local agent executes

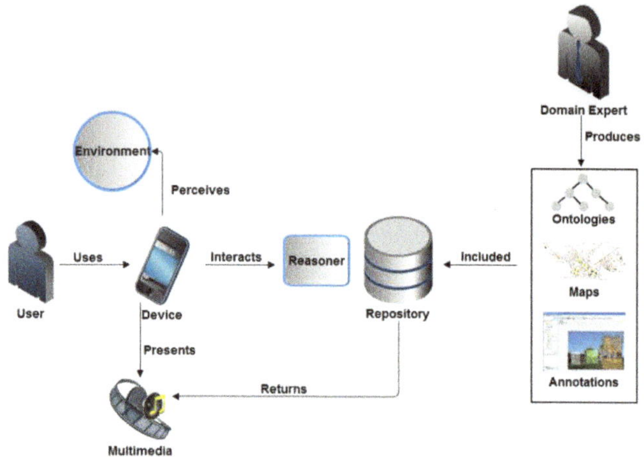

Fig. 1: Architecture and roles

autonomously and proactively in order to support the user's activity within the environment where he is moving. It discovers surrounding objects, it uses them to update the representation of the user's knowledge, reacts using the local knowledge to organize and propose the available contents and facilities by an interactive interface. If the connection works the device can access remote services, which can access a wider knowledge and have greater reasoning capabilities to look for additional contents and applications.

Remote applications implement context aware services. They use personal devices to collect perceptions and for content delivery. An ontology implements the representation of the global knowledge that is necessary to share a common dictionary and to describe the relationships among the entities/objects, which are part of the model. In our model a common ontology include all the general concepts, which are useful for a description of a pervasive environment where mobile users are moving, using their devices and interacting with available facilities and other users. The general ontology is complemented with a domain ontology that is designed by an expert of the specific application field. Experts of the application field define the domain ontology for the specific case study. They use or design a map to represent the environment. They add POIs to the map to geo-refer multimedia contents and can link them to concepts of the ontology. Furthermore they select relevant contents and annotate them using concepts and individuals of the ontology.

Concepts of the ontology are used at client side to describe a representation of the reality as it is perceived by the user. On the back-end the ontology is used to annotate digital resources like point of interests, contents, applications. It is also used to support reasoning. User's behaviors, information from pervasive devices or from other users, device properties, external events are heterogeneous data that are perceived by the device and that are used to build a dynamic changing representation of the user knowledge about the reality, within which he is moving. The applications are knowledge driven. The user's knowledge can be used by the application that is running on the device to adapt its logic locally, and it is updated remotely to improve the awareness of services at server side. Application are events based. Events can be updates of the user's knowledge or can be explicit service requests raised by the user. At each invocations semantic queries, that depend on the user's knowledge, are built and processed to get the action to be performed and the contents to be delivered. Results of the query are individuals of the ontology that are described by semantic annotation. The user's knowledge is composed of many semantic concepts with *static*, *dynamic* or *locked* properties. Components of the knowledge are:

-- *Device technology and capability*. Among the static properties here we means hardware resources, on board peripherals, display size, total memory and storage.
Dynamic ones can be power level, available memory and storage and bandwidth ;

-- *User's position*. It is a dynamic property that can change over the time and can be evaluated using different techniques and technologies, depending on the devices, on the available infrastructures and the kind of environment (indoor or outdoor);

-- *Pervasive objects*. They are dynamically discovery by the device. They can be sensors which provide information about the environment or can be used by the services itself by the device or eventually by thorough network if they are connected;

-- *Time information*. We intend the current time at user side and the time that the user is spending, or has spent within the environment.

-- *User's interests*. This part of the knowledge could be dynamically changed by the application according to the user's behaviors and to his feedbacks. The user could choose to start from an empty or a standard profile, to change it or to lock some properties interactively.

Semantic techniques are used for intelligent content and application discovery and delivery. Knowledge representation, ontology and annotations of digital resources are used to filter, organize and deliver contents and software to the device. Different techniques for reasoning can be experimented such as graph matching, description logics, neural networks or more simple ones like SPARQL queries. Here we integrate semantic techniques with discrete optimization methods to take into account constraints such as user's position and available time for exploitation.

V. CASE STUDIES

Here we present two different case studies, which have been chosen to test the proposed model and the developed framework.

A. Cultural Heritage

Exploitation of archaeological sites can be very difficult because of a lack of supporting infrastructures and because of the complex recognition and comprehension of the relevant ruins, artworks and artifacts[10]. The availability of personal devices can be used to plan and support the tourist by suggesting him the itineraries, the points of interest and by providing multimedia contents in the form of digital objects which can semantically augment the perceived reality. In this context relevant issues are the profiling of the user, the discovery and the delivery of the contents that can improve the user's satisfaction, new models of interactions with reality. Three case studies have been chosen to test the approach and the framework [12]. The S. Angelo in Formis Basilica, in Campania, near S. Maria Capua Vetere, the ancient town of Norba and the amphitheater of Capua. In these sites we cannot install complex infrastructures and they are difficult to be understood without a tourist guide. For the presented case studies we need to provide a technological solution that does not need infrastructures for letting the software know the user location and his feeling about the environment. It means that Bluetooth, RFID, GPS, electronic compass, camera, network connection and others are the technologies which have to be used together or independently to get information about the user perceptions and to augment his exploitation of the archaeological site. The first thing to be considered when modeling this kind of problem is the environment. In fact we have to model the environment where the user is moving and to reconstruct the perceptions of the user himself in order to get his particular vision about what is surrounding him. A real representation of the environment is necessary to identify landmarks and possibilities of intervention using pervasive actuators and sensors whose input will be stored in the memory of the agent and will be updated as the environment changes. The environment will be modeled as a geo-referred map with itineraries, landmarks and points of interest. The user will be able to download the map of the area to be visited at home, before to leave, or on site, if the network will be available. The map will include all the points of interest that identify the relevant objects of that area and different cultural itineraries which could be exploited on site. Also contents can be discovered and downloaded in advance according to the device capabilities. On board software and remote services will assist the cultural visit by augmenting the reality by the user's personal device. A software agent executes on the user's device to support services exploitation. It percepts the surrounding environment using the on-board peripherals and executes plans which are chosen by an ad-hoc reasoning to optimize the user's satisfaction. The knowledge of the environment acquired by the agent represents part of its own beliefs. Another set of believes describes user's knowledge about the surrounding environment that can be acquired by using peripherals of his personal mobile device, by recording and evaluating user actions or explicitly asking for user feedbacks. Some examples are user's position, interest, nearby objects, landscape, etc. Interaction with the environment and presentation of contents can be done again using user's device and his peripherals. Of course the way of localization of users and objects depends on the device technology, the available infrastructures and the kind of environment.

Indoor or outdoor localization can be implemented using heterogeneous technologies, and often absolute localization can not be performed, but it is only possible to detect nearby landmarks or objects. Besides as the device is a PDA, it has limited battery, storage and computation capacities so also the agent running on it has limited capabilities.

In order to extend the agent's capability another remote intelligent agent interacts with the local one to achieve individual or shared goals. This two-agent system is part of a loosely coupled network of problem-solver entities that work together to find solution to problems that are beyond the individual capabilities or knowledge of each entity [6]. In fact the agent running on the local device is able to collect perceptions, to communicate believes and to execute actions locally. The remote one can access a wide knowledge base and can perform a more complex reasoning. Of course different components of the user's knowledge could have different weights according to their relevance or the time of perception. We could consider the knowledge at the current time, in a time windows, or its complete history. In order to augment user's knowledge and capability to interact with the environment, services have to choose, according to their context awareness[11]; i) what content and application it has to deliver; ii) when it needs to present the content; iii) how this should be done. Finally another set of believes is the set of contents in the remote knowledge base which are available for delivery, to be used for augmenting the reality perceived by the user. The maximization of user satisfaction represents the goal to achieve. With this aim agents have to make a decision about the optimal set of contents and their organization according to user's profile, device and time availability.

B.Smart Market

The second proposed case study is an e-commerce service that profiles the customers and supports their purchases recommending products and shops [13]. On the other side the service is used by vendors, who upload information about their offers, shops and sales. The choice of this application field is due to the fact that the rising usage of mobiles as the primary route for consumer interaction with brands, and its importance as a catalyst for content sharing and recommendation, are still at odds with the level of focus and budget currently afforded to mobile in the market field. Additional advantages in this field brought by the technology are the traceability, which will help the consumers to verify the origins of the products. Besides, in case of a food related disease, the traceability of the purchased food will enable faster detection of the origin of the infection and thus contain its impact better and faster. A QRCode or an electronic tag can be read by the device and may carry out information related to drugs, making it easier for the customer to be acquainted with adverse effects and optimal dosage.

Besides, in case of an accident because of drug abuse or misuse it will be possible to quickly identify the taken drug by asking the smart device, which may also inform about the right antidote and dosage to enable the emergency treatment to react faster and better, and thereby saving lives.

Let us imagine users of different ages and habits, and with different purposes, in a market. Each of them has a smartphone with a smart application that perceives the environment, provides simple facilities, and accesses the e-service. All the customers continuously update, by the smart application, their profile in the system. They can insert information about themselves such as name, sex, age, home country, e-mail address, languages spoken, level of education, lists of interests, etc. They also set a few preferences about what information the device can autonomously communicate, such as position, nearby objects, user's behaviour. Customers use a grocery list by their smartphone updating their shopping chart. Vendors add or remove their marketplaces and products, and update their description in the knowledge base, also by semantic annotations. Just to simplify our requirements, we suppose that customers cannot add, describe and recommend products on their own. Using the autonomic behavior of the service, the objective of the system is to propose the best set of recommendations, which can help the user to improve his own utility, exploiting the available information about products, about the environment within which customers are moving and about the customers themselves.

VI. Context Aware Semantic Discovery

For sake of simplicity, in both the proposed case studies we suppose that the perceptions notified by the device to the remote services are the user's location, his/her available time, an concepts of the ontology. A common goal is the maximization of the user's satisfaction by proposing the best set of contents, which are relevant to the user's interest. Each time a new perception has been received, the user's profile is evaluated in order to eventually trigger the discovery service that updates the optimal set of contents to recommend.

The semantic discovery service provided by our framework returns a set of digital objects related to POIs in the pervasive environments. Each content is annotated by concepts from the ontology and can be discovered by a SPARQL query to the content repository. The result of the query is a set of N instances of digital objects whose relevance to the user context has not been considered yet. The context awareness of the services is exploited by computing the relevance of each content to the profile p, which is the semantic representation of the context, including interest, position, etc.. The annotation and the user's profile are represented using the Vector Space Model (VSM). VSM is a model for semantic representations of contents as vectors of items created by G. Salton [9]. In our case $\vec{a_j}$ is the vector of concepts $c_{i,j}$ of the domain ontology.

-- $\vec{a_j} = < \{c_{1,j}, o_{1,j}\}, \{c_{2,j}, o_{2,j}\}, \ldots, \{c_{l,j}, o_{l,j}\} >$

$\forall j = 1 \ldots N$

-- $\vec{p} = < c_1, c_2, \ldots, c_m >$

Sizes l and m are the number of different concepts that appear in the annotation and in the profile . If a term occurs in the annotation, $o_{k,j}$ is the number of occurrences of that term in the annotation.

Sizes l and m are the number of different concepts that appear in the annotation $\vec{a_j}$ and in the profile $\vec{p}$. If a term occurs in the annotation, $o_{k,j}$ is the number of occurrences of that term in the annotation.

We defined a score $A(\vec{a_j}, \vec{p})$ to measure the relevance of an annotation $\vec{a_j}$ to the profile $\vec{p}$ by the following formula (1)

$$w_j = w(\vec{a}_j, \vec{p}) = \sum_{k=1}^{l} r_k \quad (1)$$

where

$$r_k = o_{k,j} * \sum_{i=1}^{m} \frac{1}{d_{k,i}+1}$$

where $d_{k,i}$ is the minimum number of edges that connect the node representing the concept $c_{k,j}$ of the annotation to c_i of the profile. In (1), for each item $c_{k,j}$ of the vector $\vec{a_j}$ the relevance to the profile $\vec{p}$ is computed by adding the relevance of that concept to each concept of the profile, and by multiplying each contribution for the number of occurrences $o_{k,j}$. The relevance between two concepts is calculated by dividing 1 by the number of edges of the ontology that connect the node representing the concept c_k of the profile to $c_{k,j}$ plus 1. As a result we have a score for each annotated item that is associated to a POIs so that it is possible to order the items and the POIs according to user's preferences.

However a user has some additional constraints.

In the case of the archaeological guide the contents to be delivered can be limited by the device technology or the available time for the experience.

In the case of smart market, constraints can be the price, the distance he wants to go away, the purchase frequency, etc. For this reason it is necessary, to select the valid recommendations excluding, those that cannot be enjoyed according the current constraints. For example it is necessary that the the budget does not exceed the amounts that is available to the user. Here we aim at recommending the best products that: i) maximize the score, ii) are compliant with the user's budget, iii)can be purchased without exceeding time and space limits.

This problem can be reduced to a *discrete optimization problem* that consists in searching the optimal value (maximum or minimum) of a function $f : \vec{x} \in Z^n \rightarrow R$, and the solution $\vec{x} = \{x_1, \ldots, x_n\}$ in which the function's value is optimal. $f(\vec{x})$ is said *cost function*, and its domain is generally defined by means of a set of m constraints on the points of the definition space. Constraints are generally expressed by a set of inequalities:

$$\sum_{i=1}^{n} b_{i,j} x_i \leq a_j \qquad \forall j \in \{1 \ldots m\} \qquad (2)$$

and they define the set of feasible values for the x_i variables (the *solutions space* of the problem). In our case:

$w_i \geq 0 \ \forall j = 1 \ldots N$, $W = \{\vec{b_1}, \ldots, \vec{b_m}\}$

where:

-- w_i represents a score and

-- B a set of constraints $\vec{b_t}$, each one composed of $N+1$ integer.

We have to compute:

$$max \sum_{k=1}^{N} w_i x_i$$

so that (3)

$$\sum_{i=1}^{N} b_{i,j} x_i \leq b_{j,N+1}$$

with $x_i \in \{0, 1\} \ \forall i = 1 \ldots N$. The goal is to maximize the value delivered. The vector $\vec{x}$ represents a possible solution: its components x_i are 1 or 0 depending on whether the object is included or not in the best set. To set the constraints we create a table with a column for each content and a row for each requirement. The rows of the matrix *B* are contain in each cell $b_{i,j} = 1$ if the requirement is necessary for the content in that column, $b_{i,j} = 0$ otherwise.

In Table II we specify price and shopping time for each product.

TABLE II
RAPPRESENTATION OF PRODUCT PROPERTIES AND USER'S CONSTRAINTS

	C_1	C_2	...	C_n	User
Price	10	20		15	35
Time	10	5		5	15

Furthermore in the last column of the table we have $b_{i,N+1} = 0$the budget and the time availability of the user. For example, the first row tells us that the user can spend not more than 30 , so it is necessary to exclude one content. The same consideration must be done for the second row. Finally we have to exclude that product that affects the affinity with a minimum penalty. Further details about the algorithm and its performance evaluation are out of the scope here.

In the following section we detail the technological choices which have been taken to implement each component of the framework.

VII. TECHNOLOGICAL IMPLEMENTATION

In the following sections, we present the techniques and technology choices, explaining their role in the overall picture. We refer to the introduced case studies to explain how the framework can be used in different fields.

A. Environment Map

To provide a description of the environment within which the user is moving we need a geo-referenced map that describes buildings, roads, bans, itineraries in the case of cultural heritage or departments, sectors, shelves, routes in the case of smart marketa and, in general, Points of Interest (POIs). We used the OpenStreetMap format to design open maps. In Fig. 2 the way we used to build the map of a supermarket is shown.

In the picture there are two different views of the JSOM tool, the image level and the geo-referenced map under construction. The user can choose to build the map following the edge of a reference image or can import other vectorial and geo-referenced formats like GPX. Paths, walls and POIs are outlined. Each point of interest represents departments, sectors, shelves, or any other entities of relevance and can be described using a list of keyvalue pairs. The tool allows to export the map in an open format that can be read and used by the client application.

A. Ontology

An ontology has been designed to describe the sites of interest and to annotate the related media. A general part includes the concepts which are common to all the class of applications that can be modeled according to the proposed approach.

Among the others the *Time* class and its properties (*CurrentTime, ElapsedTime)*, allow to organize and assist the user taking into account time information and handling time constraints. *Position* class and its properties allow to localize the user and objects around him. An application specific part of the ontology includes the concepts that belong to the domain in question.

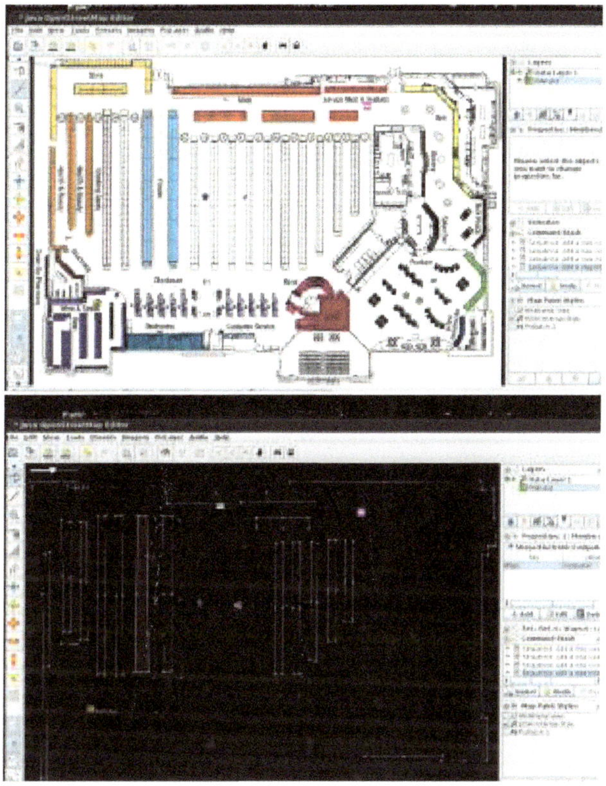

Fig. 2: Marketplace map

In the construction of ontology for the case of the archaeological guide it has been developed by experts of the cultural heritage, and we avoid here to go in deep on details which are out of our knowledge. In the case of the smart market case study, we used the following classification based on the buying habits of consumers [7]. In fact an appropriate classification allows to make better recommendations, since it is possible to identify the type of purchase made by the customer, the frequency he buys a certain product, and on the other hand it is possible to make more accurate cross-sell suggestions.

-- *Convenience products* are purchased frequently by the consumer, they do not require any major effort in the process of choice and any risk of making the wrong choice. Often, their purchase is systematic, and affordability. This also facilitates impulse buying by the individual. They are commonly illustrated by commodities, *unsought* (emergency) items, and impulse products. Examples of consumer goods that fall into the convenience category include fresh produce and grocery staples, umbrellas, gum, and batteries.

-- *Preference products* are distinguished from the previous item primarily on the basis of a perceived greater risk in their purchase. The individual, mainly because of marketing efforts, however, tends to address the problem of the choice of the brand, trying to find one that can better meet his needs. The most prominent examples of preference products are in the consumer package goods industry.

-- *Shopping products* are goods that the individual acquires after careful considerations, and after a prolonged comparison among different brands. These assets may be slightly different from the others, making the price the main driver of choice, or differentiated, making their attributes to play a key role in purchasing decisions. Examples of shopping goods are clothes, and furniture for end consumers, and equipment and components parts for industrial users.

-- *Speciality products* are goods with special features, often characterized by prestigious brands and very high prices. Examples of speciality goods include vintage imported wines, expensive sports cars, and paintings
by well-known artists.

-- *Unknown products* are not known by the consumer , or the consumer would not show any interest if it were not informed about them through advertising.

In [8], three product types are identified: *convenience*, *shopping*, and *speciality* and three steps of purchase/consumption are considered: *pre*, *during*, and *post* transaction. In Table 1 each stage is further divided into areas, in which consumers may seek benefits. Information discovery is more important in the case of shopping, rather than convenience products. Each cell in the upper half of the table depicts whether that particular benefit is important in the context of the given product type.

TABLE I
PRODUCT CLASSIFICATION: THE MATCH BETWEEN PRODUCT TYPES AND MEDIA ATTRIBUTES

	Pre-Transaction		During Transaction	
	Information Search	**Product Trial**	**Payment Options**	**Immediate Delivery**
Convenience products	*Not Important*	*Not Important*	*Important*	*Important*
Shopping Products	*Important**	*Important*	*Important*	*Moderately Important*
Speciality Products	*Important***	*Important*	*Important*	*Moderately Important*

	Post-Transaction			
	Installation Service	**Consumption**	**Return**	**After Sales Enhance-ment**
Convenience Products	*Not Important*	*Important*	*Not Import ant*	*Not Important*
Shopping Products	*May be Important*	*Important*	*Import ant*	*May be Important*
Speciality Products	*May be Important*	*Important*	*Import ant*	*May be Important*

Since the goal here is not to design a complete profile, for the creation of the ontology we focused on general concepts that make it possible to achieve the successful personalization we aim for. We identified two main disjoint classes: *Product* and *Customer*. The *Customer* class contains personal information and has two disjoint subclasses: *Member* and *Non member* to distinguish the users that are regular customers enrolled from the others. Personal information, inserted as datatype properties, consists of information common for all users that is useful to identify and partially segment users. Datatype properties included are *hasName* (type String), *hasAge* (type int), *hasEmail* (type String) and *hasGender* (type Man or Woman). In Fig. 3 a picture the relevant concepts of the ontology for a simple use case of Smart Market are shown.

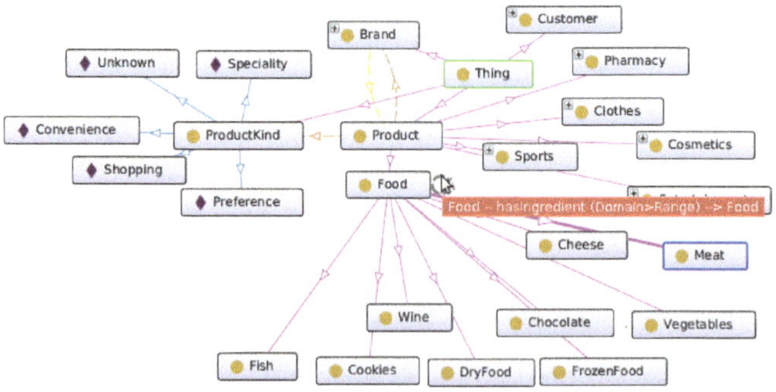

Fig. 3: Smart Market Ontology

The *Product* class has six subclasses: *Food*, *Cosmetics*, *Sports*, *Clothes*, *Entertainment*, *Pharmacy*.

The *isProductType* object property supports a classification based on the purchasing habits of consumers according to the model explained before. Besides *Food* class has the data properties: *isGlutenFree (type boolean) indicates the presence of gluten, hasExpirationDate* (type Date Time) registers the sell-by date expiration and *isFrom* (type string) provides information about the production chain. Another part of the ontology describes the type of purchase: *Convenience* , *Preference*, *Shopping* or *Speciality*. In particular Convenience products are characterized by the *hasPurchaseFrequency* that can be instantiated for each couple *(product, user profile)*.

B. Annotation

The ontology is used also for annotating the multimedia contents. To annotate texts, images and any kind of contents we chose the AktiveMedia tool.

In Fig. 4 a picture of the Amphitheater of S. Maria Capua Vetere is annotate with the *Column* and the *Arc* classes which are part of this kind of building. The output produced by the annotator is an RDF file with concepts and properties of the AktiveMedia ontology and of the domain ontology. The same has been done for the smart-market, where images show those products to be recommended.

C. Digital Repository

The Fedora repository is used to store digital objects and supports their retrieval. Into the Fedora repository a digital object is composed of a set of files which are:

-- *object metadata*: used by the client application to understand how to deliver the content;

-- *binary streams*: which are images, video, text ... any kind of raw information to be delivered;

-- *RDF annotation*: that describe the semantic of the object according to the ontology;

-- *disseminations*: filters to be eventually used for adapting the object according to the target client.

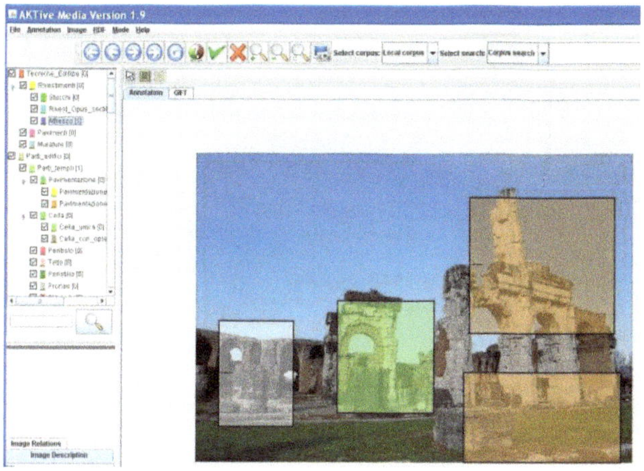

Fig. 4: The annotator

We loaded the Aktive-Media ontology and the domain ontology into the Fedora repository in order to exploit its embedded SPARQL engine that is used to select the optimal set of individuals (i.e. contents). Multimedia contents are automatically stored into the repository after the annotation phase. The RDF output is automatically processed using an XSL transformation to make it compliant with the model used by the Fedora repository.

D. Content's Delivery for the Mobile Archaeological Guide

The diversity of media content to deliver and their quality, is especially relevant in the case of the archaeological guide. In this case we have defined the following content models.

-- Multiple images whose transparency can be graduated by the user to compare changes in different periods. In the same way real pictures can be compared with paintings. Old picture can be compared with what is seen by the camera.

-- Parts of the image acquired by the camera are recognized and linked to related multimedia contents.

-- Virtual reconstructions which are synchronized with the camera output or the detected RFIDs.

-- Text, audio, video and composite media. A content descriptor is attached to every digital object. It is used by the device when the content is being delivered. The descriptor defines the right player for that media, configuration parameter and necessary input. In Fig. 5 an example of delivered content is shown. The user focuses the camera on a particular view of the S. Angelo in Formis Basilica and sees the original temple of *Diana Tifatina*. In particular the perspective viewed by the camera is the same shown by the video.

E. The Client for Mobile Archaeological Guide

At client side we extended the android version of an open source software navigator called Navit. The navigator provides basic facilities for map visualization and to guide the user along some itineraries by using the on-board GPS receiver, or other technologies. By new extension the guide is able to sense the environment by the available peripherals, to understand the situations according to which the visit will be adapted, and to enhance user's experience.

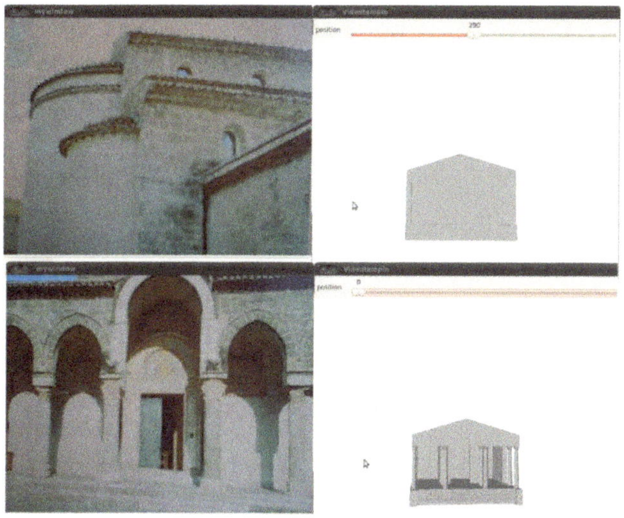

Fig.5: Synchronized video

Even if many experiences on augmented reality are proposed nowadays, usually the exploitation of vision is used only to overlap real and virtual images. Our extensions provide the following functionalities:

-- access to device peripherals to sense the environment and to update the representation of the user's knowledge about the reality around him;

-- access to remote services to update the user knowledge and to ask for available contents

-- a local cache of objects for exploiting the visit without connection;

-- a limited reasoner that is able to organize the content by itself when the connection does not work.

Perceptors are implemented by:

-- GPS positioning to localize the user in open spaces and to guide him on cultural itineraries;

-- RFID for positioning and detection of nearby POIs. This technology can be used to alert the user but also for his positioning in indoor environments;

-- CODBAR recognition to get information about artifact, monuments, ... when a RFID reader is not available;

-- image recognition by search by sample techniques which are speed-up and improved using a position based filtering;

-- monitoring of device resources and configuration;

-- collection of user interest by feedback and by an analysis of his behavior;

-- time monitoring.

On the other hand a list of functionalities are provided to deliver contents and to guide the visit. Content management (discovery, organization and fruition) is supported both at client side and at server side. Knowledge visualization and management are provided to allow the explicit specification of user's own interest. The output of the camera is used as a component of the user's knowledge as well as a map on which semantic additions are anchored, not simply superposed. The user will be able to ask for and exploit available multimedia contents, which are related to points of interest, or to personal interests expressed by semantic concepts. Multimedia contents will be adapted at server side according to the device/user/session profile to provide to the user the best quality of service. Fig. 6 shows the result of the content discovery service. The service invocation starts the camera that is used to take a photo of the landscape or of a subject of interest. In background, the client uploads the image and waits for the content retrieval. The discovery of relevant contents and the download of retrieved information runs in background meanwhile the user is interacting with other facilities provided by the client. At server side the picture is received together with the user's position. The received sample image is processed to extract a list of properties, which are invariant with the color, the zoom and little rotations by the OpenCV library. The server owns a list of images of landscapes and relevant points of interest the user can shoot. These images have already been processed by the same OpenCV library. The user's position is used to perform a filtering of the points of

Fig. 6: The Client

interest close to him, in order to speed-up the comparison of sample features with the ones of candidate images. The best matching images are returned. If they have been annotated, annotations are used to suggest other contents that are semantically related. Furthermore frames of the images, which have been semantically annotated, can be discovered and pointed in the picture that has been just taken. Other events, which open dialog for suggesting actions, itineraries, POIs or media can be related to new perceptions.

F. *The Client for Smart Market*

To access the smart market the user will exploit, by his smartphone, an Android application using an interface shown in Fig. 7 that is able to get and communicate his position and to localize POIs or entities close to him. Furthermore the user is able, by the application, to search for products in a grocery list and to add the products to be bought in his shopping cart. Let us suppose that the user wants to buy cookies. Its device updates periodically a remote agent about the user's position. Each time this component of the user's profile changes, the agent checks for a POIs that is linked to a product semantically related to the cookies concept.

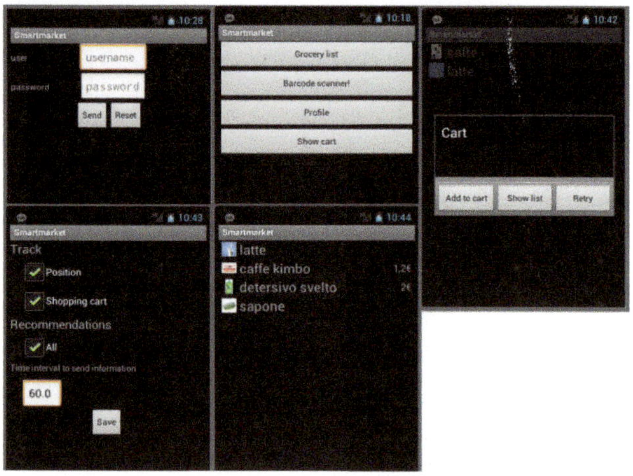

Fig. 7: The Android Client

If the POIs is found the user is notified. Starting from the POI the user can browse the list of products, which are sold at that position. In the same scenario we can imagine that the user has bought the product. Another client application of our smart service is able to recognize the user, by a fidelity card, or by any kind of personal identification mechanism, without getting any additional private information. Scanning the barcode of the purchased product the smart market service can be informed about new convenience goods in the user's habit. A new timer will be created to alert the user at the sellby expiration date of the cookies. The purchase frequency is not known and can be put equals to the expiration date. When the same, or an equivalent product, will be bought, the frequency timer will be updated and set again. Finally a list of concepts have been added to the user's profile for personalizing the discovery of similar products to recommend. In the provided example we will have *glutenFree*, *chocolate*, the *cookies brand*. At the expiration of the timer the user will be alerted about the eventual opportunity to buy the same cookies, but alternative products will be recommended according to the context aware semantic discovery strategy described in the following section. In particular among the available products the system recommends the one annotated with the *cookies* concept, the *hasBrand* property, the *glutenFree* property, and a *hasIngredient* property. Annotated texts and images are currently presented by the web interface shown in Fig. 8. In particular you have on the left the list of relevant concepts, and the picture of the product on the right.

At this point the user can move the mouse over the concepts and see the annotation on the picture. He can also move the mouse over the picture and see a tool-tip that explains the annotation.

Concepts from the list, or from the tool-tip, can be easily added to the user profile in order to let the system learning about the user's preferences for a personalized recommendation. Using the mouse button or the form is also possible to search for other products, which are semantically related to the selected concepts, in order to choose an alternative brand or similar kinds of cookies.

Fig. 8: Semantic browsing of a recommended product

VIII. CONCLUSION

In this paper, we proposed a proactive semantic framework for the intelligent delivery of services in pervasive environments. In such environments, users need to have access to the most relevant and interested services within a rich and dynamic context. The main goal is the exploitation of pervasive devices to maximize the expected satisfaction of the user by adapting applications and contents to be delivered and by augmenting and improving his utility in several contexts. Personal devices are used to interact with the environment, to run interactive applications and to present contents to the user. A smart application runs on the user device, uses the peripherals to perceive the environment, provides context awareness to remote discovery services. We described a framework that implements the proposed approach and described its application for cultural heritage and shopping recommendation. Future works will address quantitative analysis of the approach and the support of collaboration among users by personal semantic based recommendation.

ACKNOWLEDGMENT

Author thank Massimiliano Diodati and Vincenzo Reccia for their technical support.

REFERENCES

[1] H. Chen, "An Intelligent Broker Architecture for Pervasive Context-Aware Systems". *PhD thesis, University of Maryland*, Baltimore County, 2004.

[2] T. Gu, H. K. Pung, D. Q. Zhang, "A service-oriented middleware for building context-aware services", Journal of Network and Computer Applications, Vol . 28, pp. 1-18, Jan. 2005.

[3] Manzaroli D., "Architectures for Context Aware Services in Smart Environments". *PhD thesis,* May 2010

[4] G. Raffa, P. H. Mohr, N. Ryan, D. Manzaroli, M. Pettinari, L. Roffia, S. Bartolini, L. Sklenar, F. Garzotto, P. Paolini, T. Salmon Cinotti, "CIMAD - A Framework for the Development of Context-aware and Multi-channel Cultural Heritage Services", in ICHIM07 September 30, 2007 Toronto

[5] S. Kurkovsky and K. Harihar, "Using ubiquitous computing in interactive mobile marketing," *Personal Ubiquitous Computing*, vol. 10, no. 4, pp. 227–240, Mar. 2006.

[6] V.R., Durfee, E.H., Lesser and Corkill, D.D., *Trends in Cooperative Distributed Problem Solving*, IEEE Transactions on Knowledge and Data Engineering, Vol. March 1989, KDE-1(1), pages 63-83.

[7] E. B. Murphy, P., "Classifying products strategically," *Journal of Marketing*, vol. 50, no. 3, pp. 24–42.

[8] B. Sindhav and A. L. Balazs, "A model of factors affecting the growth of retailing on the internet," *Journal of Market-Focused Management*, vol. 4, pp. 319–339, 1999.

[9] G. Salton and M. E. Lesk, "Computer evaluation of indexing and text processing," *J. ACM*, vol. 15, no. 1, pp. 8–36, Jan.1968.

[10] A. Amato, B. Di Martino, S. Venticinque "BDI Intelligent Agents for Augmented Exploitation of Pervasive Environments". In: Proceedings of the 12th Workshop on Objects and Agents. Università della Calabria, 2011, vol. 741, p. 81-88, ISBN/ISSN: 1613-0073

[11] S. Venticinque, A.Amato, B. Di Martino Ontology driven Augmented Exploitation of Pervasive Environments. In: AFIN2011: The Third International Conference on Advances in Future Internet. Nice/Saint Laurent du Var, France, August 21-27IARIA, p. 41-48, ISBN/ISSN: ISBN: 978-1-61208-148-9

[12] A. Amato, B. Di Martino, S. Venticinque "Semantically Augmented Exploitation of Pervasive Environments by IntelligentAgents". In: Mue 2012: The Sixth International Conference on Multimedia and Ubiquitous Engineering. Madrid, Spain, July 10 – 13, ISBN/ISSN: ISBN: 978-0-7695-4701-5/12

[13] A. Amato, B. Di Martino, S. Venticinque (2012) A Semantic Framework for Delivery of Context-Aware Ubiquitous Services in Pervasive Environments. In INCoS 2012: 4-th International Conference on Intelligent Networking and Collaborative Systems. Bucharest, Romania, September 19 – 21, 2012, ISBN/ISSN: ISBN: 978-0-7695-4808-1/12

[14] Aversa, R., Di Martino, B., Venticinque, S. Distributed agents network for ubiquitous monitoring and services exploitation (2009) Proceedings - 12th IEEE International Conference on Computational Science and Engineering, CSE 2009, 2, art. no. 5284221, pp. 197-204.

13

Framework for Computation Offloading in Mobile Cloud Computing

Dejan Kovachev and Ralf Klamma
Department of Information Systems and Databases
RWTH Aachen University

Abstract **— The inherently limited processing power and battery lifetime of mobile phones hinder the possible execution of computationally intensive applications like content-based video analysis or 3D modeling. Offloading of computationally intensive application parts from the mobile platform into a remote cloud infrastructure or nearby idle computers addresses this problem. This paper presents our Mobile Augmentation Cloud Services (MACS) middleware which enables adaptive extension of Android application execution from a mobile client into the cloud. Applications are developed by using the standard Android development pattern. The middleware does the heavy lifting of adaptive application partitioning, resource monitoring and computation offloading. These elastic mobile applications can run as usual mobile application, but they can also use remote computing resources transparently. Two prototype applications using the MACS middleware demonstrate the benefits of the approach. The evaluation shows that applications, which involve costly computations, can benefit from offloading with around 95% energy savings and significant performance gains compared to local execution only.**

Keywords **— Application Virtualization, Middleware, Pervasive computing, Software Design, Android**

IV. Introduction

RESOURCE-DEMANDING multimedia applications such as 3D video games are being increasingly demanded on smart phones. Even if mobile hardware and mobile networks continue to evolve and to improve, mobile devices will always be resource-poor, less secure, with unstable connectivity, and with constrained energy. Resource poverty is major obstacle for many applications [14]. Therefore, computation on mobile devices will always involve a compromise. For example, on-the-fly editing of video clips on a mobile phone is prohibited by the energy and time consumption. Same performance and functionalities on mobile devices still cannot be obtained as on desktop PCs or even notebooks when dealing with high resource-demanding tasks.

Recently, the combination of cloud computing [11], wireless communication infrastructure, ubiquitous computing devices, location-based services, and mobile Web, has laid the foundation for a novel computing model, called mobile cloud computing [9]. It provides to users an online access to unlimited computing power and storage space. The cloud abstracts the complexities of provisioning computation and storage infrastructure. The end user uses them as utility and in reality they can be far-away data center or nearby idle hardware.

Offloading has gained big attention in mobile cloud computing research, because it has similar aims as the emerging cloud computing paradigm, i.e. to surmount mobile devices' shortcomings by augmenting their capabilities with external resources. Offloading or augmented execution refers to a technique used to overcome the limitations of mobile phones in terms of computation, memory and battery. Such applications, which can adaptively be split and parts offloaded [6, 18], are called elastic mobile applications. Basically, this model of elastic mobile applications gives the developers the illusion as if they are programming virtually much more powerful mobile devices than the actual capacities. Furthermore, elastic mobile applications can run as stand-alone mobile applications, but use also external resources adaptively. Which portions of the application are executed remotely is decided at runtime based on resource availability. In contrast, client/server applications have static partitioning of code, data and business logic between the server and client, which is decided in the development phase.

Our contributions include integration with the established Android application model for development of "offloadable" applications, a lightweight application partitioning and a mechanism for seamless adaptive computation offloading. We propose Mobile Augmentation Cloud Services (MACS), a services-based mobile cloud computing middleware. Android applications that use the MACS middleware benefit from seamless offloading of computation-intensive parts of the application into nearby or remote clouds. First, from the developer perspective, the application model stays the same as on the Android platform. The only requirement is that computation-intensive parts are developed as Android services, each of which encapsulates specific functionality. Second, according to different conditions/parameters, the modules of program are divided into two groups; one group runs locally, the other group is run on the cloud side. The decision for partitioning is an optimization problem according to the input conditions of the cloud and devices, such as CPU load, available memory, remaining battery power on devices, bandwidth between the cloud and devices. Third, based on the

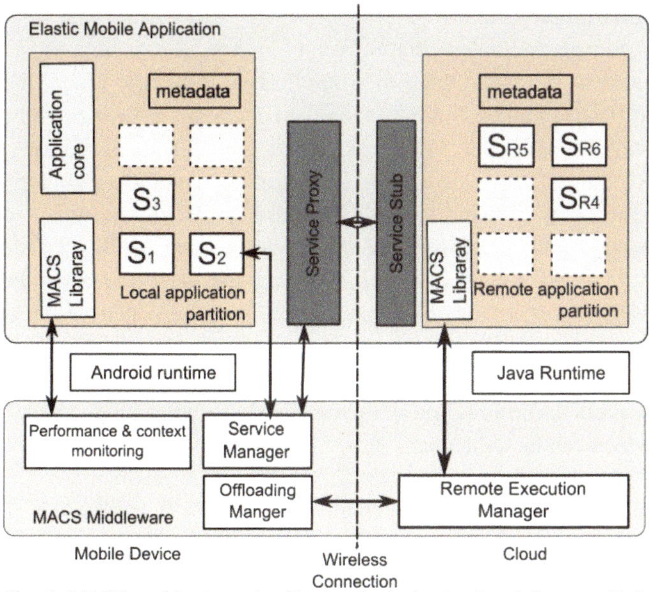

Fig. 1 MACS architecture. Application logic is structured from multiple Android services (S_i). Some of them can be offloaded into the cloud (S_{Ri}).

solution of the optimization problem, our middleware offloads parts to the remote clouds and returns the corresponding results back. Two Android applications on top of MACS demonstrate the potential of our approach.

In the rest of the paper, we first review related research in work mobile cloud computing in Section 2. Then we describe our MACS middleware with detailed descriptions of the implementation (Section 3). We explain the offloading model in our middleware in Section 4. In Section 5 we introduce two use case applications, the setup of the evaluation and the corresponding evaluation. After that, we discuss the results in Section 6. Finally, we draw conclusions and refer to the future work.

V. Related Work

Previous work has proposed many mechanisms that address the challenges of seamless offloaded execution from a device to a computational infrastructure (cloud). The encapsulation of the mobile device's software stack into a virtual machine image and executing it on a more powerful hardware can be considered as a "brute force" approach to offloading, such as proposed by Chun and Maniatis [1] or Satyanarayanan et al. [14]. More recently, Kosta et al. [8] further improved this idea. Although such virtualized offloading can be considered as simple and general solution, it lacks flexibility and control over offloadable components. Therefore, we consider that application developers can better organize their application logic using the established Android service design patterns and benefit from the MACS middleware.

Ou et al. [13] propose class instrumenting technique, i.e. a process to transform code classes into a form which is suitable for remote execution. Two new classes are generated from the original class, one is an instrumented class which has the real implementation and the same functionality as the original class, the other is a proxy class, whose responsibility is only to call the function written in the instrumented class. Then, the instrumented class can be offloaded to remote cloud, and the call will be invoked from the instrumented in the remote cloud. In MACS we adopt similar idea, but unlike Ou et al. [13] we use standardized language for the proxy interfaces which is already widespread in the Android platform. The Cuckoo framework [6] and MAUI system in [2] implement a similar idea. Our MACS middleware is inspired by these solutions. However, MACS middleware does extra profiling and resource monitoring of applications and adapts the partitioning decision at runtime.

An important challenge in partitioned elastic applications is how to determine which parts of code should be pushed to the remote clouds. The graph based approach to model the application has been used in several works. Giurgiu et al. [3] use "consumption" graphs and decide which part should be running locally or remotely. It finds a cut of the consumption graph with a goal function, which minimizes the total sum of communication cost, transmitting cost and the cost of building local proxies. The AIDE platform [4] uses a component-based offloading algorithm, which mainly focuses on minimum historical transmission between two partitions. The $(k+1)$ partitioning algorithm, introduced by Ou et al. [13], is applied to a multi-cost graph representing the class-based components. A similar approach is done by Gu et al. [4, 5]. Zhang et al. [19, 17] use a general Bayesian inference to make the partitioning decision. However, executing constantly executing graph or inference algorithms on the mobile device takes significant resources on the constrained device. We use an integer linear optimization model to describe the offloading so that it is not only easy to implement, but it can also be independently solved if the remote clouds are not available temporarily.

VI. Mobile Augmentation Cloud Services

The goal of our MACS middleware is to enable the execution of elastic mobile applications. Zhang et al. [17] consider elastic applications to have two distinct properties. First, an elastic application execution is divided partially on the device and partially on the cloud. Second, this division is not static, but is dynamically adjusted during application runtime. The benefits of having such an application model are that the mobile applications can still run independently on mobile platforms, but can also reach cloud resources on demand and availability. Thus, mobile applications are not limited by the constraints of the existing device capacities.

MACS architecture is depicted on Figure 1. In order to use MACS middleware, the application should be structured using established Android services pattern. Android is already established as the most prominent mobile phone platform. Additionally, its application architecture model allows decomposition of applications into service components which can be shared between applications. A MACS application consists of an application core (Android activities, GUI, access to devices sensors) which cannot be offloaded, and multiples services (S_i) that encapsulate separate application functionality (usually resource-demanding components) which can be offloaded (S_{Ri}). The services communicate with the application through an interface defined by the developer in the Android interface definition language (AIDL).

As service-based implementation is adopted, for each service we can profile following metadata:

- type: whether can be offloaded or not
- memory cost: the memory consumption of the service on the mobile device
- code size: size of compiled code of the service
- dependency information on other services, for each related module, we collect following:
    - transfer size: amount of data to be transferred
    - send size: amount of data to be sent
    - receive size: amount of data to be received

Metadata is obtained by monitoring the application execution and environment.

Android services are using Android inter-process communication (IPC) channels for RPC. The services are registered in the Service Manager, and a binder maintains a handle for each service. Then an application, that wants to use a service, can query the service in the Service Manager. Upon service discovery, the Android platform will create a service proxy for the client application. All the requests to access the service will be sent through the service proxy, and then forwarded to the service by the binder. After processing the requests, results are sent back to the service proxy on the client application through the binder. Finally, the client gets the result from the service proxy. From client's point of view, there is no difference between calling a remote service or calling a local function.

The offload manager determines the execution plan, and then the services to be offloaded are pushed to the cloud. The results are sent back to the application upon completion. Our approach is similar to the Cuckoo framework [6], however, MACS allows dynamic application partitioning at runtime, where Cuckoo only enables static partitioning at compile time. MACS monitors the execution of the services and the environment parameters. Whenever the situation changes, the middleware can adapt the offloading and partitioning.

The main goal of MACS is to enable transparent computation offloading for mobile applications. Therefore, our middleware tries to fit the usual Android development process and bring the developers an easier way to offload parts of their applications to remote clouds in a transparent way. MACS hooks into the Android compile system, makes modifications of generated Java files from AIDL in the pre-compile stage. Developers need to include MACS SDK libraries into their Android project.

Since our implementation wants to bring the developers an easier way to distribute their application to remote cloud, the low-level implementation should be transparent to them. The way to hide the low-level implementation is as follows. Recalling the Android compile system and combining with the idea of using the services on Android, the possible way to make modification of generated Java file from AIDL is in the pre-compile stage. Our code is embedded there to realize the transparency to the developers. Each time while the developers compile their projects by using Ant tool, our code will be embedded without notice. The way to add a customized process while building with Ant tool is to write a new target, which can be treated as a task.

At the cloud side, the MACS middleware handles the offload requests from the clients, installs of offloaded services, their initialization and method invokes (s. Figure 9 in the Appendix). The cloud-side MACS middleware is written with pure J2SE so that it can run on any machines with installed J2SE.

MACS middleware monitors the resources on the mobile execution environment and available clouds. It then forms an optimization problem whose solution is used to decide whether the service which contains the called function should be offloaded or not. When the service is determined to be offloaded to the remote cloud, our middleware tries first to execute the service remotely. If there is no such service on the remote clouds, our framework transmits the service code (jar file) to the cloud, and the corresponding results after the service execution are returned to the mobile device. The cloud caches the jar files for subsequent executions.

Except for the computation offloading, our framework also features simple data offloading. If files are needed to be accessed on the remote cloud, MACS file transmission (MACS-FTM) transfers automatically the non-existing files from the local device and vice versa. Basically, the middleware at pre-compile time inserts a line of code after a file object is created by using the File object in Java. This code snippet retrieves file information of the file at runtime. When no such file exists in the remote cloud, MACS-FTM throws an exception which is caught by the middleware which in turn obtains the file from the device.

VII. Adaptive Computation Offloading

The proposed model and corresponding algorithm are supposed to be applied for scenario which is computationally intensive [7]. The requirements for the developer are that the code should be structured in a model in advance.

The developer should also provide or use extra tools to extract meta-information from given modules and then tag each module with some parameters. The tagged parameters are used for deciding on code partition later.

Let us suppose that we have n number of modules which can be offloaded, S_1, S_2 ... S_n. Each of the modules has several properties described as metadata, i.e. for specific module i, its memory cost mem_i, code size $code_i$. Let us consider the k number of related module which can be offloaded. For each of them, we denote the transfer size $tr_1, tr_2 \ldots tr_k$, send size $send_1, send_2 \ldots send_k$, receive size $rec_1, rec_2 \ldots rec_k$, where $\{1..k\} \subseteq \{1..n\}$ and $send_k + rec_k = tr_k$. Meanwhile, we introduce x_i for module i, which indicates whether the module i is executed locally ($x_i = 0$) or remotely ($x_i = 1$). The solution $x_1, x_2 \ldots x_n$ represents the required offloading partitioning of the application.

The cost function is represented as follows:

$$\min_{x \in 0,1}(c_{transfer} * w_{tr} + c_{memory} * w_{mem} + c_{CPU} * w_{CPU}) \quad (1)$$

where

$$c_{transfer} = \sum_{i=1}^{n} code_i * x_i + \sum_{i=1}^{n}\sum_{j=1}^{k} tr_j * (x_j \text{ XOR } x_i) \quad (2)$$

$$c_{memory} = \sum_{i=1}^{n} mem_i * (1 - x_i) \quad (3)$$

$$c_{CPU} = \sum_{i=1}^{n} code_i * \alpha * (1 - x_i) \quad (4)$$

There are three parts in the cost function. The first part depicts the transfer costs for the remote execution of services, including the transfer cost of its related services which are not at the same execution location. The latter part of Eq.(2) implicitly includes the dependency relationship between modules, i.e. if the output of one module is an input of another. The c_{memory} contains the memory costs on the mobile device, and c_{CPU} the CPU costs on the mobile device is, where α is the convert factor mapping the relationship between code size and CPU instructions, which is taken to be 10 based on [12]. w_{tr}, w_{mem}, w_{CPU} are the weights of each costs, which can lead to different objectives, for example lowest memory costs, lowest CPU load or lowest interaction latency.

The three constraints are expressed as the following:

Minimized memory usage. First, the memory costs of resident service can not be more than available memory on the mobile device, i.e.

$$\sum_{i=1}^{n} mem_i * (1 - x_i) \le avail_{mem} * f_1 \quad (5)$$

where $avail_{mem}$ can be obtained from the mobile device, f_1 is the factor to determine the memory threshold to be used, because the application can not occupy the whole free memory on the mobile device.

Minimized energy usage. Second, for the offloaded services, the energy consumption of offloading should not be greater than not offloading [10], i.e.

$$E_{local} - E_{offload} > 0 \quad (6)$$

The local energy consumption can be expressed using the number of local instructions to be executed I_{local}, local execution speed S_{local} and the power used of local execution P_{local} [10]. At the first decision time, I_{local} is estimated according to the code size. After the first decision, this number is collected from our framework (by calling the statistic method provided from Android API). Obviously, there is relationship between the instruction number and the power used for that instruction while doing power profiling.

$$E_{local} = \frac{P_{local} * I_{local}}{S_{local}} \quad (7)$$

The energy costs of offloading some parts to remote cloud can be expressed as the sum of energy consumption during waiting for the results from the cloud E_{idle}, transferring (including sending E_s and receiving E_s) the services to be offloaded [10] and also the additional data which may be needed on the remote cloud E_{extra}.

$$E_{offload} = E_s + E_{idle} + E_r + E_{extra} = P_s * (t_s + t_{extra}) + P_{idle} * t_{idle} + P_r * t_r$$

The idle time of the mobile device waiting for the result from cloud can be treated as the execution time of remote cloud, so the formula becomes

$$E_{local} - E_{offload} = \frac{P_{local} * I_{local}}{S_{local}} - \frac{P_{local} * I_{local}}{S_{cloud}} - \frac{P_s * (D_s + D_{extra})}{B_s} - \frac{P_r * D_r}{B_r}$$

where D_s and D_r are the total data sizes to be sent and received, D_{extra} is the size of extra data needed because of offloading, which is determined at runtime, B_s and B_r are the bandwidths of sending and receiving data, and S_{cloud} is the remote execution speed. Additionally,

$$I_{local} = \sum_{i=1}^{n} code_i * type_i * x_i \quad (10)$$

$$D_s = \sum_{i=1}^{n} send_i * type_i * x_i \quad (11)$$

Fig. 2. Snapshot of prototype application of face detection and recognition using MACS middleware.

$$D_r = \sum_{i=1}^{n} rec_i * type_i * x_i \quad (12)$$

where $type_i \in \{0,1\}$ represents whether a service is offloadable or not.

Minimized execution time. Third, the third constraint is enabled when the user prefers fast execution, i.e.

$$t_{local} - t_{offload} > 0 \quad (13)$$

The local execution time can be expressed as the ratio of CPU instructions to local CPU frequency, meanwhile, the remote execution time consists of the time consumed by CPU, file transmission and the overhead of our middleware.

$$t_{local} = \frac{I_{local}}{S_{local}} * x_i \quad (14)$$

$$t_{offload} = (\frac{I_{local}}{S_{cloud}} + \frac{D_{extra}}{B_s} + t_{overhead}) * x_i \quad (15)$$

where $t_{overhead}$ is the overhead which our framework brings in. According to the constraints above, we now transform the partitioning problem to an optimization problem. The middleware determines the execution location by solving a linear integer optimization problem. The decision maker receives the input parameters and execution constraints. It then returns the corresponding running locations. The solution of $x_1, x_2 .. x_n$, is the optimized partitioning strategy. By using integer liner programming (ILP) on the mobile device, MACS gets a global optimization result. Whenever the the parameters in the model change, such as available memory or network bandwidth, the partitioning is adapted by solving the new optimization problem.

MACS middleware defines the abstraction of a decision maker so that we can apply different decision makers which determine the execution location of each "offloadable" service. In our experiments we used Cream[1], an open-source class library for constraint programming in Java. It provides enough features to run the decision solver on an Android platform, with acceptable solving speed in the order of tens of milliseconds. During the calculation, as an objective function is taken the sum of transmission cost, memory cost and CPU cost.

Although MACS introduces the overhead because of using a proxy for communication between offloaded services and the mobile application, the overhead is relatively small, which is shown in the evaluation. MACS also profiles each offloaded module/service to dynamically change its execution plan and adjust the partitioning.

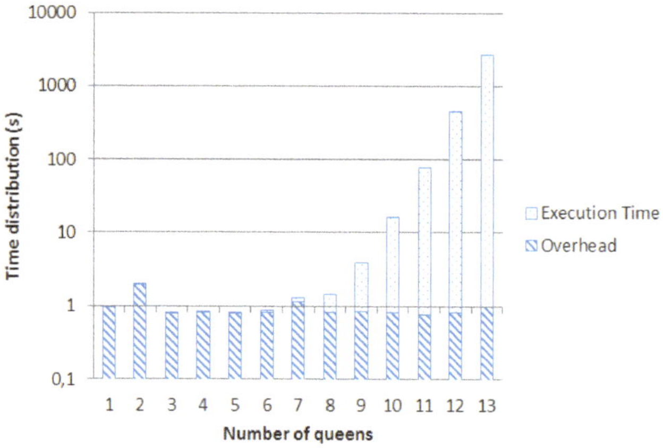

(a)

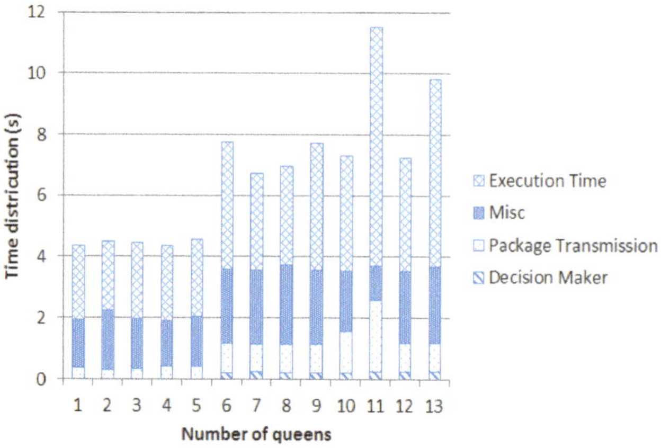

(b)

Fig. 4. Total time distribution of N-queens: (a) local execution and (b) remote execution.

VIII. Experimental Evaluation of Offloading

We evaluate our MACS middleware with two use case smartphone applications. The first application implements the well-known N-Queens problem. It is chosen because the performance bottleneck represents a pure computation problem. This use case can easily show the overhead introduced by MACS middleware. The second application involves face detection and recognition in video files. This use case involves lots of computation, but also requires much more memory resources to process and obtain results. Table 8 shows the problem space in terms of N.

[1] http://bach.istc.kobe-u.ac.jp/cream/

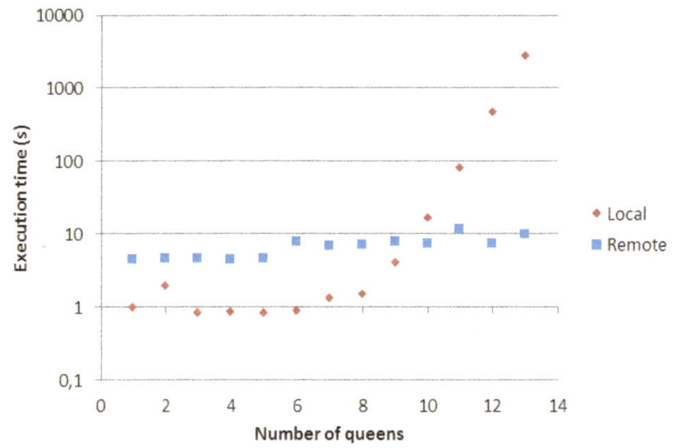

Fig. 3. Snapshot of prototype application of face detection and recognition using MACS middleware.

TABLE I
HARDWARE COMPONENTS OF MOBILE DEVICES AND DESKTOP COMPUTER

Hardware Component	Android Milestone	PC
Processor	ARM A8 600MHz	QuadCore 2.83GHz
Memory	256MB	8GB
WLAN	Wi-Fi 802.11 b/g	N/A
OS	Android 2.2	Windows XP x64

The second case can process a video file, and detect faces from the video file, cluster them and provide the time point cues for video navigation. The results can then be used for faster video navigation on small screen devices (Figure 2). The video file is processed with OpenCV[2] and FFmpeg[3] libraries. We use FFmpeg to open video files, and scan it frame by frame. Face detection in video files is done by detecting faces in video frames. In the processing, faces in the video file are detected by the existed implementation in OpenCV, and then the detected faces are recognized by the method proposed by Turk and Pentland [15], and after that, the faces are clustered.

In the implementation we used JavaCV[4] for video processing. When the application gets the results from the processing, it shows all detected faces as a clustered view. The user can select a cluster, and then navigate to the time points where that face occurs in the video. Thus, the application can accelerate navigation in a video based on persons that occur within.

A. Setup of the Evaluation

Hardware. The hardware we are using in the evaluation is as follows. A Motorola Milestone mobile phone based on Android platform 2.2 is used in the evaluation. A desktop computer which includes quad-core CPU acts as a cloud provider that can host the offloaded computation. The details about the hardware components for the mobile device and desktop computer are shown in Table 1.

TABLE II
ESTIMATED ENERGY CONSUMPTION OF A MOBILE DEVICE

Hardware Component	Estimated Energy Consumption (W)
Processor	0.4 (Idle: 0.05)
WLAN	0.75 (Low: 0.03)
LCD	0.9

Network Topology. While offloading services to the remote cloud, the mobile phone connects to a nearby access point. Since the wireless local area network is encrypted with Wi-Fi Protected Access 2 (WPA2) security protocol, the data speed is not as fast as non-encrypted considering of the overhead introduced by the security protocol. The desktop computer is connected to the Internet directly by network cable, whose bandwidth is 100 Mbps.

Energy Estimation Model. We adopt a method as

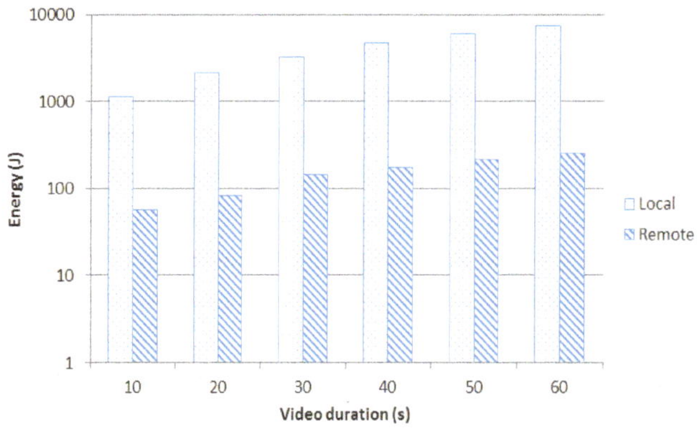

Fig. 7 Energy consumption of face detection.

the one proposed by Zhang et al. [16], a power model for an Android phone and a measurement application for the energy consumption on the Android-based mobile device on the fly. Using their software, the energy consumption of each hardware component of the Motorola Milestone such as LCD, CPU and Wi-Fi can be measured separately (see Table 2).

B. Results of the Use Case 1

We use the algorithm by Sedgewick and Wayne[5]. The basic idea is to use recursion and back-tracking to enumerate all possible solutions. Although it is not the best algorithm, it is often used for solving the N-queens puzzle. It is clear that with the increase of N, much more steps are spent to find solutions, which is extremely time-consuming for the mobile device.

We run the N-Queens on the local device and offload to the remote cloud separately, for $N=1$ to $N=13$. For $N=14$, it will take hours to finish on the local device, it is not realistic not to be offloaded while doing computation after $N=14$. Table 8 in the Appendix provides overview of the problem space in terms of N.

[2] http://opencv.willowgarage.com
[3] http://ffmpeg.org
[4] http://code.google.com/p/javacv

[5] http://introcs.cs.princeton.edu/java/23recursion/Queens.java.html

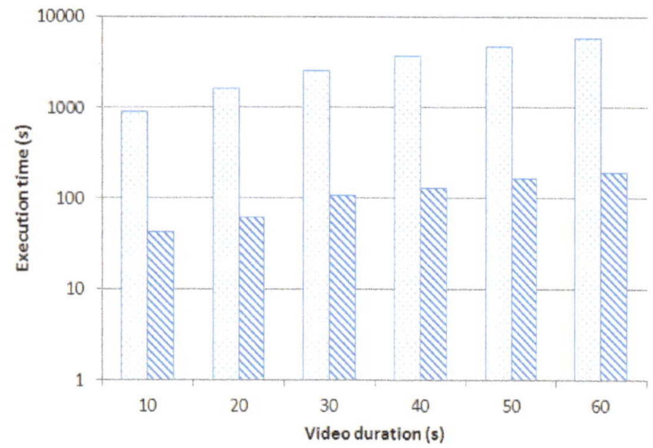

Fig. 6. Execution time of face detection

Figure 3 shows the time duration of execution of the specific calculation service. From $N=1$ to $N=9$, the execution speed on the local device is acceptable compared with the remote speed and to run the method locally is better, but after $N=10$, the remote speed dominates to be the better option as the computation time dominates the total time in the rest cases, and the remote execution speed is also relative stable, there is no huge variation for remote speed.

Figure 4 shows the different times, which are made up of the total spent time. With the increase of the queens number, the local execution time increases outstandingly, especially from $N=9$, the execution time of calculating solution occupies more than half of the total time. Meanwhile, the overhead, our framework brings in, remains constant. As for the remote execution, the overhead is broken down into three parts, one is the package offloading time, one is the decision making time, the rest is the residual overhead. It shows that our decision model costs only little time to finish the determination. The transmission time of remote package occupies also few periods of total time, since the remote package is small. The execution time of solving the N-queens is relative stable, except for $N=11$, which is a deviation during the execution and measurement.

The last Figure 5 shows the results of consumed energy with and without offloading. As for the local execution, most of the time is spent on computation, since our energy model involves CPU and LCD, and the LCD is always on while computation, so that the energy consumption of CPU and LCD dominates the total energy consumption of local execution. The execution time is significantly increased from $N=9$ compared to the remote execution, which leads to the highest energy value. In contrast to those, the remote execution time is nearly stable, so that the consumed energy is almost at the same level.

C. Results of the Use Case 2

Six video files are used in the evaluation. All of them belong to a same original video file with different length of time, 10, 20, 30, 40, 50 and 60 seconds (see Table 3). The video resolution is 720 pixels × 480 pixels, the fps is set to 30, the overall bit rate is 1500 Kbps and the video is compressed with MPEG-4 format, 3GPP Media Release 5 profile. The audio codec is AAC, and the bit rate for audio is

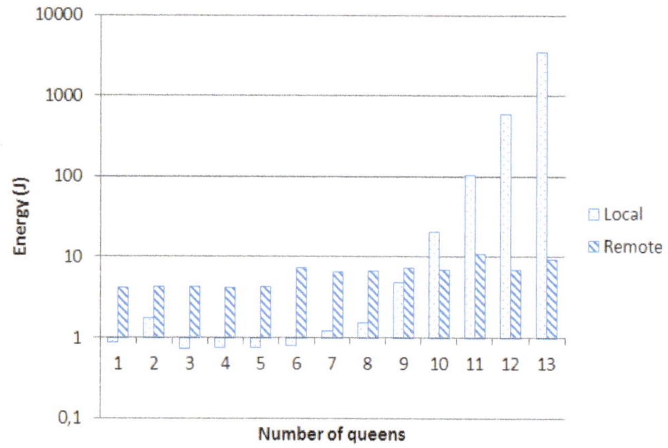

Fig. 5. Energy consumption of the N-queens app.

30 Kbps.

TABLE III
VIDEO DURATION AND FILE SIZE

Duration (seconds)	File size (Bytes)
10	1864984
20	3864612
30	5827420
40	7754219
50	9633240
60	11584020

In order to get a more accurate estimation of execution time which is used in the model, we first run the face detection services locally, and keep track of the spent time (second) and the file size (bytes), and then a linear regression is used to reflect the relationship between the spent time and the file size. Considering the number of CPU instructions provided by Android API, it can only be used to make estimation on the execution which involves no native calls, we don't directly use that count, but focus on the execution time. The regression shows that,

$$Time = 0.0005 * FileSize - 246.09 \tag{16}$$

We use this heuristic equation in our model to determine the execution time.

On Figure 6 can be seen clearly that the execution time is reduced hugely while offloading compared with the local execution. Even dealing with the 10 seconds video file, the local device spends more than 15 minutes on processing and detecting, but the corresponding remote offloading takes only less than 1 minute. Each time the computation is offloaded to the remote cloud, the execution speed can be reduced by more than 20 times, see Table 4. It is absolutely not acceptable to let the CPU of local mobile device 100% load for such long time, and it confirms that the video processing task is still a huge burden for the mobile device.

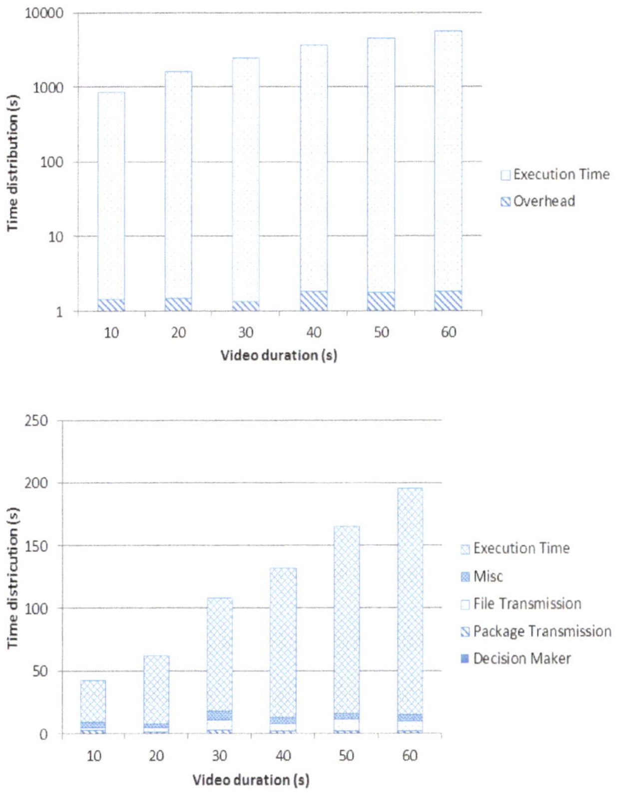

Fig. 8. Total time distribution of face detection: (a) local execution and (b) remote execution.

TABLE IV
VIDEO DURATION AND SPEED UP OF FACE DETECTION IN VIDEO FILES

Duration (seconds)	10	20	30	40	50	60
Speed up	x20	X26	X23	X28	X28	X29

With the huge difference between local and remote execution time, we can conclude that the local energy consumption is worse than the remote ones, because most of the time is spent on CPU and LCD, which are the top two energy consuming components (s. Figure 7). Table 5 also depicts the energy saving situation while offloading. Energy can be saved by more than 94% thanks to the offloading.

Figure 8 describes the composition of the local and remote total spent time in details. As the execution time increases with the bigger video file size, the overhead our framework brings occupies about only 0.1%, which can be nearly omitted. Regarding to the remote execution, the total time spent consists of execution time, transmission time of needed files, package transmission (service offloading) time, and decision making time. With the increase of the video file size, the file transmission time also rises, but compared to the total time, it is not significant. The decision making does its determination in less than 1 second, which is only 1% of the total time spent. The total overhead our framework brings is about 5% of the total time, which is acceptable considering about the speed up and energy save above.

The face clustering can only be done on the remote cloud because of the software limitations on the local mobile device. Detailed results can be seen in Table 6 in the Appendix. Most of the execution time is spent on building/rebuilding the training set. If the training set is already available before the remote execution, then the estimated execution time can be significantly reduced.

TABLE V
VIDEO DURATION AND SPEED UP OF SAVED ENERGY

Duration (seconds)	10	20	30	40	50	60
Saved Energy (%)	94.98	96.07	95.59	96.37	96.33	96.55

IX. DISCUSSION

Offloading perhaps is not suited for every mobile application, but from the results of the two use cases, we see that when an application uses complex or time consuming algorithms such as recursion, by offloading those parts into the cloud, time and energy consumption are reduced, so that the local execution time is reduced to an acceptable level. Offloading can lower the CPU load on a mobile device significantly. It can also save lots of energy, which indicates that the battery lifetime can be increased compared to the local execution, as shown in the second use case, where more than 90% of energy is saved and the calculation speed is up to 20 times over local execution.

The results also prove that the overhead of our framework is small and acceptable with the increase of needed computation, it is better to push those computations which cost considerable resources to the remote cloud. But for the small N in the N-Queens problem, the overhead occupies almost half of the total execution time because of the needed computation is small so that it takes only little time to obtain the results. This shows a clear advantage of local execution over remote offloading when the needed computation is not much. In a word, the more computation is needed, offloading has more advantages. Since we use Wi-Fi in the evaluation, the time of sending files and receiving results has small proportion, but if 3G or GPRS are used, the offloading time will surely increase.

X. CONCLUSIONS AND FUTURE WORK

The results show that the local execution time can be reduced a lot through offloading, which is sometimes not acceptable for users to wait for, and by pushing the computation to the remote cloud can lower the CPU load on mobile devices significantly thanks to the remote cloud, since most of the computations are offloaded to the remote cloud. Meanwhile, lots of energy can be saved which indicates users can have more battery time compared to the local execution. The results also prove that the overhead of our framework is small.

Our framework supports offloading of multiple Android services. If there are multiple services in one application and all of those services can be offloaded to the remote clouds, our resource monitor natively supports this situation and can make the corresponding allocation determination, so that some of the services should be offloaded and the rest of the services should be run locally.

The next steps are to enable parallelization of the offloaded services. Additionally, we can extend the current middleware so that it supports automatic partitioning arbitrary mobile applications. A great challenge is how to estimate the characteristics of an application depending on different input parameters, which is precisely the relationship between the input of the invoked method and the execution time. We could characterize the relationship between execution time and input parameters by running the target application several times and adapt the offloading algorithm accordingly.

APPENDIX

Figure 9 presents the registration flow of "offloadable" services (a) and the optimization process (b).

In Table 6, the data on the left side of the slash sign is for local execution, whereas on the right side is for remote execution.

TABLE VI
EVALUATION RESULTS OF USE CASE 1 IN DETAILS

Num-ber	Execution Time (s)	Overhead (s)	CPU Instructions	Energy (W)
1	0.973 / 4.383	0.97 / 1.97	155986 / 847523	0.877 / 4.188
2	1.953 / 4.534	1.95 / 2.269	162317 / 847189	1.759 / 4.329
3	0.82 / 4.486	0.817 / 2.012	155925 / 847421	0.739 / 4.285
4	0.847 / 4.378	0.839 / 1.943	164070 / 847368	0.766 / 4.183
5	0.827 / 4.602	0.811 / 2.062	164021 / 847941	0.751 / 4.397
6	0.883 / 7.786	0.84 / 3.629	201058 / 848234	0.812 / 7.513
7	1.318 / 6.752	1.178 / 3.592	383161 / 848363	1.242 / 6.531
8	1.473 / 7.006	0.841 / 3.779	1008483 / 848173	1.579 / 6.770
9	3.994 / 7.746	0.868 / 3.611	4475993 / 848379	4.845 / 7.475
10	16.431 / 7.337	0.836 / 3.549	22916437 / 849381	21.026 / 7.094
11	80.055 / 11.527	0.794 / 3.72	128838160 / 850019	103.754 / 11.112
12	459.055 / 7.256	0.832 / 3.571	780925799 / 849810	596.439 / 7.013
13	2766.508 / 9.839	1.015 / 3.685	6937052630 / 849590	3596.054 / 9.468

Detailed results of use case 2. The data of Table 7 in the left side of slash is local execution, the middle one is remote execution of face detection, the last one is remote execution of face detection with face recognition.

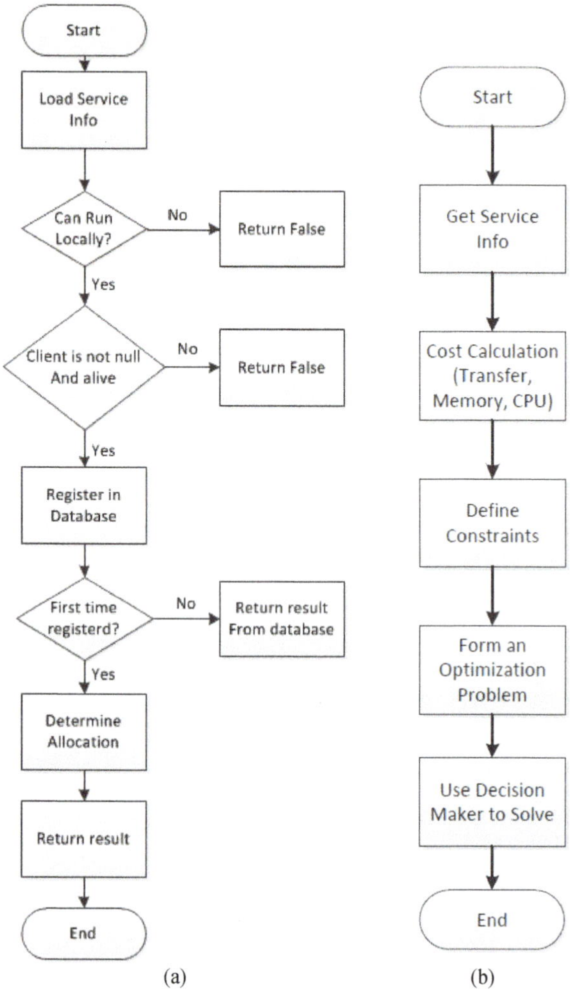

Fig. 9. Flow charts: (a) process registration and (b) allocation determination.

TABLE VII
EVALUATION RESULTS OF USE CASE 2 IN DETAILS

Duration (s)	Execution Time (min)	Overhead (s)	CPU Instructions (Million)	Energy (kW)
10	14.8 / 0.7 / 1.9	1.5 / 5.0 / 4.6	7.4 / 1.8 / 2.1	0.9 / 0.1 / 0.2
20	27.6 / 1.0 / 12.3	1.5 / 2.9 / 5.9	7.2 / 2.6 / 5.5	1.7 / 0.1 / 1.0
30	42.6 / 1.8 / 25.0	1.4 / 7.3 / 5.9	8.3 / 4.1 / 7.6	2.6 / 0.1 / 2.0
40	62.5 / 2.2 / 70.0	1.9 / 5.5 / 5.8	10.1 / 5.7 / 8.4	3.7 / 0.2 / 5.5
50	77.7 / 2.8 / 122.5	1.8 / 5.1 / 7.2	11.8 / 7.8 / 9.6	4.7 / 0.2 / 9.6
60	96.7 / 3.3 / 152.5	1.9 / 5.8 / 6.5	13.3 / 9.0 / 11.1	5.8 / 0.3 / 11.9

Table 8 shows the number of possible solutions of the N-Queens problem in terms of N.

TABLE VIII
N-QUEENS PROBLEM SPACE

N	4	6	8	10	12	14
No. Distinct Solutions	2	4	92	724	14200	365596

ACKNOWLEDGMENT

This work is supported by the Excellence Initiative of German National Science Foundation (DFG) within the research cluster Ultra High-Speed Mobile Information and Communication (UMIC) and in part by NRW State within the B-IT Research School. We thank Tian Yu for the prototype implementation.

REFERENCES

[1] B.-G. Chun and P. Maniatis, "Augmented Smartphone Applications Through Clone Cloud Execution," in *Proceedings of the 12th Workshop on Hot Topics in Operating Systems (HotOS XII)*, pp. 1-8, Monte Verita, Switzerland, 2009. USENIX.

[2] E. Cuervo, A. Balasubramanian, D. Cho, A. Wolman, S. Saroiu, R. Chandra, and P. Bahl, "MAUI: Making Smartphones Last Longer with Code Offload," In *Proceedings of the 8th International Conference on Mobile Systems, Applications, and Services (ACM MobiSys '10)*, pp. 49–62, San Francisco, CA, USA, 2010. ACM.

[3] I. Giurgiu, O. Riva, D. Juric, I. Krivulev, and G. Alonso, "Calling the Cloud: Enabling Mobile Phones as Interfaces to Cloud Applications," In *Proceedings of the 10th ACM/IFIP/USENIX International Conference on Middleware (Middleware '09)*, pp. 1–20, Urbana Champaign, IL, USA, November 2009. Springer.

[4] X. Gu, A. Messer, I. Greenberg, D. Milojicic, and K. Nahrstedt, „Adaptive Offloading for Pervasive Computing," *IEEE Pervasive Computing*, vol.3, pp.66–73, Jul. 2004.

[5] X. Gu, K. Nahrstedt, A. Messer, I. Greenberg, and D. Milojicic, "Adaptive Offloading Inference for Delivering Applications in Pervasive Computing Environments," in *Proceedings of the First IEEE International Conference on Pervasive Computing and Communications (PerCom 2003)*, pp. 107–114, Dallas-Fort Worth, TX, USA, 2003. IEEE.

[6] R. Kemp, N. Palmer, T. Kielmann, and H. Bal, "Cuckoo: a Computation Offloading Framework for Smartphones," in *Proceedings of the 2nd International ICST Conference on Mobile Computing, Applications, and Services (MobiCASE 2010)*, pp. 59-79 Santa Clara, CA, USA, October 2010.

[7] R. Kemp, N. Palmer, T. Kielmann, F. Seinstra, N. Drost, J. Maassen, and H. Bal, „eyeDentify: Multimedia Cyber Foraging from a Smartphone," in *Proceedings of the 11th IEEE International Symposium on Multimedia (ISM 2009)*, pp. 392–399, 2009.

[8] S. Kosta, A. Aucinas, P. Hui, R. Mortier and X. Zhang, "Unleashing the Power of Mobile Cloud Computing using ThinkAir," *CoRR, abs/1105.3232, 2011*, informal publication.

[9] D. Kovachev, Y. Cao, and R. Klamma, "Mobile Cloud Computing: A Comparison of Application Models," *CoRR, abs/1107.4940*, 2011.

[10] K. Kumar and Y.-H. Lu, "Cloud Computing for Mobile Users: Can Offloading Computation Save Energy?," *Computer*, vol. 43, no. (4), pp. 51–56, April 2010.

[11] P. Mell and T. Grance, "The NIST Definition of Cloud Computing," 2009.

[12] J. K. Ousterhout, "Scripting: Higher Level Programming for the 21st Century," in *IEEE Computer*, vol. 31, pp. 23–30, 1997.

[13] S. Ou, K. Yang, and J. Zhang, "An Effective Offloading Middleware for Pervasive Services on Mobile Devices," *Pervasive Mob. Comput.*,vol.3, pp.362–385, August 2007.

[14] M. Satyanarayanan, P. Bahl, R. Cáceres, and N. Davies, "The Case for VM-Based Cloudlets in Mobile Computing," *IEEE Pervasive Computing*, vol. 8, no. 4, pp. 14–23, October 2009.

[15] M. Turk and A. Pentland. "Eigenfaces for Recognition," *J. Cognitive Neuroscience*, MIT Press, vol. 3, pp. 71–86, January 1991.

[16] L. Zhang, B. Tiwana, Z. Qian, Z. Wang, R. P. Dick, Z. Morley Mao, and L. Yang, "Accurate Online Power Estimation and Automatic Battery Behavior Based Power Model Generation for Smartphones," in *Proceedings of the Eighth IEEE/ACM/IFIP International Conference on Hardware/Software Codesign and System Synthesis*, pp. 105–114. ACM, 2010.

[17] X. Zhang, S. Jeong, A. Kunjithapatham, and S. Gibbs, "Towards an Elastic Application Model for Augmenting Computing Capabilities of Mobile Platforms," in *Third International ICST Conference on Mobile Wireless Middleware, Operating Systems, and Applications*, pp.161-174, Chicago, IL, USA, 2010.

[18] X. Zhang, A. Kunjithapatham, S. Jeong, and S. Gibbs, "Towards an Elastic Application Model for Augmenting the Computing Capabilities of Mobile Devices with Cloud Computing," Mobile Networks and Applications, vol. 16, pp. 270–284, 2011.

[19] X. Zhang, J. Schiffman, S. Gibbs, A. Kunjithapatham, and S. Jeong, "Securing Elastic Applications on Mobile Devices for Cloud Computing," in *CCSW '09: Proceedings of the 2009 ACM Workshop on Cloud Computing Security*, pp. 127–134, Chicago, IL, USA, November 2009. ACM.

14

A Client-Server System for Ubiquitous Video Service

[1]Ronit Nossenson, [2]Orit Yudilevich, [2]Omer Marlowitz
[1]*Faculty of Computer Science,* Jerusalem College of Technology, Jerusalem, Israel
[2]*School of Computer Science,* The Interdisciplinary Center, Jerusalem, Israel

***Abstract* — In this work we introduce a simple client-server system architecture and algorithms for ubiquitous live video and VOD service support. The main features of the system are: efficient usage of network resources, emphasis on user personalization, and ease of implementation. The system supports many continuous service requirements such as QoS provision, user mobility between networks and between different communication devices, and simultaneous usage of a device by a number of users.**

***Keywords* — Seamless Content Delivery, Ubiquitous Multimedia Service, Personal Multimedia Delivery, Live Video Transmission, VOD Transmission.**

I. Introduction

A new generation of distributed services ranging from entertainment services such as live video streaming, video on demand and on-line gaming, to life- saving applications such as medical services and monitoring, are being deployed in heterogeneous and ubiquitous environments. To be accepted by both users and network operators, these ubiquitous services must deliver continuous service, as well as adaptive and satisfactory Quality-of-Service with a minimum overhead of network resources. Providing ubiquitous services entails a number of complex issues, such as supporting the required QoS during a session, seamless handovers between different radio access technologies (RATs), supporting user mobility, etc.

In this article we introduce a simple client-server architecture and algorithms for live video and Video On Demand (VOD) ubiquitous services. To achieve satisfactory continuous service with minimum overhead, collaboration and coordination between small number of agents uses several communication methods including wireless or cellular connections. This article is an extended version of our previous results [13].

The main features of the architecture are as follows:

1. Efficient usage of network resources complying with the required/preferred QoS.
2. User- driven architecture which enables easy personalization.
3. Ease of implementation.

Previous work on ubiquitous multimedia services has focused on middleware solutions (see, for example, [1-7, 9, 10]). These ideas are good and effective but they require the cooperation of network operators. Since the conventional business model is defined only between the content provider and its content consumers, the readiness of operators to deploy such solutions is limited.

As illustrated in Fig. 1, our novel architecture relies on both the user's communication devices (for example, smart phone, PDA or laptop) and the continuous service/content provider system; no changes or extensions are needed in the operator network. The system can completely handle a range of continuous service requirements: QoS provision, user mobility between RATs and between different communication devices, simple user interface and personalization, simultaneous usage of the same device by a number of users (while protecting privacy) and so on. The description focuses on ubiquitous live video and VOD services, but the system can be easily extended to other services as well.

The architecture is based on the "best k" algorithm to ensure efficient use of network resources [8]. This algorithm provides high quality live-video transmission by using few agents and proposes ways to minimize the usage of network resources. Experimental results show that by using the best-k algorithm, high quality video can be delivered with an overhead factor of 1.65%.

This paper is organized as follows. In the next section the

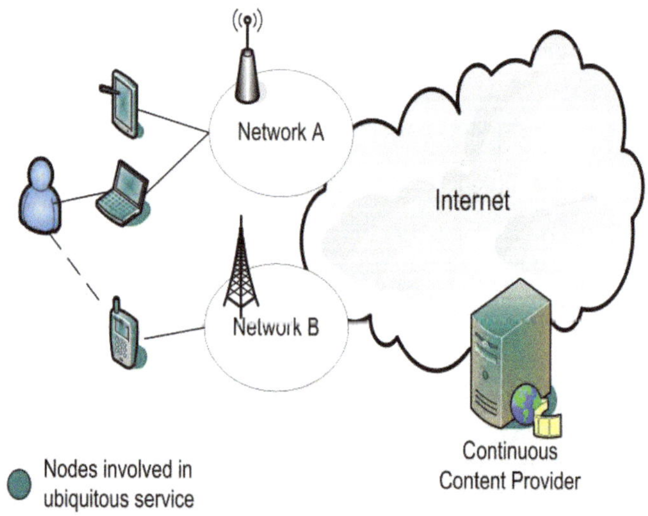

Fig. 1. High level description of a client-server system for ubiquitous service. The subscriber can receive a service via laptop, smart phone, etc.

ubiquitous system requirements are outlined. In section III, we describe the specifications of the new system by topic. These include the system building blocks, state machines and essential procedures. Next, some experimental results are described in Section IV. Finally, we suggest future research directions.

II. System Requirements

Providing ubiquitous service raises a number of issues that must be addressed. In this section we specify the technical requirements for a ubiquitous service system. First we define the terms used in this work. Next, we list the requirements according to their functionality: general, usability, QoS, multiple users' and privacy.

A. Definitions

Content provider functions as the server side. Its responsibilities include user management, QoS management and content provision. *Service* is a video service such as live video or VOD. *User* is a person who subscribes to the service. *Client* is the software which provides a special service on a user device. A user has one client on each device, per service. For example, a user who subscribes to 3 services via 4 devices has 12 clients. The client is responsible for communication between the user and the system, measuring and reporting QoS, agent management and combining the received data when necessary.

Agent is the software that is responsible for receiving and transmitting data for the service via a specific communication interface/technology. For example, a device with cellular, WiFi and BT connections has one client and up to three agents per service and user (see Fig. 2). Two or more users can use the same service via one device simultaneously (watching a movie together, for example). In this case the first user who activated the service is the *primary* user and the other users are referred to as *secondary* users. The primary user and secondary users together are referred to as the *service users' group on a device*. The primary user manages the service users' group that is using the service.

B. General Requirements

G1. The system supports ubiquitous service for live video transmission and video on demand (VOD) applications.

G2. The system attempts to provide QoS as close as possible to the preferred quality (see requirement Q1 below), with minimal user intervention and with minimal overhead for network resources.

G3. Scalability requirements: The system supports up to A potential agents per user. The system supports up to B activations of service per minute. The system supports up to C simultaneous active services. A user may have up to D active services and up to E paused services simultaneously. The parameters A, B, C, D and E can be extended by simple hardware extension.

C. Electronic Image Files (Optional)

U1. Service is supplied via one client at a specific time.

U2. The system provides a convenient user interface.

U3. The user needs to configure a set of agents for each service for each client (device).

U4. For a service, at least one device (client) and one agent must be registered.

U5. The user can alter its set of agents at any time using a convenient interface.

U6. For each service specified in requirement G1 the system supports the following operations: Start, Stop, Pause, and Resume.

U7. The "Start" operation is used for service activation the first time as well as for service re-activation after a "Stop" action. It can be used for inactive service only.

U8. The operation "Stop" is used for termination of the active or paused service.

U9. The operation "Pause" is used for temporary halts of an active service, up to a predefined timeout. The timeout can be interrupted by user operation ("Resume"). Otherwise the service is terminated.

U10. The "Resume" operation is used to continue the service after the "Pause" operation, depending on the type of service, assuming that the timeout has not expired.

U11. The outcome of the "Resume" operation for VOD

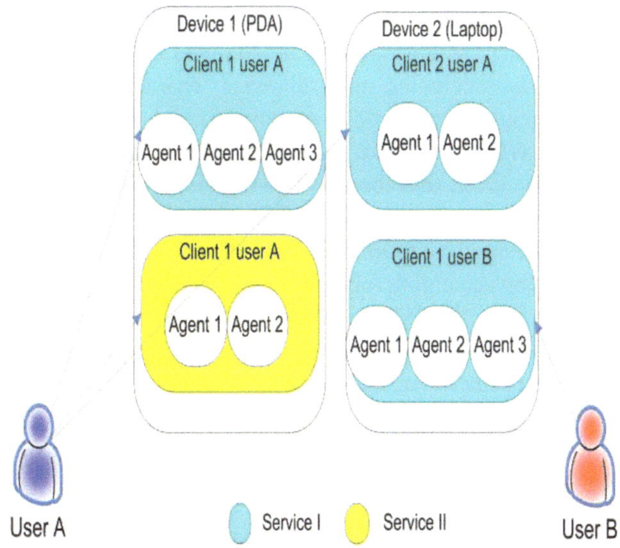

Fig. 2. User-Device-Client-Agent-Relationship. A user can have several devices and can be subscribed to several services. For each (service, user, device) triple there is one client. Each client can manage several agents. Two users can share a device, for the same service or for different services.

application is continuation of the video transmission from the point where it was paused. For the Live Video application it corresponds to resumption of the on-line video at the current time.

U12. For an active service, at least one agent is active.

U13. Changes of agents, without a change in client (switching, adding and subtract agents) are done transparently without the user's intervention.

U14. A paused service can be resumed from any subscribed client (device continuously).

U15. There can be only one active service per device.

U16. A user can have several active services, on several devices.

U17. New service activation on a device that already has another service active on it, is subject to the approval of the primary user of the active service and is equivalent to pausing/ termination of the previous service.

D. QoS requirements

Q1. The user can define *Preferred QoS parameters* and *Required QoS parameters* per service. In addition, the system has default values for these parameters per service.

Q2. The system aims to provide the user with QoS as close as possible to the preferred QoS, with minimal user intervention. If this is impossible the system aims to provide QoS above the required QoS. If the QoS falls below the required level the user is informed and the service is terminated.

Q3. The system provides continuous service for the user as long as there is at least one active client and active agent capable of providing QoS above the required level.

Q4. The QoS parameters are defined for each service separately, including the following characteristics as a minimum: bandwidth, delay and jitter.

Q5. When the QoS parameters are higher than the required level, but below the preferred thresholds, the system attempts to improve the service in the following way: (1) if the preferred QoS can be achieved using the current client (by changing agent/s), the change is performed transparently; (2) if the preferred QoS can be achieved only by another client (device), the transition can be performed, subject to the user's approval; (3) if the preferred QoS cannot be achieved using any other client (device), the system provides the best QoS possible via the current client (device).

Q6. If the QoS is below the required threshold, the system tries to improve the QoS via the current client. If this is impossible, the user is advised to move to another device. If the required QoS cannot be met the service is terminated.

Q7. For each user and active service, the system maintains a set of potential agents as a function of QoS parameters, environmental changes, user preferences, and agent availability.

E. Multiple User Requirements

M1. An active service has one primary user.

M2. There can be several secondary users for an active service.

M3. A user can join a service on a specific device which is managed by a different primary user, subject to both users' approval.

M4. A secondary user can disjoin a service; this action is equivalent to pausing or stopping the service to the specific user.

M5. A primary user can be replaced by another user, subject to both users' approval. The previous primary user becomes a secondary user in this case.

F. Privacy Requirement

P1. A user cannot access information on services that another user is subscribed to, even if they co-exist on the same device (see Fig. 2), unless the user is a primary user who is aware of the secondary users of the same service.

III. System Specifications

In this section we elaborate on the technical problems and provide specifications and algorithms for the system. The listed specifications provide a feasible solution for the pre-defined requirements. This section is divided into the following sub-sections: service state machine and QoS specification, user interface specification, agent state machine, multi user specification and user mobility issues are discussed in the final sub section.

A. Service State Machine and QoS Specifications

According to the requirements, each service can be in one of three states per user: Not Active, Active (A, B, C and D sub-states) or Paused. Fig. 3 depicts the transitions between the states, showing all the valid transitions, their triggers and actions.

1) From *"Not Active Service" to "Active Service"*: The actions that take place in this case are: (i) the system establishes a connection with the agent that sent the "Start" command before it starts to transmit the content; (ii) A handshake procedure is performed with each user's agents, and a list of available agents is generated. The handshake procedure and the management of the available agent list are described in the agent state machine below.

2) From *"Active Service" to "Not Active Service"*: This transition can occur in two cases: upon a user "Stop" command or if the QoS falls below the required level (see requirements Q2, Q6). In these cases the data flow to/from the client is terminated, a termination message is sent to all

available agents, so they can move to the "not active" state, all session data are dropped on both server and client sides.

3) From *"Paused Service" to "Not Active Service"*: A service can go from "Paused" to "Not Active" in two cases: by a user "Stop" command or by a pause timeout expiration, see requirement U9. In these cases, the system sends a termination message to all clients to turn the available agents to "not active". Service- session related information is dropped from both the server and client sides.

4) From *"Active Service" to "Paused Service"*: The user can implement this transition in several ways: by sending pause command, activation of another service on the same device, approval of client swapping proposed by the system and switching from primary user to secondary user in the multiple user service mode (see requirements U9, U17, M3). In these cases all the service session data are saved in both server and client. Service paused messages are sent to the active agents, so they can go into the "available" state. Data transmission is stopped. On the server side, additional data are stored, such as last active client, pointer to the last transmitted I frame, last packet sequence number, file offset (the location in the movie for VOD service), and last decoding format in use. In addition, the timer for maximum paused time is activated.

5) From *"Paused Service" to "Active Service"*: In case the service is resumed on the same client it was paused on, all the data exist on both the client and the server sides and the service is simply resumed. Otherwise, if the service is resumed on a different client (device) a format adaptation is performed if needed. The last I frame is sent to the client together with the following P frames and a specific notification for the video player film offset. This enables VOD service to resume from the same point it was paused on the previous device. For live video service, the service is resumed according to the current time.

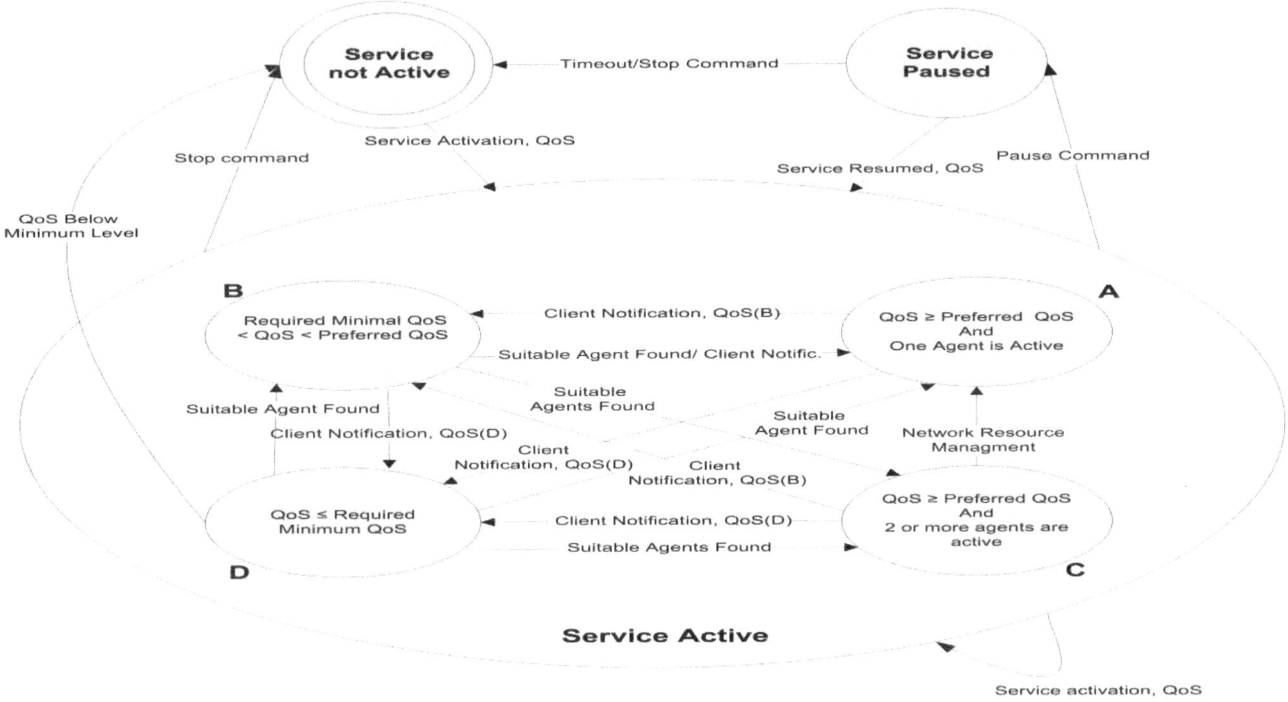

Fig. 3. Service State Machine and QoS management state machine for active service.

6) From *"Active Service" to "Active Service"*: If the "resume" command is initiated on the active client, the command is discarded, otherwise the command is equivalent to pausing the service on a current client and resuming it on a new one.

Other state transitions are server internal and are related to QoS provision. Specifically, it meets requirements G2, U12, U13, Q2, Q3, Q5, Q6 and Q7 above.

The essential server functions are:

- Ensuring an acceptable QoS level via agent management.
- Providing maximum transparency to the user.
- Ensuring efficient usage and minimum overhead of the network resources.

The essential client functions are:

- Providing the user interface to the system.
- Monitoring the QoS parameters for active services and informing the server if needed.

- Combining and synchronizing the data, using [8] or a similar algorithm.

The following information is stored on the server per active user of a specific service:

- The current QoS parameters.
- The current active client and its set of active agents.
- A list of available clients with their available agents.
- Session associated information.
- List of clients and agents used in the last time interval.
- List of forbidden client transitions.

During active service, the server runs the state machine as presented in the “Active Service State” in Fig. 3. In sub-state A the user receives its preferred QoS by one agent; thus the system does not have to improve its QoS or to reduce network resource overhead associated with it.

Sub-state B is characterized by the QoS between the preferred and required levels, thus the service can be continued along with system efforts to improve the QoS according to requirements G2 and Q5. The requirements define the following priorities: QoS, minimum user intervention and network resources (see requirements G2, U13, Q5, Q6); hence, in state B, the following procedure is performed periodically:

```
StateBProc(PreferredQoS, CurrentClient,
      AllAvaliableClients):
Begin
  Bool IsPreferredQoSPossible = false;
  Bool IsPreferredQoSReached = false;
  IsPreferredQoSPossible =
    BestQoS(PreferredQoS, CurrentDevice,
    AllAvaliableClients);
  If (IsPreferredQoSPossible)Then
    Bool IsPreferredQoSReached =
    SwapAgentbyQoS (PreferredQoS, QoSList);
  End
  If (Not IsPreferredQoSReached)Then
       ImproveCurrentClientQoS(CurrentQoS,
              CurrentDevice);
  End
End
```

The function `BestQoS` is described below, it is responsible for generating the list of respective agents and possible QoS by staging a competition between the agents (for details regarding agent competition see [8]). The QoS list includes: (i) Best QoS that can be reached by one agent of the current client (device) (`BestQoSSingleAgentCurrentClient`) and corresponding agent; (ii) Best QoS that can be reached by two or more agents on the current client (device) (`BestQoSMultiAgentCurrentClient`) and corresponding agents; (iii) Similar data about other available clients and agents (iv) Best possible QoS for each client achieved by one or more agents. The inputs to the function are: requested QoS threshold, current device, full list of

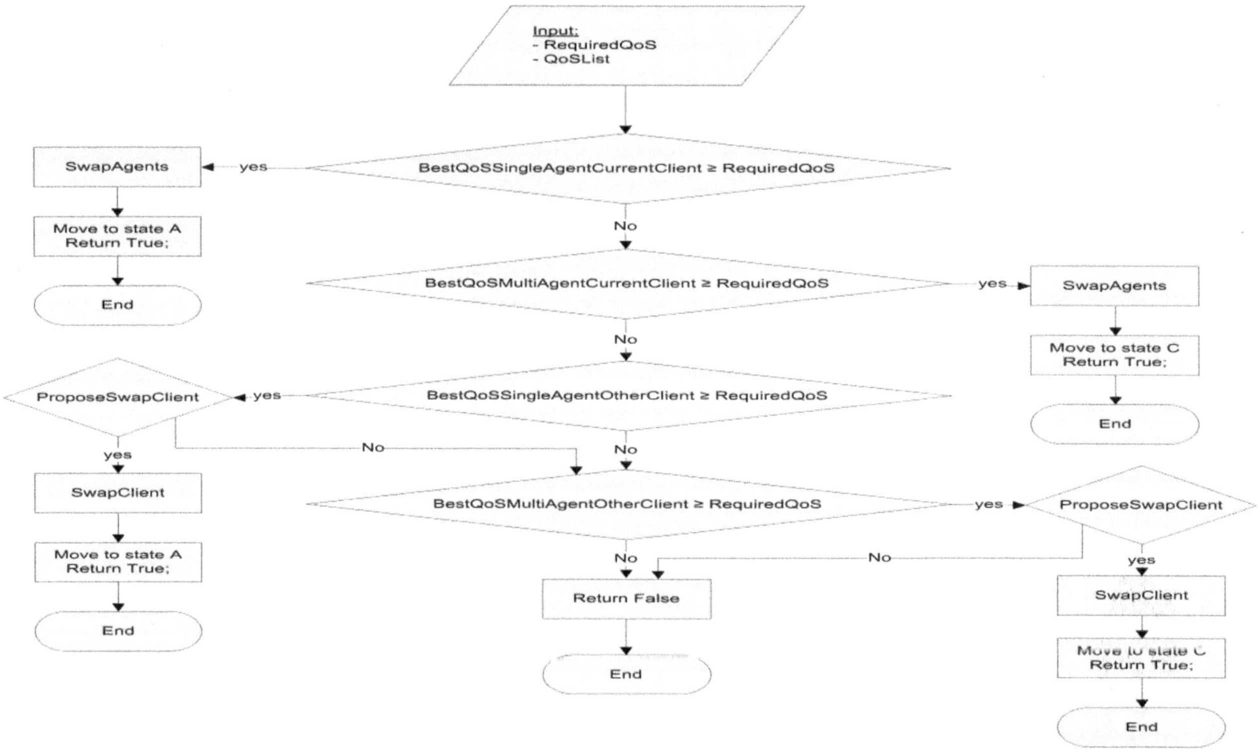

Fig. 4. Swap agent by QoS procedure flow chart

devices and agents.

The pseudo code of the function `BestQoS` is as follows:

```
BestQoS(QoSThreshold,CurrentClient,
                  AllAvaliableClients):
Begin
Competition(Current Client's Agents);
Set BestQoSSingleAgentCurrentClient, Agent;
Set BestQoSMultiAgentCurrentClient, Agents;
If((BestQoSSingleAgentCurrentClient ≥
  QoSThreshold)Or
  (BestQoSMultiAgentCurrentClient ≥
  QoSThreshold)) Then return true;
/* Needed QoS can't be reached by current
client, rest of the clients are checked */
Competition(Available Clients);
Set BestQoSSingleAgentOtherClient, AgentList;
Set BestQoSMultiAgentOtherClient,
AgentMatrix;
maxQoS = max(BestQoSSingleAgentOtherClient,
BestQoSMultiAgentOtherClient);
return (maxQos  ≥ QoSThreshold)
End
```

When agent/s that supply the required QoS is/are found, the system attempts to swap to this/these agent/s. This is performed by the `SwapAgentbyQoS()` procedure. The flow chart of this procedure is presented in Fig. 4. This procedure scans the possible agents according to a pre-defined order (requirements G2, U13, Q5, Q6). The first choice is one agent on the current client, then, multiple agents on the current clients, finally single and multiple agents of other clients. This procedure updates the QoS state machine, as required.

The system should introduce the user to the full list of other clients that can provide the preferred QoS. Client switching is always subject to user approval. In addition, the system should maintain a list of clients whose transitions are forbidden (see user interface specification for details), to avoid undesirable proposals to the user (requirement G2, Q5).

If the preferred QoS cannot be reached the system should try to improve the QoS on the current client (requirement Q5), to the best possible QoS. This is performed by the `ImproveCurrentClientQoS()` procedure. It uses the current QoS and current client as inputs. This procedure does not affect the QoS state machine, since after its termination the QoS level is still between the preferred and the required thresholds. The pseudo code of this procedure is as follows:

```
  ImproveCurrentClientQoS(CurrentQoS,
     CurrentClient)
  Begin
  MaxQoS
=max(BestQoSSingleAgentCurrentClient,

BestQoSMultiAgentCurrentClient);
   If  ((MaxQoS > CurrentQoS) And
  (MaxQoS==BestQoSSingleAgentCurrentClient))
  Then
    SwapAgents(CurrentAgents, NewAgent);
  Else if (MaxQoS > CurrentQoS)
  Then
    SwapAgents(CurrentAgents, NewAgents[]);
  End
```

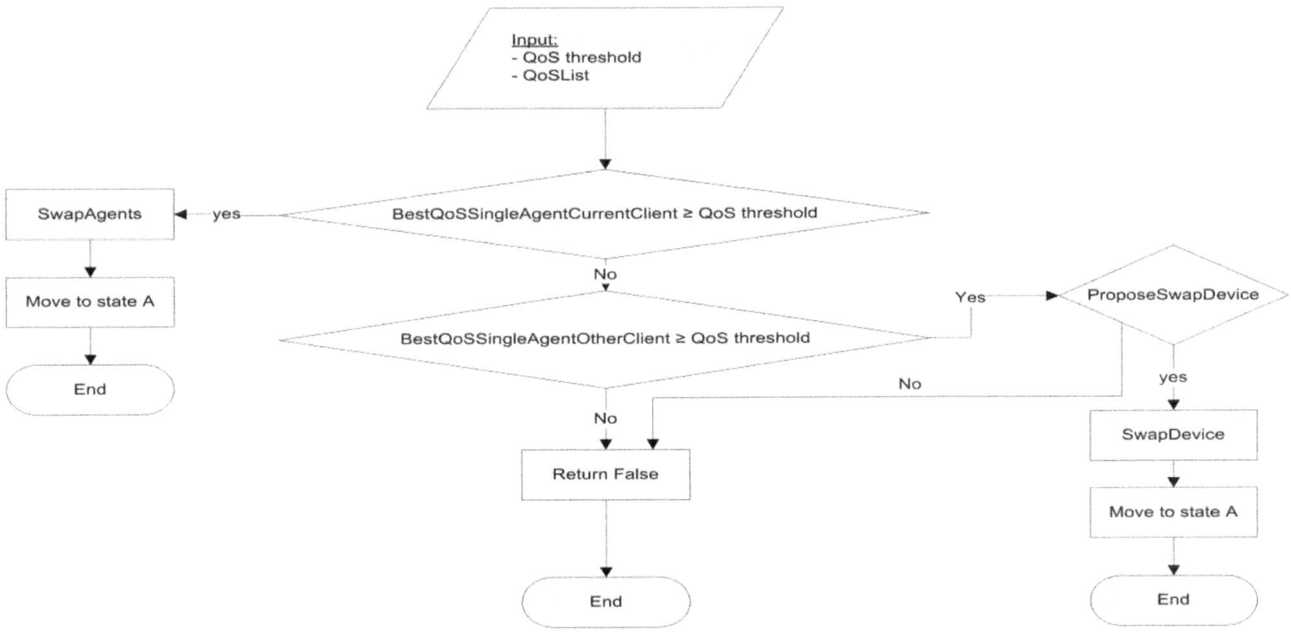

Fig. 5. Swap agent by overhead procedure Flow Chart

In sub-state C the QoS is equal or higher than the preferred QoS, but is supplied by more than one agent. In this case the system should try to reduce network resource overhead, by attempting to supply the preferred QoS using one agent if possible (requirement G2). The system should try to reduce the number of active agents. This is done by the `SwapAgentbyOverhead()` procedure that runs periodically. The flow chart for this procedure is presented in Fig. 5. The procedure updates the QoS state machine accordingly.

In sub-state D the QoS is below the required threshold. In this case the system attempts to improve the QoS; if this attempt fails the service is terminated (see Q2 and Q6.). The pseudo code of this procedure is as follows:

```
StateDProc():
Begin
Bool IsPreferredQoS, IsRequiredQoS = false;
IsPreferredQoS=SwapAgentbyQoS(PreferredQos)
;
If (PreferredQos) Then
    Move to State A;
    Return;
IsRequiredQoS                                  =
SwapAgentbyQoS(RequiredQos);
If (!IsRequiredQoS) Then
    Notify user of service termination;
    Exit Active Service State Machine;
    Service state = Not Active;
    Terminate service;
Else
  ImproveQoSCurrentClient;
  Move to State B;
End
```

B. Agent State Machine

In this sub-section we introduce the agent state machine (Fig. 6). As mentioned earlier, for active service, one or more agents can be used for service supply (they are in the "active" state), while the other agents for this service are in the "available" state. Upon service activation the system sends activation messages to all user clients (for the current service). The clients instruct their agents to move to the "*available state*". The clients and agents must respond to the activation message. This process is referred to as the "handshake procedure". The outcome of this "handshake procedure" is a list of all available agents for the current service. In order to keep the available agents list updated, all agents must send (by client) keep-alive messages. When a client recognizes that agent/s becomes available after unavailability (for example device turn on) a notification is sent to the system. If the service is active, an activation command to the agent/s is sent. Similarly, when agents become unavailable for active or paused service, they should inform the system if possible. Once an agent does not send a keep-alive message for a specific period of time it is removed from the available agents list ("*not active*" state). Agent that has completed the handshake procedure is regarded as in *available* or *active* state until service termination.

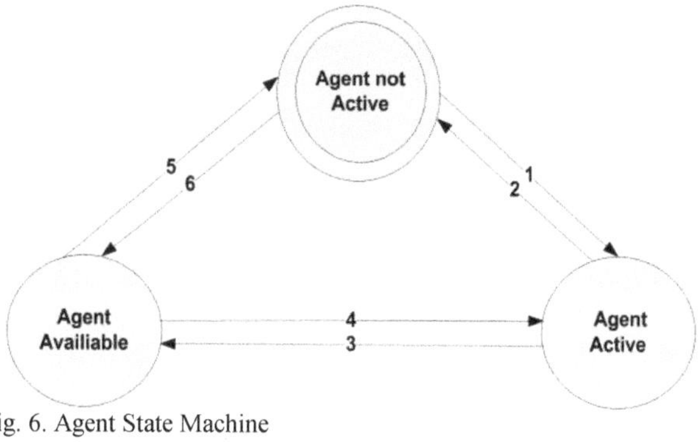

Fig. 6. Agent State Machine

Below are the detailed agent state transitions of Fig. 6.

1) From "*Not Active" to "Active"*. This transition occurs only upon service activation on the specific device.

2) From *"Active" to "Not Active"*. This transition takes place when the agent was one of the agents that provided the service and the service is terminated due to a user's command or due poor QoS (see sub-section QoS specification above).

3) From *"Active" to "Available"*, this transition occurs in the case of agents/client switching for QoS reasons (see QoS specification sub-section above) or a service state change from active to paused (see service state machine sub-section above).

4) From *"Available" to "Active"*, this can occur if the agent is selected by the system for service transmission for QoS reasons or when the service state moves from "Paused" to "Active" by command from this agent.

5) From *"Available" to "Not Active"*, this transition takes place for available agents of a service when the service state is changed to "Not Active" (from "Active" or "Paused" states).

6) From *"Not Active" to "Available"*, this transition takes place for all agents upon service activation command.

In "Not Active" state the agents have no connection to the system. In the "Available" state the agents are connected to the system and keep-alive messages are exchanged as described above. An agent is "Active" if it is one of the agents that supply the service.

C. User Interface Specifications

User interface has to meet requirements G3, U3, U4, U5, U6, Q1 and P1. In order to meet privacy requirements (P1) the interface to a service should be a remote web page and it should be password protected, for example.

The actions that can be performed by the user interface are:

1) *Join the service:* First, the user has to register and to accept a username and password (requirement P1) to access the system. The next step is setting the preferred and required QoS (requirement Q1). The user can choose the default values. The use of these values is described in the QoS specification sub-section above. Then, the user should register the devices; this process is described below.

2) *Client and agent management:* Client and agent management is divided into three sub-processes: client/agent addition, client/agent subtraction and client transition management. As a guideline the user does not have to address agents. The user chooses to alter the client setting; the system requests the data (list of connection interfaces) from the client and presents the options to the user. These procedures can be implemented at any time according to requirement U5.

(i) *Client/Agent addition*: When subscribing a device or agent to a service, a user can choose to connect from the current device ("Add this device") or to identify the subject device by a unique identification, an IP address for instance. After the device is chosen the system queries the device about possible agents (connection interfaces). All possible agents are presented to the user and the user chooses which agents to install. In case of agent addition the user should choose "modify client settings" and the system will return all the options. After the user confirms the choice the profile at the client side is updated. If the number of potential agents exceeds the A parameter from requirement G3, the system displays the message to the user and client/agent addition is aborted.

(ii) *Client/Agent subtraction*: The procedure can be performed from any device. In agent subtraction, as mentioned above, user chooses "modify client settings" and the system shows all agents. In the case of client subtraction the system presents the user list of all of its clients (devices) and the user chooses the client/s to subtract. The system does not allow subtraction of the last client/agent (requirement U4). After the user confirms the choice the profile at both server and client sides is updated.

3) *Client transition management:* In the QoS specifications sub-section, it was noted that in case of insufficient QoS or inefficient use of network resources the system can suggest switching clients to the user. As mentioned earlier the system should try to minimize the use of this option, to meet the requirement for minimal user intervention (G2). For this reason there should be a minimum period of G1 minutes between two successive proposals to switch devices. Additionally, the system should store recently (for G2 minutes) used agents and clients, in order to avoid "ping-pong" transitions between clients. For the user's convenience the system should avoid proposing invalid transitions. For this reason the system stores a list of forbidden client transitions, per user. The user should manage this list. This list can be updated in two ways: by the user interface or when the system displays the list of optional clients for transition, the user can assign "Do not propose this transition again" to one or more clients. The system also should not propose a switch to a device in active service (of any user) to comply with requirement U15. In summary, client transition can be proposed to a user if: this is the first proposal for G1 minutes and there exists a potential device for transition that has not been used in the last G2 by this user and the service, does not appear on the forbidden transitions list, does not run an ubiquitous service to any user and meets the QoS threshold. Upon the user's approval of the transition client, a swapping procedure is performed. Client swapping is similar to pausing and resuming the service on a new client; the only difference is that the "pausing" is triggered by the user's approval of the client switching and not by the pause command. Pausing and resuming service is described in the service state machine sub-section above. Further work could be done to elaborate the list of proposed clients for transition, by studying users' preferred transitions, for example.

4) *Activation, stopping, pausing and resumption of a service:* These actions are described in the service state machine and the QoS sub-sections. For privacy (requirement P1) the activation and resumption of a service should be password protected operations.

5) *Multiple user service management:* The interface is described in the multiple user service management sub-section below.

6) *Disjoin the service:* The user should choose to unsubscribe to the service. After the user confirms the choice, all associated session information on both the client and server sides are removed.

D. Multiple User Specifications

One aspect of ubiquitous service is that several users may start to consume a service together and then want to continue to consume it separately at a different time and place. The requirements for multiple users (M1 – M5) specify this case.

In order to support the multiple user mode the system and the client need to support the following actions:

- Add user to service user group.
- Remove user from service user group.
- Primary user substitution.

All of the above only apply to active service.

1) *Add user to service user group:* To add a user to a service group, the primary user should choose the "Add user to the current service" option and specify the user. The system asks for the new user's confirmation, according to

requirement M2, by requesting a password. Once the subject user is approved to join the service, the system performs the following steps: the user is added to the service's user group, a connection is established with all the users' clients to move agents to the "available" state and to generate a list of available agents. The service state for the new user is "paused". Note that the user must be subscribed to the service, but the device can only be subscribed by the primary user.

2) *Remove user from service group:* Confirmation by the subject user is performed as above. If the primary user wants to leave the service a new primary user should be chosen according to requirement M5 (below). Then the departing user should chose to stop or to pause the service (requirement M3).

3) *Primary user substitution:* The following steps should be performed: (i) upon user request to change primary user, the system should generate a list of potential primary users. The new primary user can be a user who is in the service's user group and subscribed the current device to the service. If there is no such user, the primary user change cannot be done. (ii) The list of potential primary users is displayed to the current primary user and he/she should choose the new primary user. (iii) The last step is to stage the competition between agents of the new primary user on the current client, and to choose active agent/s for the service [8]. The previous primary user becomes the secondary user and stays in the service's user group.

E. User Mobility

User mobility is a focal issue in ubiquitous service. In this sub-section we show that the mobility issue is resolved in our system. User mobility can have two negative effects: the user needs to switch devices and/or the QoS degrades due to coverage or load issues. If the user needs to switch devices this should be done by pausing the service on the old client and resuming it on the new one. The case of QoS degradation is discussed above. Thus our specifications resolve the user mobility issue without having to take any location associated actions, by enabling a high level of customization and addressing QoS as a general issue that is not related solely to mobility.

IV. Performance Evaluation

In the performance evaluation of our method we consider the following additional competing methods: single transmission (that is, the way live video is transmitted today) and simple (not controlled or coordinated) multiple transmission of 2-5 agents. In simple multiple transmissions, the agents transmit the video streams without coordination with the server and the client joins the streams using the simple join function of the minimal arrival time. For every packet sequence number, the client considers the first instance to arrive. That is, the resulting arrival times are the minimum arrivals times of every packet.

Due to the short distance between the agents, we cannot assume that they are statistically independent. Therefore, our method for evaluating the suggested solution is by measurements of real traffic, rather than theoretical analysis. First, we transmit video using several agents in various conditions in order to collect the data. The transmission of the agents was done using LU60 of LiveU [11], using one to five cellular modems connected to three different cellular networks. Each agent has a different connection to the internet. Next, a feasible solution for splitting and joining is implemented. We record the received data with LiveU's server (LU1000) [11] and also using 'Wireshark' software [12]. We collect data which is relevant to parameters such as delay, jitter and retransmission ratio. Therefore, we record for each packet in each transmission from each agent the

TABLE III
PACKET LOSS RATIO

Process	Average	Worst case
1 agent	2.56%	17.0%
2 agents	0.04%	1.22%

TABLE IV
AVERAGE PERCENTAGES OF JITTER CONDITION VIOLATION

Cond.	Best (A)	Best (B)	Best (C)	1 agt.	2 agt.	3 agt.	4 agt.	5 agt.
>0	66%	79%	83%	80%	73%	70%	69%	69%
>1	61%	75%	79%	76%	69%	66%	64%	64%
>2	58%	71%	75%	75%	67%	63%	61%	60%
>3	55%	67%	71%	73%	65%	60%	57%	56%
>4	45%	49%	51%	69%	56%	48%	41%	37%
>5	42%	44%	46%	67%	54%	44%	37%	31%
>6	37%	37%	38%	65%	50%	39%	31%	24%
>7	35%	35%	35%	63%	48%	37%	28%	22%
>8	33%	33%	33%	60%	45%	34%	26%	20%
>9	26%	27%	28%	53%	38%	29%	22%	17%
>10	13%	14%	16%	39%	26%	19%	15%	12%
>11	7%	8%	11%	32%	20%	14%	11%	9%
>12	6%	7%	10%	30%	18%	13%	10%	8%
>13	5%	7%	9%	29%	17%	12%	9%	8%
>14	5%	7%	9%	28%	17%	12%	9%	8%
>15	5%	6%	9%	28%	16%	11%	9%	7%
>16	5%	6%	8%	27%	16%	11%	8%	7%
>17	4%	6%	8%	26%	15%	10%	8%	6%
>18	4%	5%	7%	25%	14%	10%	7%	6%
>19	4%	4%	6%	21%	12%	8%	6%	5%
>20	3%	3%	4%	16%	9%	6%	5%	4%
>25	2%	2%	2%	11%	6%	4%	3%	3%
>30	2%	1%	1%	9%	5%	4%	3%	3%
>35	2%	1%	1%	8%	4%	3%	3%	2%
>40	2%	1%	1%	8%	4%	3%	3%	2%
>45	2%	1%	1%	7%	4%	3%	3%	2%
>50	2%	1%	1%	6%	3%	2%	2%	2%

Packet Sequence Number and Time of Arrival. Finally, we evaluate the method's potential performance using the data collected at the beginning.

The recording is done throughout the day including both peak (busy hour) and off-peak hours. Each experiment consists of 5 samples of video transmissions using one to five simultaneous agents. The experiment is repeated 9 times with long video files (about 15 minutes, 30,000-65,000 packets each). In addition, the experiment is repeated twice with short video files (five minutes) to validate that the observed statistic behavior also fit short transmissions. Overall, the recording trace includes statistics of ~ 6 million real packets.

The analysis was performed three times, with jitter requirements of 13 msec. ("Condition A"), 25 msec. ("Condition B"), and with jitter requirement of 40 msec. ("Condition C"). The considered performance parameters are overhead factor, packet loss ratio and jitter.

Regarding the average overhead, processes with one agent naturally have no overhead (factor 1), processes with two agents have overhead of factor 2 (every packet is transmitted twice), and so on. Table I summarizes the overhead factor of our method relative to a single agent process. Each line in the table specifies the average overhead factor, the observed minimum overhead factor and the observed maximum overhead factor. The best k process with parameter 13 has an average overhead factor of 3.08, the best k process with parameter 25 has an average overhead factor of 1.91 and the best k process with parameter 40 has an average overhead factor of 1.65. The differences in the overhead factors are due to the fact that fewer competitions are generated when the requirement from the jitter is less demanding. In a competition all the potential 5 agents transmit two segments, thus, the overhead increases with the number of competitions. Interestingly, the observed minimum overhead factor of the best k process with parameter 40 is 1.09 which is an excellent result. In this observation, only 46 competitions were generated by the algorithm (92 segments) out of 4000 segments in the total transmission and all other 3908 segments were transmitted by a single agent only (97.7%).

To understand the source of the overhead results, Table II plots the percentages of segments transmitted by a number of agents in each algorithm. For example, when using the best k process with parameter 40, an average of 83.7% of the segments were transmitted by only one agent, 0.1% of the segments were transmitted by exactly two agents and 16.2% of the segments were transmitted by all 5 agents (during competitions). Table II illustrates that the best k algorithms with parameters 25 and 40 chose most of the time to use only one transmitting agent, but kept replacing it when its performance decreased. These insights imply that

TABLE I
OVERHEAD FACTOR

Process	Average	Minimum observed	Maximum observed
Best k (A)	3.08	2.05	3.93
Best k (B)	1.91	1.10	2.79
Best k (C)	1.65	1.09	2.72

TABLE II
AVERAGE PERCENTAGES OF TRANSMITTING AGENTS

Process	% using 1 agt.	% using 2 agt.	% using 3 agt.	% using 4 agt.	% using 5 agt.
Best k (A)	45.5%	3.2%	0.0%	0.0%	51.3%
Best k (B)	77.0%	0.4%	0.0%	0.0%	22.6%
Best k (C)	83.7%	0.1%	0.0%	0.0%	16.2%

the overhead can be reduced significantly by developing a different mechanism to replace/select the transmitting agents other than a competition.

Regarding the packet loss ratio, all best k processes and all multiple transmission processes with 3 agents or more have 0.0% average packet loss ratio. The measurements of the processes with one and two agents are presented in Table III. Generally, the networks are reliable and usually the packet loss ratio is very low. However, a very high packet loss ratio of up to 17% packet loss ratio was observed for a single agent in some cases. Naturally, using additional agent reduces the packet loss ratio dramatically, and using more agents or more sophisticated algorithms reduce the packet loss ratio to 0.0.

In order to evaluate the impact of the statistics on the actual user experience we study the function 1-CDF (Cumulative Distribution Function). It represents the average percentages of times that the arrival process violates the corresponding jitter condition. Table IV presents these jitter statistic of the competing methods. Each line describes the average percentages of times that the arrival processes violate the corresponding jitter condition. That is, the packets inter-arrival time is larger than the specified threshold. For example, in line number twelve, the jitter condition is "smaller than 11", and the process "best k with parameter 25" violates this condition in 8% of the samples on average while the process that uses simple multiple transmissions of three agents violates this condition 14% on average. As can be seen from this table, starting from jitter condition "smaller than 11" the best k processes with parameters 13 and 25 outperform the other processes with a significant small number of condition violation. The best k process with parameter 40 behaves very similar to the process with five multiple agents starting from jitter condition "smaller than 13". Note that all best k processes perform at least three times better than single transmission.

All above mentioned results imply that there is no need to require a strong performance condition to improve the performance significantly. The performance of the algorithm under different performance requirements is similar. However, the overhead increases with the performance condition strength. Thus, selecting normal to weak performance condition is recommended.

V. FURTHER WORK AND CONCLUSION

This article described a novel approach to ubiquitous multimedia service - client/server architecture. The system incorporates detailed requirements, specifications and algorithms. We address all known issues related to ubiquitous service: QoS management, efficient usage of network resources, limited overhead, simultaneously usage of a device by a number of users, user mobility and user interface. An additional advantage of our system is that it is network independent, and thus can use any RAT technology. Obviously, it can coexist with other ubiquitous service architectures.

REFERENCES

[1] P. Bellavista, A. Corradi, and L. Foschini, "MUM: a middleware for the provisioning of continuous services to mobile users", In IEEE Sym.. on Computers and Communications, Vol. 1, pp. 498-505, 2004.

[2] P. Bellavista, A. Corradi, and L. Foschini, "Context-aware handoff middleware for transparent service continuity in wireless networks", Pervasive Mob. Comput. vol. 3(4), pp. 439-466, August 2007.

[3] P. Bellavista, A. Corradi, and C. Giannelli, "Mobility-aware management of internet connectivity in always best served wireless scenarios", In Mobile Networks and Applications, Vol. 14, pp. 18-34, Kluwer Academic Publishers, 2009.

[4] Y. Cui, K. Nahrstedt, and D. Xu, "Seamless User-Level Handoff in Ubiquitous Multimedia Service Delivery" Multimedia Tools Appl. vol. 22(2), pp. 137-170, February 2004.

[5] R. Ferrus, O. Sallent, and R. Agusti, "Interworking in heterogeneous wireless networks: comprehensive framework and future trends", IEEE Wireless Communications, vol. 17(2), pp. 22-31, April 2010.

[6] Yong-Ju Lee, Choon-Seo Park, Jin-Whan Jeong, Hag-Young Kim and Cheol-Hoon Lee, "UMOST : Ubiquitous Multimedia Framework for Context-Aware Session Mobility", Int. Conf. on Multimedia and Ubiquitous Engineering (MUE 2008), pp. 3-8, April 2008.

[7] K. Nahrstedt, D. Xu, D. Wichadakul, and B. Li, "QoS-aware middleware for ubiquitous and heterogeneous environments", IEEE Comm. Magazine, vol 39(11), pp. 140-148, November 2001.

[8] R. Nossenson and O. Markowitz, "Using coordinated agents to improve live media contents transmissions", Int. Conf. on Systems and Networks Communications (ICSNC 2011), pp. 167-170, Oct. 2011.

[9] M. Takemoto, T. Oh-ishi, T. Iwata, Y. Yamato, Y. Tanaka, K. Shinno, S. Tokumoto, and N. Shimamoto, "A service-composition and service-emergence framework for ubiquitous-computing environments", International Symposium on Applications and the Internet Workshops, pp. 313-318, 2004.

[10] M. Takemoto, H. Sunaga, K. Tanaka, H. Matsumura, and E. Shinohara, "The Ubiquitous Service-Oriented Network (USON) - An Approach for a Ubiquitous World Based on P2P Technology", in Proc. Peer-to-Peer Computing, pp.17-24, 2002.

[11] LiveU web site: http://www.liveu.tv/ accessed: January 2012.

[12] Wireshark web site: http://www.wireshark.org/ accessed: January 2012.

[13] R. Nossenson, O. Yudilevich and O. Markowitz, "Client-Server Architecture and Algorithms for Ubiquitous Video Service", The 6th International Conference on Multimedia and Ubiquitous Engineering (MUE 2012), Madrid, Spain, July 2012.

15

Smart Indoor Positioning/Location and Navigation: A Lightweight Approach

José Antonio Puértolas Montañés[1], Adriana Mendoza Rodríguez[2], Iván Sanz Prieto[3]
[1]Bizeu Consulting, [2]University Complutense of Madrid, [3]AXA Tech

Abstract — **In this paper a new location indoor system is presented, which shows the position and orientation of the user in closed environments, as well as the optimal route to his destination through location tags. This system is called Labelee, and it makes easier the interaction between users and devices through QR code scanning or by NFC tag reading, because this technology is increasingly common in the latest smartphones. With this system, users could locate themselves into an enclosure with less interaction.**

Keywords — **Indoor location system, NFC technology, QR codes.**

I. Introduction

INDOOR Positioning shall be defined as any system which attempts to provide an accurate positioning inside of a covered structure; it is used to locate people or required objects in large buildings and in closed areas/spaces. At present, there are several types of location-sensing systems, each having their own strengths as well as limitations. [1]Indoor location sensing systems have become very popular in recent years, because many applications need to know the physical location of objects. These systems provide a new layer of automation called automatic object location detection[2] and use wireless concepts, optical tracking or ultrasonic techniques.[3]

The relevance of context-awareness for mobile users has been demonstrated in several applications which have been summarized in applications of context-aware computing, and have included fieldwork, museums, route planning, libraries and tourism.[4] There are three technologies commonly used for indoor location systems –ultrasonic, infrared and radio frequency. These can be supplemented by inertial systems which are generally used for prediction. [5]

One of the most well-known location-based systems is the Global Positioning System (GPS), which is widely used to track moving objects located outdoors and determines the position (altitude, longitude and latitude) of an object in any point of the planet by triangulation [6]. However, GPS, as it is satellite dependent, has an inherent problem of accuracy, when determining the location of objects inside buildings. Active Bat and Cricket, are cited as [7] use ultrasonic pulse TDOA to enable high precision and dimensional positioning in indoor environments. Although, ultrasonic positioning system requires additional hardware to send and receive ultrasonic pulse, it can determine 3D position of indoor objects with an accuracy of a few mm to a few cm. [1]

Indoor location technologies may be also classified according to the location positioning algorithm, the physical layer or location sensor infrastructure. Locations sensing approaches can be classified as follows: location fingerprinting (scene analysis), triangulation, trilateration, hyperbolic lateration, proximity, and dead reckoning.[8]

In this paper/article, Indoor positioning systems are classified according to the technology on which they are based. There are systems which use Wireless networks, Satellite-based techniques, infrared systems, ultrasonic systems, Ultra Wide Band (UWB), RSS, hybrid systems, GPS indoor, Bluetooth…

Technical deployment of indoor positioning systems involves significant investment. Also these systems have a lot of problems like multipath effects, they need a specific infrastructure, require a Line Of Sight[1] (LOS) etc. also they consume battery of mobile devices when obtaining the information of indoor location.

We present an indoor positioning and orientation system, which optimize the user mobility in closed spaces. This technology employs a location codes system. It is a simple and inexpensive solution to obtain the indoor location and orientation.

This paper is organized as follows. Section 2 presents relevant overview of existing indoor location systems, section 3 explains the most common problems of indoor location systems, section 4 describes the indoor position system based in QR codes. Finally, section 5 ends the paper with a discussion about future work and conclusions.

II. Overview of Existing Indoor Location Systems

Physical location class can be broken down into three subcategories: describe locations, spatial locations and network locations:

- Location related to geographic objects such as mountains, lakes, cities, roads, countries or other structures that have a description such as name, identifier, or number is known as a descriptive location.

[1]Line of sight is a type of propagation that can transmit and receive data only where transmit and receive stations are visible each other without any sort of an obstacle between them.

- Spatial location is used more in professional applications where a descriptive location does not provide enough details.
- Network location refers to a location based on the topology of a communications network. A user´s device position in a network is achieved based on its Internet Protocol (IP) address. However in mobile networks a network location is achieved from the base stations used by the mobile terminal. [8]

The techniques used to estimate location in a cellular network have been successfully used in wireless Local Area Networks (WLANs), Bluetooth, Ultra-wideband (UWB), Ultrasound, Radio Frequency Identification (RFID), Hybrid technologies, and QR scanning method.

- *Systems based on WLANs:*

In the last few years, WLAN radio-signal-based positioning systems, supported by underlying Radio Frequency (RF) and Infra-Red (IR) transmission technologies has experimented an enormous expansion and is expected to continue this trend due to the fact that it is an economical solution providing convenient connectivity and high speed links, and can be implemented with relative ease in software.

Wi-Fi RTLS (Real Time Location Systems) enables position sensing by using existing wireless LAN routers [9]. Additionally, WLAN covers a large area and is not restricted by line of sight issues. A WLAN can support a large number of nodes and vast physical areas by adding access points to extend coverage. Therefore, WLAN allows users to be truly mobile as long as the mobile terminal is under the network coverage area. [4]

WLAN is appealing because it allows enhanced connectivity and is particularly useful when mobile access to data is necessary. Additionally, user flexibility and probability can easily be reconfigured while requiring no cable infrastructure.[4]

The Wave LAN system triangulates 2D position of an object within a building by using either empirical data or a mathematical model of indoor radio propagation. Advantages of this system are such that it requires few base stations and it uses the same general wireless networking in the buildings. [3]

A special case of this technology is a systems based on fingerprinting. Location fingerprinting refers to techniques that match the fingerprint of some characteristic of the signal that is location dependent. The fingerprints of different locations are stored in a database and matched to measured fingerprints of different locations are stored in a database and matched to measured fingerprints at the current location of an MS (Mobile Station). [10]

The fingerprinting technique is relatively simple to deploy compared to the other technique such as angle-of-arrival (AOA) and time-of-arrival (TOA). Moreover, there is no specialized hardware required at the mobile station (MS). Any existing wireless LAN infrastructure can be reused for this positioning system. The most common algorithm to estimate the location computes the Euclidean distance between the measured RSS vector and each fingerprint and database. The coordinates associated with the fingerprint that provides the smallest Euclidean distance is returned as the estimate of the position. [11]

- *Systems based on Bluetooth:*

The Bluetooth wireless technology is a technology originally designed as a short-range connectivity solution for personal, portable, and handheld electronic devices.

Bluetooth is lighter standard than WLAN, highly ubiquitous (embedded in most phones, personal digital assistants (PDAs), etc.) and supports several other networking services in addition to IP. Bluetooth tags are small size transceivers. As any other Bluetooth device, each tag has a unique ID. This ID can be used for locating the Bluetooth tag. [2]

The position estimation occurs in the mobile terminal without the need of changes in the already fixed installed network topology. The developed system is based on the known and already published triangulation methods using the received signal power strength of the surrounding Bluetooth access points. These points must exist previously. For precise position estimation, the dependence between the distance and the received signal strength has to be determined. [12]

- *Systems based on UWB:*

Ultra Wide band (UWB) impulse radio signals are employed for location and tracking. This system platform was constructed using standard electronics components, therefore, it allows fast performance evaluation and to estimate the time of arrival (TOA) of received pulse signals. Systems based on UWB use receivers deployed in a building to track users carrying small tags which emit ultra-wideband signals. Finally, these signals are transferred to a server where the location of the transmitter is calculated [12].

The main advantage of the UWB signals is to have high temporal resolution and to provide accurate TOA measurements in multi path environments. The transmitted signal has a sequence of short pulses. These are propagated in the media and received by at least four receivers placed in known positions [3].

- *Systems based on Ultrasounds:*

Indoor ultrasonic location sensing systems frequently use a number of ceiling-mounted transmitters to measure the position of a user in a room, and they process data from time of flight between transmitters and this user. Usually, these systems are composed by a hardware, consisting of transmitters or beacons which emit signals; software-based algorithms of location and at least one receiver that the user must carry. These receivers can take several forms depending on the system used.

For example, the Cricket system has an accuracy of 1 m to 3 m and supports multiple target discoveries. The user´s position is not shared unless the user accepts advertising that information. Thus there are no privacy issues due to the location being estimated on the mobile device.

Cricket´s limitations are the absence of a centralised management and monitoring functions. The computational requests are also quite high, which means high power consumption in order to synchronise the ultrasound pulse an RF data.

Another system based on Ultrasounds is called Bat system. This system consist that users wear small badges which emit an ultrasonic pulse when are radio-triggered by a central controller. The system determines the time of flight of the pulses from the badges to a network of receivers on the ceiling, and calculates the 3D positions of the badges using a multilateration algorithm. [13]

- *Systems based on RFID:*

Radio Frequency Identification (RFID) is a rapidly developing technology for automatic identification of objects [12]. The RFID position estimation is based on electromagnetic communication between RFID readers and tags, so RFID readers can read data emitted from RFID tags. Also, it is becoming more common that mobile devices have Near Field Communication (NFC) reader based on RFID technology, which allows reading these tags [13].

NFC technology is a bidirectional short range, wireless communication technology. Therefore, the communication occurs between two nearby devices: on one hand, a mobile phone acting as a NFC reader, and on the other hand, RFID tags, which are processed by the device. Systems based on NFC technology orient the user by gathering his destination point; this way the mobile device collects the current position from NFC tags, so the system computes the optimal route. Finally, another factor to keep in mind is that this technology can operate in card emulation, reader / writer, and peer-to-peer (P2P) modes [14].

RFID tags are categorized as either passive or active. Active tag contains embedded batteries in order to provide a much broader scanning range. The major disadvantage of using active RFID tag based solutions is the high cost of the active tags. [14]. In contrast, passive tags operate without a battery and are mainly used to replace the traditional barcode technology. They are much lighter and less expensive than active tags, offering a virtually unlimited operation lifetime. However, their read ranges are limited, offering a less range than active tags [7]. The employed signal technologies include RF (radio frequency) such as ultrasound, infrared, vision-based systems and magnetic fields[15].

The main advantages of this technology, is that non-contact is required and not need direct vision (N-LOS) [3]. Therefore, RFID systems have been widely adopted as an attractive technology for many significant applications such as asset tracking and industrial automation [13].

- *Systems based on hybrid technologies:*

Hybrid systems combine two or more technologies in order to improve the accuracy and precision of the location estimation. Depending on used technologies by these systems, they will have certain advantages or disadvantages. These technologies have been described in a bellow section.

An example of hybrid systems is the system called indoor GPS. In this system, user position tracking is mainly defined by at least four (or more) transmitters and a receiver. A battery operated transmitter uses laser and infrared light to transmit one-way position information and elevation from the transmitter to the receiver. The receiver has photodiodes inside its module and senses the transmitted laser and infrared light signals. The signal is transferred through a wireless network connection providing mobility to users. The 3D position of the optical receiver is then calculated by the process of triangulation. Triangulation is used if the angles to known (α and β) are given. With two known locations, the absolute position in 2D can be determinate. The two angles are used to determinate the line-of-sights (LOS) for each of the known locations. Indoor GPS, unlike UWB, depends on a clear line of sight and some calibration points are required. However, both technologies offer centimetre level positioning accuracy with Indoor GPS positioning offering significantly higher precision.[4]

- *Systems based on QR scanning method:*

QR-Codes are a type of visual tag, like a 2D barcodes, since they are more efficient than traditional tags and have a much higher information capacity. Due to the open QR-Codes specification, these codes can be employed without restriction and for all applications. Many mobile services rely on them; indeed it has been proposed that they should be embedded in images, such as those used on printed paper tickets. [16]

QR-Maps, is a solution to obtain indoor user locations using QR-Codes (Quick Response Codes) and Google Maps API. Since a smartphone is available, a QR-Code with a short text can be decoded. This text is sent to a location server, which returns an URL showing the map centred in the specific location[16]. Thus, no special hardware is required, simply a smartphone with a camera capture QR-Codes and Internet connection to access the map.[17]

There is another system which combines all previously mentioned features (QR scanning, do not required any specific API installed; do not need any transmitters and receiver, etc.). This system is called Labelee and it based on scanning QR codes, which allows the location and position in indoor environments.

The main improvement regarding the system described above is that Labelee allows showing the position at indoor environments, therefore users knows their position and it will be much easier to know where they would like to go.

Indoor position systems could be classified in active or passive systems. Passive location represents the estimation of a person´s position without require of a physical device, for example, a tag or sensor. However, most of location techniques are part of the active systems class, due the need of people who want to locate themselves into an enclosure to carry an electronic device or another object which allows estimating their position.[8] Table 1, summarizes the main characteristics of the technologies described above:

TABLLE 1: VARIABLES OF THE TECHNOLOGIES USED ON INDOOR POSITIONING SYSTEMS

					Labelee	
	WiFi	Bluetooth	UWB	Hybrid S./I. GPS	RFID/ NFC	Scanning/QR
Outdoor location	✗	✗	✗	✗	✓	✓
Indoor location	✓	✓	✓	✓	✓	✓
Location and Orientation in Indoors	✗	✗	✗	✓	✓	✓
Guided indoors	✓	✓	✓	✓	✓	✓
Accuracy	3 m	2 m	10 cm	1/5 m	0 m	0 m
Range	Room Scale	10 m	15 m	Gobal	Scale according enclosure	Scale according enclosure
Signal	RF	RF	RF	RF	RF	___
Necessary Infrastructure	WLAN router	Specific Trans.	Specific Trans.	Specific Trans.	NFC Tag	QR Tag
Battery consumption/Energy Spending	Medium	Medium	Medium	High	Low	Low
Specific Application	✓	✗	✗	✓	✗	✗
Cost	High	High	High	High	Low	Low

III. REVIEW OF THE MOST COMMON PROBLEMS

Since GPS is not available indoor location, many approaches have been proposed to facilitate indoor location based on wireless technologies such as RFID (radio frequency identification), Bluetooth or IEEE 802.11. All these solutions require the deployment of a wireless infrastructure and, in some cases, specific technologies seldom found in smartphones.[16]

The GPS signal [7] is not usually available in indoor environments. [18] Satellite-based navigation systems require an unblocked line of sight between the satellite and the receiver to function properly. If this prerequisite is not met, satellite navigation is degraded; different methods and technologies are needed.

One way for locating a mobile device in two dimensions requires the measurement of the Line Of Sight (LOS) distance between the mobile phone and at least three participating BSs (Base Stations). Each distance measurement generates an area which is focused at the measuring base station and has a radius equal to the mobile-to-base distance. So, the three regions unambiguously determine the location of the device.

However this may become a problem if the LOS is not continuous. The usage of signals that can penetrate building materials may overcome this problem in the near future[15]. A mobile phone positioning system is accurate to around 100 m. at indoor, short-range (approximately 10 m radius) wireless cell may be used to determine positions based on cell identification, but requires many stations. [18]

When the range measurements are corrupted by noise, then the location of the mobile station can be determined by finding the solution that is most consistent with the measurements. However, in dense urban environments, there may be no direct path from the mobile to the BS. Due to reflection and diffraction, the propagating wave may actually travel excess path lengths on the order of hundreds of meters and the direct path is blocked. In order to mitigate the effect of the measurement BS, it is necessary to develop location algorithms that are robust to the NLOS (Non Line Of Sight) error.[19]

Another factor that affects the indoor positioning systems is called multipath. This is the propagation phenomenon that results in radio signals reaching the receiving antenna by two or more paths. The multipath transmission (multi path fading) includes large scale fading and small scale fading. Large scale fading corresponds to average signal power attenuation, small scale fading is due to dramatic alterations in amplitude and phase. Almost all indoor position systems have these effects of multipath systems that include constructive and destructive interference, and phase shifting of the signal. Destructive interference cases fading[20]. It is important to note that, RSS measurements are affected by multipath effects or attenuation when passing through walls, distorting them from the expected free space values. [21]

A solution to compensate the multipath effects on received signal, is equalized these effects by using Equal Gain Combining Algorithm further correlating the received signal with reference signal so that we can eliminate other unwanted signals. The problem is that obtained signal has more noise power, so signal may not sufficient to locate Indoor positioning in centimetre level or millimetre level.[20]

Moreover, the cost of a positioning system may depend on many factors. These factors include money, time, space, weight, and energy. The time factor is due to installation and maintenance, also, mobile units may be having tight space and weight constraints. Sometimes, installation of indoor positioning systems may not have to result in additional costs, since previously there could exists hardware being used for another purposes. Energy is another important cost factor. For example, some mobile units like RFID tags are completely energy passive since they only respond to external fields and, thus, could have an unlimited lifetime, unlike most devices with rechargeable battery, which have a lifetime of several hours without recharging [2].

Another factor to consider is complexity of a positioning system. This can be attributed to hardware, software and operation factors. Most of the mobile units lack strong processing power and long battery life; so, positioning algorithms with lower complexity would be preferred. Usually, it is difficult to derive the analytic complexity formula of different positioning techniques; thus, the computing time is considered. Another factor to keep in mind is location rate, which is an important indicator for complexity. The dual of location rate is location lag, which is the delay between a mobile target moving to a new location and reporting the new location of that target by the system.[2]

Apart from the general problems mentioned above, each indoor positioning system, according to the technology on which they are based, has its specific problems, which will be analysed below.

- *Inconvenient of WiFi based systems:*

In order to enable the user to locate himself in enclosures, it is necessary that the mobile phone and the network are synchronized; therefore, the device should calculate its distance from the access point (AP).

The only available information is the signal strength received from each AP. Indeed, the received signal strength is measured. Such information is available because the APs send signals periodically. Mobile devices use those signals to handle the roaming inside the network. Given this consideration, it is possible to get a list of the received power coming from all the APs covering the area where the mobile phone is moving.

The easiest approach for locating a mobile device in a WLAN environment is to approximate its position by the position of the access point received at that position with the strongest signal strength. However the main drawback is its large estimation error. The accuracy is proportional to the density of access points; which is in the range of the 25 to 50 meters for indoor environments. [22]

In order to configure a wireless navigation system in an indoor environment it is required individually encoded transmitters and the setup of a sufficient number of receivers to provide at least room-level resolution. Also, for the navigation environment a map is required, which holds the RSSI at strategic points in that area, usually in form of a grid. The process, during which the RSSI is recorded at each grid point, is called fingerprinting and it is a very time consuming one. It is usually only valid for a particular antenna/access point pair and is very sensitive even to the slightest change in the surroundings (e.g. the removal of a particular item of furniture). Some problems of fingerprinting are that such techniques require specialized hardware in every base station, or access point to correlate the multipath characteristics. In WLANs, an easily available signal characteristic is the received signal strength (RSS) and this has been used for fingerprinting. The RSS is a highly variable parameter and issues related to positioning systems based on RSS fingerprinting are not understood very well. This may not very costly for small areas, but it becomes an increasingly important component as the area to be covered and the number of users becomes large. [10]

We must also take into account a problem with the multi floor environments, because two or more very different locations could potentially have the same RSS location-fingerprint. Instead of storing average RSS values, the joint or marginal distributions of the RSS are used in fingerprinting locations. The experiments are once again conduced in hallways and the accuracy rages between 5 and 20 feet in most cases.

- *Inconvenient of Bluetooth based systems:*

This system was not designed taking into account privacy and security topics, due to the installation of Bluetooth beacons that will send advertisements to mobile devices which are within its scope. Also, device address data sent from a Bluetooth Sensor to the Ad Server is not encrypted. [1]

Another disadvantage is the Bluetooth range which depends on the Bluetooth Class. The maximum range is approximately 100m for Bluetooth Class 1. The range can be extended with directional antennas and signal amplifiers but this increases the cost to deploy a Bluetooth-based localisation system [7]. Furthermore, the position accuracy is dependent on the amount of cells used [14]

- *Inconvenient of UWB based systems:*

Indoor location systems are based on UWB, and have some disadvantages produced by the effects of metallic and liquid materials, these effects can be reduced with a good placement of the UWB readers. UWB technology is suitable for 2D and 3D location estimation. Most common methods used in a UWB localisation system are Uplink-Time Difference of Arrival (TDOA) and Angle of Arrival (AOA). A combination of TDOA an AOA can reduce the number of sensor required for a system using just TDOA.[8]

The problems of using this ultrasonic technique are the requirement of large number of receivers across the ceilings and their placements across the ceiling which needed quite sensitive alignments. [3]These systems are loss of signal due to obstruction; false signals due to reflections; and interference from high frequency sounds.[5]

- *Inconvenient of Ultrasound based systems:*

The main issue of using ultrasonic techniques is the requirement of a large number of receivers across the ceiling. Their placement needs some calibration [3]. Other disadvantages is that need a single-user access, if multiple collocated transmitters send at the same time, their signals can interference with one another, and make it difficult for the receiver to distinguish between them. Also, these systems presents a poor performance in the presence on in band noise, this means that, people create ultrasonic noise trough their everyday actions, although they are rarely aware of it because ultrasound is inaudible. [23]

- *Inconvenient of RFID based systems:*

A drawback to consider in systems based on RFID technology is the type of label used. Active tags are more accurate, hence their cost is much higher, however, passive tags are more economical, and yet have a lower precision. Also, It must be taken into account the greater the number of labels installed, a larger area will be covered and therefore, the cost will be higher.

- *Inconvenient of Hybrid based systems:*

All previously described inconvenient are mixed on hybrid based systems, so it uses these technologies.

- *Inconvenient of scanning QR Codes:*

The main disadvantage of methods based on QR codes scanning, is that the determined position is always the position of the marker. So the used navigation device must be placed closed to the marker. The range depends on the resolution of the used camera; the accuracy depends on the accuracy of the measured marker position. [18]

QR-Codes have been specifically designed for fast information transfer. QR-code data retrieval depends on factors such as camera quality, picture angle or luminosity, as well as on the area of the code. Thus area depends on the size of the data. In our case data size is minimal, just a short text identifying the map and the location, which corresponds to a small QR-Code that can be placed on any wall or displayed on any screen with negligible aesthetic impact.[16]

IV. NEW APROACH TO CREATE A CHEAP, PLATFORM INDEPENDENT AND LIGTH WEIGHT INDOOR POSITIONING SYSTEM

A recent investigation of Baltimore University shows that our position in known environments is represented in a short of interior map located in cerebral hippocampus, formed by neuronal groups called "places-cells". When we make a walk that we had make before, places cells are activated sequentially in our interior map, showing the right way.[24] Working in a similar way, Labelee employ location tags which contains a QR code, which are distributed at indoor environments, acting as a places-cells, so users only needs to scan the QR code using their devices to find the way to go. These location tags act the same form as location cells, because they encode their position and orientation to allow guidance and navigation.

Labelee is an indoor positioning and orientation system, which optimizes the client/user mobility inside the building. Labelee employs the LC (location codes) system, which includes a QR (quick response) code and a NFC chips that acts as an information point for the user: it stores its geographic position and leads him to its target.

The user only has to take a picture of one of those QR labels or read the NFC chips with his "smartphone", and then the system will place him and lead him in the map, without the need of downloading or installing any app in the mobile phone. Figure 1 shows how Labelee works. Since a user arrives at a closed environment, and wishes to know where he. Users can locate other points of interest in the map.

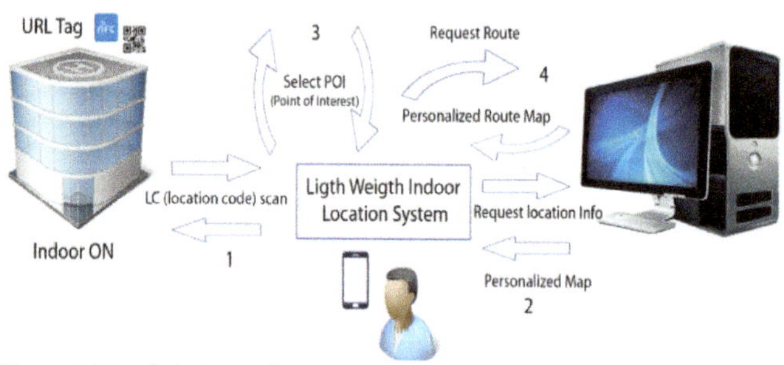

Figure 1:How Labelee works

To know these points, users need to scan the nearest location tag. Once a point of interest is selected, Labelee indicates the current position and the selected point of interest. If the selected point of interest is located on a different floor, Labelee shows the closer available stairs or elevator and returns a map displaying the shortest way to change floors.

To navigate on indoor environments with Labelee, it is necessary to have all the indoor space information previously processed, in order to calculate all possible routes. Depending on the extent of the environment, or the complexity of these, there will be more or less location tags that correspond to the placement of the labels. Those location labels will be strategically distributed all over the place, when the user scans any location tag, Labelee will indicate his current position and will guide him step by step to his destination. This system works with any smartphone or tablet, even if the smartphone have any camera, just entering the URL which appears besides the QR code. In the figure 2 shows this process.

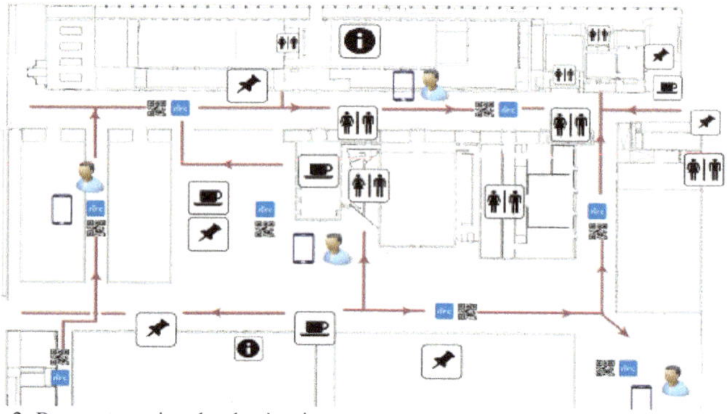

Figure 2: Routes to arrive the destination

Therefore, applications of Labelee are varied and can be installed in any type of area where it needs to have reference plans. This system is very useful, for environments due to the functions they perform, attract different people daily.

In outdoor spaces, Labelee is an alternative in urban historical centres or any place where, due to the morphology or the density of habitants, the GPS signal is not optimal despite the fact is an open space. Also, this solution enables visitors the location, orientation and guidance in public parks, city centres, amusement parks and tourist attractions, apart from showing other routes of various kinds.

Moreover, Labelee is very helpful choice, because it does not rely on any wireless network antennas or transmitters; hence, the required installation is minimal, so it is suitable for shopping centres, parking, airports, museums, hospitals, data processing centres, buildings and office centres, special events (fairs, festival, congresses, conferences, etc.), stadiums, holiday resorts, cruises, etc.

V.CONCLUSIONS AND FUTURE WORK

This paper has presented an indoor location system based on QR codes and recently RFID technology, through NFC readers, more common in the latest smartphones; users could locate themselves into the environment with less interaction. Moreover, it would be interesting to try to regulate the transmission of location according to the bandwidth, in order to get a greater energy savings and a less data consumption.

As future work, we intend to use smartphone compasses to show the user how to go the desired destination. With the of help of accelerometers, it will be also possible to estimate walking distances as additional information. Also the management module could be improved to get information from the public which uses different browsers to find the target audience

Some improvements will be shown, for example plant aerial view of buildings, different levels of detail within the map, zoom in the area where the user is situated-so you could get a clearer idea of where is located.

A cartographic base with vector maps is being created at this moment, which will be useful, for example, to remember the position where our car is parked, offer custom routes and maps, different levels of detail inside the map, show distance and time to our destination using 3D maps.

Another important topic is regarding optimize the file sizes, in order to get a better transfer. Also is being done. Also it is being created a test procedure for different mobile platforms, which will serve as the basis for the development of an easy-to-use application which will have continuous integration, so eventual mistakes will be corrected very quickly. Incidents resolution and detection will be faster, because events, queries and errors are being defined and stored in a specific database. Location could be also shared, using instant messaging applications, social networks, or just share with tourist information webpages, entertainment, etc. A new web portal with maps will be developed, which will have maps classified by categories from all areas where the system is installed. Therefore users could interact, share or point their places of interest.

Overall, a direct relationship exists between the accuracy and the cost in positioning systems. However, the system described in this article, is considered a fine-grained system, which means that its accuracy can be estimated with a high degree of resolution, with a low cost.

REFERENCES

[1] [1] L. Escobés, "Summary of the State of the Art in indoor location systems," no. December, pp. 1–3, 2009.

[2] [2] H. Liu, S. Member, H. Darabi, P. Banerjee, and J. Liu, "Survey of Wireless Indoor Positioning Techniques and Systems," vol. 37, no. 6, pp. 1067–1080, 2007.

[3] [3] H. Koyuncu and S. H. Yang, "A Survey of Indoor Positioning and Object Locating Systems," vol. 10, no. 5, pp. 121–128, 2010.

[4] [4] H. M. Khoury and V. R. Kamat, "Evaluation of position tracking technologies for user localization in indoor construction environments," Automation in Construction, vol. 18, no. 4, pp. 444–457, Jul. 2009.

[5] [5] C. Randell, "Low Cost Indoor Positioning System ⋆ ," no. 15986.

[6] [6] R. G. Crespo, G. G. Fernandez, and O. S. Martínez, "In Premises Positioning – Fuzzy Logic," pp. 284–291, 2009.

[7] [7] Y. Fukuju, M. Minami, H. Morikawa, T. Aoyama, I. Science, and F. Sciences, "DOLPHIN : An Autonomous Indoor Positioning System in Ubiquitous Computing Environment Yasuhiro FUKUJU *, Masateru MINAMI *, Hiroyuki MORIKAWA **, and Tomonori AOYAMA * * Graduate School of Information Science and Technology , The University of Tokyo ** Gr."

[8] [8] G. Deak, K. Curran, and J. Condell, "A survey of active and passive indoor localisation systems," Computer Communications, vol. 35, no. 16, pp. 1939–1954, Sep. 2012.

[9] [9] U. Rueppel and K. Stuebbe, "BIM-Based Indoor-Emergency-Navigation-System for Complex Buildings," Tsinghua Science & Technology, vol. 13, no. October, pp. 362–367, Oct. 2008.

[10] [10] P. Prasithsangaree, P. Krishnamurthy, P. K. Chrysanthis, T. Program, and I. Section, "ON INDOOR POSITION LOCATION WITH WIRELESS LANS," pp. 720–724, 2002.

[11] [11] K. Kaemarungsi and P. Krishnamurthy, "Modeling of indoor positioning systems based on location fingerprinting," Ieee Infocom 2004, vol. 2, pp. 1012–1022.

[12] [12] S. Feldmann, K. Kyamakya, A. Zapater, and Z. Lue, "An indoor Bluetooth-based positioning system : concept , Implementation and experimental evaluation."

[13] [13] M. Hazas and a. Hopper, "Broadband ultrasonic location systems for improved indoor positioning," IEEE Transactions on Mobile Computing, vol. 5, no. 5, pp. 536–547, May 2006.

[14] [14] B. Ozdenizci, K. Ok, V. Coskun, and M. N. Aydin, "Development of an Indoor Navigation System Using NFC Technology," no. April, pp. 25–27, 2011.

[15] [15] R. Mautz, "Overview of current indoor positioning systems," Geodesy and Cartography, vol. 35, no. 1, pp. 18–22, Jan. 2009.

[16] [16] E. Costa-montenegro, F. J. González-castaño, D. Conde-lagoa, A. B. Barragáns-martínez, P. S. Rodríguez-hernández, and F. Gil-castiñeira, "QR-Maps : an Efficient Tool for Indoor User Location Based on QR-Codes and Google Maps," pp. 928–932, 2011.

[17] [17] Y. Xue, G. Tian, R. Li, and H. Jiang, "artificial object mark in complex indoor environment," pp. 6648–6653, 2010.

[18] [18] C. Lukianto and H. Sternberg, "Overview of Current Indoor Navigation Techniques and Implementation Studies Overview of Current Indoor Navigation Techniques and Implementation Studies," no. May 2011, pp. 18–22.

[19] [19] M. P. Wylie and J. Holtzman, "The non-line of sight problem in mobile location estimation," Proceedings of ICUPC - 5th International Conference on Universal Personal Communications, vol. 2, no. 3, pp. 827–831.

[20] [20] P. Veeranath, D. N. Rao, S. Vathsal, and N. Bhasker, "Reducing Multipath Effects in Indoor Channel for Analysis of GPS / Pseudolite Signal Acquisition," vol. 3, no. 2, pp. 1–6, 2013.

[21] [21] Y. Á. López, M. E. D. C. Gómez, J. L. Álvarez, and F. L.-H. Andrés, "Evaluation of an RSS-based indoor location system," Sensors and Actuators A: Physical, vol. 167, no. 1, pp. 110–116, May 2011.

[22] [22] F. Evennou and F. Marx, "Advanced Integration of WiFi and Inertial Navigation Systems for Indoor Mobile Positioning," EURASIP Journal on Advances in Signal Processing, vol. 2006, pp. 1–12, 2006.

[23] [23] M. Hazas and A. Ward, "A Novel Broadband Ultrasonic Location System," pp. 264–280, 2002.

[24] [24] J. Sampedro, "SOCIEDAD Un GPS en el cerebro," pp. 4–5, 2013.

Dynamic, ecological, accessible and 3D Virtual Worlds-based Libraries using OpenSim and Sloodle along with mobile location and NFC for checking in

Rubén González Crespo, Sergio Ríos Aguilar,
Pontifical University of Salamanca in Madrid, Spain
Roberto Ferro Escobar, Nicolás Torres
Distrital University Francisco José de Caldas, Bogotá, Colombia

***Abstract* —This paper proposes the implementation of a 3D virtual library, using open platforms such as OpenSimulator and Sloodle, applied to the integration of virtual learning environments. It also proposes their application to the creation of open libraries to share and disseminate the new dynamic nature of knowledge, in the understanding that 3D virtual worlds may contribute to the future of libraries as part of green initiatives to achieve an ecologic and sustainable planet.**

***Keywords* — 3D Virtual Library, OpenSim, Sloodle, NFC, GPS check-in, Grid.**

I. Introduction

TODAY, one of the most relevant topics in computer-based content management research efforts is the way of presenting and visualizing content, stepping ahead from simple static web delivery of documents, books and publications via HTTP on the Internet. Nowadays, users demand better ways of accessing new sources of information, and an evidence of this is the current evolution of Digital Libraries, making huge efforts in order to make large collections of documents available to general public.

Anyway, current prevailing paradigms show that changes are needed in the way these contents are presented to the information consumers, who are looking to interact with Web contents in a more friendly manner using richer interfaces. And provided that recent evolution in office and domestic broadband access to Internet has improved the performance of networks interconnections and hence the globally perceived speed of Internet traffic, this change begins to come true thanks to virtual worlds simulators like Open Simulator [1].

This simulator was created under the guidelines of Secondlife, released in 2007 by Linden Labs [2], and placed under the GPL license [3]. OpenSim meets the general client-server model and has a modular structure written in C# and supports multiple viewers, such as the HIPPO viewer [4] , RealXtend viewer [5] and the Secondlife official viewer itself

The dynamic nature of Web 3.0 [6] combines artificial intelligence techniques, geospatial mobility, mashups, interactivity, multimedia and 3D transformations to build a semantic database able to learn. And everything is accomplished in 3D virtual worlds, where the user can be embodied in avatars that allow travelling to and wandering many worlds constantly arising, built using freely available design tools. Secondlife was initially used as a game in a virtual world built on a commercial server, and after a few years, due to the many worlds that can be created, Linden Labs decided to release the code to allow for an open construction of virtual worlds and private islands by programmers.

One of the most promising research areas related to 3d virtual worlds technology is the study of the integration of these 3d virtual worlds with current virtual learning environments and content management systems, given the powerful emergent properties arising from such a combination.

Another prevalent trend is the use of mobile broadband in order to access any kind of information using a personal mobile device, and one promising technology now close to reaching the necessary critical-mass is Near Field Communications (NFC). This technology provides a proximity-based interface to a wide array of services, including mobile check-ins, in special check-ins to virtual worlds using smart tags deployed at strategic places in the real world.

Also, when NFC is not available, and we need to promote (or restrict) the access to the virtual world Library from within certain known premises, it is always possible to get the real-world location of the user proving his/her presence via location based services (ie GPS, A-GPS...etc), used as an alternative check-in method. Current Android and iOS mobile devices (>90% market share) provide us with location services with enough accuracy (meeting E-911 and E-112 strict requirements with a wide margin)

This paper proposes the fundamental basis for the development of a 3d virtual world based Library using OpenSim, and facing the integration of a virtual learning environment (VLE, LMS) such as Moodle, with the help of the Sloodle platform.

Fig. 1. View of an open virtual world simulator with geospatial database connectivity and information sharing in real time.

II. TOOLS FOR CREATING METAVERSES

A. OpenSim 3D modeller

OpenSim is a 3D application server that uses the same standards found in SecondLife to communicate with their users. This modeller emerges from the analysis of the structure of SecondLife viewer's source code, released early in 2007 by Linden Labs under GPL. It was reverse engineered in order to create a 3D application server, and is characterized by being free software, having a modular structure written in C # and its support for multiple viewers; to date, OpenSim is in version 0.7.4

Among the most promising features of OpenSim is the ease with which universities can customize their LMS (Learning Management Systems) [8] integrating the legacy systems with OpenSim databases and creating content that suit the needs and methodologies of each institution.

OpenSim is being currently used to design, create and easily implement 3D virtual worlds by a huge base of programmers around the world. With this tool, it is possible to create content targeted to different areas, such as e-learning, virtual tours, simulations, among others. Moreover, being a free software tool leads itself to low-cost development and allows developers to contribute to the development of new features.

B. Moodle for Web courses

Moodle (Modular Object Oriented Dynamic Learning Environment) is an open source software package for creating courses and web sites based in PHP and designed to work with different database engines, especially MySQL [9]. Moodle aims to help educators to create and easily manage online learning communities, and is designed to foster the construction of knowledge by students.

Its first version was published in August 2002 and to date the current version is 2.3. Moodle currently has over 6.856. 000 courses offered through 72.164 registered sites in 223 countries, with Spain holding the second place in World Ranking of registered courses, and Colombia in the eighth position.

C. Sloodle project

The SLOODLE project (Simulation Linked Object Oriented Dynamic Learning Environment) [10] is a successful code that integrates multi-user virtual environments like Second Life and OpenSim with learning management platform MOODLE. Sloodle provides a wide range of tools (some still in development) that enable virtual learning worlds and immersive education, and are fully integrated with a web-based learning management system, used by thousands of educators and students worldwide.

D. Ubuntu operating system

Ubuntu is an operating system that uses a Linux kernel and based on Debian [10], is maintained by the British firm Canonical and the developer community. This distribution is intended to improve the end user experience, making it easier to interact with the operating system through an improved GUI. It is one of the most popular distributions today, with approximately a 50% share among Linux distributions.

For the development of this project, it should be noted that Linux is widely known as an operating system designed from the beginning as multithreaded as multiuser, so it does manage very safely the permissions for each user on processes and files, preventing users from accessing each other's information. Ubuntu has default user "root" disabled, so that the system repeatedly asks the user's password to install components and forces him to be especially careful with permissions to modify files and folders.

III. INSTALLATION OF METAVERSE IN OPENSIM

Below is a brief summary of the steps needed in the Ubuntu distribution in order to assemble a metaverse in OpenSim, and linking it to the Moodle platform.

A. Installing XAMPP Platform

Given that we need a web server to distribute the contents held by Moodle and that it is advisable to configure MySQL as the default database engine for a OpenSim server, we will use the XAMPP [11] Apache distribution, that includes the installation of MySQL. The XAMPP installation files for linux are available on the official website "Apache Friends". Once downloaded the file, you must deploy it uncompressed in the "opt" directory. Finally you must start the server using the command "/opt/lampp/lampp start".

One of the most powerful utilities found in XAMPP is the php-based administration tool for MySQL databases "phpMyAdmin", which can be accessed typing in a browser "http://localhost".

By default, the admin page does not have access control enabled. So, anyone with access to the server is able to change these valuable data. We recommend setting passwords. The XAMPP documentation describes how to do this.

B. Installing the Metaverse world with OpenSim

On the OpenSim download page we can find the necessary files for the installation of the server. It is advisable to mount the server on a 32-bit Linux version, since 64-bit versions often have compatibility issues with OpenSim and its dependencies.

OpenSim is designed to run on Microsoft's .NET platform. As the code and scripts used in the metaverse are based on C#, we should install the Mono platform on Linux, which has the libraries necessary to compile such languages. The required files for installing Mono on Linux are available at their official download page or they can be downloaded directly using the terminal installation commands (eg Ubuntu: sudo apt-get install mono-complete). After having installed mono opensim unzipped the file, we must open a Terminal. To start the server, just type "mono OpenSim.exe" command, with root permissions.

If there are no compatibility issues, the server should start gracefully, loading one by one all the necessary dependencies to the metaverse. If this is the first time you run OpenSim on the computer, it does not have any metaverse to begin with, so the simulator requests the creation of one, asking for the following information:

- New region name[ ]: alphanumeric character. Enter the desired name for the region.
- RegionUUID [default_UUID]: UUID. Unique identifier for this region. OpenSim proposes a default UID, so you should just press enter unless you prefer a specific UUID.
- Region Location [1000.1000]: x, y coordinates. It is the point in a coordinate plane you want for the region. Mode must be specified for non-grid overlay on existing regions region. Otherwise you can accept the default one point by pressing the enter key.
- InternalAddress [0.0.0.0]: Wildcard Mask (IP address). This address is actually a wildcard mask that allows you to restrict the range of addresses that can access the server. In most cases, should be left with default value, so that all hosts are allowed to access our server.
- InternalPort [9000]: # port (0-65535). Internal server port to be used for server access.
- AllowAlternatePorts: True or False. This feature is still in experimental stage. Default is false.
- ExternalHostname: ip address. This is the address used by clients to access the server, which should take into account some other considerations that are discussed later. To start, you can leave "SYSTEMIP".
- Do you wish to join an existing region to Udin estate (yes / no)? [Yes]: This is used to attach the newly created region to an existing state. If there is a previously created state which wants to join the region you should write its name, or on the contrary, type " no "and proceed to create one.

The parameters that are required to create one are:

- Estate name [MyEstate]: alphanumeric character. Any name.

The following data creates a new user on the server and will be the first user to access it:

- Estate owner first name [Test]: alphanumeric character. The name to use for the estate owner.
- Estate owner last name [User]: alphanumeric character. The name to be used for the estate owner.
- Password: alphanumeric character. User Password created.
- Email [ ]: alphanumeric character. Email of the newly created user.

Now we have a region, a state and a user. So, we can run the server and it can be accessed from any viewer. These parameters can be modified at any time by accessing the Regions.ini located in the bin/Regions directory within OpenSim. Text file appears as shown in Figure 2.

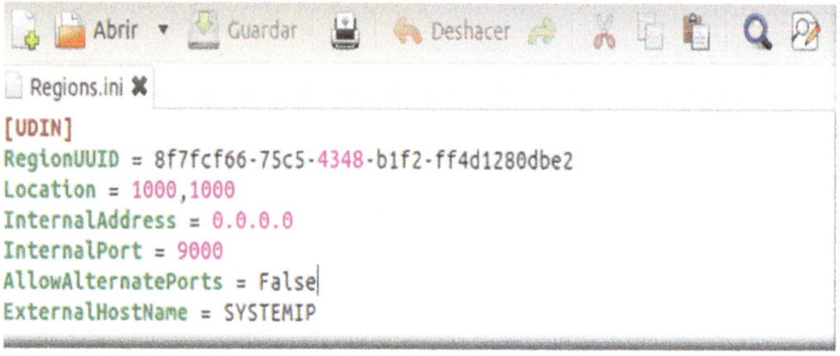

Fig. 2. View of Regions with location, internal address, port and the name UDIN server, Regions ID it is compose for a hexadecimal number unique in the world.

C. OpenSim settings for different network topologies.

OpenSim is a simulator that runs on a server, ie it is not a local program as a video game in which the user starts a process in his computer and it ends when he finishes using it. The situation resembles that of an interactive website in which the user makes requests to the server and starts or finishes sessions, accesses databases or performs specific actions in which processing occurs on the server (which may be in another location or on the same computer from which you connect the client). If you have a server, you must have a client with the necessary application to access its services. In case you have OpenSim Hippo viewers, like OSV, RealXtend or just the Second Life Viewer, they allow users to connect to servers running the simulator. Because the client and server can be located on the same network or on separate networks, we'll go on providing a brief explanation of the basic configurations of the above parameters, so that clients are able to access to servers.

Depending on the topology, we can have multiple address mappings (static or dynamic –depending on if the server has DHCP enabled or not-)

Changes on the server are made in the file Regions.ini, located in the bin/Regions folder within OpenSim. The only parameters that must be modified to access are "ExternalHostname" and "Internal Address".

It is possible to change the client directly off the display. The only parameter important to change is Login URI (the socket through which you access the server), it also allows viewers to save configurations and to assign names and other descriptive parameters that are not of interest here. Here are some settings for the server and the client, and network topologies.

- *Topology 1: Client and server are hosted on the same computer.*

The client and server configuration shown in Table I:

TABLE I
SETTING PARAMETERS FOR TOPOLOGY 1

Client	InternalAddress	0.0.0.0
	ExternalHostName	SYSTEMIP
Server	Login Uri	http://localhost:port

The network topology is shown in Figure 3.

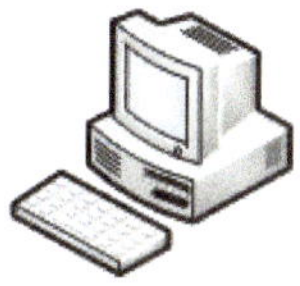

Fig. 3. Topology 1, this computer hosts server and client software, the virtual worlds development builder is inside.

- *Topology 2: Client and server on the same network.*

In this case we have the client and server using the same network serving different IP addresses (usually within a network with private addresses). The configuration is shown in Table II:

TABLE II
SETTING PARAMETERS FOR TOPOLOGY 2

Cliente	InternalAddress	0.0.0.0
	ExternalHostName	SYSTEMIP
Servidor	Login Uri	http:// public_address:server

The network topology is shown in figure 3.

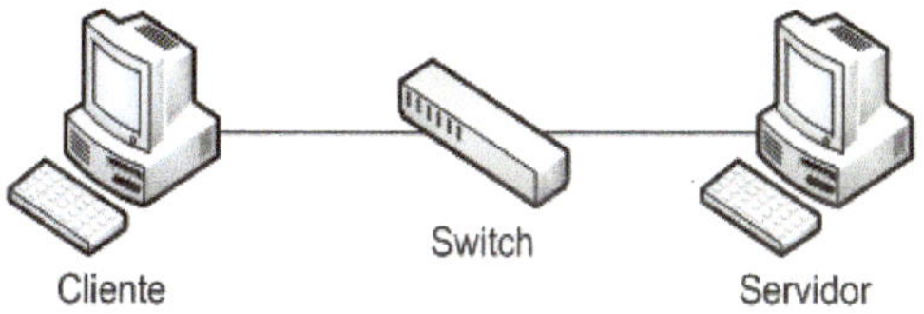

Fig. 4.. Topology 2, the client and server using the same network and serving different IP addresses

- *Topology 3: Client and server connected via the Internet, where the server has a public address.*

This configuration occurs when the server is connected to the internet through a broadband modem provided by an ISP and obtains a public address. The server configuration is the same as in the previous case. On the client, we can use this server's public address as part of the login URI. Figure 5 shows this network topology.

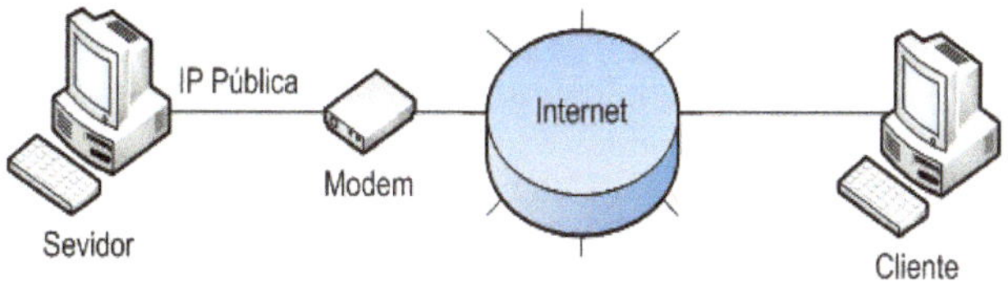

Fig. 5. Topology 3 Client and server connected via the Internet, where the server has a public address over the Web.

- *Topology 4: Client and server connected via the Internet, where the server has a private address.*

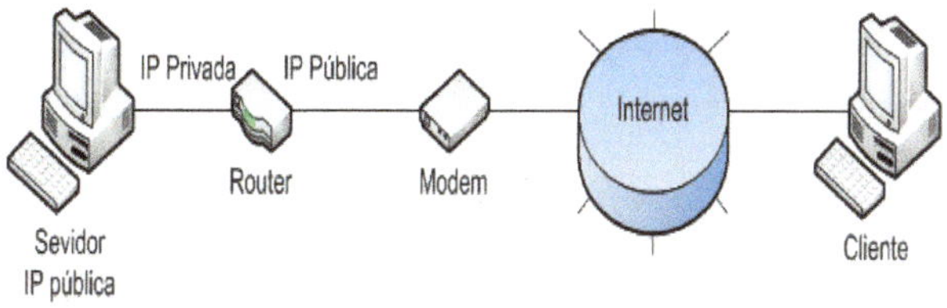

Fig. 6. Topology 4 Client and server connected via the Internet.

This configuration occurs when the server is connected to internet through a router connected to the ISP modem and therefore has a private address, which is translated by the router (NAT) to get out to the Internet. In this case, in addition to the above settings, we should enable port forwarding in the router for transport protocols TCP and UDP.

TABLE III
SETTING PARAMETERS FOR TOPOLOGY 4

Cliente	InternalAddress	private _address
	ExternalHostName	public_ address
Servidor	Login Uri	http:// public_address:port

To allow the client to connect to the server in the latter topology is also necessary to make port forwarding in the router. In current routers this function is usually called "virtual server" and is used to specify which ports redirect the public address to a private address.

D. Construction of objects in OpenSim

Creating and modifying objects in OpenSim is the basis for the construction of a metaverse. The first thing you should do is increasing the size of the island, this is accomplished using the "terrain fill" command long with the number of the desired size. The region may have a maximum size in the grid, because if set to a larger size, it would overlap with other regions in grid mode.

Different tools within OpenSim are:

- Vision (ctrl + 1) is used to change the angle of view, the approach and orbit of the camera.
- Move (Ctrl + 2): With this option you can move the different objects in the region.
- Edit (Ctrl + 3): this is one of the most important tools; this tool can easily modify position, rotation and size of objects.
- Create (ctrl + 4): This tool creates different primitive objects like cubes, cylinders, cones, spheres, prisms, toroids and others.
- Ground (ctrl + 5): With this tool you can raise, level, smooth down or reset the selected field. Using this tool you can also change the properties to specific "plots", including the type of texture that will be used to play media.

Fig. 7. OpenSim viewer, basic tools menus and tabs for designing objects in the virtual worlds.

When you open any of the aforementioned tools, you can make specific modifications to the objects just created, using the tabs at the bottom of the popup window.

The tabs available in the tools menu are:

- General: here you can make a description and edit the permissions and ownership of objects, and define the action performed by default when an avatar touches it.
- Purpose: In this tab you can set "physical" properties of the object in the metaverse, such as position, rotation, size, or even if the object is material, ghost, etc. The detailed description of the management of client tools for building an OpenSim metaverse are not the subject of this document.
- Features: edit the characteristics associated with the physics engine simulator. OpenSim accepts different physics engines that allow more realistic simulations of objects, but for the purpose of this project are not relevant.
- Texture: each object may also have different textures associated, which create more realistic environments and are also used for multimedia playback within the metaverse.

The complexity of building a metaverse starting from scratch using basic objects and the time it takes will depend on the user's ability using these tools. The detailed description of the metaverse construction is not the subject of this document.

IV. Linking the Metaverse Created with Moodle

If you have a Moodle platform installed on a server accessible to the client, this can be linked to the OpenSim metaverse created using Sloodle tools, so that users can make use some of Moodle functions directly from within the metaverse. For this project we installed Moodle on the same machine hosting an OpenSim server, in order to perform some tests.

A. Installing Sloodle

The files required for Sloodle operation must be deployed on the server and on the user/client machine. The files of the server are available for download at the official project wiki. The decompressed folders must be copied into the following subfolders. It is important to remove any previous version of Sloodle directories installed in Moodle, before copying the new ones.

- Module (required): You must copy the "Sloodle" folder into the "mod" folder within Moodle.
- Block (Optional): You must copy the "sloodle_menu" folder into the "blocks" folder of Moodle.
- Plugin Sloodle object allocation (Optional): You must copy the "sloodleobjet" folder into the "mod/assignment/type" folder within Moodle.

The user files for OpenSim virtual objects are required to make use of the functionality of Moodle within the metaverse. This is called the Sloodle Set. Once downloaded, it must be loaded into the metaverse via the "load oar" command using the command line in the OpenSim server. Keep in mind that this loads a new region, so before you do so, it is advisable to save the region on which you are working.

Once SLOODLE oar has been loaded, the objects appear. Sloodle objects therefore must be stored in the user's inventory. With the above, the modules are installed and Sloodle is ready for use with OpenSim.

B. Moodle in OpenSim

Before you can access Moodle from OpenSim using a user with admin role, you must enter edit mode through the "Enable Editing" button located in the upper right of the course and proceed to click on the "Add activity" drop down menu and then the "SLOODLE Controller". This leads to a page designed to setup the data controller, of which we only need to supply the name, so after filling this field, so you can click Save and return to course.

This controller handles all communications between Moodle and OpenSim and is therefore necessary to use any other Sloodle tools.

Some of the Sloodle tools, such as "Presenter", must have a Moodle driver object that is added the same way as the main controller. These tools added from Moodle page are required for objects that are added to Sloodle metaverse and they have some functionality, otherwise you cannot access the contents of the Moodle database. Other tools such as the "Sloodle Choice" or "Quiz Chair" can interact directly with native Moodle tools as "Choice" or "Quiz" respectively.

In this case the example will be a "Presenter" because you need to make video presentations in the metaverse. The detailed explanation of each one of the Sloodle tools is not the purpose of this paper and is left to the reader. As previously stated, the driver is first added to Moodle using "Add Activity" and then "Presenter". At this point, you must add the metaverse object. All Sloodle client tools can be added through the metaverse's "Sloodle Set", added above the metaverse and kept in inventory.

Before performing any task in Moodle using OpenSim, you must link the SLOODLE set located on the client host with Sloodle Controller located on the server. The steps for doing this are:

- Step 1: Click the SLOODLE Set
- Step 2: Write in Moodle site address, for example http://localhost/moodle or http://200.56.46.103:54661/moodle.
- Step 3: A dialog box appears in the viewfinder. Continue clicking on "Go to page".
- Step 4: To authorize objects Sloodle opens a browser which in turn shows the viewer . It should come with admin role, or shows the course in which you added the controller . Select "authorize previously added to Moodle", and post the form, after which you can close the browser.
- This should be done only once for the main controller Sloodle set, and once for each object you add to SLOODLE set in the metaverse.
- Step 5: Finally you must click again on the Sloodle Set, and then click Download configuration, so that the object has authorized access to the specified Moodle platform.

At this point you have a connection between OpenSim and Moodle but still have no functionality. Functions are added by SLOODLE set objects. The steps to add a Sloodle tool are:

- Step 1: Click the Sloodle set.
- Step 2: Click "Rez an object"
- Step 3: Click the object number to be added to the metaverse, in this case "7" representing the Presenter.
- Step 4: Click the object and then "Go to page" to authorize the new object in Moodle, as explained above for the controller and then close the page.
- Step 5: Click on the object and then select "Download configuration"

If everything is ok, the object is linked to the Moodle platform. These changes are required in Moodle as Sloodle objects do not store information, all the information is in the Moodle database.

- Module (required): You must copy the "Sloodle" folder into the "mod" folder within Moodle.
- Block (Optional): You must copy the "sloodle_menu" folder in the "blocks" folder within Moodle.
- Plugin Sloodle objects allocation (Optional).

V. LIBRARY DESIGN FOR VIRTUAL WORLDS

The use of digital libraries has changed over the years and even more with the advent of the Internet, from that moment the way knowledge is accessed really changed because in cyberspace there are countless sources of information: text, video and more. The change is so important that new generations of students prefer to search the web before going to a library; therefore arises the need for more innovative spaces that encourage children, youth and adults to enter virtual worlds and visit this new interaction spaces and new models of accessing knowledge. The ACL [12] make rules for Distance Learning Library Services, since 2000.

The fundamental basis for the creation of a virtual library are similar to a real repository but the interface is implemented by Sloodle, and therefore it must have intelligent access protocols. This is achieved using Web-based architectures, OWL, XML [13] [14], Ontologies, Semantic Services for RDF presentation [15] , data management, Grid repositories [16], databases compatible with XAMPP, connection to external sources of scientific data, social, geographic and more. All of them can be achieved by using Service Oriented Architectures (SOA) [17].

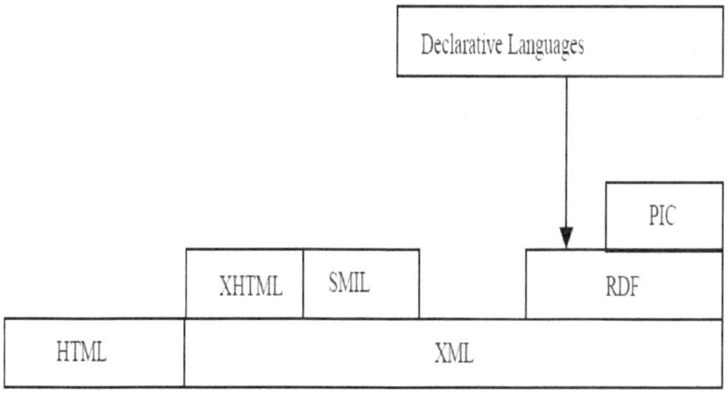

Fig 8. Language Layers on the Web, the Semantic Web, with machine processable information defined for their semantic content. Such semantic interoperability thanks of the W3C, XML/XML Schema and RDF/RDF Schema [13].

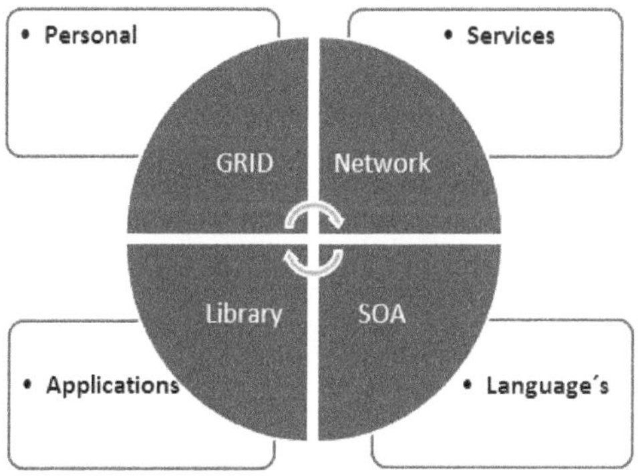

Fig. 8. Example using SOA for building virtual libraries over virtual worlds

One interesting and promising way of accessing the virtual Library at known places where we would like to promote it, is through publicly accessible NFC smart tags. In this scenario, when a user touches (or approaches) these tags with an NFC-enabled mobile phone, an event fires up an app which grabs the necessary URI that allows the user to access the virtual world.

This use of NFC technology can also act as a check-in method, or proof of attendance at known places where the Virtual Library is being promoted.

As seen, it is necessary to involve quite a lot of human resources, as researchers, Engineers, software developers, university students, research groups in the field of network architecture and broadband access to high speed Internet, Grid computing [18] and SOA services and a community of architects builders and 3D metaverse designers [22].

VI. RESULTS

The main result is a metaverse based on the engineering faculty of the "Francisco Josė de Caldas" Distrital University [19], hosted on a server accessible from the internet, and tested for Video Streaming [20] using OpenSim textures and mobile networks inside the University campus .

In Figure 10 you can see the photo of the real Engineering faculty building, and at the right, OpenSim built the building as proposed for the operation of a virtual 3D library. When the user logs in and reincarnates as an Avatar, is able to visit the site and see all kinds of information, all of the above accomplished by using Sloodle. In the future, we expect to further develop this initiative in collaboration with security staff trained to feed the virtual world.

Fig. 9. Metaverse example created in the Distrital University, with access to virtual library, Left real Building, Right virtual 3D Building in Opensim.

Also, there are objects -within the metaverse- linked to the Moodle platform installed on the same server.

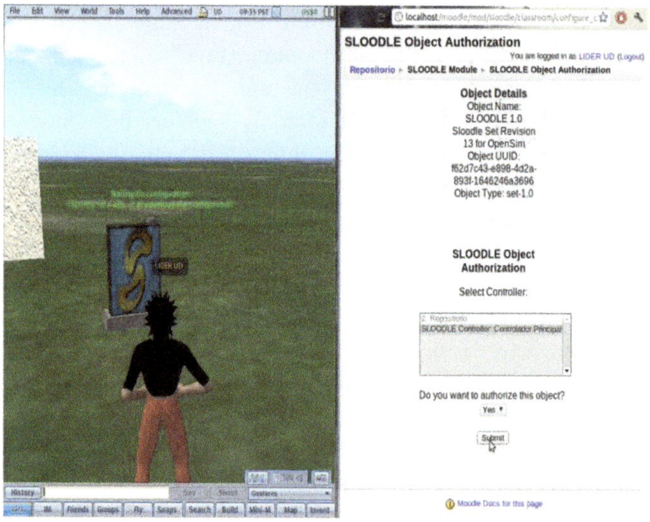

Fig 10. View window Sloodle controller with access from the metaverse.

VII. CONCLUSION

These kind of metaverses may target the educational virtual environments and the dissemination of scientific material, even in simulation labs, thanks to the programming tools here presented, which allow the control of the characteristics of the objects in these metaverses. Different research groups of the Distrital University are working with free tools like OpenSim and finding their application in specific fields of knowledge, and also other universities have already made significant progress: a good example of application is the development of 3D virtual libraries in these worlds.

One promising research area is the mobile broadband access to these virtual digital libraries, just making the knowledge ubiquitous, and special interest should be paid to recent proximity based interfaces such as NFC, which will be prevalent in mobile devices in the very near future, and a invaluable tool for check-ins as a gate between real and virtual worlds.

The use of virtual digital libraries help the environment as they save resources as printed paper, physical visits to sites, reduction in greenhouse gas emissions, electricity, time of travelling, encourages telecommuting from home and an endless resources that help to improve our world.

ACKNOWLEDGEMENTS

This proposal was carried out in the network of advanced technology research in Francisco José de Caldas University in Bogotá Colombia, as part of the Ph.D. in Engineering, within the "LIDER" [21] research group. Special thanks to Dr. José Nelsón Perez for his valuable support to develop these important projects that allow the creation new scientific horizons. With his support, we expect to continue moving forward with new developments in the research group.

REFERENCES

[1] G. P. Fishwick, "An introduction to Opensimulator and virtual environment agent-based M&S applications," in Simulation Conference (WSC), Proceedings of the 2009 Winter, Dec. 2009, pp. 177 –183.64.

[2] Linden Research, Inc, Available: http://lindenlab.com

[3] M. Barbulescu, M. Marinescu, O. Grigoriu, G. Neculoiu, V. Sandulescu, and I. Halcu, "GNU, GPL in studying programs from the systems engineering field," in Roedunet International Conference (RoEduNet), 2011 10th, June 2011, pp. 1 –4.

[4] Hippo OpenSim Viewer, Available: http://mjmlabs.com/viewer

[5] RealXtend Viewer, Available: http://realxtend.org

[6] M. Pattal, Y. Li, and J. Zeng, "Web 3.0: A real personal web! More opportunities and more threats," in Next Generation Mobile Applications, Services and Technologies, 2009. NGMAST '09. Third International, Conference on, Sept. 2009, pp. 125 –128.

[7] T. Kuroiwa and S. Bhalla, "Dynamic personalization for book recommendation system using web services and virtual library enhancements," in Computer and Information Technology, 2007. CIT 2007, 7th IEEE International Conference on, Oct. 2007, pp. 212 –217.

[8] M. Phankokkruad and K. Woraratpanya, "Web services for learning management systems: Communication architecture, in Communications (MICC), 2009 IEEE 9th Malaysia International Conference on, December. 2009, pp. 403 –408.

[9] M. Di Giacomo, "MySQL: Lessons learned on a digital library," Software, IEEE, vol. 22, no. 3, pp. 10 – 13, May-Jun 2005.

[10] S. Spaeth, M. Stuermer, S. Haefliger, and G. Von Krogh, "Sampling in open source software development: The case for using the debian gnu/linux distribution," in System Sciences, 2007. HICSS 2007. 40th Annual Hawaii International Conference on, Jan. 2007, p. 166.

[11] D. D. Dvorski, "Installing, configuring, and developing with Xampp," March 2007.

[12] Content and Design of Academic Library Web Sites for Distance Learners: An Analysis of ARL Libraries, Association of College & Research Libraries. Guidelines for distance learning library, December 2001.

[13] Hannes Bohring, Soren Auer, "Mapping XML to OWL Ontologies" June (2005) Available: http://www.informatik.unileipzig.de/~auer/publication/xml2owl.pdf.

[14] Brewster, C. O. (2004). "Knowledge Representation with Ontologies: The Present and Future". IEEE, 72-81.

[15] Simons, Gary F., William D. Lewis, Scott Farrar, D. Terence Langendoen, Brian Fitzsimons and Hector Gonzalez. 2004b. The semantics of markup: Mapping legacy markup schemas to a common semantics. Proceedings of 4th Workshop on NLP and XML: RDF/RDFS and OWL in Language Technology. Association for Computational Linguistics, Barcelona, Spain, July 2004. Online: http://emeld.org/documents/SOMFinal1col.pdf

[16] K. Czajkowski, I. Foster, N. Karonis, C. Kesselman, S. Martin, W. Smith, and S. Tuecke. "A resource management architecture for metacomputing systems". In IPPS/SPDP '98. Workshop on Job Scheduling Strategies for Parallel Processing.

[17] R. Fracchia, G. Jeney, M. Mazzotti, E. Piri. (2011). Optimization of Multimedia over wireless IP links via X-layer design" University of Bologna.

[18] Niaye NM, et. al.. "Performance comparison of hierarchical checkpoint protocols grid computing". International Journal of Interactive Multimedia and Artificial Intelligence. 2012, v 1, n 5, (pp 46-53).

[19] M. Erdmann, D. Fensel, I. Horrocks , M. Klein, S. Melnik. "The Semantic Web on the respective Roles of XML and RDF". IEEE, In Conference publication in the Third Workshop on Workflows in Support of Large-Scale Science, On 17-17 Nov. 2008, pp 1-10.

[20] Francisco José de Caldas Distrital University 1948 in Bogotá Colombia, Available: http://udistrital.edu.co

[21] Julian Camargo, Roberto Ferro,. (2009). "Research Laboratory for Development Electronic and Networks" LIDER-UD". Available: http://201.234.78.173:8080/gruplac/jsp/visualiza/visualizagr.jsp?nro=00000000005315.

[22] Ndiaye, N. M., P. Sens, and O. Thiare, "Performance comparison of hierarchical checkpoint protocols grid computing", International Journal of Interactive Multimedia and Artificial Intelligence, vol. 1, issue Special Issue on Distributed Computing and Artificial Intelligence, no. 5, pp. 46-53, 06/2012 DOI: 10.9781/ijimai.2012.156

M.Sc. Roberto Ferro Escobar, Electronic Engineer, M.Sc. in Networks 2004, became a Member of IEEE in 1999, a Senior Member in 2011, teaching at the "Francisco José de Caldas" Distrital University since 2006 in Bogotá Colombia, ex-manager of the Research Network of Advanced Technology "RITA", former manager program of Electronic Engineering 2009-2011, is currently developing his Ph.D thesis on Computer Engineering from the Universidad Pontifical of Salamanca in Madrid Spain.

Ph.D. Rubén González Crespo, Ph.D. in Computer Engineering from the Pontifical University of Salamanca in Madrid Spain, lecturer in Sciences and Engineering in the same university.

M.Sc. Eng. Sergio Ríos Aguilar. MSc in Telecommunication Systems, Computer Engineer, and PhD Candidate, lecturer in Sciences and Engineering at Pontifical University of Salamanca in Madrid, Spain Researcher on advanced mobile services.

Eng. Nicolas Torres Electronic Engineer became a Member of IEEE in 2010, student at the "Francisco José de Caldas" Distrital University in Bogotá Colombia 2006, Active member of LIDER research group.

17

A System for Personality and Happiness Detection

Yago Saez[1], Carlos Navarro[1], Asuncion Mochon[2] and Pedro Isasi[1]

[1]*University Carlos III of Madrid, Computer Science Department, Madrid, Spain*
[2]*Applied Economics Department, UNED, Madrid, Spain*

***Abstract* — This work proposes a platform for estimating personality and happiness. Starting from Eysenck's theory about human's personality, authors seek to provide a platform for collecting text messages from social media (*Whatsapp*), and classifying them into different personality categories. Although there is not a clear link between personality features and happiness, some correlations between them could be found in the future. In this work, we describe the platform developed, and as a proof of concept, we have used different sources of messages to see if common machine learning algorithms can be used for classifying different personality features and happiness.**

***Keywords* — personality detection, Android OS, happiness, written text, machine learning, classifying algorithms**

I. Introduction

Since Hans Jürgen Eysenck in 1947 defined the pillars, or traits, that form personality [1], numerous studies have been conducted and many works have been written about the subject, see Section II. These works have supported his theory of individual differences between humans with regards to personality. This theory is also known as the PEN model because of the three traits on which it is based: Psychoticism, Extroversion and Neuroticism. The theory provides a direct way to obtain a score for each component by using questionnaires, specifically the EPQ-R questionnaire. Each of the three personality traits has a biological basis, so the scores obtained for the traits represent different brain processes.

Researchers have tried to obtain information about the personality of human beings through direct means such as the EPQ-R questionnaire, but they have also used indirect methods. Because personality is considered to be stable over time and throughout different situations, specialized psychologists are able to infer the personality profile of a subject by observing the subject's behavior.

One of the sources of knowledge about the behavior of individuals is written text. According to research in this field, it is reasonable to expect that different individuals will have different ways of expressing themselves through the written word, and these differences will correspond to their individual personality profiles, as well as their moods.

This type of reports offers numerous advantages for researchers because a substantial amount of information about a subject's personality profile can be obtained without their presence or any additional specific effort on the subject's part.

A. Multidisciplinary work

Although research on personality profiling and analysis of the written word is part of psychology, collaboration with other disciplines, such as computer science, is necessary for certain purposes. Even with a solid psychological theoretical foundation, it is also necessary to be able to use quantitative methods to analyze large amounts of information. Such methods are especially applicable when analyzing large amounts of written text.

It is thus necessary to undertake this type of research with a multidisciplinary team, in which social sciences researches and computer scientists combine their knowledge to create efficient tools for the analysis of human personality. Computer science provides the tools necessary to collect, process and classify text samples of psychological interest in a systematic fashion, based on the principles of software engineering and artificial intelligence.

A tool with the aforementioned characteristics will be of great interest for the economy and human happiness. For example, if a system that could recognize the personality traits of a criminal in a matter of minutes with a high degree of confidence was available to law enforcement, a more efficient handling of critical situations could be achieved.

The remainder of this article is structured in the following manner. The following section describes the most relevant works related to this research. Section III describes the objectives and answers to common questions. Section IV depicts Eysenck`s theory of personality background. After Section IV, we describe the proposed platform, (Section V), the classifier module (Section VI) and the preliminary results (section VII). Finally, the main conclusions are presented in Section VIII.

II. State of the Art

The U.S. Army War College has shown an interest in predicting and controlling the behavior of an individual or group of individuals based on knowledge of their personalities. They believe that a system capable of this would have important applications in State security, competition in the labor market, political elections, or simply in the acquisition of knowledge about any person whose behavior might be of interest, see [2].

To perform a strategic personality simulation, they recommend taking into account the intersection between internal and external elements as well as external situational factors and personal influences.

Professors of computer science Gill and Oberlander conducted [3] a study on the recognition of the "extroversion/introversion" personality trait based on written text. They based their work on the Eysenck model [4]. For this purpose, they asked subjects with known scores on the EPQ-R questionnaire to write two e-mails to a fictitious friend. They subsequently analyzed these e-mails with a text analysis program called LIWC (*Linguistic Inquiry and Word Count*) and with the psycho-linguistic database MRC. They generated bigram profiles according to the degree of extroversion of the subjects (high or low). The results showed differences between the two sub-types of samples. Based on these differences, it was found that extroverts use more punctuation and exclamation signs, produce texts with more words, make more references to social situations, and use a greater number of positive words. Introverts, in contrast, are more likely to use the first-person singular, express themselves using more emotionally negative words, and use more coordinating conjunctions. The researchers also made lists of frequently used bigrams for both groups.

With their results, both authors conclude that the personality dimensions have relevance and validity for working with human-computer communication and computer learning.

Young presents in 2003 a geographical profiling, which consists of the profiling of criminals based on questions such as "when" or "where," instead of based on their motivations, age, gender, or other indicators [5]. With this approach, the need to incorporate computer science into the profiling process is emphasized to analyze large databases and prevent people from overlooking important information or connections between crimes. This type of analysis becomes imperative in the case of serial killers, who may commit crimes in different states that involve victims who do not know each other. The proposal coincides with the nature of this project in that it warns about the need for interdisciplinary work and highlights the importance of computer science for the processing of data that individual psychologists would not be able to analyze manually.

In this article [6], the principle of geographic profiling is presented. Geographic profiling is an attempt to obtain a wide body of information about criminal cases to provide a general psychological description of an unknown subject (UNSUB) — a possible suspect. After going into detail about the description of geographic profiling, the author presents several programs for collecting the essential information for this purpose. First, the Violent Criminal Apprehension Program (VICAP) is presented, which is used by the FBI to efficiently analyze the connections between existing criminal cases. Second, Kim Rosso's Criminal Geographic Targeting (CGT) is exhibit. This computer program produces a topographic map by performing many calculations that group together similar crimes, and it takes into account human movement patterns. Lastly, the Predator system, developed by Dr. Grover and M. Godwin, is described. This system uses multivariate analysis to carry out geographic profiling and produces a 3D, color-coded map to classify different areas according to the probability that the perpetrator lives or operates in them.

The word done by F. Mairesse and M. Walker may be considered to be the most important antecedent of the System for Personality Detection (SPD) project [7]. The researchers attempted to automatically identify personalities based on pieces of recorded conversations. Their personality analysis was based on the Five Factor Model (see [8]), which, is closely related to the personality traits of the PEN model used in the present project. In addition to confirming previous studies, the authors reached conclusions about personality. For example, they found that correlations between linguistic indicators and personality traits are higher in informal spoken dialog; this conclusion has stimulated the use of informal language in SPD. They also concluded that the most complex trait to analyze is "neuroticism," whereas "agreeableness" and "conscientiousness" provide the best results. Prosodic indicators were found to be the most accurate predictors for "extroversion." Finally, they concluded that their hypothesis, which proposes that it is possible to automatically detect personality through language, is confirmed, and they find that their procedure is applicable to a variety of fields.

The work of T. Polzehl, S. Moller, and F. Metze shows the results of implementing a personality evaluation paradigm for spoken input, and it compares human and computer performance in carrying out this task [9]. For this investigation, a professional speaker wrote speeches corresponding to different personality profiles, in accordance with the Five Factor Model questionnaire NEO-FFI. Then, human judges who did not know the speaker estimated the five personality factors. Recordings were also analyzed by using methods based on acoustic and prosodic signals. The results were very consistent between the acted personalities (as evaluated by the judges) and the initial classification of the results. Based on this, the authors concluded that they had made a first step toward the use of personality traits in conversations for future human-machine communication.

The study of A. V. Ivanov, G. Riccardi, *et al.* focused on personality prediction in the context of human spoken conversation [10]. For that purpose, once again, the Five Factor Model was used as a reference. The authors' final goal is to create a machine called the *Personable and Intelligent Virtual Agent*, which is capable of adjusting its linguistic behavior as required by the human with whom it converses. This would facilitate human-machine communication. During this research work, a simulated tourist help agent was created, which gathered linguistic and acoustic information from the subjects taking part is a role-playing game. These individuals volunteered their scores in the Big Five (Five Factor Model) questionnaire, and they were classified by their traits in a binary fashion: high or low. The results showed that machines can be trained to automatically predict personality traits based on conversations. In addition, statistically significant data were presented for the prediction of traits such as "conscientiousness" and "extroversion."

Linguistic Inquiry and Word Count (LIWC) is private software that analyzes text and calculates the degree to which an individual uses words from different categories, see [11]. A wide variety of sources are used, such as e-mails, transcripts of conversations, speeches, and poems. With LIWC, it is possible to obtain, for example, information about the number of emotionally negative words or self-references used, among many other dimensions of language.

Research on the topic of personality is often focused on one trait in particular: extroversion/introversion. Researchers in this field strive to find personality indicators, with the goal of creating simulated human-machine conversations, instead of focusing their discoveries on the creation of tools for personality profiling and happiness analysis. It is worth mentioning that, with the exception of the works [3], [12] and the LWIC2007 package (2007), all investigations were carried out based on spoken conversations and not on written text, in contrast with this work. In any case, existing research focused on the inference of personality and happiness based on the analysis of written text does not make use of mobile devices as a platform.

Regarding the research works that do focus on the creation of profiling tools, they are all centered on geographic profiling; they do not include personality as a factor in the profiling of the subject. Despite this, these works emphasize the need to combine disciplines to produce their tools. That is the spirit of this project.

III. Objectives

The main goal of this project is to develop a prototype system that is capable to collect information in written Spanish from different sources of interpersonal communication on a mobile device.

The project consists of a module in which a client application is developed for mobile devices running the Android operating system. This application is in charge of compiling and sending information about the user to a server application, which stores the information as it is received.

Independently of the goals set for this work, and according to advances in joint research with a team of criminologists from the Institute of Forensic Sciences and Security (ICFS), work will begin on a prototype for a classifier module that, by processing the collected data, will search for markers to classify the user according to Eysenck's theory of personality.

For this purpose, a system will be created to classify the user based on previously established principles of analysis and natural language processing.

Why mobile devices?

According to a study carried out by CISCO Systems (2013), in 2016, there will be more mobile devices than people, which means that there will be a large number of potential users for the system. In addition, it is worth mentioning that many of the most commonly used means of communication are concentrated on these devices.

Why Android systems?

There are many reasons to implement this project on Android devices, the first of which is that the Android OS provides programmers with more flexibility for the development of applications because it allows for free access to all device resources: an indispensable requirement for the development of the proposed system.

Additionally, the percentage of mobile devices running Android rose to 84.1% by the middle of 2012, according to a study by the consulting company Kantar, i.e., more than four out of five people in Spain who possess a mobile device have one that runs Android. This allows for wider distribution of the application.

Nevertheless, not all Android devices are useful to us, or at least not all of them can provide us with the same sources of information. Because of this, we will focus on smartphones, the devices through which most interpersonal communication takes place.

Why in Spanish?

For the purpose of analyzing the conduct of an individual through their writings, knowing and being able to analyze the language in which the individual expresses himself or herself is paramount, from a psychological point of view. The mere fact that someone uses certain specific words or expressions gives structure to the subject's personality profile. Because of this, a single language must be selected for the development of the application. For the application to be used by people in other countries, it would need to be adapted to the appropriate socio-linguistic context.

This project is being developed in Spain, so the native language (Spanish) of the potential users has been selected.

IV. Theoretical Background

The theory of personality by Hans J. Eysenck [1] is based on multidimensional taxonomies of personality. From this point of view, there exist personality traits that allow for the description, and therefore prediction, of human personality and conduct, see [13].

Eysenck recognizes three personality traits: psychoticism, extroversion and neuroticism, giving rise to the acronym in PEN theory. These traits manifest themselves in different types of human behavior:

TABLE I
CHARACTERISTICS THAT DEFINE THE THREE PERSONALITY TRAITS OF THE PEN MODEL.

Extroversion	Neuroticism	Psychoticism
Sociable	Irrational	Aggressive
Dominant	Inhibited	Cold
Assertive	Taciturn	Egocentric
Active	Emotional	Impersonal
Lively	Tense	Impulsive
Boastful	Anxious	Antisocial
Daring	Depressed	Creative
Carefree	Feeling guilt	Unfeeling
Adventurous	Low self esteem	Harsh

These traits cannot be understood categorically because they are not mutually exclusive. A subject's personality is composed of three independent traits, which must be understood from a dimensional point of view, [13].

Hence, it is important to understand that the three traits are independent, but together, they determine a personality profile corresponding to the idiosyncrasies of the subject. The potential of their combinations cannot be disregarded.

With this model, an underlying biological basis of the three traits is provided. Eysenck believed that the Extroversion-Introversion trait corresponds to cortical arousal. Specifically, it is controlled by the Ascending Reticular Activating System (ARAS). According to the author, extroverts possess a lower degree of cortical arousal, meaning that they present low cortical activation. In contrast, introverts are *a priori* expected to be highly activated. Given the low "internal" activation of extroverts, they would require external and more intense stimulation, whereas introverts are over-activated and do not require external stimulation to maintain a high level of arousal [14].

The Neuroticism-Stability trait is related to the autonomous nervous system, or the limbic system, which is in charge of regulating emotional impulses. Therefore, a highly neurotic individual will have an unstable autonomous nervous system, leading to intense reactions to stimuli. This would explain the variability of mood and anxiety in neurotic subjects. In stable subjects, the exact opposite would be found, [14].

Psychoticism is the most complicated trait within Eysenck's theory, and only recently has some light been shed on its biological nature. Psychoticism has been found to be related to the vulnerability to psychotic disorders, although this does not mean that people with high scores on this trait are certain to suffer from such personality disorders [14]. The Eysenck Personality Questionnaire-Revised (EPQ-R) [4] is currently used to evaluate the traits proposed by Hans. J. Eysenck.

Lastly, it is worth mentioning the relationship of Eyseck's theory with another multi-trait personality model, which is highly favored by the scientific community: the Five Factor Model. This model, also known as "The Big Five" model [8], is based on five fundamental personality traits: Extroversion, Neuroticism, Openness to Experience, Agreeableness and Conscientiousness [13]. These traits are to be evaluated via the NEO Personality Inventory-Revised (NEO PI-R) [15], or the Big Five Questionnaire (BFQ) [16].

Extroversion and Openness to Experience correspond to the Extroversion trait in PEN theory, Neuroticism has a homologous trait in Eyseck's theory, and Psychoticism would be inversely correlated with Conscientiousness and Agreeableness.

V. TECHNICAL PROPOSAL

In this section, the architecture and design of the system to be developed is presented and the different components of the system are explained.

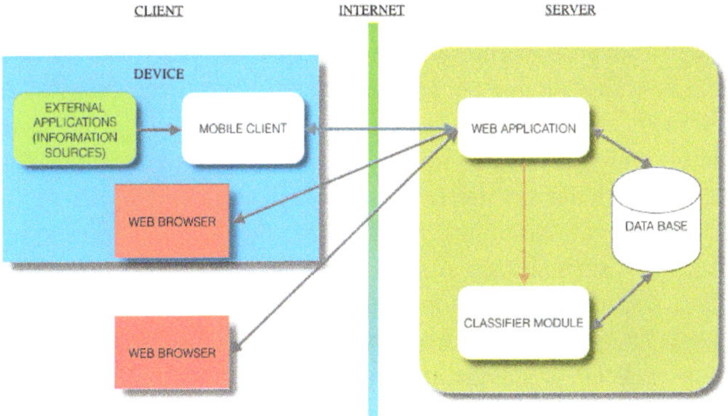

Fig. 1: System architecture

The model to be implemented corresponds to a distributed computer system, which will be composed of numerous devices. Existing classical architectures for distributed systems include the client-server (C/S) architecture and peer-to-peer (P2P) architecture. The C/S architecture is employed when there is a dependency relationship between the devices, which are interconnected in a computer network. This occurs when some functions are performed on the server, and it is the client that communicates with and requests a response from it. In the P2P architecture, every device may function as both client and server.

In the SPD project, there is a logical split within the application. Due to the restrictions described in the non-functional requirements, the system is spread across different computers (physical separation). Only one of the computers—or a group of them functioning as one— will provide services to the rest, thus becoming the "server," the others will submit requests to it, thus becoming "clients." Thus, the chosen architecture is the C/S architecture.

The elements included in the architecture of the SPD system are the following:

- **Client:** software in charge of interacting directly with the user and communicating with the server to submit requests to the system. It will consists of the following:
  - **Mobile device:** the equipment owned by the user, which contains the following elements:
    - External applications: an indispensable aspect of the functioning of the system is that the user has a set of applications for interpersonal communication installed on the device, which will serve as the source of information. The

text samples needed for personality profiling will be obtained from these applications. These sources of information will have to be accessible to the mobile client.

- Mobile client: application that will be developed in this project, which allows the interaction with the user. It will mainly be in charge of gathering information and communicating with the server to classify the personality traits. It will also work as a client to access the information provided by the external applications.

- **Web client:** some actions will have to be carried out through a web client external to the system. It may be located within the mobile device itself, or on any other device with basic web navigation capabilities.

- **Server:** the software to be implemented will run on a computer that is accessible to the clients installed on various user terminals via the Internet. It will itself be in charge of the task of communicating with the clients, providing services such as registration within the system, setting up user accounts, etc. It contains the following elements, which must be differentiated from the server itself:
  - **Web application:** responsible for communicating with both types of clients, mobile and web-based.
  - **Database:** element that will store all user access accounts and all information compiled from the mobile client.
  - **Classifier module:** module that will take care of processing and determining, from the information stored in the database about a given user, which personality profile defines said user.

In addition, both the client and the server will employ a Model-View-Controller (MVC) architecture. This is an architecture typically used for graphical interfaces such as web pages. It separates the components into different layers for the reuse of code, and its decoupling facilitates the development, expansion and maintenance of the application. In this project, a pure implementation of the model will not be used because it will not always be the user who carries out the actions and requests between clients and server.

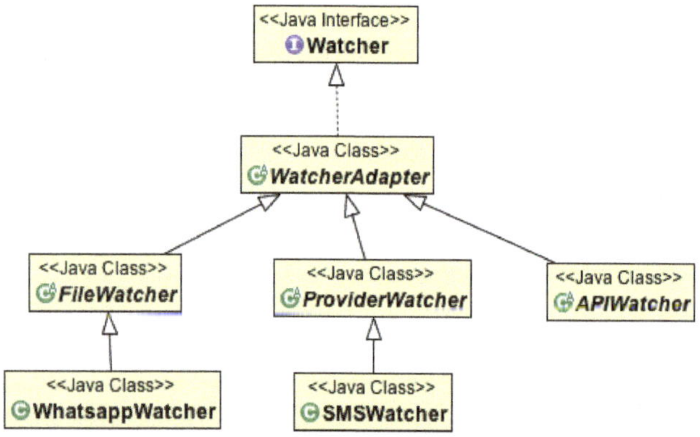

Fig. 2: Details of the hierarchy of implementation of the recollectors.

A. Alternatives

With regard to possible alternatives, we would like to mention that a similar software tool with the same goals as the one presented in this work does not exist, or at least is not publically known. Therefore, this project cannot be implemented using some existing alternative scheme. Notwithstanding, it would be reasonable to carry out an analysis of existing technologies, such as tools or programming languages, to determine which could be useful and to explain why some should be chosen instead of others. We would like to mention that there exists a real-time monitoring software package for Android called *MobileSpy*, which overlaps with SPD as it also collects data. This software, in contrast to the proposed SPD software, collects data from additional sources, such as pictures taken and websites visited by the user. Such information is not necessary for this project, as it is oriented toward espionage.

VI. Classifier Module

In accordance with what was explained in Section III, a prototype classifier module has been programmed as we proceed with the psychological research on the classification and quantification of personality traits through written text.

The classifier is composed of two parts:

First, there is the main process. Its functions include communicating with the database to procure the user data to be processed. In addition, this first part will be in charge of calling the functions of the second part— functions that search for matches of personality indicators— for the evaluation of the established parameters in the compiled user entries. Finally, the main process will have the task of saving in the database the results obtained by the matching functions (for future access, without the need to repeat the analysis on data that have already been processed), and it will responsible for analyzing the results to evaluate the scores of the different personality traits.

The analysis process is still being developed by the ICFS team and, thus, has not been implemented yet. Basically, they are the search functions that have the task of individually finding and counting every match identified in the data entries, according to the indicators established by the group of psychologists. This group of professionals built a series of lists of indicators to identify personality traits. For example, to identify high scores in the "neuroticism" trait, a list of "emotionally negative words" has been set as an indicator. These indicators are not simply based on the contents of the text, as in the previous example; they also take into account the structure of the text, e.g., lexical density.

Regarding the technical aspects of the classifier, the first decision was to use a language processor to link the analysis to the theoretical guidelines developed by the psychologists. A reference in this field is the *Natural Language Toolkit* (NLTK), but it does not have enough resources to function in the Spanish language. After taking into account many possible solutions and testing a variety of resources to analyze text, we found that the open source toolkit Freeling was the most approriate for this project. Freeling offers a wide range of functionalities similar to those of NLTK, but with more resources for Spanish. Furthermore, there exists a translation of the WordNet— a lexical database— called the *Multilingual Central Repository* (MCR), which is compatible with *Freeling* and is on par with numerous resources created by other groups. This helped with the automation of the analysis of text using *Freeling*. Thus, the steps needed to implement the language processor have been determined, and this work may proceed as soon as the psychological part of the research allows for it.

VII. First Prototype: Preliminary results

In this work, the aim is to try to classify messages with within the personality features described in Table 1 using supervised machine learning algorithms. The key idea is that these messages have words that will be preprocessed and clustered in order to see whether its possible to match somehow the obtained clusters with the personality features model described before.

However, although the mobile application described before is already developed and ready to deploy we don't have enough information to analyze nor prove the complete method with Whatsapp and SMS messages. There are not many public datasets available in Spanish, so what we have done as a proof of concept is to work with one public dataset with real messages and two datasets generated from messages collected from different websites. For the public dataset we have used a subset of a corpus made of 63.017 *Twitter* messages released by the end of 2013, 17. This dataset has been analyzed for sentiment/opinion analysis with different techniques by several research institutions (IMDEA, LSI-UNED, Elhuyar Fundazioa, L2F, and SINAI-Universidad de Jaen). Other available dataset in Spanish is the 15M Dataset from Complex Systems and Networks Lab (http://cosnet.bifi.es/research-lines/online-social-systems/15m-dataset). This large dataset of *Twitter* messages from the Spanish 15M movement is not useful for this purpose because it is mainly made of hashtags. The other two datasets are made of messages collected manually from several blogs focusing on collecting different lengths of messages (no rights for publishing). For datasets #1 and #2 a classification for extraversion, neuroticism and psychoticism has been made. Personality (based on Eysenck's theory) could be considered an objective measure, but happiness is usually considered opinion. Therefore, there is not a direct translation between the personality features such as extraversion, neuroticism, and psychoticism into happiness. We know that in PEN model, personality it is measured as a weighted combination of features. If written text can express personality features we think that it can express happiness as well, and for future studies we will try to find if any of these combinations can be somehow correlated into an acceptable happiness classification. For these reasons we have decided to include a happiness classification for dataset #3.

Table II
Type of dataset used for testing the clustering algorithms

Dataset	Nº messages	Avg. words / Message	Description
#1	1772	6.53	Twitter
#2	314	10.62	Blogs
#3	200	50.02	Blogs

It is important to remark that the main objective of these experiments is to check if it is possible to identify and classify different types and lengths of messages with common machine learning techniques. In order to do this, we are testing different datasets with several classifying algorithms. First common stage when analyzing texts is to preprocess information. This step will analyze words trying to identify language (English, Spanish, etc.), common lingo, symbols, emoticons, words without semantics and prepositions, articles and conjunctions.

After this initial step, then a stemming analysis takes place. The stemming is a wide used process in information retrieval that reduces or derives words to their stem or root form (if possible). Once this stage finishes, the cleaned dataset is used for clustering.

The first experiments were conducted with WEKA. When working with so many parameters, many different setups can be used. For replication purposes here is a summary of the actions taken, the parameters used and the algorithms tested:

- Preprocessing stage (applied directly to the dataset):
  - Filtering stage (importing data in WEKA):
    - Euclidean distance
    - StringToWordVector filter
    - IDF Transform = True
    - All keywords in lowercase
    - Remove all special characters, punctuation, accents
    - Stopwords: prepositions, articles, conjunctions
    - Eliminate those words that appear less than 5 times.
    - Stemming algorithm: Snowball
    - Deleting keywords with less than 3 letters (before stemming)
    - Deleting numbers
  - Used algorithms (WEKA):
    - J48, binary splits, 0.25 confidence factor
    - Random Forest, 30 trees, no depth limit
    - Support Vector Machine (SVM), LibLINEAR

The Table 3 shows the preliminary results with the three different algorithms and random cross-validation (10 folds).

Table III
Results for all scenarios and algorithms tested

	J48	Random Forest	SVM
#1	Accuracy: 0.42 Kappa: 0.18	Accuracy: 0.57 Kappa: 0.37	Accuracy: 0.56 Kappa: 0.38
#2	Accuracy: 0.65 Kappa: 0.53	Accuracy: 0.58 Kappa: 0.44	Accuracy: 0.65 Kappa: 0.53
#3	Accuracy: 0.64 Kappa: 0.27	Accuracy: 0.64 Kappa: 0.28	Acurracy: 0.82 Kappa: 0.65

The first dataset has in average very short messages for an adequate classification. The accuracy of the tested algorithms in the best case is a 57% (SVM), with many false positives mainly between neuroticism and psychoticism classes (see Table 4 in Appendix). Other issue is the nature of the messages as *Twitter* it is not usually a social media for sending personal messages. Furthermore, most of the selected messages are classified into psychoticism and neuroticism, due to the fact that the dataset comes from primer elections, which bias the type of message collected. These messages are probably really useful for opinion mining but excessively focused on primer elections and short to extract what our psychologists are looking for. The second and third datasets have clearly more promising results. We think that is because messages are longer than the ones from the *Twitter* dataset (see average words per message in Table 2) and because there are not only focused on one specific topic.

VIII. CONCLUSIONS

Returning to the goals laid out in Section II, we have created a mobile application for the Android OS that allows for the acquisition of information in the form of written text and allows for communication with a server. In our system the information is acquired from two sources: the *Whatsapp* application and the SMS service. We have also created a server that receives this information from the mobile clients and stores it. Finally, we have also begun work on the creation of a classifier for the gathered information. A prototype has been developed that will continue to be improved.

With regard to the objectives established previously, it can be concluded that all the requirements have been successfully met, except for one: the requirement for the presentation of results in the mobile application. To meet this requirement, it will be necessary to fully develop the classifier and the prototype but this work is still in an early stage although it is a great starting point for future analysis. A lot of work needs to be done prior having more conclusive results. However, so far, we are proud to say that the first objective is achieved. A lot of time has been needed for collaborating with psychologists, understanding the possible ways of identifying personalities and their link with happiness. Finally, developing an application able to collect the *Whatsapp* messages has been a hard task but will allow us the possibility of building the first public corpus of *Whatsapp* messages up to the date. The future of this dataset depends on the overcoming of other problem encountered: the reticence of users to install the application and their non-objective use when installed.

A. Future lines of work

In the short term, we have several goals for continuing the work already done in this project.

First of all, we consider very important to be able to carry out an initial research, so the preliminary results can guide our team and the psychologists to refine the whole system. This first step is already set to be carried out in the near future, the application will be installed on the mobile phones of experimental subjects who will have taken the EPQ-R personality test beforehand. After that, information will be collected and classified by using the tools that have already been developed, which by then will have been improved. The appropriate statistical analysis will be carried out to compare the results of the personality test with the application information. With these results, we will be able to determine how fine-tuned the classifier and the theoretical psychological guidelines are, and we will see what aspects need to be improved so that the personality profile obtained from the application can be trusted. Additionally, we may introduce Artificial Intelligence techniques, such as genetic algorithms, neural networks or automatic learning techniques, as they may help improve the classification.

As mentioned in the previous section, it will be necessary to complete and improve the classifier; this task is intimately related to the research to be carried out. The classifier is a part of the system that cannot be considered finished with only a first build. It will be necessary to verify the proper functioning of the global system and to perform certain adjustments in response to the results. Hence, the system will be improved due to advancements in psychological practice as well as future research that may be carried out.

We are determined to refactor this application and its principles beyond the criminological and clinical fields and to extend the technology to personal use. The field in which we are very interested is human-machine communication. If a machine were able to automatically identify personality from the text provided by a person, it could adapt to the needs and tastes of the person. As [18] mentioned, the moment we are able to identify personality, we are one step closer to being able to simulate it. That is, we could be closer to achieve artificial intelligence.

IX. ACKNOWLEDGMENTS

The authors wish to acknowledge Álvaro Ortigosa Juárez (CEO at ACC -Agencia de Certificaciones de Ciberseguridad – and CNEC – Centro Nacional de Excelencia en Ciberseguridad), Manuel de Juan Espinosa (CEO at ICFS – Instituto de Ciencias Forenses y de la Seguridad), Carlota Urruela Cortés (project leader), Irene Gilpérez López and Pilar González Villasante (Main Psychology Team at the ICFS) for their support and collaboration during an important part of this work.

X. APPENDIX

Detail of confusion matrix obtained with different algorithms and datasets.

TABLE IV
CONFUSION MATRIX FOR TESTS DONE WITH DATASET #1

J48 Confusion Matrix: Dataset #1

Act \ Pred	Extraversion	Neuroticism	Psychoticism	Unknown
Extraversion	**62**	184	48	18
Neuroticism	85	**354**	150	48
Psychoticism	70	138	**247**	115
Unknown	25	19	129	**80**

Random Forest Confusion Matrix: Dataset #1

Act \ Pred	Extraversion	Neuroticism	Psychoticism	Unknown
Extraversion	**114**	187	10	1
Neuroticism	46	**507**	82	2
Psychoticism	1	192	**365**	12
Unknown	1	13	210	**29**

SVM Confusion Matrix: Dataset #1

Act \ Pred	Extraversion	Neuroticism	Psychoticism	Unknown
Extraversion	**199**	102	9	2
Neuroticism	130	**394**	106	7
Psychoticism	22	162	**312**	74
Unknown	1	14	147	**91**

TABLE V
CONFUSION MATRIX FOR TESTS DONE WITH DATASET #2

J48 Confusion Matrix: Dataset #2

Act \ Pred	Extraversion	Neuroticism	Psychoticism	Unknown
Extraversion	**51**	7	22	0
Neuroticism	8	**47**	8	13
Psychoticism	30	12	**30**	7
Unknown	12	14	8	**45**

Random Forest Confusion Matrix: Dataset #2

Act \ Pred	Extraversion	Neuroticism	Psychoticism	Unknown
Extraversion	**52**	6	17	5
Neuroticism	0	**58**	5	13
Psychoticism	32	16	**26**	5
Unknown	9	18	5	**47**

SVM Confusion Matrix: Dataset #2

Act \ Pred	Extraversion	Neuroticism	Psychoticism	Unknown
Extraversion	**49**	3	21	7
Neuroticism	0	**56**	5	15
Psychoticism	16	10	**43**	10
Unknown	2	16	5	**56**

TABLE VI
CONFUSION MATRIX FOR TESTS DONE WITH DATASET #3

J48 Confusion Matrix: Dataset #3

Act \ Pred	Happy	Unhappy
Happy	**670**	330
Unhappy	403	**597**

Random Forest Confusion Matrix: Dataset #3

Act \ Pred	Happy	Unhappy
Happy	**702**	298
Unhappy	417	**583**

SVM Confusion Matrix: Dataset #3

Act \ Pred	Happy	Unhappy
Happy	**844**	156
Unhappy	193	**807**

REFERENCES

[1] H. J. Eysenck, Dimensions of Personality, 1947.
[2] N.D. Livergood, Strategic Personality Simulation: A New Strategic Concept. U.S. Army War College, 1995
[3] Gill, A. and Oberlander, J. [2002] Taking care of the linguistic features of extraversion. In Proceedings of the 24th Annual Conference of the Cognitive Science Society, pp363--368. Washington DC, August 2002
[4] Eysenck, H.J. and Eysenck, S. B. G., 1985. Revised Eysenck Personality Questionnaire (EPQ-R).
[5] G. Young, "Mapping Mayhem: The Geography of Crime," Computoredge, 2003.
[6] K. Rossmo, Profiling and Geography, ISBN-0849381290, 1999.
[7] F. Mairesse, M. and Walker, M., 2006. Automatic Recognition of Personality in Conversation. In: Association for Computational Linguistics ed., Proceedings of the Human Language Technology Conference of the North American New York: Association for Computational Linguistics. Automatic Recognition of Personality in Conversation, pp. 85
[8] Goldberg, L. R. (1981). Language and individual differences: The search for universals in personality lexicons. In L. Wheeler (Ed.), Review of Personality and Social Psychology, Vol. 2. Beverly Hills, CA: Sage.
[9] T. Polzehl, S. Moller and F. Metze, Automatically Assessing Acoustic Manifestations of Personality in Speech, 2010.
[10] A.V. Ivanov, G. Riccardi, A.J. Sporka and J. Franc, "Recognition of Personality Traits from Human Spoken Conversations," Annual Conference of the International Speech Communication Association, 2011.
[11] J. W. Pennebaker, R.J. Booth and M.E. Francis, Linguistic Inquiry and Word Count (LIWC2007), 2007.
[12] M. de Juan-Espinosa and A. Ortigosa Juárez, Detección Automática De La Personalidad En Comunicación Vía Ordenador. Madrid: Universidad Autónoma de Madrid, n.d.
[13] B. Moreno, Psicología De La Personalidad. Procesos. International Thomson Editores, 2008.
[14] M. de Juan-Espinosa, Personalidad y Criminología, Apuntes De Psicología Criminológica. Madrid: Universidad Autónoma de Madrid, 2005.
[15] Costa, P.T. and McRae, R.R., 1990. Revised NEO Personality Inventory (NEO PI-R).
[16] Caprara, G.V., Barbarranelli, C. and Borgogni, L., 1993. Big Five Questionnaire (BFQ).
[17] Villena-Román, J., García-Morera, J., Moreno-García, C., Ferrer-Ureña, L., Lana-Serrano, S., Carlos González-Cristóbal, J. Westerski, A., Martínez-Cámara, E., García-Cumbreras, M.A., Martín-Valdivia, M.T., Alfonso Ureña-López, L., 2012, TASS-Workshop on Sentiment Analysis at SEPLN, Workshop on Sentiment Analysis, Sociedad Española para el Procesamiento del Lenguaje.

[18] M. de Juan-Espinosa, Personalidad Artificial: Hacia Una Simulación De Las Diferencias De Personalidad En Situaciones De Interacción. Madrid: Universidad Autónoma de Madrid, 1997.

18

Intelligent Position Aware Mobile Services for Seamless and Non-Intrusive Clocking-in

Sergio Rios-Aguilar,

Pontifical University of Salamanca en Madrid, Spain

Abstract **– This paper analyzes the viability of the use of employees smartphones as a valid tool for companies in order to conduct presence control. A Mobile Location Aware Information System is also proposed for a non intrusive Presence Control using exclusively terminal-based reactive location technologies, meeting cost minimization, and universal access criteria. The focus is providing trust to the employees, so that they feel safe and in control of when the location data is gathered while satisfying the control needs of the employer. LAMS platform is a state-of-the-art framework for synchronous mobile location-aware content personalization, using A-GPS terminal-based/network assisted mobile positioning techniques and UAProf data processing at the origin server.**

Keywords **- Clocking in, Presence Control, Terminal Based Positioning, Network Assisted Positioning, A-GPS**

I. Introduction

There is enough evidence suggesting that use of Information Technology can play a quite important role on the growth of small and medium size businesses. In this line of thought, IT can be employed to bring about increased competitiveness if it enables businesses to create new jobs, increase productivity and sales achieving new levels of administrative efficiencies. It goes without saying that these outcomes can be achieved through measurable improvements of key performance indicators [1].

Tardiness, or –worst case- absenteeism, are two of the most persistent obstacles to productivity, profitability and competitiveness. They cause overtime, late deliveries, dissatisfied customers and a decline in employee morale amongst workers who are expected to cover for an absent or late coming employee. Early recognition and rapid intervention are key when it comes to managing absence in the workplace and can prevent absenteeism from becoming a long-term problem. [2] [18].

As for performance measurement, companies should collect and systematize all the information available so that they signal or allow the execution of their strategy to be successful in their business and remain in business. Once a company acknowledges tardiness and absenteeism as key problems, it should try to collect information detecting such behaviors.

Workplace surveillance and business organizations go hand in hand, and that employee monitoring is nothing new. The implication is that surveillance at work is, first, a necessity, and second, a normal, taken-for-granted element of working life. Employees expect to have their performance reviewed, objectives set, and information gathered on their activities and whereabouts – indeed, this is seen as good management practice. [18].

Controversies generally arise when employee monitoring goes beyond what is reasonable or necessary (i.e. when employers use what employees perceive as intrusive monitoring, gathering precise information as to how employees use their time) or when the application of monitoring negatively affects existing levels of autonomy and trust. [18][19]

So, when employers intend to control tardiness of absenteeism, they generally make use of clock-in systems. But they prove to be ineffective when applied to remote workers (ie. Sales force, field force…etc). [20]

Here we will be presenting LAMS platform, a Location Aware Mobile System for non intrusive clocking in. Field force employees make use of their own (or company supplied) smartphone to clock-in, and they are in full control of the positioning process and can trust that the system only gathers location data when they allow it to do so.

II. Location Technologies and LAMS Overview

One of today's key technologies related to advanced mobile services development is the physical positioning of mobile terminals requesting services. Moreover, the use of terminal based and network assisted solutions in order to achieve higher precision and lower acquisition times is the current direction of the state of the art research being conducted on this field.

In this paper, at first we will be presenting the fundamentals of Location Services and current location technologies, and we will go on with a description of the proposed platform designed to assist the employee clocking in process, with full HTML5/xHTMLMP-compliant contents personalization, together with integration services for spatial analysis, in terminal based and network assisted mobile services.

This implementation currently uses W3C Location API code within the terminal, in order to obtain the local position measurement from the embedded positioning device and the A-GPS data from a SUPL-enabled network. [14]. This information is then feeded to the JavaEE core of the platform, which build XML-based spatial queries and forwards them to the available GNSS servers, processes their XML responses (GML, PoIX) through XSL transformations to the final HTML5/JQueryMobile contents delivered to the requesting client terminal.[1],[2]

III. Positioning Technologies

Currently available technologies for physical determination of the position of a mobile terminal fall into two broad areas: terminal based technologies and network based ones.

In the former case, the positioning intelligence resides in the mobile terminal or in its SIM/USIM card. Within these technologies we have those based/dependant on GNSS systems (Galileo, GPS, Glonass), those which use the mobile network operators (MNOs) infrastructure (i.e E-OTD, Enhanced Observed Time Difference) and finally, those hybrid solutions which constitute the main focus in this paper: terminal based and network assisted positioning, currently represented by A-GPS.[9],[11]

In the latter case, network bases solutions don't require the integration of the positioning intelligence within the mobile handset. So, this kind of positioning services can be provided to all existing handsets with no distinction, as there are no change sin hardware required. The tradeoff is the relative lack of precision, comparing to the aforementioned terminal based solutions. Representative technologies in this area are CGI+TA (Celle Global Identity + Time Advance) and UL-TDoA (Uplink Time Difference of Arrival) [5][7]

A. A-GPS and Enhanced A-GPS

The Assisted GPS technology appeared recently and represents a key turning point. The technology enables a powerful hybridization between a worldwide location means – GPS– and a mass-market communications means – GSM/UMTS.

Moreover, Assisted GPS comes in handy mixing the best of the two worlds, since it compensates for the major faults of GPS and GSM/UMTS: a purely network- based technology does not provide sufficient accuracy (80 meters at best), and pure GPS solutions suffer from long delays before position delivery (typically several minutes). The principle of Assisted GPS consists of coupling satellite positioning and communication networks, sending assistance data to the receiver integrated in the handset to improve its performance [7][10]

Compared with standard GPS, Assisted GPS offers (1) very short latency to get a position, by sending satellite data much more rapidly, than GPS itself; (2) Hence, very low power consumption; and (3). Increased sensitivity, therefore increased availability of the location service, particularly in dense urban area and indoor environments.

Enhanced A-GPS is an improvement of the classic A-GPS positioning with new satellite technologies: EGNOS (European Geostationary Navigation Overlay Service), and Galileo. The Enhanced A-GPS provides EGNOS-based assistance data to GPS-enabled mobile phones, via GSM/GPRS or UMTS networks. This is obtained by incorporating an EGNOS/GPS reference receiver in the Enhanced A-GPS server, which receives the EGNOS correction messages.[11]

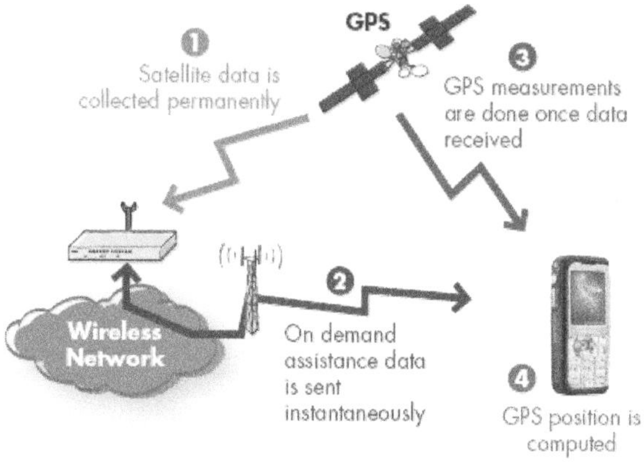

Fig 1. A-GPS Operation Flow

EGNOS is a system consisting of three geo-stationary satellites and 34 ground stations in Europe, the Americas and Asia. EGNOS improves GPS & GLONASS position accuracy and availability by 'adding' measurements from its three geo-stationary satellites locally visible in Europe to the Galileo, GPS & GLONASS, and by providing ionosphere, orbitography and clock corrections.

IV. Location Aware Mobile Services (LAMS) Platform

The LAMS Platform has been developed as a proof-of-concept of the proposed service architecture, acting as an universal LCS (LoCation Services) client.(See Fig.3)

Once the platform gets the positioning info obtained using local AGPS data through W3C Location API, sends it to the server , which in turn, trans codifies this information to XML, using an specific XML application (LAMSX). With this information, the platform sends a spatial query to a GIS server in order to obtain reverse geocoding data which carries the desired Points of Interest (POIs) also in native XML format. [4],[5],[6]

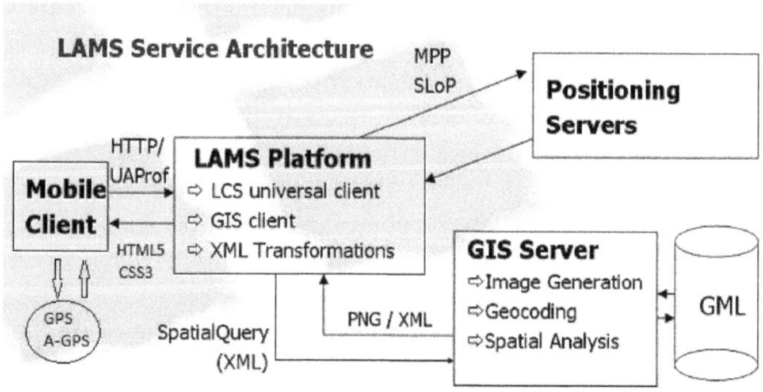

Fig 2. LAMS Platform Service Architecture

At this point, the platforms enters a dynamic content generation phase, personalizing the overall response to the make, model and capabilities of the requesting handset, so that the employee gets the best experience in the clocking in process, no matter which phone he or her has been using to complete the task [3]

Finally, the outcome of this content generation phase is the end HTML5 / JQuery Mobile or xHTML/MP page targeted to the handset.

V. Conclusions

In this paper has been presented the practical use of Location Services, applied to non intrusive field force clocking in, allowing the employee for the control of their position data gathering times.

This knowledge can be used along with the also described UAProf framework to achieve full personalization of the contents delivered to the end client. In this case, we are able to dynamically generate location-dependent, device-dependent and user preferences-aware contents.

Also, this paper demonstrates the feasibility of a time-constraint service provision when using terminal based technologies with the most up to date enablers, such as W3C Location API, OMA SUPL and A-GPS at application level.

References

[1] De Assis Lahoz, M., & Camarotto, J. A. (2012). Performance indicators of work activity. Work (Reading, Mass.), 41 Suppl 1, 524–31. doi:10.3233/WOR-2012-0207-524

[2] Ball, K. (2010). Workplace surveillance: an overview. Labor History, 51(1), 87–106. doi:10.1080/00236561003654776

[3] Franlin Reynolds, Johan Hjelm et al. Composite Capability/Preference Profiles (CC/PP): A user side for content negotiation. World Wide Web Consortium http://www.w3.org/TR/NOTE-CCPP

[4] Ríos Aguilar, S. Generación dinámica de contenidos WAP para terminales móviles, Libro de Ponencias del Congreso "Mundo Internet 2000", February 2000

[5] Ríos Aguilar, S. Position-aware WAP Contents Personalization. Wrox Wireless Developer Conference, Amsterdam, July 2000

[6] Ríos Aguilar, S. Interfacing to a Mobile Positioning Center. Wireless Developer Network (www.wirelessdevnet.com), September 2004

[7] Schiller Jochen, Voisard Agnes. Location Based Services. Elsevier-Morgan Kaufman 2004

[8] Spinney Jonathan, Wireless Location Uses in the User Plane and Control Plane. http://www.lbszone.com/content/view/148/45 Visited in May 2007

[9] Stojmenovic I. Handbook of Wireless Networks and Mobile Computing. John Wiley & Sons, New York, 2004

[10] Katz, J.E. , Aakhus M. (eds). Perpetual Contact. Mobile Communications, Private Talk, Public Performance. Cambridge University Press, Cambridge, UK 2005

[11] AGILE project. http://www.galileo-in-lbs.com/index.php?id =398. May 2007

[12] OMA-TS-MLP, "Mobile Location Protocol", Open Mobile Alliance, http://www.openmobilealliance.org/

[13] OMA-RD-SUPL-V2_0, "Secure User Plane Requirements v2.0",Open Mobile Alliance, http://www.openmobilealliance. org/

[14] JSR-179 Location API for JavaME. Sun Microsystems Java Community Process http://jcp.org/aboutJava/communityprocess/final/jsr179/index.html

[15] Wu Shiow-Yang, Wu Kun-Ta, Effective Location Based Services with Dynamic Data Management in Mobile Environments, Wireless Networks 12, 369-381. Springer Science, 2006

[16] Küpper Axel, Treu Georg, Efficient Proximity and Separation Detection among Mobile Targets for Supporting Location-based Community Services. Mobile Computing and Communications Review, Vol 10, No. 3. 2008

[17] Xu Toby, Cai Ying. Location Anonymity in Continuous Location-based Services. Proceedings of the 15th International Symposium on Advances in Geographic Information Systems. ACM GIS 2007

[18] Ball, K. (2005). Organization, Surveillance and the Body: Towards a Politics of Resistance. Organization, 12(1), 89–108. doi:10.1177/1350508405048578

[19] Axtell, C. (2011). The Well-being of the Mobile Workforce. iPass 2011 Report.

[20] M. López-Fernández et al.. "Control of attendance applied in higher education through mobile NFC technologies". Expert Systems With Applications Vol 40, Issue 11, September 2013

19

Efficient Measurement of the User Experience of Interactive Products. How to use the User Experience Questionnaire (UEQ).Example: Spanish Language Version

Maria Rauschenberger, *MSP Medien-Systempartner, Germany*; Martin Schrepp, *SAP AG, Germany*
Manuel Pérez Cota, *University of Vigo, Spain*; Siegfried Olschner, *DATEV eG, Germany*
Jörg Thomaschewski, *University of Applied Science Emden/Leer, Germany*

***Abstract* — Developer, manager and user feedback is needed to optimize products. Besides the basic Software qualities – usability and user experience are important properties for improving your product.**

Usability is well known and can be tested with e.g. a usability test or an expert review. In contrast user experience describes the whole impact a product has on the end-user. The timeline goes from before, while and after the use of a product. We present a tool that allows you to evaluate the user experience of a product with little effort. Furthermore the tool is available in different languages and we are using the new Spanish Version. We show how this tool can be used for a continuous user experience assessment.

***Keywords* — Software Quality, User Experience, Questionnaire, Usability, Test, Development**

I. INTRODUCTION

IS your redesign of the website better than the old version? Has the development effort spent to increase user experience really paid off? If you want to answer such questions you need a quantitative method to measure user experience [1]. An efficient and inexpensive method to do such measurements is the usage of rigorously constructed and validated questionnaires.

The concept of user experience combines well-known aspects like efficiency and effectiveness with additional criteria like aesthetics, joy-of-use or attractiveness. The first group of criteria is often referred as pragmatic quality aspects [2], while the second group is called hedonic quality aspects. Another often-used terminology to distinguish both classes of quality criteria is usability goals versus user experience goals [3]. The dependency of pragmatic and hedonic quality is presented in Fig. 1.

One well investigated research question is the relationship of pragmatic and hedonic quality. Empirical evidence proves that products, which are perceived to show a high level of hedonic quality, are also perceived as easy to use [4], [5], [6]. These and similar observations cause some authors [7] to state that 'What is beautiful is usable'. In contrast other studies point out [8], [9] an opposite dependency. The perception of the aesthetic value of a user interface increased when the number of concrete usability problems decreased. Thus, in this study a 'What is usable is beautiful' effect was observed.

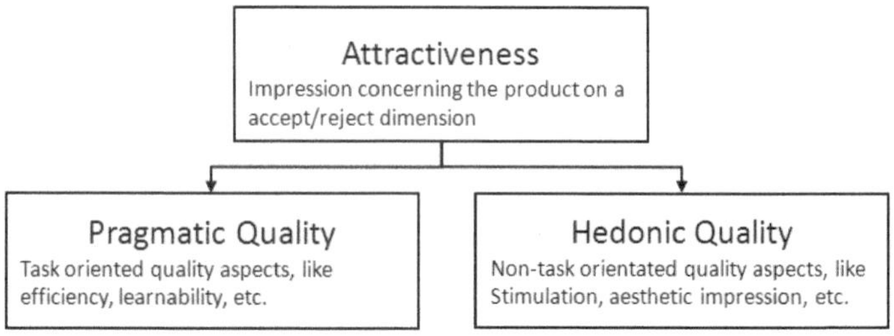

Fig. 1. Grouping of different quality attributes.

Why are perceived hedonic and pragmatic quality aspects associated? As possible explanation for this connection halo-effects [10], mediation by the mood of the user [11] or mediation by other variables [6] have been suggested. Since it is quite difficult to separate these effects experimentally [8] it is currently unclear which of these hypotheses are able to explain this effect.

These results indicate that it is necessary to consider both pragmatic and hedonic aspects if we want to measure how satisfied users are with a given product.

This is the underlined idea of constructing the User Experience Questionnaire (UEQ) [12], [13] that is described in this paper. In the context of the questionnaire user experience is understood as the overall impression of a user when he or she interacts with a product, i.e. covers both pragmatic and hedonic quality aspects.

The UEQ allows a quick assessment of the user experience for any interactive product. The scales of the questionnaire are designed to cover a comprehensive impression of user experience. The questionnaire format supports the user response to immediately express feelings, impressions, and attitudes that arise when they use a product.

If a new product is rolled out or if an existing product is evaluated the first time typical questions are '*Does the product create a positive user experience?*' or '*How do users feel about the product?*'. To answer such questions it is sufficient that a representative sample of users of the new product fill out the UEQ. 30 answers are usually enough to get a valid impression. For example, the answers can come from participants of a usability test or pilot users.

Another application is the continuous quality assessment of a software product within a development process [14]. In this approach a measurement with the UEQ is collected with each new version of the software. Thus, we can directly see if new versions bring an improvement in user experience if the scale values for the six scales of the UEQ increase with the new version (for an example on the concrete implementation of such a process, see [14]). An application of the UEQ in the process of idea and innovation management is described in [15].

User experience is not only a snapshot of the present usage a product has. It is an entire impression a product makes on the user. Even more, the user's judgement starts before touching and using a new product. In addition the change of impression carries on during and after the usage [1]. The UEQ is able to present the distinct results over time for the result analysis.

The UEQ is a semantic differential. For such questionnaires it is especially important that users see the items in their native language. So far the UEQ was available in German, English, French and Italian. We present in this paper the Spanish language version of the questionnaire.

We describe in the following how the UEQ was constructed and validated. In addition, the structure of the questionnaire and the meaning of the subscales are explained. We then show, how the UEQ should be applied in a company and how the results can be analyzed. Besides, the DATEV eG a big business software company is presenting their design process with the UEQ. Finally, we describe the creation of the Spanish language version of the UEQ.

II. Construction and Validation of the User Experience Questionnaire (UEQ)

The items and scales of the UEQ were created by a data analytical approach. First, a set of 229 potential items was built as a result of several brainstorming sessions with usability experts. Second, this set was reduced to an 80 items raw version by an expert evaluation. Third, the eighty items raw-version of the questionnaire was used in several studies focusing on the quality of interactive products, including e. g. a statistics software package, cell phone address book, online-collaboration software, or business software. In total the data of 153 participants were collected for the initial data set. Finally, the scales and the items representing each scale were extracted from the data by factor analysis (principal components, varimax rotation). Six factors resulted from this analysis. Details concerning the process can be found in [12], [13].

The reliability (i.e. the scales are consistent) and validity (i.e. the scales do really measure what they intend to measure) of the UEQ scales was investigated in several studies (in 11 usability tests with a total number of 144 participants and an online survey with 722 participants). A review of all available studies showed that reliability (Cronbach's Alpha was used for an estimation of internal consistency) of the scales was sufficiently high. In addition, the validity of the scales was investigated in a number of studies [12], [13], [14]. Results indicate good construct validity.

III. Structure of the Questionnaire

The user experience questionnaire contains 6 scales with 26 items in total:

1) **Attractiveness**: General impression towards the product. Do users like or dislike the product? This scale is a pure valence dimension. *Items: annoying / enjoyable, good / bad, unlikable / pleasing, unpleasant / pleasant, attractive / unattractive, friendly / unfriendly*
2) **Efficiency**: Is it possible to use the product fast and efficient? Does the user interface looks organized? *Items: fast / slow, inefficient / efficient, impractical / practical, organized / cluttered*
3) **Perspicuity**: Is it easy to understand how to use the product? Is it easy to get familiar with the product? *Items: not understandable / understandable, easy to learn / difficult to learn, complicated / easy, clear / confusing*
4) **Dependability**: Does the user feel in control of the interaction? Is the interaction with the product secure and predicable? *Items: unpredictable / predictable, obstructive / supportive, secure / not secure, meets expectations / does not meet expectations*
5) **Stimulation**: Is it interesting and exciting to use the product? Does the user feel motivated to further use the product? *Items: valuable / inferior, boring / exiting, not interesting / interesting, motivating / demotivating*
6) **Novelty**: Is the design of the product innovative and creative? Does the product grab users attention? *Items: creative / dull, inventive / conventional, usual / leading edge, conservative / innovative*

The dependency of the UEQ scale is presented in Fig. 2.

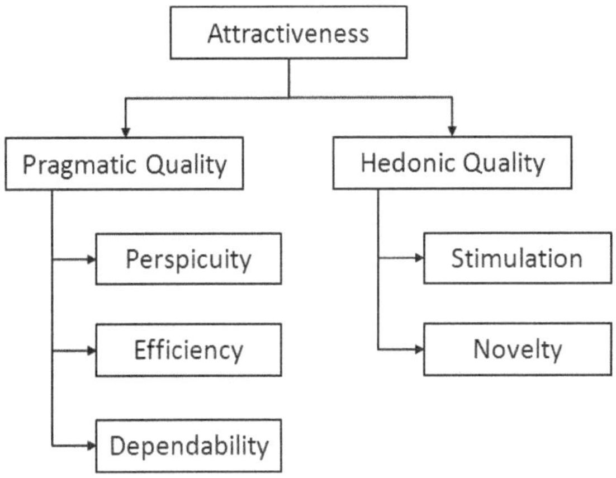

Fig. 2. Scale structure of the UEQ questionnaire.

For the specific questionnaire the order of the items and their orientation (starting with the positive or the antonym statement) is randomized. The specific English questionnaire is shown in Fig. 3 and the Spanish questionnaire is shown in Fig. 7.

	1	...	7		
annoying	O	...	O	enjoyable	1
not understandable	O	...	O	understandable	2
creative	O	...	O	dull	3
easy to learn	O	...	O	difficult to learn	4
valuable	O	...	O	inferior	5
boring	O	...	O	exciting	6
not interesting	O	...	O	interesting	7
unpredictable	O	...	O	predictable	8
fast	O	...	O	slow	9
inventive	O	...	O	conventional	10
obstructive	O	...	O	supportive	11
good	O	...	O	bad	12
complicated	O	...	O	easy	13
unlikable	O	...	O	pleasing	14
usual	O	...	O	leading edge	15
unpleasant	O	...	O	pleasant	16
secure	O	...	O	not secure	17
motivating	O	...	O	demotivating	18
meets expectations	O	...	O	does not meet expectations	19
inefficient	O	...	O	efficient	20
clear	O	...	O	confusing	21
impractical	O	...	O	practical	22
organized	O	...	O	cluttered	23
attractive	O	...	O	unattractive	24
friendly	O	...	O	unfriendly	25
conservative	O	...	O	innovative	26

Fig. 3. English version of the UEQ

IV. How to Apply the Questionnaire

For a successful application of the UEQ the acceptance of following two groups are needed: Users and Managers. To achieve a high user acceptance of the UEQ you should take following points into account:

- Background and benefits of the method should be clear to the user
- A personal contact should be available for the responders
- The time interval between repeated measurements should be long enough

To achieve acceptance by product managers consider to:

- Provide help during the interpretation of the UEQ factor values. Define your range of good, medium and bad and explain the theoretical background
- Combine old and new UEQ values in one picture and show the changes during the development in order to increase the intelligibility of the measurements
- Search for other user feedback that supports the interpretation of the UEQ outcome and integrate it into your report
- Enhance the UEQ results with concrete enhancement suggestions based on user experience expertise and use this as a base for further discussions about the next development goals

V. Analyzing Result

After collecting the answers from the users a three step analysis as presented can follow. To reduce the effort for data analysis an MS Excel file is created, doing all the necessary calculations. Only the raw data of the questionnaire results have to be entered into the tool. The tool then calculates the scale values, creates a bar chart to visualize the results and calculates some basic statistical indicators necessary for an interpretation of the data, for example confidence intervals for the scales. Fig. 4 presents an example of a result and Fig. 45 shows an example of a comparison of two product versions.

A. Verifying the validation

The first step is to confirm the Cronbach's Alpha data, which describes the consistency of the items of the scales (i.e. if all items in the scale measure the same quality). It is calculated automatically for each study in the excel sheet which can be downloaded from www.ueq-online.org.

If the Alpha value for a scale is small this is an indication that some of the items in this scale are possibly misinterpreted or interpreted in a direction that does not reflect their intention in the context of the UEQ. In this case it is questionable if this specific scale can be interpreted for the final result.

There are two well-known effects that can cause a small value of the Alpha-Coefficient for a scale. First, it is possible that the context in which the questionnaire is applied yields to a misinterpretation of some items in the scale. For example, in a study with informatics students the item 'secure/not secure' was referred from the users to the security (i.e. absence of malware or spyware) of the web-service and not to the dependability of the interaction.

Second, a scale may be simply irrelevant in the context in which the questionnaire is applied. Thus, the participants may have problems to interpret the items of the scale properly, which lowers the correlations between the items of the scale and thus decreases the Alpha-Coefficient.

If the alpha coefficient is higher or equal than 0,7 the scales show high consistency, i.e. all items in a scale measure the same aspect and it is unlikely that one of the items is misinterpreted in the given context.

But it can also happen that all items in a scale are influenced by a context specific effect, i.e. one of the scales differs highly from the other scales due to a special target group.

In a study with 20 participants the scale novelty had low results caused by a target group with different age. The VoIP-Software Skype was evaluated. The younger group had no enthusiasm about the technology, because they had known it for a long time. It was not exciting anymore. Elsewise the older group did not know Skype or any similar product. It was their first contact with this technology and they found it very fascinating. The consequence was that one group perceived Skype very stimulating and the other not.

After exanimating the Alpha value next step is the interpreting of the overall result as descripted in Chapter B.

B. Intepretate the overall result

The items are scaled from -3 to +3. Thus, -3 represents the most negative answer, 0 a neutral answer, and +3 the most positive answer. When analyzed the following aspect should be considered. Scale values above +1 indicate a positive impression of the users concerning this scale, values below -1 a negative impression. Due to well-known answer effects, like the avoidance of extremes, observed scales means are in general in the range of -2 to +2. More extreme values are rarely observed, so a value near +2 represents a very positive near optimal impression of participants.

Fig. 4 shows an example for an overall result for a product. The graphic is automatically generated by the data analysis sheet (Excel) that can be downloaded together with the questionnaire.

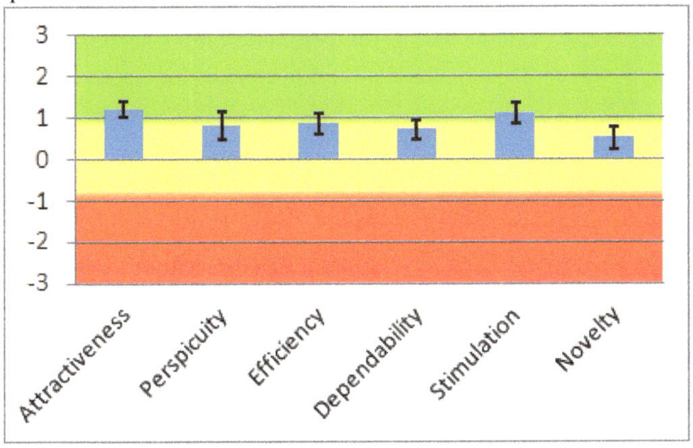

Fig. 4. Example of an overview result.

Thus, this particular product created a slightly positive impression concerning Attractiveness and Stimulation, but is judged neutral concerning the other 4 scales. The error bars represent the 5% confidence intervals for the scale means, i.e. the probability that the true value of the scale mean lies outside this interval is less than 5%. The width of the error bars depend on the number of respondents and on the level of agreement between the respondents. Thus, the more the participants that filled out the questionnaire agree concerning their evaluation of the product the smaller are typically the width of the error bars. Thus, if there are many respondents to the questionnaire and the error bars are still wide, this can be an indication that there are different sub-groups of participants with quite opposite options about the product.

Two different products or product versions can thus easily be compared concerning their user experience by comparing the scale means. See Fig. 5 for a comparison of two product versions concerning the observed scale means.

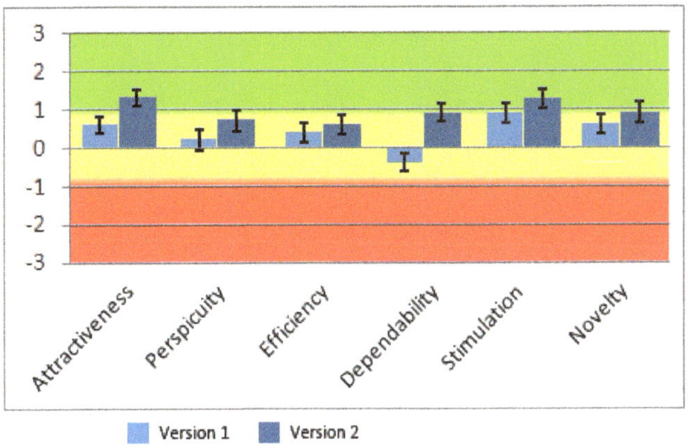

Fig. 5. Example of a comparison of two product versions concerning the UEQ scales.

In this example version 2 is much better concerning Attractiveness, Perspicuity Efficiency and Dependability. Concerning the hedonic scales Stimulation and Novelty both versions seems to be comparable.

To find out if the difference concerning the scale values is significant on the 5% level (or any other level you choose) it is necessary to apply a statistical test that compares the scale means (for ex. a t-test). It is not sufficient to check if the error bars do not overlap. If they do not overlap it can be concluded that the difference is significant at 5% level. But the opposite is not true. The error bars can overlap and the difference may still be significant!

The scales can be grouped into three categories. Atractiveness is a pure valence dimension. The scales efficiency, perspicuity and dependability describe the pragmatic quality of the product. The scales stimulation and novelty describe the hedonic quality of the product.

C. Analyzing the results of the individual items

After the overview the details have to be examined. First if you have two software versions with the UEQ results the items results are placed opposite each other. Items with extreme differences give a hint which areas have been improved or not. These way product versions can be compared easily and exact with one another. Also the detail analyzing shows, which areas should be improved for the next release (See Fig. 6). If it is the first product release see if some items show extreme results compared to other in the same UEQ results.

While analyzing each item the target group could give hints about what caused the significant distinction. Therefore the basic demographic data has to be collected with the UEQ results as well.

The UEQ exists in different languages which are tested reliably. Nevertheless, because of the complexity of language, it is also possible that translation deviance the results.

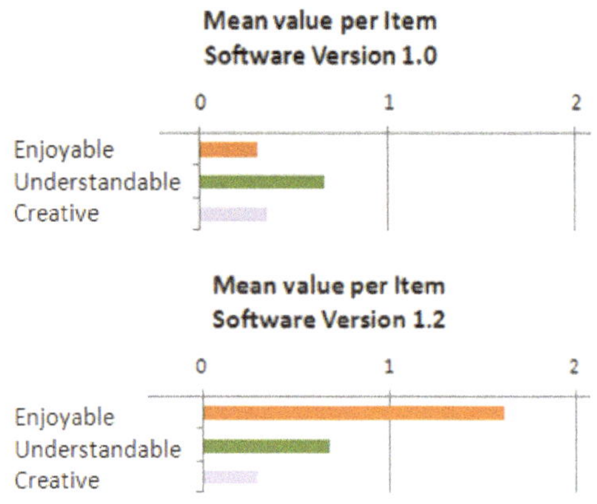

Fig. 6. Example for the detail analyzing of the results from the UEQ-Excel-Sheet (a specimen of the first three items)

VI. Application in the Company DATEV eG

This part presents an example how the UEQ is applied for benchmarking in a big business software company. A general impression of a process is presented in [16].

A. About DATEV eG

The cooperative DATEV eG, Nuremberg (Germany), is a software company and IT service provider for tax consultants, auditors and lawyers as well as their clients. Roughly 5800 employees produce more than 220 applications and provide service for about 39800 cooperative members.

B. Usage of UEQ within a defined Design Process

The concept of user centered design is meanwhile part of the official DATEV eG software development model and the UEQ is an integral component among other UCD methods like classical usability testing, focus groups, persona development and heuristic evaluation. The questionnaire is used to get user feedback at different development stages and all UEQ data are collected in one database.

C. Scenarios of use

One major goal is to perform a regular standardized survey with our users in consultant companies and enterprises. The challenge here is the integration into software release plans and market research activities. The UEQ is currently used successfully in three scenarios:

- Evaluation of new beta versions by selected beta testers
- Assessment of released software by randomly selected users
- At the end of a classic usability test to evaluate a new prototype

In the last scenario it is not the primary goal to get an accurate assessment, but the outcome will give an orientation whether the new software design will bring a significant improvement compared to the DATEV eG benchmark and previous measurements for the tested application. Of course one must be cautious, the tasks in a laboratory test do not represent the entire application and the demonstrated improvements in some parts will perhaps have no effect on the overall user experience of the complete application.

A current project is the test of the combination of online questionnaire and focus group. The outcome of the online-UEQ should be the base for questions in asynchronous online focus groups. Another example how to use the UEQ is described in an article concerning user experience for business software [16].

Because of the special form of the UEQ it is important that participants fill out the questionnaire in their natural language. Thus, it is for companies that use the UEQ on multi-national level important to have language versions of the questionnaire available.

VII. Creation of a Spanish Language Version

First, the German version of the UEQ was translated into Spanish by a native speaker and a bilingual person. After that the Spanish version had been retranslated into German. If the words turned out to match the original words the translation was declared to be successful. Otherwise the process was repeated until all words matched. To demand a one-to-one translation from one language into another is not entirely possible. The reason for that are the different meanings of one word, which make it difficult to find synonym in any language.

The translator was open minded and didn't know the questionnaire before. For more information see [17].

	1 ... 7		
desagradable	O ... O	agradable	1
no entendible	O ... O	entendible	2
creativo	O ... O	sin imaginación	3
fácil de aprender	O ... O	difícil de aprender	4
valioso	O ... O	de poco valor	5
aburrido	O ... O	emocionante	6
no interesante	O ... O	interesante	7
impredecible	O ... O	predecible	8
rápido	O ... O	lento	9
original	O ... O	convencional	10
obstructivo	O ... O	impulsor	11
bueno	O ... O	malo	12
complicado	O ... O	fácil	13
repelente	O ... O	atractivo	14
convencional	O ... O	novedoso	15
incómodo	O ... O	cómodo	16
seguro	O ... O	inseguro	17
activante	O ... O	adormecedor	18
cubre expectativas	O ... O	no cubre expectativas	19
ineficiente	O ... O	eficiente	20
claro	O ... O	confuso	21
no pragmático	O ... O	pragmático	22
ordenado	O ... O	desordenado	23
atractivo	O ... O	feo	24
simpático	O ... O	antipático	25
conservador	O ... O	innovador	26

Fig. 7. Spanish version of the UEQ

VIII. Validation Studies for the Spanish Version

The Spanish Language version of the UEQ is already validated in two bigger studies.

In the first study 94 students evaluated the user experience of the Amazon web-shop (www.amazon.es). The scale means and confidence intervals are shown in Fig. 8.

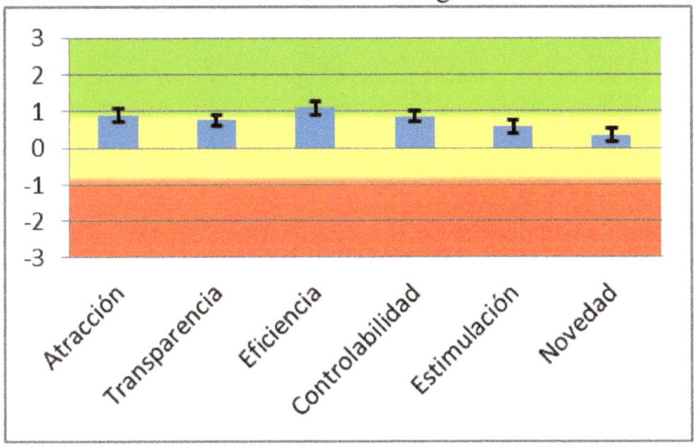

Fig. 8. Result for Amazon web-shop.

Thus, overall the participants had a slightly positive or neutral impression concerning the user experience of the Amazon web-shop. The impression concerning the pragmatic quality (Perspicuity, Efficiency and Dependability) is clearly higher than the impression concerning the hedonic quality (Stimulation, Novelty).

An analysis of the Cronbach Alpha coefficient showed that the single scales showed high consistency values (Attractiveness: 0.85, Perspicuity: 0.59, Efficiency: 0.74, Dependability: 0.48, Stimulation: 0.75, Novelty: 0.64). This is an indicator that the scales are sufficiently consistent.

In a second study 95 students evaluated the user experience of Skype. Again scale means and confidence intervals are shown in Fig. 9.

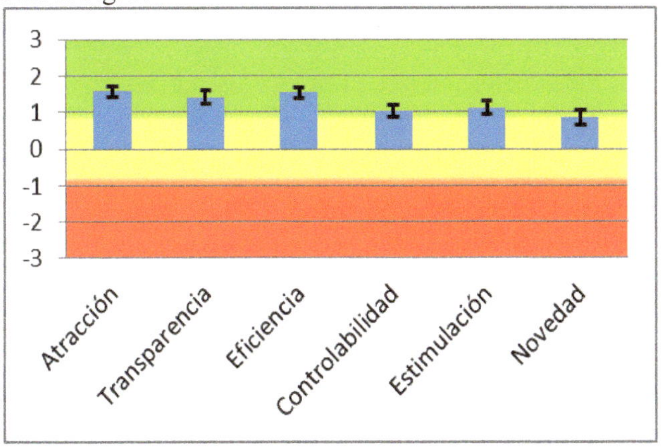

Fig. 9. Result for Skype.

The impression concerning the Skype user experience is quite positive. Again pragmatic quality is judged better than hedonic quality aspects. If we compare these evaluations to the results for the Amazon web-shop we clearly see that Skype creates a better user experience.

As in the first study alpha coefficient for the scales shows high values (Attractiveness: 0.83, Perspicuity: 0.71, Efficiency: 0.72, Dependability: 0.55, Stimulation: 0.78, Novelty: 0.71) again indicating sufficient scale consistency.

Of course further studies are necessary to finally judge if the psychometric properties of the Spanish version are identical to the existing and well-evaluated German and English version. But these first results are positive.

IX. Availability

The UEQ questionnaire can be used free of charge. The questionnaire itself, a data analysis tool and literature describing the construction of the questionnaire can be downloaded from www.ueq-online.org. The questionnaire and the analysis tool are available in several languages. Currently German, English, French, Italian and the Spanish version are available. It is worked on a Portuguese Version as well.

X. Summary

We described the construction, the result analyzing and the validation studies of the Spanish language version of the User Experience Questionnaire. This questionnaire allows a fast evaluation of the user experience of interactive products. It measures not only usability aspects like efficiency, perspicuity and dependability, but also user experience aspects like stimulation or originality.

Since the UEQ has the form of a semantic differential, it is quite important that participants can rate a product in their natural language. Thus, the new language version allows the application of the UEQ in Spanish speaking target groups.

The first available validation studies suggest that the scale quality of the Spanish version is sufficient to apply the questionnaire in projects to collect feedback about user impressions.

References

[1] DIN EN 9241-210, 2011-01, Ergonomics of human-system interaction – Part 210: Human-centred design for interactive systems, Beuth, Berlin, 2011.

[2] Hassenzahl, M., "The effect of perceived hedonic quality on product appealingness", Internatinal Journal of Human-Computer Interaction, 13, pp. 481-499, 2001.

[3] Preece, J., Rogers, Y., Sharp, H., "Interaction design: Beyond human-computer interaction". Wiley, New York , 2002.

[4] Kurosu, M., Kashimura, K., "Apparent usability vs. inherent usability: experimental analysis of the determinants of the apparent usability". In: Conference Companion of human factors in computing systems, pp. 292--293. Denver, Colorado, 1995.

[5] Tractinsky, N., "Aesthetics and Apparent Usability: Empirical Assessing Cultural and Methodological Issues". CHI'97 Electronic Publications http://www.acm.org/sigchi/chi97/proceedings/paper/nt.htm 1997.

[6] Hassenzahl, M., Monk, A., "The influence of perceived usability from beauty". Human-Computer Interaction, 25(3), pp. 235-260, 2010.

[7] Tractinsky, N., Katz, A.S., Ikar, D., "What is beautiful is usable", Interacting with Computers, 13, pp.127--145, 2000.

[8] Ilmberger, W.; Schrepp, M. , Held, T., "What kind of cognitive process causes the relationship between aesthetics and usability". In: Holzinger, A. (Ed.): USAB 2008, LNCS 5298, pp. 43-54, 2008.

[9] Tuch, A.N., Roth, S.P., Hornbaek, K., Opwis, K. & Bargas-Avila, J., "Is beautiful really usable? Toward understanding the relation between usability, aesthetics, and affect in HCI". Computers in Human Behavior 28 (5), pp. 596–1607, 2012.

[10] Hartmann, J., Sutcliffe, A., DeAngeli, A., "Towards a theory of user judgement of aesthetics and user interface quality". ACM Transactions on Computer-Human Interaction, 15, 2008.

[11] Norman, D., "Emotional design. Why we love (or hate) everyday things". New York: Basic books, 2004.

[12] Laugwitz, B.; Held, T., Schrepp, M. "Construction and evaluation of a user experience questionnaire". In: Holzinger, A. (Ed.): USAB 2008, LNCS 5298, pp. 63-76, 2008.

[13] Laugwitz, B.; Schrepp, M., Held, T. "Konstruktion eines Fragebogens zur Messung der User Experience von Softwareprodukten [Construction of a questionnaire to measure the user experience of software products] ". In: A.M. Heinecke, H. Paul (Eds.): Mensch & Computer 2006 – Mensch und Computer im Strukturwandel. Oldenbourg Verlag, pp. 125 – 134, 2006.

[14] Laugwitz, B.; Schubert, U.; Ilmberger, W.; Tamm, N.; Held, T., Schrepp, M., "Subjektive Benutzerzufriedenheit quantitativ erfassen: Erfahrungen mit dem User Experience Questionnaire UEQ [Quantitative measurement of subjective user satisfaction: Experiences with the user experience questionnaire UEQ]". In: Brau, H.; Diefenbach, S.; Hassenzahl, M.; Kohler, K.; Koller, F.; Peissner, M.; Petrovic, K.; Thielsch, M.;Ulrich, D., Zimmermann, D. (Eds.); Usability Professionals 2009, pp. 220 – 225, 2009.

[15] Winter, D., Pietschmann, J., "UX in den frühen Phasen des Innovationsprozesses. User Experience von Anfang an bedenken[UX in the early phases of the innovation process. Consider User Experience from scratch.]". In: Brau, H.; Lehmann, A.; Petrovic, K.; Schroeder, M. (Eds.); Usability Professionals 2012 , pp. 274-279, 2012.

[16] Rauschenberger, M., Hinderks, A.; Thomaschewski, J., "Benutzererlebnis bei Unternehmenssoftware: Ein Praxisbericht über die Umsetzung attraktiver Unternehmenssoftware [User experencing of Busnisssoftware: A practical report how attractive Busnisssoftware could be developt.]". In: Brau, H; Lehmann, A.; Petrovic, K.; Schroeder, M.C. (Eds.); Usability Professionals 2011, pp. 158 - 163, 2011.

[17] Rauschenberger, M., Schrepp, M.; Olschner, S.; Thomaschewski, J.; M. P. Cota, "Measurement of user experience: A Spanish Language Version of the User Experience Questionaire (UEQ) ". In: Information Systems and Technologies (CISTI), Vol. 1, pp. 471-476, 2012

Recognizing Human Activities User-independently on Smartphones Based on Accelerometer Data

Pekka Siirtola and Juha Röning, *Department of Computer Science and Engineering*

P.O. BOX 4500, FI-90014 University of Oulu, Finland

Abstract — **Real-time human activity recognition on a mobile phone is presented in this article. Unlike in most other studies, not only the data were collected using the accelerometers of a smartphone, but also models were implemented to the phone and the whole classification process (preprocessing, feature extraction and classification) was done on the device. The system is trained using phone orientation independent features to recognize five everyday activities: walking, running, cycling, driving a car and sitting/standing while the phone is in the pocket of the subject's trousers. Two classifiers were compared, *k*nn (*k* nearest neighbors) and QDA (quadratic discriminant analysis). The models for real-time activity recognition were trained offline using a data set collected from eight subjects and these offline results were compared to real-time recognition rates, which are obtained by implementing models to mobile activity recognition application which currently supports two operating systems: Symbian^3 and Android. The results show that the presented method is light and, therefore, suitable for be used in real-time recognition. In addition, the recognition rates on the smartphones were encouraging, in fact, the recognition accuracies obtained are approximately as high as offline recognition rates. Also, the results show that the method presented is not an operating system dependent.**

Key words—**Activity recognition, classification, mobile phones**

I. INTRODUCTION AND RELATED WORK

HUMAN activity recognition using wearable sensors, such as accelerometers, has been widely studied during the recent 20 years. Despite several years of study and promising recognition results, not many commercial products, besides pedometers, exploiting these results are available. There are some exceptions, however, such as Polar Active [17] and ActiGraph [1], which can be used to detect the intensity of the activity. Nevertheless, in overall, it seems that companies and people have not been willing to spend money on technology enabling activity recognition. Now, things are about to change: Smartphones are sold more and more every year (smartphone shipments: 2009: 169 million, 2010: 295 million [23]). Smartphones include a wide range of sensors, such as accelerometers, magnetometers, gyroscopes, and GPS, all of which are sensors used for activity recognition in the past studies. Therefore, people already have the technology enabling activity recognition and mobile application stores (AppStore, Nokia Store, Android Market, etc.) can be used to distribute activity recognition applications directly to end-users.

In this article real-time user-independent human activity recognition is presented. The presented method uses only orientation independent features and it is used to recognize five every day activities. Article compares the performance of two classifiers (QDA and *k*nn) in offline and real-time scenarios. Unlike in most of the other studies, in this study the recognition models are implemented to mobile phone to see how models work in real-life, outside laboratory conditions. In addition, models are tested using two different mobile phones: Nokia N8 running Symbian^3 operating system and Samsung Galaxy Mini which is running Android 2.2.1 operating system. It is shown that the method presented in this study enables accurate recognition results not only when the acceleration data is studied offline but also when the whole recognition process (preprocessing, feature extraction and classification) is done in real-time on device. What is more, it is shown that the method is operating system independent.

Human activity recognition using accelerometers has been carried out in various studies, such as [2], [5], [25], [26]. These studies were done using accelerometers build for research use. Therefore, based on these results, it is not straightforward to build a commercial product. There are also some articles where activity recognition using mobile phones has been studied ([3], [14], [16], [19], [24], [27]).

In each of these studies, the data is collected using a mobile phone and the activity recognition is done afterwards on PC,

This work was done as a part of MOPO study (ClinicalTrials.gov Identifier: NCT01376986). Pekka Siirtola would like to thank GETA (The Graduate School in Electronics, Telecommunications and Automation) for financial support. The authors would slso like to thank Infotech Oulu and the Finnish Funding Agency for Technology and Innovation for funding this work.

P. Siirtola is with the University of Oulu, Department of Computer Science and Engineering (e-mail: pekka.siirtola@ee.oulu.fi).

J. Röning is with the University of Oulu, Department of Computer Science and Engineering (e-mail: jjr@ee.oulu.fi).

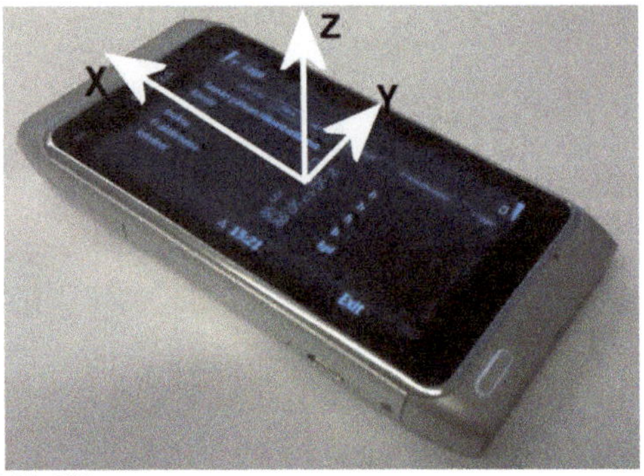

Fig. 1. Nokia N8.

based on collected data. Thus, the activity recognition algorithms are not implemented on the phone, and the classification is not done in real-time on a mobile phone as in our study.

An activity recognition system running purely on a smartphone is presented in [7]. The presented system can be trained on the device and it also does the classification in real-time on the device. The recognition is based on features calculated using geometric template matching and support vector machine (SVM) is used as a classifier. Unfortunately, the article does not include recognition rates: thus, the evaluation of the system is difficult. However, the smartphone application is available from Android Market. The system described in [13] can also be found from Android Market. It seems to recognize activities with high accuracy, but all the features used are not orientation independent. In addition, personalized mobile activity recognition system for Android phones is presented in [8]. In this application user can select which activities he wants application to recognize but it requires training data collection gathered by the user.

Activity recognition using mobile phones has some limitations. Because smartphones are expensive products, people do not want to carry a phone while performing activities where there is a danger to break it. Therefore, it is not necessary to recognize most of the sports activities such as playing football or swimming. Thus, this study concentrates on recognizing five everyday activities, *walking*, *cycling*, *running*, *idling (=sitting/standing)* and *driving/riding a car*. The importance of the everyday activity has been shown in several studies, for instance, in [4] it is shown that there is a relationship between moderate intensity lifestyle activity and cardiometabolic health.

Although the latest smartphones are equipped with processors enabling huge calculation capacity, the activity recognition algorithms must nevertheless be light. The mobile phone can be running several applications simultaneously and the activity recognition algorithms are not allowed to use the whole processing power, nor disturb other applications. Therefore, the recognition must be done using light methods.

The paper is organized as follows: Section II describes sensors and data sets. Section III introduces the techniques and methods used in this study. Models trained using offline data and their accuracy are presented in Section IV. Sections V and VI evaluates the accuracy of the activity recognition based on models trained using offline data, when detection is done in real-time on a mobile phone running Symbian^3 and Android operation systems. Finally, conclusions are discussed in Section VII.

II. DATA SET

The data for training the models were collected using a Nokia N8 smartphone [15] running Symbian^3 operating system, Figure 1. N8 includes a wide range of sensors: tri-axis accelerometer and magnetometer, two cameras (12 MP and 0.3 MP), GPS, proximity sensor, compass, microphones and ambient light sensor.

The models used in this study were trained based on activity data collected from eight healthy subjects. The trousers' front pocket was fixed as the phone placement, but the subject was allowed to determine whether the phone was placed in the left or right pocket. The participants performed five different activities: walking, running, cycling, driving a car, and idling, that is, sitting/standing. The total amount of the data collected was about four hours.

These activities were selected because normal everyday life consists mainly of these five activities. Walking and running are different from the other three because everyone has a personal walking and running style. Other activities are not personal, for instance, while cycling, the movement trajectory is predefined. Therefore, the models to recognize walking and running are most challenging to train.

The real-time classification using Nokia N8 was tested by seven subjects, three of whom were subjects whose data were not used to train the recognition models. These subjects carried the phone in their trousers' front pocket and performed from one to five activities.

In addition, the real-time classification was tested using Samsung Galaxy Mini smartphone running Android 2.2.1 operating system. Galaxy Mini is a low budget smartphone having tri-axis accelerometer, proximity sensor, compass and 3.15MP camera. It uses 600MHz ARMv6 processor. Galaxy Mini was tested by six subjects who carried the phone in their trousers' front pocket and performed from one to five activities.

In this study, only the tri-axis accelerometer was used in this study to detect activities. Accelerometers were running at full speed, which is a phone model dependent feature. However, all the samples were not used in activity recognition process. The latest value from accelerometer was called every 25 milliseconds. Therefore the used sampling frequency was 40Hz, which is much less than the maximum sampling frequency. The highest possible frequency was not used because it varies between devices and it also varies depending what functions of the phone are used. The used method enables the same sampling frequency to any smartphone, making recognition less phone model dependent.

The training data were collected by subjects whose age

varied from 25 to 34 years (average 29 years) and height from 1.65 to 1.90 meter (average 1.78 meter) and real-time classification was tested by subjects whose age varied from 27 to 34 years (average 30 years) and height from 1.65 to 1.90 meter (average 1.75 meter). They performed activities outside the laboratory. Subjects walked inside and outside, mainly on flat surface but also in a staircase. Streets where subjects walked, run, drove a car, and cycled were normal tarmac roads, and the route and speed were determined by subjects themselves. Partly the same roads were employed in offline and real-time tests. The roads used for collecting driving a car data included motorways, as well as roads at the city center. Idling consists mostly of office working but includes also standing.

III. Activity recognition

In this study, two different activity recognition classifiers were compared: quadratic discriminant analysis [10] (QDA) and *k* nearest neighbors [6] (*k*nn). In addition, the recognition was performed using three different settings: (1) offline recognition, to compare different features, classifiers and to evaluate models for online recognition, (2) online recognition on the device, to test the models in realistic real-life conditions, and (3) on device recognition on different phone models to test operating system dependency of the models.

In each case, the raw data were processed in the same way to obtain comparable results.

A. Preprocessing and feature extraction

The purpose of this study was to develop a user-independent activity recognition method that runs purely on a smartphone and gives accurate recognition results also when the system is used in non-laboratory conditions. The recognition was supposed to work when the mobile phone is placed in the trousers' front pocket. However, the mobile phone can lay on the pocket in numerous different orientations. There are two ways to eliminate the effect of the orientation: (1) by recognizing the orientation of the phone, or (2) by eliminating the orientation information. On the other hand, the orientation is impossible to recognize using only accelerometers. Therefore, the effect of orientation had to be eliminated. In the preprocessing stage, the three acceleration channels were combined as one using square summing to obtain the magnitude acceleration, which is orientation independent. Moreover, the orientation of the phone has limitations, the screen or the back of the phone is always against the user's leg when the phone is in the pocket. Therefore, it was tested if features extracted from a signal where two out of three acceleration channels were square summed would improve the classification accuracy.

The online activity recognition was done using a sliding window technique. The signals from the sensors were divided into equal-sized smaller sequences, also called windows. From these windows, features were extracted and finally the classification of the sequences was done based on these features. In this study, the windows were of the length of 300 observations, which is 7.5 seconds, because the sampling

TABLE I
The results of offline recognition using QDA.

Subject/ Activity	Idling	Walking	Cycling	Driving	Running
Idling	94.3%	1.2%	0.3%	4.2%	0.0%
Walking	1.0%	95.6%	2.3%	0.0%	1.3%
Cycling	0.4%	3.4%	94.3%	1.9%	0.0%
Driving	3.7%	0.0%	2.2%	94.2%	0.0%
Running	0.0%	0.0%	0.0%	0.0%	100.0%

TABLE 2
The results of offline recognition using *k*nn.

Subject/ Activity	Idling	Walking	Cycling	Driving	Running
Idling	94.5%	1.1%	0.0%	4.3%	0.0%
Walking	1.0%	90.2%	8.6%	0.0%	0.3%
Cycling	0.3%	1.7%	94.6%	3.4%	0.0%
Driving	4.2%	0.0%	2.1%	93.8%	0.0%
Running	0.0%	0.4%	0.0%	0.0%	99.6%

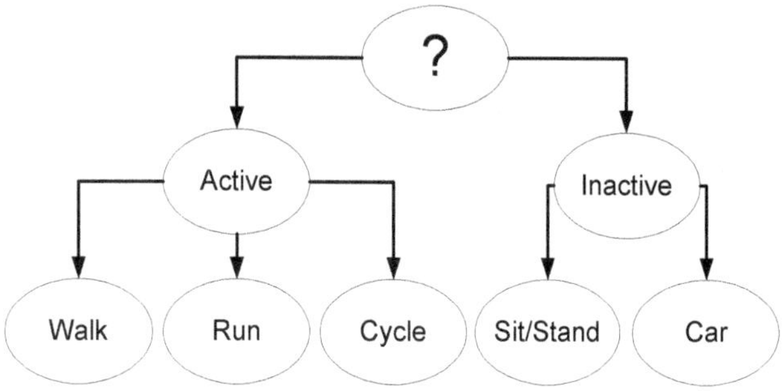

Fig. 2. The decision tree obtained to recognize the type of activity

frequency was 40Hz. In offline recognition, the slide between two sequential windows was 75 observations, while in online recognition, the slide was set to 150 observations. To reduce the number of misclassified windows, the final classification was done based on the majority voting of the classification results of three adjacent windows. Therefore, when activity changes, a new activity can be detected when two adjacent windows are classified as a new activity. For instance, if the slide is 150 observations, a new activity can be detected after 450 observations, which is around eleven seconds if the sampling rate is 40Hz.

The total number of 21 features was extracted from magnitude acceleration sequences. These features were standard deviation, mean, minimum, maximum, five different percentiles (10, 25, 50, 75, and 90), and a sum and square sum of observations above/below certain percentile (5, 10, 25, 75, 90, and 95). The same features were also extracted from the signals where two out three acceleration channels were square summed together. It was noted that the combination of x and z axis signal channels improved the classification most. Therefore, from each window, the total number of 42 orientation independent features were extracted, 21 features from the magnitude acceleration signal and 21 features from the signal where x and z were square summed.

B. Classification

The classification result was obtained using the decision tree presented in Figure 2, which classifies activities using a two stage procedure. In the first classification stage, a model is trained to decide if the studied subject is currently active (walking, running or cycling) or inactive (driving a car or idling). In the second stage, the exact activity label is obtained. One model has to be trained to classify an active activity as walking, running or cycling, and the other to classify an inactive activity as idling or driving.

The models were trained offline using the collected data. These models were implemented to a smartphones (Symbian^3 and Android) and also used in online tests. To compare different classifiers, the classification was performed using two different classification methods, *k*nn and QDA. The most descriptive features for each model were selected using a sequential forward selection (SFS) method [9]. QDA classifiers were trained using the whole training data set, similar to *k*nn classifier for the offline recognition. However, because of the limited computational power of the smartphone, the activity recognition on the device was performed using only a limited number of randomly chosen instances from training data.

IV. Model Training and Offline Recognition

The purpose of the offline recognition is to build and test accurate models that can later be implemented on a mobile phone to enable user-independent and operating system independent real-time recognition of the activities on the device. Models were trained for *k*nn and QDA classifiers based on the data collected from eight persons.

A. Results

To obtain reliable user-independent results, the training was performed using the leave-one-out method, so that each person's data in turn was used for testing and the other seven sequences were employed for model training.

The results are shown in Tables 1 and 2.

B. Classification

The offline recognition results show that the both classifiers, QDA and *k*nn, enable accurate results. The average classification accuracy using QDA is 95.4%, while *k*nn enables

TABLE 3
The Results of Online Recognition on Device Using Nokia N8 and *k*nn

Subject/ Activity	Idling	Walking	Cycling	Driving	Running	Average
Subject 1	91.5%	99.9%	89.2%	91.4%	87.2%	91.8%
Subject 2	99.9%	99.9%	93.7%	87.8%	92.4%	94.7%
Subject 3	76.3%	99.9%	89.6%	-	92.8%	89.7%
Subject 4	-	-	-	97.6%	-	97.6%
Subject 5	95.6%	99.9%	89.5%	89.4%	97.9%	94.5%
Subject 6	94.1%	99.9%	93.8%	-	99.9%	96.9%
Subject 7	83.3%	99.8%	98.1%	-	99.9%	99.4%

TABLE 4
THE RESULTS OF ONLINE RECOGNITION ON DEVICE USING NOKIA N8 AND QDA.

SUBJECT/ ACTIVITY	IDLING	WALKING	CYCLING	DRIVING	RUNNING	AVERAGE
SUBJECT 1	98.5%	65.6%	95.7%	99.6%	91.5%	90.2%
SUBJECT 2	99.9%	97.6%	91.2%	88.5%	99.2%	95.3%
SUBJECT 3	99.9%	97.6%	91.3%	-	99.8%	96.9%
SUBJECT 4	-	-	-	87.2%	-	87.2%
SUBJECT 5	98.2%	99.3%	97.6%	98.1%	99.9%	98.6%
SUBJECT 6	99.9%	96.0%	93.8%	-	99.9%	97.4%
SUBJECT 7	99.9%	99.8%	98.1%	-	99.9%	99.4%

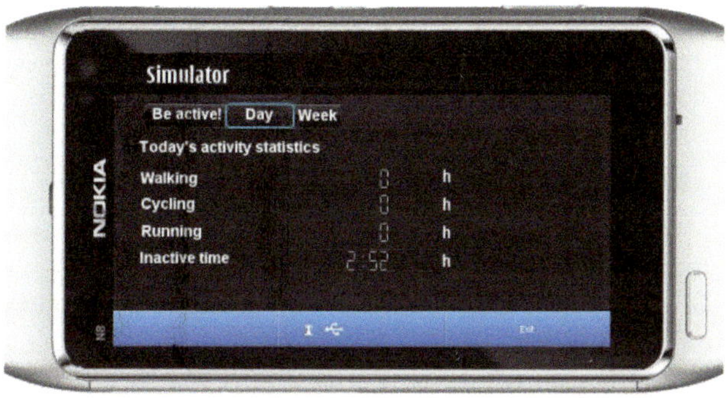

Fig. 3. Activity recognition application for Symbian^3 smartphones.

an accuracy of 94.5%. It should be noted, however that this difference is not statistically significant according to paired *t*-test. Also, each of the five activities are recognized with high accuracy.

V. REAL-TIME EXPERIMENTS ON DEVICE USING NOKIA N8

An activity recognition application for Symbian^3 devices was build using Qt [18] programming language, Figure 3. Every Nokia phone running a Symbian^3 operating system has the same kind of accelerometers, and therefore, the results presented in this section can be obtained using any Nokia Symbian^3 phone. The application uses the activity recognition models that were trained using the data presented in Section II. It should be noted that offline recognition employing *k*nn uses the whole training data set to recognize activities from the test data, making the classification process complex. As mentioned before, because of the limited computational power of the smartphone, the recognition on the device was performed using only a limited number of randomly chosen instances from training data. In this study, eight instances from each activity per subject were chosen as instances of *k*nn-based recognition model. QDA -based real-time classification results were obtained using the very same models used in offline recognition.

A. Results

The application and the real-time classification were tested by seven persons carrying Nokia N8 smartphone on their trousers' front pocket. Three of these were different from the eight subjects that collected the data for training the recognition models. The recognition results are shown in Tables 3 and 4. Both classifiers were running on the device in parallel; thus, the results are comparable.

B. Discussion

The real-time experiment showed that the application and models are running smoothly on the device. When activity recognition is done using QDA classifier, the application uses under 5% of the CPU's (680Mhz ARM11 processor) capacity. Therefore, the application can be employed alongside other applications, such as games. The usage of *k*nn as a classifier uses slightly more CPU capacity. In addition, the recognition rates on the device are around as high as offline. The average recognition rate using QDA is 95.8%, while using *k*nn it is slightly lower, 93.9%. According to paired *t*-test, this difference is not statistically significant, however.

Online recognition was tested by subjects (subjects 1, 2 and 3) whose data was not used for training as well as subjects (subjects 4, 5, 6 and 7) whose data was used for training. In both cases, the average recognition rate is high. However, there are two cases where user-independent classification has not succeeded very well. Walking activity of Subject 1 was recognized only with the rate of 65.6% when QDA is used as a classifier and cycling of Subject 3 using *k*nn was recognized correctly only in 76.3% of the cases. In both cases, cycling and walking were mixed together. It seems that inner class variation of these activities is too low causing misclassification. As mentioned above, walking is one of the most difficult activities to recognize user-independently, because every subject has a personal walking style. In addition, not the whole training data were used to train the *k*nn model to keep to recognition process light, which may have caused the weak recognition rates with Subject 3's cycling activity. In overall, the recognition on the device works well, however. It seems that the first phase of the recognition, where observations are classified as active or inactive, appears to work almost perfectly. Therefore, to make recognition even more accurate, the second phase of the classification should be improved.

The models used in online recognition are user-independent and it also seems that they are "car-independent". Two different cars were used in the data collecting phase. Although during the online test, subjects 1 and 2 used a car not used to train the models, the recognition rate is still high. On the other hand, to make sure that the models are car-independent, more tests should be carried out using different cars and road conditions.

VI. Real-time Experiments on device using Android-phone

According to the results of the previous section, both tested classifiers produce as good results. However, QDA is simpler than *k*nn and, therefore, it is lighter and more suitable for be used in an application that is supposed to run at the background all the time. Android-version of the activity recognition application was built after Symbian^3-version using Java programming language. Therefore, based on the experiences gathered using Nokia phones, it was decided that Android-version would use QDA as a classifier.

A. Results

Real-time classification on Android device was tested by five subjects, again, carrying the phone on their trousers' front pocket. Subject 1's data was not used in to train the models. The data of other five subjects were used to train models but data is different to the one used in this section. The results are shown in Table 5. The purpose of these experiments was to show that the presented activity recognition method is operating system independent.

B. Discussion

The results show that activity recognition models are accurate also when they are running in Android-based smartphone. Therefore, the presented method is not an operating system dependent.

Noticeable is the weak recognition accuracy of the activity driving a car. Only 67.8% of the cases were recognized correctly when Subject 3 was using the application. However, Subject 3 had to stop several times during the data collection session because of the red traffic lights, and in this study, these stops were considered as driving a car. If these stops are considered as idling and therefore removed from driving results, the recognition rate of driving would be almost 100.0%. On the other hand, in the case of Subject 4, driving was recognized with a rate of 98.8%. This subject did not have to wait at traffic lights.

However, driving was the only activity recognized with a low rate. All the other activities were recognized with really high accuracy and the average recognition rate is 96.5%. For instance, running is recognized perfectly with a rate of 100.0%. Moreover, the recognition accuracy of walking is also nearly 100%. The cycling did mix up with walking a little but still the average recognition rate of cycling was as high as 94.5%. In fact, the average recognition rate on a device using Samsung Galaxy Mini is higher than offline recognition rate and recognition rate using Nokia N8. However, these results are not fully comparable because they are based on separate data collections.

VII. Conclusions

Orientation independent real-time activity recognition of five everyday activities using a mobile phone was introduced in this study. The whole classification process, including preprocessing, feature extraction, and classification, was done on the device. Recognition accuracies were tested using two classifiers (*k*nn and QDA) to compare different classification algorithms. In addition, real-time activity recognition using QDA as a classifier was tested using two different phones (Nokia N8 and Samsung Galaxy Mini) running different operating systems (Symbian^3 and Android). These operating systems were chosen to this study, because not only they are popular, but also they enable real multitasking, unlike some other smartphone operating systems, such as iOS and Windows Phone 7. User-independent models for online recognition were trained offline using a data set collected by eight subjects.

When the recognition rates of on device recognition using Nokia N8 are studied in detail it can be seen that using the models trained offline, the recognition rates on Nokia N8 device are around as high as offline recognition results. In the offline case, the average classification accuracy based on the

TABLE 5
THE RESULTS OF ONLINE RECOGNITION ON DEVICE USING SAMSUNG GALAXY MINI AND QDA.

Subject/ Activity	Idling	Walking	Cycling	Driving	Running	Average
Subject 1	92.5%	97.1%	100.0%	--	100.0 %	97,4%
Subject 2	91.7%	100.0%	98.9%	--	100.0%	97.7%
Subject 3	100.0%	100.0%	90.9%	67.8%	100.0%	91.7%
Subject 4	--	--	--	98.8%	--	98.8%
Subject 5	96,7%	97.8%	87.5%	--	100.0%	95.5%
Subject 6	98.5%	100.0%	95.2%	--	--	97.9%

data used to train the models using QDA is 95.4%, while *k*nn enables an accuracy of 94.5%. While performing online recognition on Nokia N8, the average recognition rate using QDA is 95.8%, while using *k*nn it is slightly lower, 93.9%. However, in some cases, user-independent real-time recognition results on Nokia N8 are not as high as expected. In order to achieve even more accurate online results with every subject, the training data should contain more variation. Now it seems that in some cases the models for online recognition are build using too homogeneous a data set, and therefore, the models are not as good as they could be. Nevertheless, the results are encouraging. Moreover, activity recognition application runs smoothly on N8. It uses under 5% of CPU capacity when QDA is employed as a classifier; thus, other applications can be run alongside. Using *k*nn as a classifier requires more CPU capacity.

Because of the accuracy and lightness of the QDA classifier based activity recognition, Android-version of the application was tested only using QDA. Android-version was tested by six subjects carrying Samsung Galaxy Mini smartphone on their trousers' pocket. The results are really good with every subject. For instance, running is recognized perfectly with rate 100.0%. All the other activities were also recognized with really high accuracy. The only exception was driving a car -activity, where detection accuracy was lower than expected with one test subject. However, the main reason for this was red traffic lights which caused long unwanted stops.

Based on the results it is clear that activity recognition works reliably operating system independently. It seems that in this study, the on device recognition results using Android-phone are a little higher than the ones gained using Symbian^3 phone. However, subjects did not carry Symbian^3 and Android phones at the same time, the real-time recognition results of Tables 4 and 5 are not fully comparable.

Real-time activity recognition on device is working reliably on both tested operating systems, Symbian^3 and Android, though there are differences in accelerometers between phone models and operating systems. The main difference of accelerometers is the maximum sampling rate. However, in this study, the maximum sampling rate was not used. Though, the accelerometers were running at full speed, but still, a new value to be used in activity recognition was called every 25 millisecond. Therefore, the used frequency was 40Hz, which is much less than the maximum frequency of most of the smartphones. Thus, the presented method can be used with every smartphone and it is not dependent on the phone model.

Real-time recognition on the device was only tested by predefined five activities and not when the subject is doing something else. Null-data recognition is not included in this study, and therefore, such activities cause incorrect classifications. Thus, to improve the accuracy of the application, null-activity recognition should be included. Also building a behavior recognition system based on the activity recognition results could reduce the number of misclassifications [12]. In addition, it should be tested how different trousers affect the results, now every test subject was wearing jeans.

The presented activity recognition application is not body position independent. The system is trained to recognize activities when the phone is placed to the subject's trousers' pocket. Although trousers' pockets are the most common place to carry a phone [11], especially among males, a body position independent approach should be considered. Body position independent recognition is naturally more difficult than position dependent, and therefore, most likely the recognition rates would not be as high as the ones presented in this study. However, high position independent recognition rates are for instance achieved in [14].

Although, the recognition accuracy on the device is excellent, there are still some remaining issues. The application uses too much battery and, therefore, even lighter methods should be used. For instance, human activities can be recognized from lower frequency signals than the ones used in this study [21]. Therefore, the sampling frequency could be reduced or the number of required classifications could be reduced by using periodic quick-test [22]. However, even now without memory and processing power optimization, the battery of Nokia N8 and Samsung Galaxy Mini lasts over 24 hours while the application is running at the background.

In this study, everything except model training is done on the device. Other option would be to send the accelerometer data to the server, perform the classification process there and send the results to a mobile phone. In this case, calculation capacity would not be an issue, but on the other hand, privacy issues should be handled. Moreover, data transfer is not free and can cause exceptionally high costs, especially when the mobile phone and application are used abroad.

ACKNOWLEDGMENT

The authors would thanks to Tero Vallius for Qt coding and all the partners of MOPO-study (ClinicalTrials.gov Identifier: NCT01376986).

REFERENCES

[1] ActiGraph: http://www.theactigraph.com/

[2] Bao, L., Intille, S.S.: Activity recognition from user-annotated acceleration data. Pervasive 2004 pp. 1--17 (2004).

[3] Brezmes, T., Gorricho, J.L., Cotrina, J.: Activity recognition from accelerometer data on a mobile phone. In: Distributed Computing, Artificial Intelligence, Bioinformatics, Soft Computing, and Ambient Assisted Living, Lecture Notes in Computer Science, vol. 5518, pp. 796--799. Springer Berlin / Heidelberg (2009)

[4] Camhi, S.M., Sisson, S.B., Johnson, W.D, Katzmarzyk, P.T., Tudor-Locke, C., Accelerometer-determined moderate intensity lifestyle activity and cardiometabolic health, Preventive Medicine, Volume 52, Issue 5, 1 May 2011, Pages 358-360

[5] Ermes, M., Pärkkä, J., Mäntyjärvi, J., Korhonen, I.: Detection of daily activities and sports with wearable sensors in controlled and uncontrolled conditions. Information Technology in Biomedicine, IEEE Transactions on 12(1), 20--26 (Jan. 2008)

[6] Fix, E., Hodges, J.L.: Discriminatory analysis: Nonparametric discrimination: Consistency properties. Tech. Rep. Project 21-49-004, Report Number 4, USAF School of Aviation Medicine, Randolf Field, Texas (1951)

[7] Frank, J., Mannor, S., Precup, D.: Activity recognition with mobile phones. In: Machine Learning and Knowledge Discovery in Databases, Lecture Notes in Computer Science}, vol. 6913, pp. 630--633. Springer Berlin / Heidelberg (2011)

[8] Gomes, J., Krishnaswamy, S., Gaber, Mohamed, Sousa, P. and Menasalvas, E. (2012) *MARS: a personalised Mobile Activity Recognition System.* In: Proceedings of the IEEE International Conference on Mobile Data Management, 23-26 July, 2012, Bengaluru, India.

[9] Haapalainen, E., Laurinen, P., Junno, H., Tuovinen, L., Röning, J.: Feature selection for identification of spot welding processes. Proceedings of the 3rd International Conference on Informatics in Control, Automation and Robotics pp. 40--46 (2006)

[10] Hand, D.J., Mannila, H., Smyth, P.: Principles of data mining. MIT Press, Cambridge, MA, USA (2001)

[11] Ichikawa, F., Chipchase, J., Grignani, R.: Where's the phone? a study of mobile phone location in public spaces. In: Mobile Technology, Applications and Systems, 2005 2nd International Conference on, pp. 1 --8 (2005)

[12] Koskimäki H, Huikari V, Siirtola P & Röning J : Behavior modelling in industrial assembly lines using a wrist-worn inertial measurement unit, accepted to Journal of Ambient Intelligence and Humanized Computing (2012)

[13] Kwapisz, J.R., Weiss, G.M., Moore, S.A.: Activity recognition using cell phone accelerometers. SIGKDD Explor. Newsl. 12, 74--82 (2011)

[14] Lu, H., Yang, J., Liu, Z., Lane, N.D., Choudhury, T., Campbell, A.T.: The Jigsaw continuous sensing engine for mobile phone applications. In: Proceedings of the 8th ACM Conference on Embedded Networked Sensor Systems, SenSys '10, pp. 71—84 (2010)

[15] Nokia N8: http://europe.nokia.com/find-products/devices/nokia-n8

[16] Peebles, D., Lu, H., Lane, N.D., Choudhury, T., Campbell, A.T.: Community-guided learning: Exploiting mobile sensor users to model human behavior. In: Proceedings of the Twenty-Fourth AAAI Conference on Artificial Intelligence, AAAI 2010, Atlanta, Georgia, USA, July 11-15 (2010)

[17] Polar Active: http://www.polaroutdoor.com/en/support/product_support?\product=29451

[18] Qt: http://qt.nokia.com/

[19] Ryder, J., Longstaff, B., Reddy, S., Estrin, D.: Ambulation: A tool for monitoring mobility patterns over time using mobile phones. In: Computational Science and Engineering, 2009. CSE '09. International Conference on, vol.~4, pp. 927 --931 (2009)

[20] Samsung Galaxy Mini: http://www.samsung.com/hk_en/consumer/mobile/mobile-phones/smartphone/GT-S5570EGAVDC-spec

[21] Siirtola P, Laurinen P, Röning J & Kinnunen H (2011): Efficient Accelerometer-Based Swimming Exercise Tracking, IEEE Symposium on Computational Intelligence and Data Mining (CIDM 2011), April 11-15, 156--161.

[22] Siirtola, P., Koskimäki, H., Röning, J.: Periodic quick test for classifying long-term activities. In: IEEE Symposium on Computational Intelligence and Data Mining (CIDM 2011), pp. 135--140 (2011)

[23] Smartphone shipments: http://www.bgr.com/2011/03/10/berg-smartphone-shipments-grew-74-in-2010/

[24] Sun, L., Zhang, D., Li, B., Guo, B., Li, S.: Activity recognition on an accelerometer embedded mobile phone with varying positions and orientations. In: Ubiquitous Intelligence and Computing, Lecture Notes in Computer Science, vol. 6406, pp. 548--562. Springer Berlin / Heidelberg (2010)

[25] Suutala, J., Pirttikangas, S., Röning, J.: Discriminative temporal smoothing for activity recognition from wearable sensors. In: Proc. of the 4th International Symposium on Ubiquitous Computing Systems (UCS07), pp. 182--195. Tokyo, Japan (2007)

[26] Van Laerhoven, K., Cakmakci, O.: What shall we teach our pants? Wearable Computers, 2000. The Fourth International Symposium on pp. 77--83 (2000)

[27] Wang, S., Chen, C., Ma, J.: Accelerometer based transportation mode recognitionnon mobile phones.Wearable Computing Systems, Asia-Pacific Conference on, 44--46 (2010)

P. Siirtola received his M.Sc. degree in Mathematics from University of Oulu, Oulu, Finland, in 2006. At the moment he does graduate studies in Department of Computer Science and Engineering where he works as a research scientist, sponsored by the Graduate School in Electronics, Telecommunications and Automation (GETA). His research interests include pattern recognition, data mining and machine learning methods and applications.

J. Röning obtained the degree of Doctor of Technology in 1992, Licentiate in Technology with honors in 1985, and Diploma in Engineering (MSEE) with honors in 1983, all at the University of Oulu in Finland. From 1983 he has been a member of faculty of the University of Oulu, where he is currently Professor of Embedded System and head of the Department of Computer Science and Engineering. Professor Röning has two patents and has published more than 200 papers. His main research interests are in intelligent systems, especially mobile robots, machine vision, and software security.

Neural Networks through Shared Maps in Mobile Devices

William Raveane, María Angélica González Arrieta

Universidad de Salmanca, Spain

Abstract **— We introduce a hybrid system composed of a convolutional neural network and a discrete graphical model for image recognition. This system improves upon traditional sliding window techniques for analysis of an image larger than the training data by effectively processing the full input scene through the neural network in less time. The final result is then inferred from the neural network output through energy minimization to reach a more precize localization than what traditional maximum value class comparisons yield. These results are apt for applying this process in a mobile device for real time image recognition.**

Keywords **— Computer vision, Image recognition, Deep neural networks, Graphical models, Mobile device.**

I. INTRODUCTION

HYBRIFD intelligent systems have consistently shown benefits that outperform those of their individual components in many tasks, especially when used along neural computing [1]. In recent years, two main areas of computer vision have gained considerable strength and support: On one side, soft computing techniques based on non-exact but very accurate machine learning models like neural networks, which have been successful for high level image classification [7]. Contrasting these systems, computer vision techniques modeled by graphical models have enjoyed great reception when performing low level image processing tasks such as image completion [6]. In this paper, we combine both of these techniques to successfully classify and localize a region of interest within an input image.

We use Convolutional Neural Networks (CNN) [3] for the classification of image content. CNNs have become a general solution for image recognition with variable input data, as their results have outclassed other machine learning approaches in large scale image recognition tasks [4]. Paired to this CNN classifier, we use energy minimization of a Markov Random Field (MRF) [8] for inference and localization of the target within the image space. Graphical models such as this have been implemented in areas of computer vision where the relationship between neighboring regions plays a crucial role [2].

We review the implementation of this system specifically within a mobile device. With the increasing use of mobile hardware, it has become a priority to provide these devices with computer vision capabilities. Due to the high computational requirements, this need has mostly been met by outsourcing the analysis to a remote server over the Internet. This approach introduces large delays and is hardly appropriate when interactivity and responsiveness are paramount. Embedded environments have intrinsic architecture constraints which require algorithms to make the best use of the available computing capacity. The proposed system exploits this specific platform by reducing the overall required memory throughput via a parallel execution approach. This is achieved by applying layer computations over the entire image space, as opposed to running smaller patches individually, as is common with the sliding window approach normally used in this type of image classification.

The structure of this work is as follows: In Section 2, some background knowledge is reviewed detailing the functionality of CNNs and window analysis in general. We then introduce in Section 3 an optimized approach for the techniques previously discussed, including the architecture constraints that must be made to implement the proposed system. Section 4 goes over the discretization of the system and the inference process for obtaining the final output result. Section 5 continues with the results obtained from the proposed method and a brief comparison with other alternative approaches. Finally, Section 6 concludes by discussing the observations made and some additional applications where this system can be used in.

II. BACKGROUND

In this section, a brief description of CNNs and their layer types is given, as well as an overview of the traditional sliding window approach.

A. Convolutional Neural Networks

The network on which our system is based upon is a standard CNN. Figure 1 depicts the layer structure of such a network, and it is the reference architecture used throughout this paper to describe the concepts of the framework presented.

In the initial stages of the CNN, a neuron consists of a two-dimensional grid of independent computing units, each producing an output value. As a result, every neuron will itself output a grid of numerical values, a data structure in $\Re^2$ referred to as a map. When applying CNNs to image analysis, these maps represent an internal state of the imageafter being processed through a connective path leading to that particular neuron. Consequently, maps will usually bear a direct positional and feature-wise relationship to the input image space. As data progresses through the network,

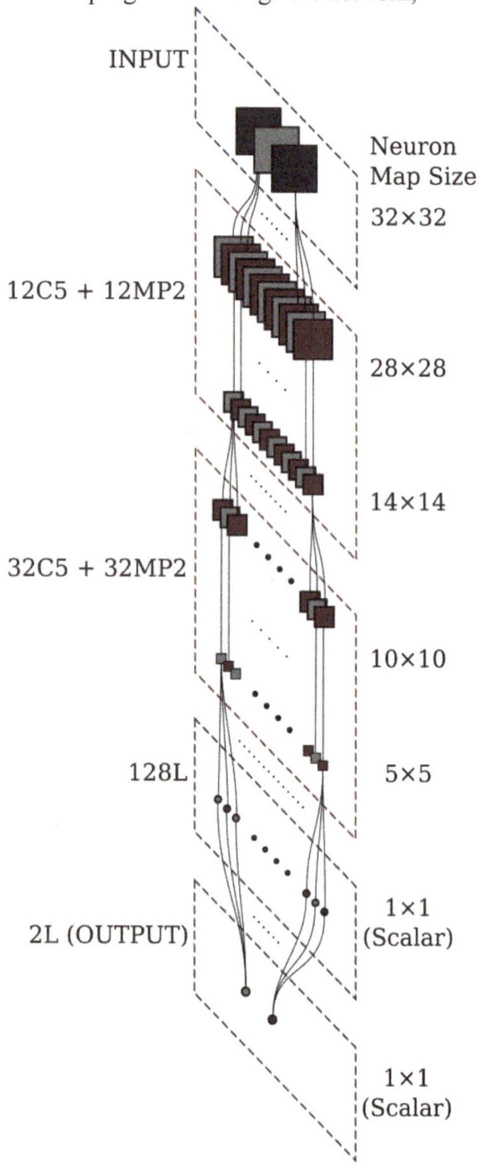

Fig. 1. A typical convolutional neural network architecture, with three input neurons for each color channel of an analyzed image patch, two feature extraction stages of convolutional and max-pooling layers, and two linear layers to produce a final one-vs-all classification output.

however, this representation turns more abstract as the dimensionality is reduced. Eventually, these maps are passed through one or more linear classifiers, layers consisting of traditional single unit neurons which output a single value each. For consistency, the outputs of these neurons are treated as1×1single pixel image maps, although they are nothing more than scalar values in$\Re^0$.

B. CNN Layer Types

The first layer in the network consists of the image data to be analyzed, usually composed as the three color channels. The notation$N_j X K_j$is used to describe all subsequent layers, where$N_j$is the neuron map count of layer$j$, $X \in \{C, MP, L\}$denotes the layer type group (Convolutional, Max-Pooling, and Linear), and$K_j$is the parameter value for that layer.

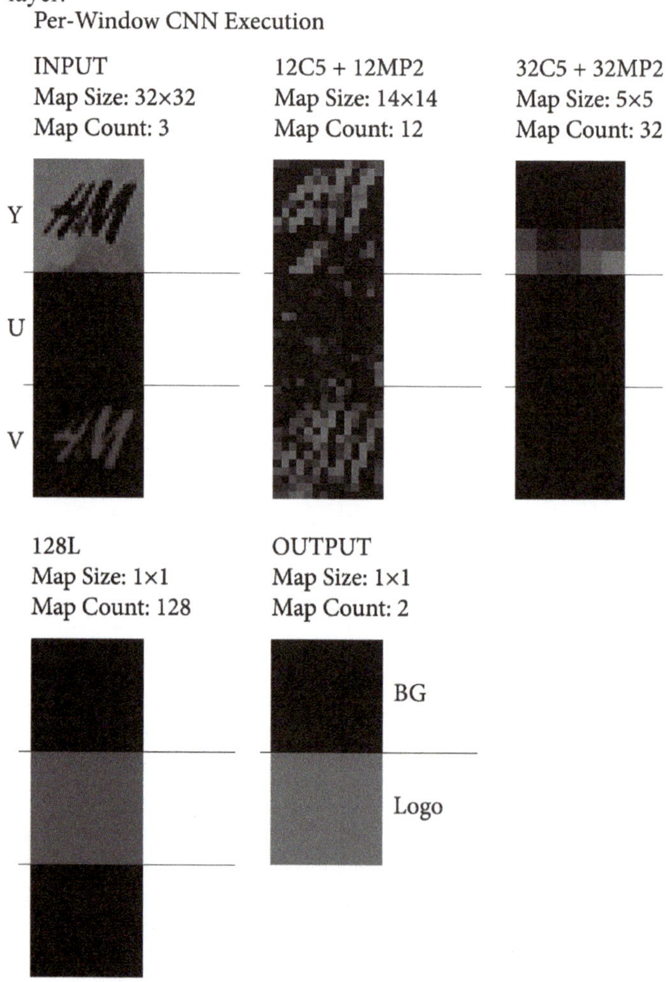

Fig. 2. Visualization of the first three neuron maps at each stage of the CNN. Note the data size reduction induced at each stage. The output of this execution consists of two scalar values, each one representing the likelihood that the analyzed input image belongs to that neuron's corresponding class. In this case the logo has been successfully recognized by the higher valued output neuron for class "Logo".

The first part of every$C \rightarrow MP$feature extraction stage is a convolutional layer. Here, each neuron linearly combines the convolution of one or more preceding maps. The result is a map slightly smaller than the input size by an amount known as the kernel padding, which arises from the boundary conditions of the valid convolution algorithm. It is defined as $(K_j/2) - 1$, where $K_j$is the convolutional kernel size of layer$j$. Therefore, the layer's map size will be given by$M_j = M_{j-1} - (K_j/2) - 1$, where $M_{j-1}$is the the preceding layer's map size.

A max-pooling neuron acts on a single map from a preceding convolutional neuron, and its task is to subsample a pooled region of size K_j. The result is a map size that is inversely proportional to said parameter by $M_j = M_{j-1}/K_j$. The data may then be passed to one or more additional $C \rightarrow MP$ feature extractors.

Linear layers classify feature maps extracted on preceding layers through a linear combination as in a perceptron -- always working with scalar values -- such that $M_j = 1$ at every layer of this type.

Finally, the output of the final classification layer decides the best matching label describing the input image. Fig. 2 shows the information flow leading to this classification for a given image patch, where the CNN has been trained to identify a particular company logo.

C. The Sliding Window Method

Recognition of images larger than the CNN input size is achieved by the sliding window approach. This algorithm is defined by two quantities, the window size S, usually fixed to match the CNN's designed input size; and the window stride T, which specifies the distance at which consecutive windows are spaced apart. This stride distance establishes the total number of windows analyzed W for a given input image. For an image of size $I_w \times I_h$, the window count is given by:

$$W = (\frac{I_w - S}{T} + 1)(\frac{I_h - S}{T} + 1) \rightarrow W \propto \frac{I_w I_h}{T^2}$$

Figure 3 shows this method applied on an input image downsampled to 144×92, extracting windows of $S = 32$ for the simple case where $T = S/2$. A network analyzing this image would require 40 executions to fully analyze all extracted windows. The computational requirement is further compounded when a smaller stride is selected -- an action necessary to improve the resolving power of the classifier: at $T = S/8$, 464 separate CNN executions would be required.

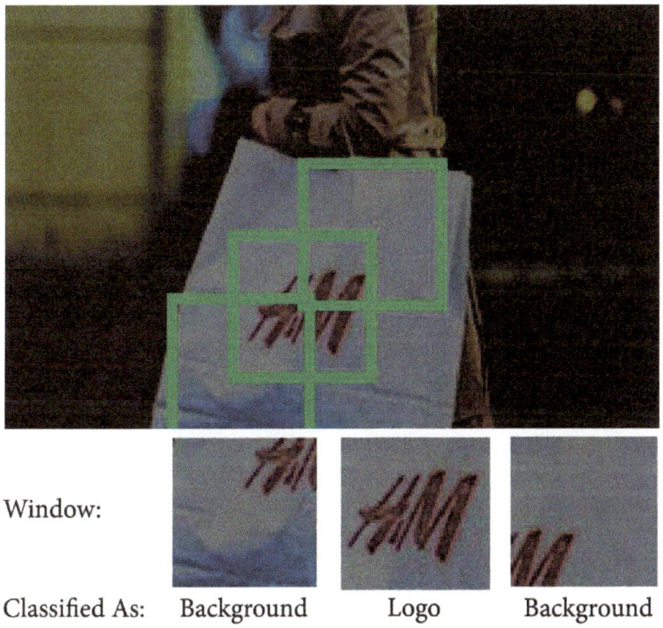

Fig. 3. An overview of the sliding window method, where an input image is subdivided into smaller overlapping image patches, each being individually analyzed by a CNN. A classification result is then obtained for each individual window.

III. Optimized Network Execution

The method proposed introduces a framework where the stride has no significant impact on the execution time of the $C \rightarrow MP$ stages, as long as the selected stride is among a constrained set of possible values. This is achieved by allowing layers to process the full image as a single shared map instead of individual windows. Constraints in the possible stride values will result in pixel calculations to be correctly aligned throughout the layers.

A. Shared Window Maps

CNNs have a built-in positional tolerance due to the reuse of the same convolutional kernels over the entire neuron map. As a result of this behavior, their output is independent of any pixel offset within the map, such that overlapping windows will share convolved values. This is demonstrated in Fig. 4.

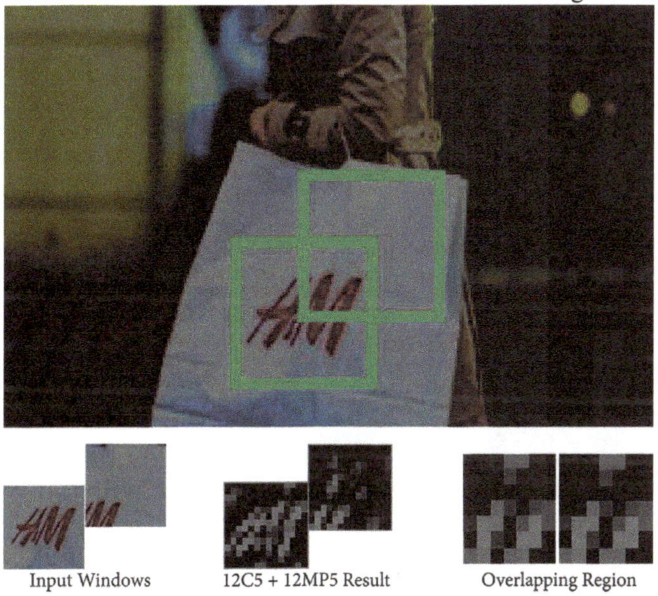

Fig. 4. Two adjacent windows extracted from an input image, passed through the 12C5 + 12MP5 feature extractor. A detailed view of the convolved maps in the overlapping top-right and bottom-left quarters of each window shows that these areas fully match.

This leads to the possibility of streamlining the feature extractors by running their algorithms over the full input image at once. Hence, each $C \rightarrow MP$ neuron will output a single map shared among all windows, where subdivisions of this map would normally match the outputs of the corresponding windows, had they been executed separately as in the traditional method. This greatly reduces the expense of calculating again convolutions on overlapping regions of each window. Figure 5 shows an overview of the shared map process, which passes the input image in its entirety through each stage of the network.

By doing this, the output layer now produces a continuous and localized class distribution over the image space, a result which contrasts greatly to that of a single classification value as was previously seen in Fig. 2. The output of this execution consists of image maps where each pixel yields the relative position of all simultaneously classified windows.

Similar to the per-window execution method, the intensity value of a pixel in the output map represents the classification likelihood of the corresponding window. Note how the relative position of the logo in the input image has been discovered after only one shared map execution of the network. An account of the window size and stride is also on

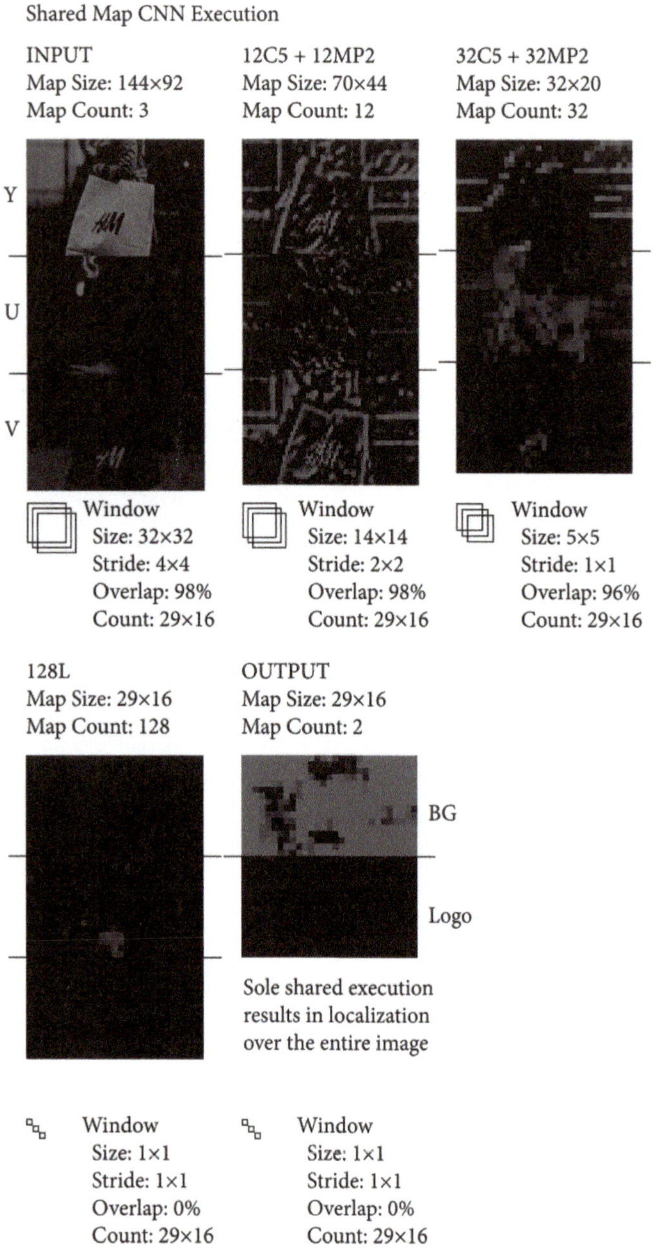

Fig. 5. The shared map execution method for a convolutional neural network, where each layer processes an entire image in a single pass, and each neuron is now able to process maps with dimensions that far exceed the layer's designed input size.

display, illustrating how it evolves after each layer, while the total window count remains the same. Here, the correspondence of each 32×32 window in the input image can be traced to each one of the pixels in the output maps.

B. Window Configuration

The operation of the shared map process relies greatly on the details of the dimensionality reduction occurring at each layer within the network. For this reason, it is necessary to lay certain constraints that must be enforced when choosing the optimum sliding window stride.

At each layer, the window size and stride are reduced until

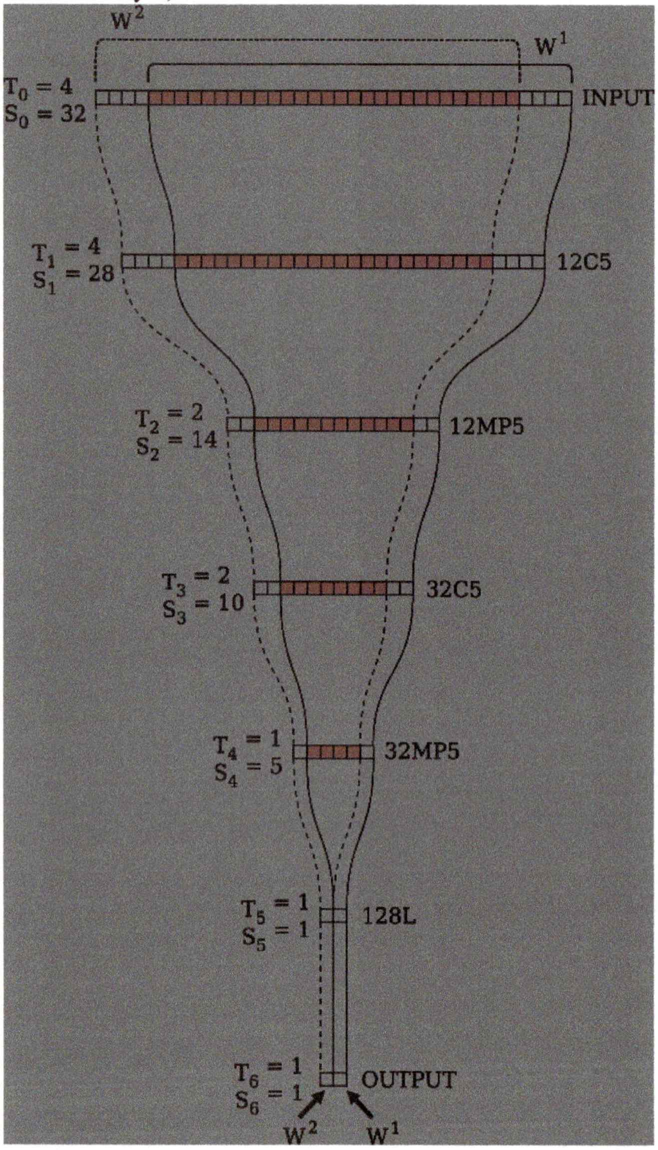

Fig. 6. The CNN layers and their effect on the window pixel space, illustrated in
one dimension for simplicity. Two successive 32×32 windows W^1 and W^2 are shown. Overlapping pixels at each layer are shaded. Starting with an input layer window stride $T_0 = 4$, the final output layer results in a packed $T_6 = 1$ window stride, so that each output map pixel corresponds to a positional shift of 4 pixels in the input windows, a relationship depicted by the column paths traversing all layers.

they eventually become single pixel values at the final linear layers. The amount of reduction at each stage varies according to the type of the layer and its parameters. All of these quantities can be found in a well defined manner as given by:

$$S_j = \{ \begin{matrix} S_{j-1} - K_j - 1 & if j \in C \\ S_{j-1}/K_j & if j \in MP \\ S_{j-1} & if j \in L \end{matrix} \}$$

$$T_j = \{ \begin{matrix} T_{j-1} & if j \in C \cup L \\ T_{j-1}/K_j & if j \in MP \end{matrix} \}$$

Where the window size S_j and its stride T_j at layer j depends on the various parameters K_j of the layer and the window size and stride values at the preceding $j-1$ layer. This equation set can be applied over the total number of layers of the network, while keeping as the target constraint that the final size and stride must remain whole integer values. By regressing these calculations back to the input layer $j = 0$, one can find that the single remaining constraint at that layer is given by:

$$T_0 \equiv a \prod_{j \in MP} K_j where a \in \mathbb{Z}$$

In other words, the input window stride must be perfectly divisible by the product of the pooling size of all max-pooling layers in the network. Choosing the initial window stride in this manner, will ensure that every pixel in the final output map is correctly aligned throughout all shared maps and corresponds to exactly one input window. Fig. 6 follows the evolution of the window image data along the various layers of the sample network architecture, showing this pixel alignment throughout the CNN.

IV. DISCRETE INFERENCE OF CNN OUTPUT

The output from the convolutional neural network as seen in Fig. 5 consists of multiple individual maps, where each one embodies a visual depiction of the relative confidence, per-class, that the system has for every window sampled.

The common practice to obtain a final classification from an output value set as seen in Fig. 5 is to identify which class has a higher output value from the CNN at each each window (here, each pixel in the output map). While efficient, results from this procedure are not always ideal because they only take into account each window separately.

Furthermore, maximum value inference is prone to false positives over the full image area. Due to their non-exact nature, neural network accuracy can decrease by finding patterns in random stimuli which eventually trigger neurons in the final classification layer. However, such occurrences tend to appear in isolation around other successfully classified image regions. It is therefore possible to improve the performance of the classifier by taking into account nearby windows.

There exist many statistical approaches in which this can be implemented, such as (i) influencing the value of each window by a weighted average of neighboring windows, or (ii) boosting output values by the presence of similarly classified windows in the surrounding area. However, we propose discrete energy minimization through belief propagation as a more general method to determine the final classification within a set of CNN output maps. The main reason being that graphical models are more flexible in adapting to image conditions and can usually converge on a globally optimal solution.

A. Pairwise Markov Random Field Model

Images can be treated as an undirected cyclical graph $G = (N, E)$, where nodes $n_i \in N$ represent an entity such as a pixel in the image, and graph edges $e_{ij} \in E$ represent the relationship between these nodes. If, for simplicity, 4-connectivity is used to represent the relationship between successive nodes in a graph; then each node will be connected to four others corresponding to its neighbors above, below, and to each side of the current element.

The output space of the convolutional neural network can therefore be represented in this manner through a graph. However, instead of describing pixel intensity values, each node in the graph represents the classification state of the corresponding window. This state takes on a discrete value among a set of class labels $c \in CL \equiv \{BG, Logo\}$ corresponding to the classification targets of the CNN. Thus, each node in the graph can take on one of several discrete values, expressing the predicted class of the window that the node represents. Fig. 7 displays the structure of such a graph.

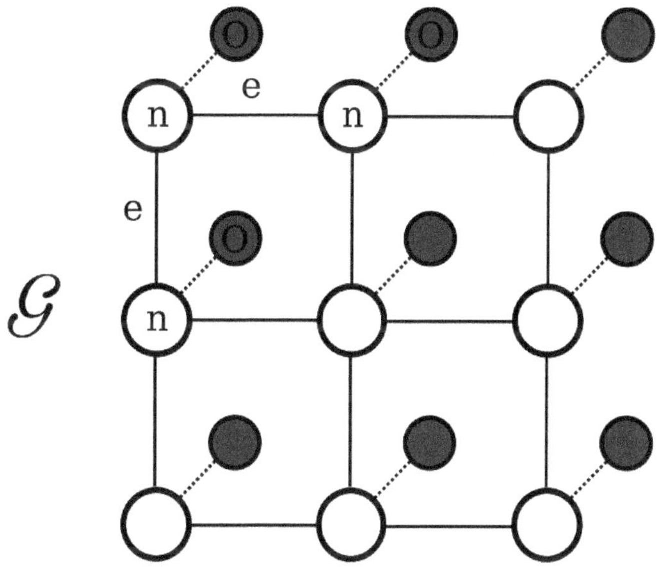

Fig. 7. A subset of the MRF graph G formed by the CNN output space, where each node n_i represents the classification state of a corresponding window analyzed with the network, whose outputs are implemented into this system as the observed hidden variables O. Nodes have a 4-connectivity relationship with each other represented by the edges e_{ij} thus forming a grid-like cyclical graph.

It can be seen that if nodes represent classification outcomes, there is a strong relationship between them. The reason is that continuity throughout a map tends to be preserved over neighboring regions due to strong local correlation in in input images. This inflicts a Markovian property in the graph nodes where there is a dependency between successive nodes. Therefore, this graph follows the same structure as an MRF, and any operations available to this kind of structure will be likewise applicable to the output map.

B. Energy Allocation

To implement energy minimization on an MRF, it is necessary to assign energy potentials to each node and edge. These energies are usually adapted from observed variables, and in this case, they correspond to the values of the output maps and combinations thereof. Therefore, MRF optimization over a graph G can be carried out by minimizing its Markov random energy R, given by:

$$R(G) = R(N, E) = \sum_{n_i \in N} \theta_i(n_i) + \sum_{e_{ij} \in E} \theta_{ij}(e_{ij})$$

Here, $\theta_i(\circ)$ corresponds to the singleton energy potential of node n_i, and $\theta_{ij}(\circ)$ is a pairwise potential between nodes n_i and n_j. Starting from the CNN output map observations, the singleton potentials can be assigned as:

$$\theta_i = \begin{bmatrix} \omega_i^0 \\ \omega_i^1 \\ \vdots \\ \omega_i^l \end{bmatrix}$$

$$\omega_i^a = \sum_{c \in C} \left\{ \begin{matrix} 1 - (O_i^c)^2 & if a = c \\ (O_i^c)^2 & otherwise \end{matrix} \right\}$$

Where l is the total number of classes in set CL (2 in the sample CNN architecture), and O_i^c is the observed CNN value for window $n_i \in N$ and class $c \in CL$. In this manner, each ω_i^a value is an MSE-like metric that measures how far off from ideal training target values did the CNN classify window n_i as. Thus, a lower potential value will be assigned to the most likely class, while a higher potential value will be given to other possible classes at this node.

Pairwise potentials can be defined as:

$$\theta_{ij} = \begin{bmatrix} \delta_{ij}^{00} & \delta_{ij}^{01} & \cdots & \delta_{ij}^{0c} \\ \delta_{ij}^{10} & \delta_{ij}^{11} & & \delta_{ij}^{1l} \\ \vdots & & \ddots & \\ \delta_{ij}^{l0} & \delta_{ij}^{l1} & & \delta_{ij}^{ll} \end{bmatrix}$$

$$\delta_{ij}^{ab} = |O_i^a - O_j^b|$$

Where each value δ_{ij}^{ab} is a straightforward distance metric that measures the *jump* in CNN output values when switching from class a to class b between windows n_i and n_j. Thus, these potentials will be small if the same class is assigned to both nodes, and large otherwise. Fig. 8 shows all energy assignments per node pair.

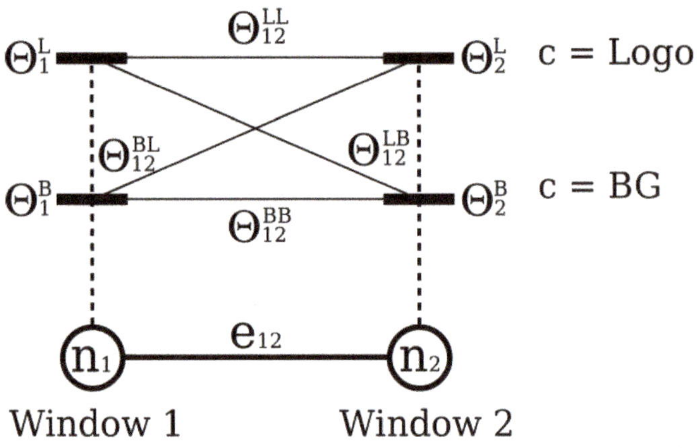

Fig. 8. A detail of the potential energies assigned to each of two nodes $\{n_1, n_2\}$ connected by edge e_{12}. The singleton potentials Θ_i^a correspond to the energy associated with node i if assigned to class a, and the pairwise potentials Θ_{ij}^{ab} are the changes in energy that occur by assigning class a to node n_i and class b to node n_j.

It is worth noting that these θ_{ij} pairwise potentials between neighboring windows (nodes) are the only feature that sets apart this process from the traditional winner-takes-all approach which would otherwise be implemented through the minimization of the energy in the θ_i singleton potentials by themselves.

C. Energy Minimization by Belief Propagation

Applying Belief Propagation [5] to find the lowest possible energy state of the graph will now yield an equilibrium of class assignments throughout the image output space.

Due to the cycles inherent of image-bound graphs, a special variation of the algorithm must be used, in this case Loopy Belief Propagation [5]. This variation requires the minimization to be run several times until the solution converges and an equilibrium is found. However, due to various existing optimizations for this algorithm, this process is very straightforward and can be solved in polynomial time.

V. RESULTS

The test application is developed for the Android mobile OS as an OpenGL ES shader which makes use of the available computing capabilities of the device GPU. The main logic of the system is placed within a fragment shader running the CNN per-pixel over a Surface Texture memory object. The test device is equipped with a quad core 1.3 GHz Cortex-A9 CPU with a 12-core 520 MHz Tegra 3 GPU. This SoC architecture embeds 1 Gb of DDR2 RAM shared by both the CPU and GPU.

The test system executes the same CNN architecture described in Fig. 1, except for the classification layer having 32 output neurons corresponding to one background label and 31 different logo labels. This network is exectued over 8 simultaneous 144×92 images forming a multi-scale image pyramid. The energy minimization technique as described in Section 4 is then applied, but over a 3D graph formed with 6-

connectivity between nodes such that each window is also aware of window classifications at the corresponding larger and smaller scale steps. Table 1 gives a summary of the results obtained from this setup.

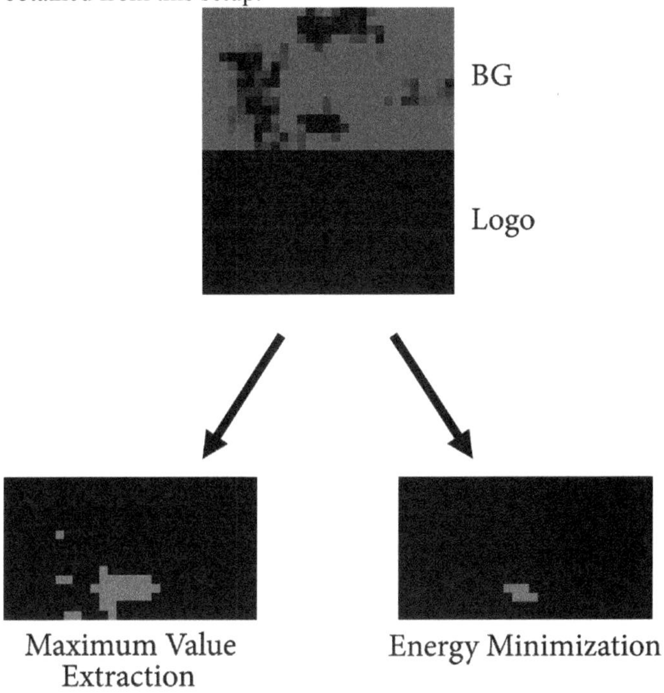

Fig. 9. Comparison of the final "Logo" classification and localization, applying the classical maximum value per class extraction vs. our proposed energy minimization inference method on the two CNN output maps introduced in Figure 5

It is of great interest to note the final 32×32 configuration. Regardless of the fact that there is no overlap at this stride, a 3.0 speedup is still observed over running the windows individually. This is due to the inherent reduction in memory bandwidth through the system's pipelined execution approach, where the entire image needs to be loaded only once per execution. This contrasts the traditional approach where loading separate windows into memory at different times requires each to be individually sliced from the original memory block -- a very expensive operation in the limited memory throughput of mobile devices.

Server platforms have a restriction in the PCIe bus speed between the CPU and GPU, but instead offer very fast local memory access within the GPU. As a result, these architectures would allow window extraction at lower relative latencies. The SoC architecture of mobile devices do not face similar CPU to GPU memory bottlenecks, as these chips are usually located within the same circuit. Their lower energy requirements, however, force local memory access to be radically slower. Therefore, this architecture favors the parallel usage of data blocks, a fact which the system we have presented exploits in full. As such, we consider it to be a mobile-first oriented algorithm, although it would offer likewise improvements in other platforms.

The results of the inference system are more of a qualitative nature, as it is difficult to objectively establish a ground truth

TABLE I
SPEEDUP RESULTS

T_0	W (L)	W (P)	OC	T (PW)	T (SM)	Speedup
4×4	464	3,712	98%	29,730	1,047	28.4x
8×8	112	896	94%	7,211	387	18.6x
12×12	60	480	86%	3,798	311	12.2x
16×16	32	256	75%	2,051	240	8.5x
20×20	24	192	61%	1,536	252	6.1x
24×24	15	120	44%	945	203	4.7x
28×28	15	120	23%	949	200	4.7x
32×32	8	64	0%	514	171	3.0x

Results of tests with several input layer stride T_0 configurations, from the closest packed 4×4 to the non-overlapping 32×32 layouts. A total window count at each pyramid level W (L), and over the full 8 level pyramid W (P), as well as the window overlap coverage OC per input map is given for each of the stride selections. An average over 20 test runs for each of these configurations was taken as the execution time in milliseconds for each of the methods described herein – the traditional per-window execution method T (PW), and our shared map technique T (SM). A speedup factor is calculated showing the performance improvement of our method over the other.

TABLE II
INFERENCE RESULTS

Algorithm	Accuracy	PPV	F_1
Maximum Value	0.942	0.341	0.498
Weighted Average	0.964	0.391	0.430
Neighbor Boosting	0.972	0.489	0.591
Energy Minimization	0.981	0.747	0.694

Results of various inference algorithms for the final classification, describing the Accuracy (TP + TN / ALL), PPV (TP / TP + FP), and F_1 (2TP / 2TP + FP + FN) metrics.

basis for such experiments. This system aims to localize classified windows, therefore it is subject to an interpretation of which windows cover enough of the recognition target to be counted as a true positive. Regardless, Table 2 gives an indicative comparison of the system against the competing techniques previousy described. Fig. 9 shows a visual comparison.

VI. CONCLUSIONS

A system for the optimization of convolutional neural networks has been presented for the particular application of mobile image recognition. The performance figures presented in Table 1 correspond to a device architecture which, at the time of this work, is a commonly available specification on end user devices. It must be noted that with the rapid growth that is being observed in mobile hardware capabilities, the effects of these optimizations are likely to grow in their significance. GPUs capable of new technology will extend the reach of the parallel-wise optimizations described. Relevant advancements in this area would include things such as heterogeneous parallel processing via OpenCL EP and zero-copy memory transfer between the camera and GPU through tighter SoC integration. General availability of such technologies will open an ever larger possibility of mobile computer vision opportunities.

Although a simple logo classification task was used here as a sample application, CNNs allow for many other image recognition tasks to be carried out. Most of these processes would have great impact on end users if implemented as real time mobile applications. Some examples where CNNs have been successfully used and their possible mobile implementations would be (i) text recognition for visually interactive language translators, (ii) human action recognition for increased user interactivity in social applications, or even (iii) traffic sign recognition for embedded automotive applications. Any of these applications could be similarly optimized and discretized by the system presented here.

In addition to the CNN classifier, the MRF model is very flexible as well and its implementation can be adjusted to domain-specific requirements as needed by each application. For example, a visual text recognizer might implement pairwise energy potentials which are modeled with the probabilistic distribution of character bigrams or n-grams over a corpus of text, thereby increasing the overall text recognition accuracy.

Furthermore, although the analysis of a single image has been discussed, this system is similarly extensible to multiple images processed together. The most common example of this is the analysis of a multi-scale image pyramid, something vital within mobile applications as variable distances between the camera and its target will cause the object to be observed at different sizes within the analyzed image. In such a case, the MRF would be extended to a 6-connectivity 3D grid, where nodes would be equally aware of window classifications at the corresponding larger and smaller scale steps.

Therefore, we believe this to be a general purpose mobile computer vision framework which can be deployed for many different uses within the restrictions imposed by embedded hardware, but also encouraging the limitless possibilities of mobile applications.

REFERENCES

[1] Borrajo, M.L., Baruque, B., Corchado, E., Bajo, J., Corchado, J.M.: Hybrid neural intelligent system to predict business failure in small-to-medium-size enterprises. International Journal of Neural Systems 21(04), 277–296 (2011)

[2] Boykov, Y., Veksler, O.: Graph Cuts in Vision and Graphics: Theories and Applications. In: Paragios, N., Chen, Y., Faugeras, O. (eds.) Handbook of Mathematical Models in Computer Vision, pp. 79–96. Springer US (2006)

[3] Ciresan, D.C., Meier, U., Masci, J., Gambardella, L.M., Schmidhuber, J.: Flexible, High Performance Convolutional Neural Networks for Image Classication. In: Proceedings of the Twenty-Second International Joint Conference on Artificial Intelligence. pp. 1237–1242 (2011)

[4] Ciresan, D.C., Meier, U., Gambardella, L.M., Schmidhuber, J.: Convolutional Neural Network Committees For Handwritten Character Classification. In: 11th International Conference on Document Analysis and Recognition. ICDAR (2011)

[5] Felzenszwalb, P.F., Huttenlocher, D.P.: Efficient Belief Propagation for Early Vision. International Journal of Computer Vision 70(1), 41–54 (May 2006)

[6] Komodakis, N., Tziritas, G.: Image completion using efficient belief propagation via priority scheduling and dynamic pruning. Image Processing, IEEE Transactions on 16(11), 2649–2661 (Nov 2007)

[7] Krizhevsky, A., Sutskever, I., Hinton, G.E.: ImageNet Classification with Deep Convolutional Neural Networks. pp. 1106–1114. No. 25 (2012)

[8] Wang, C., Paragios, N.: Markov Random Fields in Vision Perception: A Survey. Rapport de recherche RR-7945, INRIA (September 2012)

William Raveane is a Ph.D. candidate in Computer Engineering at the University of Salamanca, Spain, currently researching mobile image recognition through deep neural networks. He has also worked for several years in private companies in various topics ranging from computer vision to visual effects.

María Angélica González Arrieta is a professor at the Department of Computer Engineering and Automation at the University of Salamanca, Spain. Her research interests are primarily in pattern recognition and neural networks. She is also a member of the BISITE research group.

22

Scene integration for online VR advertising clouds

Michael Kalochristianakis, Markos Zampoglou, Kostas Kontakis, Kostas Kapetanakis, Athanasios Malamos

Department of Informatics Engineering, Technological Educational Institution, Heraklion, Crete

Abstract — **This paper presents a scene composition approach that allows the combinational use of standard three dimensional objects, called models, in order to create X3D scenes. The module is an integral part of a broader design aiming to construct large scale online advertising infrastructures that rely on virtual reality technologies. The architecture addresses a number of problems regarding remote rendering for low end devices and last but not least, the provision of scene composition and integration. Since viewers do not keep information regarding individual input models or scenes, composition requires the consideration of mechanisms that add state to viewing technologies. In terms of this work we extended a well-known, open source X3D authoring tool.**

Keywords — **Virtual reality, web advertising, 3D technologies**

I. INTRODUCTION

ADVERTISING through the world-wide web has been gaining attention during the last decade in order to cover the needs of enterprises for promotion via the cyberspace. As 3D technologies become more ubiquitous, virtual reality (VR) advertising becomes an appealing possibility. VR allows full control over digital worlds that is, the potential for creativity without limits. Although this type of creativity has boosted fields relating to multimedia it is still rather immature as far as marketing is concerned. There has been research on the field such as evaluations that designate the advantages of enabling VR technologies. For instance, the impact of allowing users to interact with the virtual representation of products is significant both in terms of appeal, penetration and general product awareness [1] [2] [3]. One of the issues regarding the general application of VR approaches in advertising is cost. 3D content is in general more expensive to produce in comparison to more traditional multimedia such as video, audio, images and text. The design of models and composite scenes, visual effects and interaction inevitably requires technical knowledge and work that cannot be allocated in terms of projects with narrow scope, besides artistic design and work. Productive commercial VR solutions are typically produced upon platforms or engines.

Our approach in the field of VR in terms of the iPromotion project aims is to provide an online infrastructure that will allow advertisers to market products and services by building and exposing experiential environments for their end users. They will be able to include ads within web page banners or as web page pop-ups in terms of traditional online marketing but they would also be capable to produce immersive experiences through other interfaces, such as interactive widescreens or touch tables. Our system is design to support both the composition and the exposure of VR content by means of cloud-enabled storage and services. The advantages the new advertising paradigms are obvious; advertisers can build and distribute their content using central, online infrastructures. Through appropriately designed 3D worlds advertising can provide a direct representation of the ideas, messages or product characteristics based on exciting interactions that may take any wanted form ranging from multimedia stimuli to games, puzzles and interaction scenarios. At the same time, users or potential customers will be given the opportunity to take part in a simulated experience regarding the advertised goods.

In order to sustain the aforementioned scenario it is essential to provide users with the means to compose any desired virtual worlds that is, not only to provide them with sufficient amounts of VR content units that is, 3D models or scenes, and also with the means to the content for the composition of complex scenes or even worlds. Complying with established standards is a means to accomplish only the former; even if the provision of significant amount of content units can be based on offering standard, royalty-free or available complying content for our platform, the ability to build more complex content cannot not supported by any. The scene composition module of our platform is designed to provide this use case by adding visual controls to content units so that they can be managed in terms of broader environments. It takes advantage of the appropriate event models to store information about the position, size, scale and rotation at least, in order to create simple representations for complex worlds so that content units are reused by accessing the cloud of our platform [9].

The rest of the paper is organized as follows: Section II presents the related work in the field of 3D scene representation, the standards and the tools. Section III briefly elaborates on the research project that produced this work, the iPromotion VR platform. The section overviews the architecture and components of iPromotion, focusing on the aspects that offer contributions to the VR field. Section IV

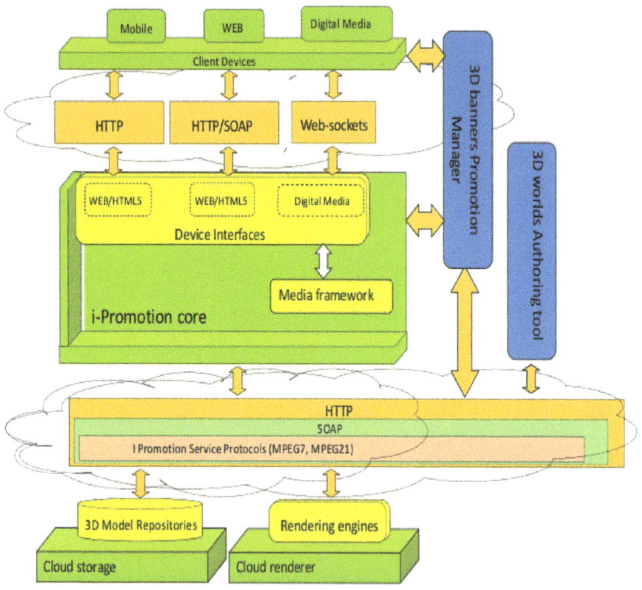

Fig. 1: the component architecture of the system

is the primary topic of this paper. Finally, section V discusses our conclusions and explores potential next steps in the development our scene composition software.

II. Procedure for Paper Submission

A. 3D technologies

The evolution of cross-platform web applications has triggered the use of XML-encoded textual files that provided a platform-independent solution for information exchange. In the field of two dimensional graphics, Scalable Vector Graphics (SVG) is an XML-based open standard that provides the necessary tools for development. SVG is capable to produce accurate results by describing vectors in XML tags. Graphics can thus be re-sized or zoomed without any loss of quality and can also be transferred in the same manner as any other XML-encoded textual files. The protocol can also utilize text annotations withing the images in order to provide supplemental functionality. Search engines or end-users can search such text annotations to retrieve content based on search criteria or individual needs. SVG is also a W3C recommendation [6] and can thus be integrated the Document Object Model (DOM) and eXtensible Stylesheet Language (XSL) besides other standards. Thus, interactivity and animation can be applied in SVG representations through object module manipulation languages such as Javascript.

X3D is a royalty free, open, ISO ratified standard for developing 3D graphics and also a run-time architecture to represent and communicate. X3D is developed by the Web3D Consortium [7] as the successor of the Virtual Reality Modeling Language (VRML). In X3D, representations use XML-encoded format and thus information is platform independent and ideal for use over the internet. X3D supports real-time communication and integration with web-service architectures, distributed networks, cross-platform applications, inter-application information transfer. It is modular by nature and is thus capable to support the concatenation or superposition of 3D components. The standard is capable to support, besides high-quality graphics, real-time interaction and audio-video sequences. The functionality of the standard is structured in layers called profiles that define the functionality subset that is to be used. The run time environment of X3D is essentially a browser engine that carries the rendering logic. The engine is released as a standalone application, as a web browser plug-in and as an open development project.

B. Scene authoring

The need to develop and combine X3D models and scenes within our platform led us to consider the available tools and their potential to be extended, maintained and used. The relevant technologies include X3D players, plug-ins, development environments and authoring platforms. A cohesive overview can be found in the pages of the WEB3D consortium. There is also an interesting comparison of the authoring and scene conversion tools in the paragraph that regards authoring. The table presented there compares most of the relevant tools that is BS Editor, SwirlX3D, X3D-Edit, Flux Studio and Vivaty Studio besides some conversion tools. The information is kept updated since it refers to the current versions of the software. One can observe a number of points. To start with, full profile is not completely supported by any authoring tool while it is "nearly supported" by BS Editor and X3D-Edit. Most tools are certified for the interchange profile and it is an easy task to retrieve this information and the detailed list of supported X3D components for each tool. Two of the authoring tools presented, Flux and Vivaty studio, are no longer active. Besides the information presented in the comparison by WEB3D consortium, a natural categorization for technologies regards their relation to the consortium that is and the details of the implementation. X3D-Edit relies on the NetBeans development platform and offers the Scene Access Interface (SAI) API in order to load and manipulate the scenegraph. BS Editor relies on a variety of technologies ranging from JAVA to proprietary ones and offers a software development kit (SDK) that can extend the capabilities of the tool. The core of the development kit also includes a scene manipulation API however, using the SDK is licensed. Blender is another interesting tool in the field of authoring released under the general public license. It offers a complete environment for model and scene manipulation, useful features for import/export, additional functionality such as game modes and full access to the source code.

C. Web advertising infrastructure

The component architecture of the iPromotion platform is presented in Fig. 1. The system implements a 3-tier serviceoriented platform composed of the following components:

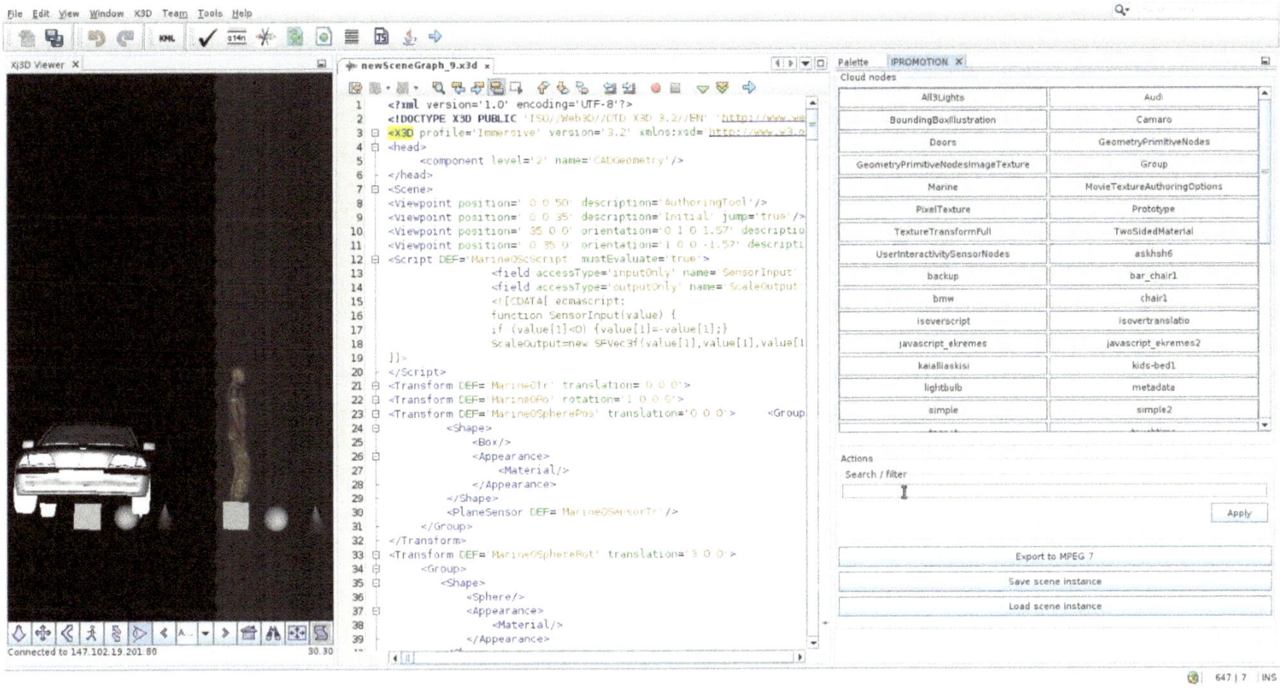

Fig. 2: the scene composition environment takes advantage of the X3D event model to implement scene management

- front-end services: based on HTTP, HTTP/SOAP and WebSockets, the system communicates with any type of device in order to adapt the rendering procedure to its capabilities using the mechanisms described in the previous paragraphs and also with any component within the architecture
- middleware container: it implements the fundamental framework that performs 3D rendering and interfaces with the presentation layer, the cloud that stores scenes, the MPEG-7 production service [8] and, last but not least, with high level tools such as 3D banners manager.
- cloud storage infrastructure: the repository of scenes and models repository is stored in a remote infrastructure implemented by as storage-as-a-service. The cloud supports transparent mechanisms for module search and retrieval over distributed processing technologies such as Hadoop and NoSQL databases [12].

Our work has been part of an online advertising platform infrastructure current under development under the iPromotion a research project. The aim of the project is to provide the all enabling framework and applications for online advertising based on VR technologies. The final outcome of the project is expected to support at least two high level use cases for online advertising. The first one will be targeting standard internet access from workstations, mobile or intelligent devices. Users will be provided with the opportunity to search online clouds of models and scenes and select any ones that interest them. Scenes will be transparently downloaded, presented transparently to the user who will be the capable to interact with the advertisement. The second use case will involve modern interaction interfaces that is, touch-screens, motion sensors such as the Microsoft KINEKT or ASUS XTION, and possibly wall screens or projectors. Users will be offered with VR environments conforming at least to the immersive X3D profile that supports 3D interactions and will thus be capable to interact with the scene using all the aforementioned interfaces. Both use cases will be supported by mechanisms for content-based search that take advantage of online MPEG-7 media location descriptors [4] for 3D scenes. MPEG-7 representation is also stored in the cloud along with the scenes. The transmission mechanism will conform with the MPEG-21 [5] standard in order to implement multimedia adaptation that is, to ensure that content can be played in the requesting devices. Finally, MPEG-4 is used to merge this information into unified 3D multimedia objects, incorporating information about digital and the actual scene description. Our infrastructures implement remote rendering of VR environments [10][11]. That is, 3D scenes are rendered on the server tier and end users receive video streams that have been adapted to the characteristics of their terminal devices as far as any characteristic is concerned (image quality, the frame rate). To this end, the open source Xj3D Browser has been ported to expose Axis2 Java Web service interface. A detailed description of the platform can be found in [13].

III. Composing Scenes

In order for the iPromotion platform to complete the cycle of 3D advertising production and service provisioning it is essential to support an authoring module that will allow scene authors or designers to produce models, scenes or worlds to be exposed by the platform. The authoring environment was required to be an autonomous desktop application environment but also to be able to exploit iPromotion services transparently that is, to use the scene repository cloud and the web services that facilitate scene creation, such as the services for scene description and retrieval. Considering parameters such as compliance to standards, technology excellence, development efficiency and organization we have chosen to use and extend the well-known X3D-Edit platform, already mentioned in section II that describes related work in the field. More specifically, X3D-Edit is released by the Web3D forum and is thus compliant with the spirit and the principle of the project. It is based entirely on JAVA and specifically on the NetBeans development platform, an important design characteristic since it offers all the benefits of the known window environment relieving the development process from the respective load while allowing efficient extension mechanisms through the creation of autonomous modules and plugins. X3D-Edit itself is modular by design since its codebase groups its functionality into distinct modules that group its functionality; KMLEditor, JOMWrapper, LookAndFeel, AppConfiguration, can be easily identified in terms of functionality and also in the code base, version 3.3, besides others. The tool is thus very easy to extend and also to compile since it exploits the build system of the NetBeans platform.

Fig. 2 presents the authoring tool extension of X3D-Edit that serves the necessary functionality for iPromotion. It is integrated within X3D-Edit as an autonomous plugin that exposes the functionality for iPromotion. The plugin implements a window viewpoint labeled "IPROMOTION" and is constructed as a separate NetBeans module, shown at the right in the figure. The panel on the left includes the port of the X3D viewer for JAVA. The main window of the tool is the X3D editor module used to compose X3D scenes and then load them in the viewer. The iPromotion panel presents a matrix containing all the scenes available from the cloud. The latter are selectable visual component so that they can be loaded. Users may also apply simple name matching in order to reduce the number of presented scenes using the input box bellow the scene matrix. Thus, after the wanted scene is located they are selected, fed in the viewer and then positioned as users will require. The new scene can be saved in the cloud transparently and can be loaded again for further processing. Each time a scene is saved, MPEG-7 indexing is transparently produced since the final scene output is submitted to the respective web service.

The main challenge behind the implementation of the aforementioned functionality is found in the inherent representation of scenes by X3D rendering modules, viewers. Such modules typically parse X3D input and do not discriminate among autonomous, integral 3D models that typically represent a single thematic object, which would be a very useful characteristic for our system as well as other ones. We have considered the alternatives for overcoming this problem; an elegant solution would be to construct a small framework that would be able to hold information regarding the state of the viewer that related to individual models being loaded even if there is no such notion in Xj3D rendering. The <x3d> tag is unique in every model and so is the <scene> tag. If the Xj3D viewer is to produce the outcome of model concatenation, they must first dispose the aforementioned tags, be processed and only then be fed to the viewer. This functionality can be implemented using both XML processing and the SAI API that supports methods for appending scenes in the one already presented by the viewer. In any case, the viewer would not be able to discriminate among the individual models it renders. Unless this functionality is included in the standard, the viewer can be set to expose information about each inserted model implicitly, event if it does not explicitly support such functionality. The solution we employed relied on visual controls in the form of basic shapes that have been inserted into the scene in order to provide feedback to the XJ3D runtime environment. Fig. 2 illustrates the visual dimension of the this mechanism; the cube, sphere and cone shapes behind the two scenes in the viewer, the car and the marine, are sensor nodes connected with the position, rotation and scale attributes of each scene with the aid of route elements in the first two cases and with the help of appropriate Javascript scripts in the latter. Thus, using standard X3D mechanisms, scenes are manageable and any environment can be built by adding more individual objects. This mechanism includes a number of details, the most important of which have to do with scene and sensor naming, listening to events, correct sensor positioning and then keeping the state of the viewer. Identifies must be named uniquely within scenes that is, all element definitions, DEFs, that relate to scene reference must be changed in real time. An efficient convention is to assign them their original names in addition to a suffix composed by the name of the scene and a number that counts how many instances of the scene have been loaded. This convention allows the authoring to maintain a structure, indexed by the unique scene instance names, holding the numbers that correspond to position, scale and rotation translations as returned by event listener mechanisms. Note however that, since the scene name is used as a key value to recover content it uses such as textures, images, etc, from the cloud, it must be maintained. Fig. 3 presents the implementation of the controls using X3D route elements that redirect numerical values from the sensors to the models they are to control using translation elements. As illustrated in the figure, the mechanism is implemented by means of X3D objects added in the scene. The names of Nodes correspond to the unique scene names suffixed with a string that identifies the sensor. Fig. 3 illustrates only the suffix for sensor node names however the implementation assigns names according to the aforementioned rule in order to implement control for any model. For the transformation of the size we employed the code illustrated in the script in order to convert vertical motion on the cone to scale information. Correct positioning of the sensors includes the computation of the maximum dimension of the scene in the Y-axis as measured from the center of the scene and the initialization of the position parameters for the sensors. This logic includes the identification of dimensions for large shapes and top level drawable objects and the transformations causing them to move, rotate, displace, etc.. The maximum dimension of the scene will define the position of the sensors so that they are visible and not hidden within space occupied by part of the object.

Fig. 4 presents parts of the output file that is stored in the cloud when a scene is produced. The format we have selected in order to incorporate composite multimedia objects based on 3D graphics is the MPEG-4 eXtensible MPEG-4 Textual format (XMT). XMT is XML-based described in MPEG-4 part 11 regarding scene description and application engine [14]. It allows us to describe the structure of multimedia scenes, by incorporating large parts of SMIL, SVG and X3D. Especially with respect to X3D, XMT uses an almost identical set of commands, making the two formats highly compatible. Besides representing simple or composite scenes, however, XMT also allows for the incorporation of more complex multimedia features, such as MPEG-7 descriptions and MPEG-21 rights management, all in the same structure. The top part of fig. 4 presents how scenes are described in terms of a world. The <es_descriptor> tag contains among other, the <streamsource> tag that defines the name of the scene which is also its index within the iPromotion cloud. Each <es_descriptor> defines a distinct scene that needs to be added in the viewer. The body of the XMT file groups all the scenes within the world within the <children> tag and keeps their position, rotation and scale as shown in the lower art of fig.4. This formalization allows scenes to be reused since complex scenes refer to simple, basic models and need not store additional X3D information. Models, that have not been created by combination of other 3D objects, do not need this kind of representation. For them the XMT information is absent from the cloud and it is never created; this is a means for separating composite and primitive scenes.

The mechanisms described in the previous paragraph are sufficient to allow the implementation of high level use cases that hide internal operations. Scenes are accessed through the iPromotion cloud API, and their manipulation relies on the SAI API. As mentioned in the previous paragraph, the implementation could rely on XML manipulation. However is an ISO standard and so is its binding to the JAVA language. We evaluated both approaches and empirically found that the former performed much slower than using the SAI API. Supplemental services, such as the creation and validation of indexing MPEG-7 information for the cloud, use relevant web services exposed by the middleware. Thus, users only deal with scene processing either graphical, or even textual through the X3D editor panel. Moreover, the tool exploits the editing characteristics X3D-Edit which supports all the X3D visual components. Another interesting characteristic is that the authoring tool is designed to allow the addition of textual meta-tags in the final scene. This information is incorporated into the MPEG-7 description and can thus be used for applying detailed search functionality. The searchable pieces of information will not be confined to textual meta-information; it will also include content descriptors for color, shape, texture and animation. As both the X3D and MPEG-7 standards are based on XML, the extraction of content description from scenes can be achieved through with XML-to-XML transformation.

IV. Conclusions And Future Work

The previous paragraphs describe a scene authoring and management tool designed to cover the needs of iPromotion, our virtual reality advertising platform. The primal use case that this software covers exceeds the basic scene creation and editing; the functionality we aim to deliver aims to provide o higher layer in 3D authoring by focusing on the concatenation of completed, autonomous models and/or scenes in order to create manageable 3D worlds. We have extended a valuable scene composition tool in the field, X3D-Edit, that relies on the Netbeans platform to offer scene manipulation functionality in moduls. We thus added a new window in the top level view of the tool that integrates iPromotion functionality transparently, more specifically, we employed the programming interfaces for cloud access and scene indexing. The former required the support of proper representations for the new types of scenes and the latter required the construction and communication with appropriate web services that extract indexing information. The core of the system however is the scene manipulation logic that keeps the state of the worlds constructed by users, that is the identification of individual scenes their normalization so that they can be fed into the XJ3D browser. Thus, a scene can be added multiple times; information about each instance is kept and all of them will be capable to access any resources originally available to them such as textures, images, audio, etc.. Scenes are controlled by visual objects properly positioned below them. These controls are accompanied with X3D and Javascript code that transforms their events to scene positioning, rotation and scale. The described work is currently in alpha stage that is, it is in progress. Future work will focus on all aspects of scene authoring. Since XJ3D browsers currently support access to online resources, scene elements using the cloud interface can be limited only to the necessary extend. The tool is following the evolution of the iPromotion VR platform and is expected to be tested against more elaborate models representing 3D environments. It will be evaluated for performance and usability.

Acknowledgments

The work described in this publication has been funded by the Programme “Cooperation” of the Operational Programme “Competitiveness and Enterpreneuship” (OPCE II) of the

Ministry of Development and Competitiveness, partly funded by the European Regional Development Fund, 2007-2013.

REFERENCES

[1] David Mazursky and Gideon Vinitzky, Modifying consumer search processes, ISSN 1860-2037 in enhanced on-line interfaces, Journal of Business Research 58 (2005), no. 10, 1299 – 1309.

[2] Kil-Soo Suh and Young Eun Lee, The effects of virtual reality on consumer learning: an empirical investigation, MIS Quarterly 29 (2005), no. 4, 673–697.

[3] Serge Baile, Sana Debbabi, and Mohamed Daassi, Effect of online 3D advertising on consumer responses: the mediating role of telepresence, Journal of Marketing Management (2010), no. 2, 1472–1476.

[4] Bangalore S. Manjunath, Phillipe Salembier, and Thomas Sikora, Introduction to MPEG-7: multimedia content description interface, John Wiley & Sons, Inc., New York, NY, USA, 2002.

[5] Ian S. Burnett, Fernando Pereira, Rik van de Walle, and Rob Koenen, The MPEG-21 book, John Wiley & Sons, the remote renderer: for higher quality scenes, to be 2006.

[6] WORLD WIDE WEB CONSORTIUM – W3C (2001), Scalable Vector Graphics (SVG) 1.0 Specification, http://www.w3.org/TR/SVG10/, (Accessed 19 May 2010)

[7] WEB3D CONSORTIUM (2004), Extensible 3D (X3D) ISO/IEC http://www.web3d.org/x3d/specifications/#x3d-spec, (Accessed 19 May 2010)

[8] P. Spala, A. G. Malamos, A. D. Doulamis, and G. Mamakis, "Extending MPEG-7 for efficient annotation of complex web 3D scenes," Multimedia Tools Appl, vol. 59, no. 2, pp. 463–504, 2012.

[9] M. Zampoglou, P. Spala, K. Kontakis, A. G. Malamos, and J. A. Ware, "Direct mapping of x3d scenes to mpeg-7 descriptions," in Proceedings of the 18th International Conference on 3D Web Technology June 20-22, 2013 San Sebastian, Spain, accepted for publication.

[10] F. Lamberti and A. Sanna, "A streaming-based solution for remote visualization of 3D graphics on mobile devices," IEEE Transactions on Visualization and Computer Graphics, vol. 13, no. 2, pp. 247–260, 2007.

[11] T. K. Capin, K. Pulli, and T. Akenine-M¨ller, "The state of the art in mobile graphics research," IEEE, Computer Graphics and Applications, vol. 28, no. 4, pp. 74–84, 2008.

[12] G. Kousiouris, G. Vafiadis, and T. A. Varvarigou, "A front-end, hadoop-based data management service for efficient federated clouds," in Cloud Computing Technology and Science (CloudCom), 2011 IEEE Third International Conference on, pp. 511–516, 2011.

[13] Markos Zampoglou, Athanasios G. Malamos, Kostas Kapetanakis, Konstantinos Kontakis, Emmanuel Sardis, George Vafiadis, Vrettos Moulos and Anastasios Doulamis. "iPromotion: A Cloud-Based Platform for Virtual Reality Internet Advertising", Big Data and Internet of Things: A Roadmap for Smart Environments, Studies in Computational Intelligence 546, DOI: 10.1007/978-3-319-05029-4_19, Springer International Publishing Switzerland 2014

[14] Fernando Pereira and Touradj Ebrahimi. The MPEG-4 book. Prentice Hall Professional, 2002.

Michael Kalochristianakis is currently an associate lecturer and researcher at the Technological Educational Institution of Crete (TEIC), at the Department of Informatics Engineering. In the past I have been an associate lecturer for the Dpt of Applied Science at TEIC, for the Dpt of Production and Engineering Management at the Technical University of Crete and for the Academic Library at the University of Crete. He has also worked for the Research Academic Computer Technology Institute and Press "Diophantus" as an IT researcher/engineer and for the software industry as a J2EE programmer. Michael holds a Doctorate degree in Computer Engineering and Informatics, a Masters degree in Computer Science and a Diploma in Electrical Engineering and Computer Technology. Pieces of my work have been published in international peer-reviewed journals and have been presented at international conferences. He is also a reviewer for several academic journals.

Markos Zampoglou was born in 1981 in Thessaloniki, Greece. He graduated from the dept. of Applied Informatics of the University of Macedonia, Greece, in 2004. He received an MSc in Artificial Intelligence from the University of Edinburgh in 2005 with a Distinction. He was awarded a Doctorate degree from the dept. of Applied Informatics, University of Macedonia, Greece in 2011. Markos is currently an associate lecturer and researcher at the Technological Educational Institution of Crete (TEIC), at the Department of Informatics Engineering.

Kostas Kontakis is a graduate student at the Technological Educational Institution of Crete (TEIC), Department of Informatics Engineering pursuing his MSC. His primary interests include semantic web and ontologies. He has been with the Multimedia Content Lab for two years. Kostas holds a BS degree from the Department of Informatics Engineering.

Kostas Kapetanakis is an experienced systems administrator and programmer working for the Multimedia Lab. He is also graduate student at the Technological Educational Institution of Crete (TEIC), Department of Informatics Engineering. He is currently pursuing his MSC degree. Kostas holds a BS degree from the Department of Informatics Engineering.

Athanasios G. Malamos received a BSC degree in Physics from the University of Crete (1992) and a PhD from the Technical University of Crete in 2000. From 1997 to 2002 he was a research assistant and a researcher in the ICCS National Technical University of Athens. Since 2002 is with the Technological Educational Institute of Crete, at the Department of Informatics Engineering, as an Assistant Professor (2002-2006) and as an Associate Professor (2006 until present). Prof. Malamos leads the Multimedia Lab. He has been running EU and National funded research projects. He has served as program committee and reviewer for several international conferences and workshops. He is a reviewer for IEEE, Springer as well as for international journals that fall within his interests and he is a member of the WEB3D consortium. His research interests include multimedia services, virtual reality, 3D modeling and multimedia semantics.

A Fault-Tolerant Mobile Computing Model Based On Scalable Replica

Meenakshi Sati[1],Vivek Vikash[2], Vishwanath Bijalwan[3] , Pinki Kumari[4],Manish Raj[5], Meenu Balodhi[6], Priya Gairola[7], Vijay Bhaskar Semwal[8]

[1,2,3] NITTR Chandigrah
[4] Bansatali Vidya Peeth rajstan
[5]IIIT Allahabad
[6,7] Uttarakhand Techincal Univesity
[8] Siemens Information System, Banglore

Abstract — **The most frequent challenge faced by mobile user is stay connected with online data, while disconnected or poorly connected store the replica of critical data. Nomadic users require replication to store copies of critical data on their mobile machines. Existing replication services do not provide all classes of mobile users with the capabilities they require, which include: the ability for direct synchronization between any two replicas, support for large numbers of replicas, and detailed control over what files reside on their local (mobile) replica. Existing peer-to-peer solutions would enable direct communication, but suffers from dramatic scaling problems in the number of replicas, limiting the number of overall users and impacting performance. Roam is a replication system designed to satisfy the requirements of the mobile user. Roam is based on the Ward Model, replication architecture for mobile environments. Using the Ward Model and new distributed algorithms, Roam provides a scalable replication solution for the mobile user. We describe the motivation, design, and implementation of Roam and report its performance. Replication is extremely important in mobile environments because nomadic users require local copies of important data.**

Keywords — **Fault-tolerant, Mobile Computing, Roam, Data Replictaion.**

I. Introduction

REPLICATION in mobile environments requires fundamentally different solutions than those previously proposed, because nomadicity presents a fundamentally new and different computing paradigm. Before mobile computing was a feasible reality, software designers simply didn't design their systems to be mobile-enabled. Computers were largely stationary, so there was no need to consider what would happen if it were easy for people to move around geographically. Consider the case of mobile-IP. When the Internet Protocol (IP) was developed, mobility was not feasible, and therefore was not considered in the design. Now that mobility *is* a reality, many researchers are trying to fix or redesign IP to support real mobility. Like the mobile-IP case, the replication system software must also be redesigned to account for mobile computing. *Roam* is built using the Ward Model, a new replication architecture that has been designed especially with mobility in mind, though of course it applies equally well to stationary environments [1]. The Ward Model provides a new replication paradigm that is neither strictly peer nor client-server; rather, it is a hybrid model of the two that allows everyone to be effectively a peer while maintaining good scalability in the number of replicas. Additionally, *Roam* provides a number of new distributed algorithms. For instance, *Roam* contains new and improved distributed algorithms for garbage collection and for dynamic management of the version vector, the main data structure behind most optimistic replication systems.

In this paper we present an approach for data replication that considers the mobility of clients: The data sets which have to be replicated to mobile clients depend often on dynamic parameters like location of the client or time. Therefore, we introduce a generic model for specifying fragments of the global database with respect to such parameters. The increasing availability of mobile devices and wireless communication technologies opens new applications which can improve services by providing real-time access to remote information sources. Examples of such applications are not only stock tickers or messenger applications but also mobile hospital information systems and car navigation systems. However, due to the mobile and wireless nature of these devices, several limitations have to be taken into account in developing and running such mobile information systems:

(1) The structure of the network as well as the availability of mobile nodes are subject to continuous changes.

(2) The bandwidth of the communication channels is – compared to traditional wired networks – relatively low and communication is much more expensive.

(3) The resources of mobile devices, e.g. memory and computing power, are strictly limited. For a mobile information system, the first two properties prohibit data access by only remote querying.

Thus, the mobile device should keep a copy of the data that is the subject of the current operations. The third issue restricts the data managed locally on the mobile device to a subset of the overall (remote) database. This raises the need for *semantic replication*: data that is relevant in a certain situation has to be transferred to the mobile device and in case of local updates the modifications are propagated back to the main database. A main issue in this context is the reduction of the replicated data in order to meet the requirements of minimal resource consumption (storage space, transfer time or volume). A possible application is an advanced car driver information system. In addition to street maps stored on a local disk the mobile system retrieves further information about local (i.e., relevant to the current location and time) traffic spots, accidents, detours etc. from a federated database integrating these information from various sources and displays it to the user or uses it for other operations, e.g. route planning. When this information becomes irrelevant, it can be removed from the local database. This scenario could be extended to support updates – a driver reports an accident or the system detects a traffic jam automatically – as well as other kinds of parameters, for example the kind of car (truck, automobile). In this paper we present a replication approach which is based on semantic extensions of the global view. Using this approach, location-based services and services requiring replication depending on dynamic features can be easily (on the global view level) implemented. It is embedded into the context of mobility. While looking at related work we cannot consider all requirements caused by the special characteristics of mobile computing. Therefore, we focus on work related to replication in mobile information systems.

Here we present a replication system designed to satisfy the requirements of the mobile user. Roam is based on the Ward Model, a replication architecture specifically aimed at mobile environments. Together with the Ward Model and new, distributed algorithms such as improved garbage collection techniques and version vector management, Roam provides a scalable replication solution for the mobile user. We describe not only the design and motivation of such a system, but its implementation and performance as well.

Roam addresses scalability with a three-pronged attack. The Ward Model addresses issues of replica management, consistency topologies, and update distribution. Dynamic algorithms and mechanisms handle the scalability of the versioning information, required for consistency maintenance. Finally, we address consistency itself with new algorithms and mobile-friendly semantics. In summary, Roam is a comprehensive replication system for mobile and non-mobile users alike. With it, users can truly compute while mobile, paving the way for both improved user productivity and new and unseen research along mobile computing avenues. The particular point in the solution space is one which we believe addresses a real problem and has wide applicability, not just for replicating files in a mobile context but more generally for a wide range of replication-related problems, including but not limited to military cases, software development scenarios, distributed database problems like airline reservation systems, and general-purpose distributed computing.

The rest of paper is organized as follows: section 2 reviews the related work; section 3 discuss the basic ward model; section 4 presents the advanced ward model; section 5 shows the architectural design; section 6 presents the results; and section 7 conclude the paper.

II. A Brief Review of Related Work

Computing in the 1970s and 1980s meant using almost completely stationary machines. Designed to be situated in one location and rarely if ever moved, these machines were heavy, awkward, and clumsy. The resulting software architectures were adequate for the time, because these machines were too clumsy to move on any regular basis. Even the emerging "portable" machines at the end of the 1980s were not really portable; they could at best be described as \luggable." However, by the 1990s the hardware industry had made rapid progress in chip and LCD technology, as well as in general miniaturization. Machines emerged that were truly portable and as powerful as their stationary cousins. Making use of this new type of machine with its attendant portability, users became increasingly mobile. In 1996, approximately one third of the computers sold were mobile-enabled: that is, portable form-factor with communications capability.

Coda [2] is an optimistically replicated file system primarily using the client-server model. Coda provides replication flexibility akin to selective replication at the clients, but not at the replicated servers. The servers run a form of peer replication. Coda clients cannot directly inter-communicate due to restrictions of the client-server model. Use of this model dramatically simplifies the consistency algorithms, at the cost of limiting the system's utility for mobility. Coda is clearly superior in the low-bandwidth scenario, having greatly optimized communications and synchronization, especially in environments with weak connectivity [3]. Some of the same ideas could be applied in Roam, though additional research would be required to incorporate them into a peer model.

The Little Work project [4] is similar to Coda, but modifies only the clients, leaving the AFS [5] servers unaltered. Congestion caused by clients' slow links is reduced in a variety of ways, including client-side modifications of AFS, Little Work's underlying RPC, and other congestion avoidance and control methods. However, clients cannot directly communicate, hindering the usability of the system in dynamic, mobile environments.

The Bayou system [6] replicates databases rather than file system objects. Like Roam, it uses the peer-to-peer model. Unlike Roam, Bayou does not attempt to provide transparent conflict detection. Applications must specify a condition that determines when a conflicting access has been made, and must specify the particular resolution process.

Ficus [7] is one of the intellectual predecessors of Roam, and Roam therefore shares many of its characteristics. Both are based on a peer model, though Roam's Ward Model scales better than the Ficus traditional peer model. Both provide selective replication control. While each maintains consistency with a periodic reconciliation process, Ficus also uses best-effort propagation at update time. Ficus is aimed at a distributed Internet environment, and works well for its target. However, it is unsuitable for mobile use, and does not scale well.

Rumor [8] is the direct predecessor of Roam; much of Roam's implementation is directly based on modified Rumor code. Rumor is in turn a descendent of Ficus, and shares many of its characteristics and problems. It is based on the traditional peer model, and relies upon periodic reconciliation to maintain consistency. While Rumor is better suited to a mobile environment than Ficus, its scaling properties are substantially the same.

An example of remote computing is illustrated in [9] applied to the geometry over internet. Resources such as executables, languages, packages, can be used from a remote computing system. They implemented a distributed system using geometry that outsources the computing-intensive tasks to remote servers that may be located in other universities or companies, linked to grids and clusters and so on. The architecture developed stresses the interoperability of the software, and a suitable high degree of decoupling between components hosted in various locations.

The definition of a level oriented business process methodology is enhanced in [10], which encourages the adaptation of the modeling notation to the modeling and technical knowledge shown by the expert. Their approach reduces the complexity found by domain experts and enables them to model their processes completely with a level of technical detail directly proportional to their knowledge. They also generate the codes on mobile plateform.

A simple client-server system architecture and algorithms is introduced in [11] for ubiquitous live video and VOD service support. The main features of the system are: efficient usage of network resources, emphasis on user personalization, and ease of implementation. The system supports many continuous service requirements such as QoS provision, user mobility between networks and between different communication devices, and simultaneous usage of a device by a number of users.

A framework for computation offloading is proposed in [12] for mobile cloud computing. Offloading of computationally intensive application parts from the mobile platform into a remote cloud infrastructure or nearby idle computers addresses this problem. They presented Mobile Augmentation Cloud Services (MACS) middleware which enables adaptive extension of Android application execution from a mobile client into the cloud.

Clustering and regression based technique is introduced in [13] for analyzing anonymized cellular network data to identify generally important locations, and to discern semantically meaningful locations such as home and work. They test it on arbitrary cellphone users, including those with low call rates.

Traditional client-server architectures are not capable of supporting the desired infrastructure and providing the required communication capabilities among replicas. While such functionality could have potentially been added to the client-server model, we expected that modifying a fundamentally client-server design to incorporate a rich communications structure would require such basic structural changes as to imply effectively starting from scratch. The goal is therefore to design a peer-based model that scales well. Replication services based on the traditional peer model, such as Rumar [8], AFS [5], and Ficus [14], all suffer from scaling problems. In response, we have designed the Ward Model. It provides a new and different form of the peer model, one based on a hybrid between client-server and peer solutions that clusters replicas into groups without affecting the underlying any-to-any communication capability between all system participants. Ward based model can also be integrated in other domains such as in fruit diseases [15], human activity recognition [16], PCA approaches [17], 3D model [18], semantic information [19], wireless sensor network [20, 21] and learning techniques [22].

III. Basic Ward Model

We describe the basic Ward Model by defining wards, and ward sets. We then explain how the model maintains consistency and how the model supports mobility.

The key idea behind wards is to group volume replicas into containers that capture the notion of common or typical communication partners. For example, given four replicas in Los Angeles and four replicas in New York, the system would perform poorly if each replica in Los Angeles typically synchronized with a replica in New York. While the topology produces correct results, each machine pays the additional cost of long- distance communication, in terms of latency, efficiency, and price per byte. Additionally, at synchronization time, there is a greater chance of a failed network connection between Los Angeles and New York than between two local partners in either location, given that generally the long distance communication depends on having the local communication operational. The obvious improved approach is to have one candidate in Los Angeles communicate with one candidate in New York; these two candidates afterwards disseminate the information among their local colleagues.

Given that synchronization should typically occur with a local partner, both for economic and efficiency reasons, we would like to group replicas together to capture the notion of synchronization locality. We therefore build wards as a collection of “nearby" volume replicas, the details of which are discussed below. The ward members are only required to be loosely connected-continual, high-quality connectivity is not necessary.

Ward set

The ward set refers to the set of replicated data stored within the ward. In the basic Ward Model, the ward set is, by definition, equivalent to the entire volume, although the definition changes in the advanced model. Like the volume itself, the ward set is dynamic in character; it changes as the volume itself changes in response to new object creations and existing object deletions. It is the ward master's responsibility to synchronize its ward set with the ward sets from other wards. In the basic Ward Model, ward-set synchronization can be accomplished simply by contacting one other ward master: since all wards store the entire volume, all ward sets are identical. In the advanced model a more complex synchronization architecture will be utilized.

Maintenance of consistency

Consistency is maintained simultaneously both within each ward and among wards. In both intra- and inter-ward scenarios, the consistency topology refers to the communication pattern used between replicas. All of our consistency algorithms are topology-independent. For example, a quadratic message complexity results from an all-pairs reconciliation topology, but a ring (using a gossip-based transfer of information) [20-25] reduces the message complexity to a linear cost. A superior messaging plan avoids the quadratic cost when inter-site communication is available, but gracefully handles degraded communication. Using two separate topologies within the ward and between wards reduces the generality of the model and increases the requirement for special-case code. Although one could potentially identify hypothetical reasons why two topologies would be required, a topology that applies equally well to both scenarios is clearly a better choice for generality and simplicity. The adaptive ring allows rarely or never-communicating sites to share data by relying on third-party replicas to gossip on their behalf. For these reasons, it is an attractive topology to consider, and one that applies equally well both within and between wards.[25-30]

For instance, assume the ward contains replicas 2, 4, and 6. If replica 5 joins but only replica 4 is knowledgeable of 5's existence, then replica 6 will synchronize with replica 4, which 6 believes to be the next replica in the ring. However, as part of 6's synchronization it learns of 5's existence, so the ring automatically "heals" itself as the new information propagates. Since the underlying algorithms are topology independent, correctness is not affected by temporary spokes on the adaptive ring. The example is illustrated graphically in Figure 1. (Ward is used generally here, and also refers to the meta-ward that contains all ward masters.). Ward masters are illustrated as solid black replicas; wards are double circles surrounding the replicas. Arrows indicate synchronization paths, although the synchronization topology adapts itself to network topology.

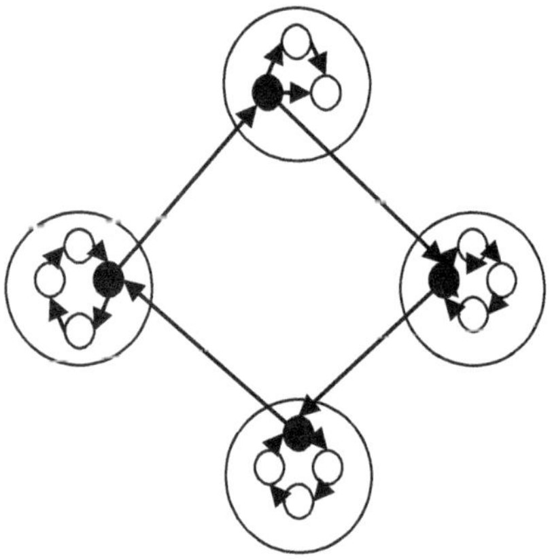

Fig. 1. The basic adaptive ring topology, both within and between wards.

Support for mobility

The model provides support for intra-ward mobility essentially for free, since everyone in the ward is a peer in the traditional sense. Intra-ward mobility occurs anytime the user is mobile within a restricted geographic area and is therefore only likely to encounter other machines from the same ward. For instance, moving within one's once from the desk to the couch demonstrates an intra-ward mobile action, as does moving around the once or within town to the local coffee shop. In all examples, the degree of motion is large enough to potentially change the set of "best" or most efficient communication partners, but not large enough to bring the user outside of his or her ward. For example, two colleagues in the once may never directly synchronize, typically relying on a set of third-party replicas to relay updates. However, if they meet at someone's home to discuss plans for the following day, they will usually want their machines to directly communicate.

IV. ADVANCED WARD MODEL

In the advanced Ward Model we introduce selective replication, the ability for a ward member to physically store only select portions of the complete volume. The details concerning how the reconciliation algorithms and controls handle selective replication are discussed in the next section. Here we describe the changes selective replication makes to the ward definitions, controls, and constructs.

Wards

The basic definition of a ward remains unchanged. The ward is still a collection of "nearby" volume replicas. However, all replicas need not store the same portions of the volume, which impacts both the ward set and synchronization topologies, as described in the following sections. The characteristics and abilities of ward members similarly remain unchanged. Specifically, any-to-any communication is still enabled between any two ward members. However, due to selective replication, any-to-any communication between all replicas may not always make physical sense for any particular layout of the file objects onto the set of volume replicas.

Ward set

The ward set is defined to be set of replicated data stored within the ward. In the basic Ward Model the ward set is equivalent to the volume. However, selective replication allows each ward member to store select portions of the volume, meaning that the set of data stored at each replica may be smaller than the entire volume. The ward set itself, therefore, may be smaller than volume, though of course it can never be larger. It is equivalent to the union of all replica sets for all replicas in the ward. The ward set changes dynamically as the set of data stored within the ward changes. For instance, if a given ward member uses the selective replication controls to locally add a new file system object, the ward set expands to include this new object. The ward set similarly decreases in size when replicas locally drop file replicas. Additionally, ward motion can change the ward set, as replicas move into and out of a given ward.

Ward masters

The ward master is responsible for the inter-ward synchronization of the entire ward set, and therefore must be able to identify the complete ward set. Use of both selective replication and ward changing, however, can change the ward set. Since any ward member can dynamically and optimistically change its replica set using the selective replication controls, and because new machines can at any time move into the ward (carrying with them their accompanying replica set), the ward set changes dynamically, lazily, and without global coordination. Without selective replication the ward set could be identified simply by naming the volume. With selective replication, however, the best mechanism for the ward master to identify its complete ward set is to individually name all objects in it.

Finally, we would like selective replication to help alleviate disk storage at the ward master. A ward member that wants to store a 100MB file should not necessarily force the ward master to also store that 100MB file solely for inter-ward synchronization purposes. We allow the ward master to store only a virtual replica-the equivalent to naming the object without storing _le data. When data is required for synchronization purposes, the ward master relays it via the physical data site.

Maintenance of consistency

Since each replica in the ward has a potentially different replica set, it follows that each ward master has a potentially different ward set. Therefore, in both intra- and inter-ward synchronization, we must use a more robust topology than a simple adaptive ring. We augment the basic adaptive ring to account for the differences between replica sets and ward sets. When the ward master is viewed as a “super-replica," the ward set appears exactly the same as a replica set for a super-replica, meaning that one approach can again be used in both intra- and inter-ward synchronization. The solution uses an adaptive ring for each file object, rather than one for the whole volume, and then coalesces multiple per-file rings into a single ring based on the intersection between replica.

Support for mobility

Selective replication introduces new difficulties for the ward motion algorithms. In the basic Ward Model, a machine moving into a new ward is guaranteed to have the same replica set as the new ward's ward set, because all participants store full volumes. Since the ward set is equivalent to the moving replica's replica set, it is straightforward for the new machine to integrate with the new ward. With selective replication, the mobile machine's replica set may differ from the new ward's existing ward set. As a result, the advanced Ward Model requires more rich and robust ward motion algorithms.

The ward master is responsible for the inter-ward synchronization of all intra-ward data. When a new machine enters the ward and brings with it data files not in the current ward set, there are two options: either the ward set expands to incorporate the new data objects, or it doesn't. In the former case the ward set changes, possibly causing changes at other ward masters since they must keep track of what is stored at the various ward masters to properly form their adaptive per-file rings. In the latter case the ward set remains unchanged, and there are no ripple effects affecting the other ward masters, but the mobile machine cannot synchronize all of its data completely within the new ward. Some of the data must be synchronized with another ward, most likely the original ward that the mobile machine came from. To properly decide which option is best, we must look at how physical motion actually occurs. Real mobility seems to occur in one of two modes:

visit: a temporary trip to a remote location, measured in hours or a small number of days

long stay: a longer stay at a remote location, perhaps for \a while," perhaps more permanently.

Each mode has accompanying expectations of cost and performance. Users expect a temporary mobile move to be lightweight and inexpensive: since they're not planning on staying very long, they don't want to pay a large up-front cost. Additionally, users will generally accept sub-optimal performance, given that they know the motion is temporary and the up-front cost is minimal. On the other hand, users moving for longer periods of time are generally willing to pay a more expensive up-front cost to gain better performance. Since they know they will remain at the remote location for a long time, they want good performance while there. The up-front cost is amortized over the length of the stay; users staying a short time do not gain significant benefit from the large cost, and are therefore generally unwilling to pay it. Real motion, of course, occurs over a continuous time spectrum, and doesn't always fall exactly into one of the two classifications. However, as a general paradigm the classifications seem to work fairly well, especially since positions in the middle of the spectrum can essentially be placed in either category.

V. System Architecture

As we discussed earlier, we are having several models to achieve services of Replication, but any of them is not able to fulfill our requirements. So we are using a new model, called Ward Model. Ward model is a hybrid model which combines the features of peer-to-peer and client server model as shown in figure 4. This model provides all the four basics qualities of replication system for mobility. In Ward Model, several new terms are used. These terms are Wards, Ward Member (MC), Ward Master (WM) and Ward set which we discussed already in previous sections. Figure 2 describes the ward model and Figure 3 shows the graphical views of ward models.

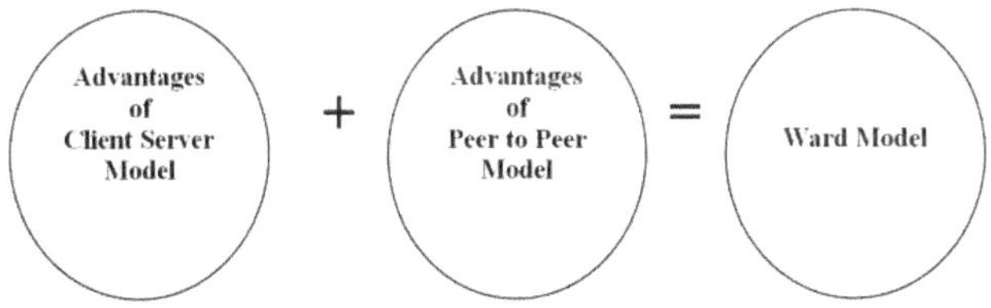

Fig. 2: Ward Model

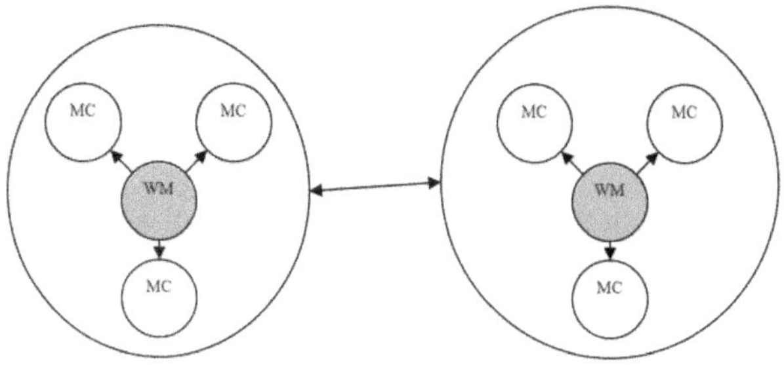

Fig. 3. Graphical view of Ward Model

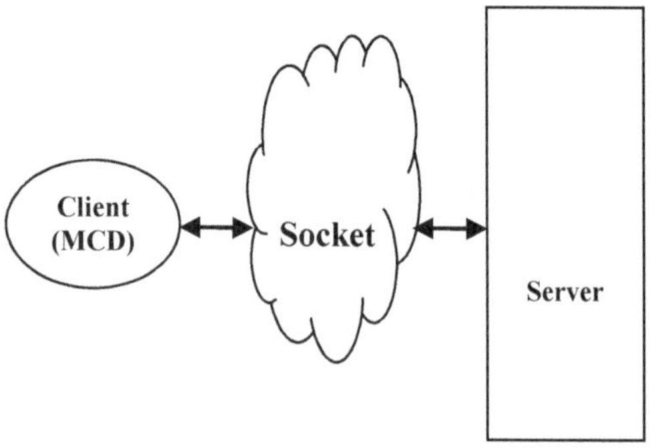

Fig. 4. Client server using socket.

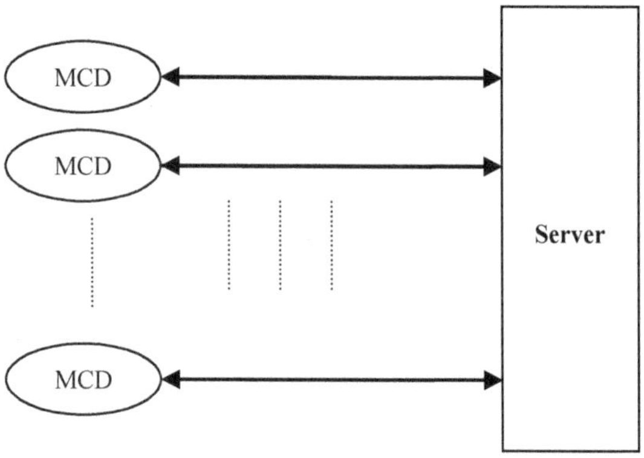

Fig. 5. Multiple Clients connected to Server.

Figure 4-9 shows the block architectures of the various phases of the approach such as, working of client server using socket, how multiple clients are connected to server, the process of client authentication, pinging of clients at server using hash table, receiving file from server and sending file to server.

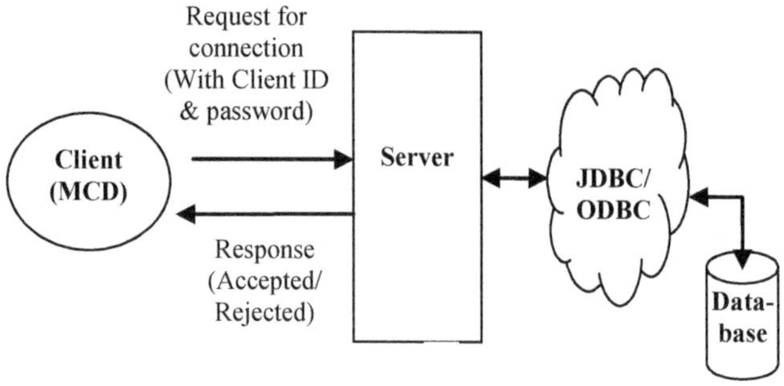

Fig. 6. Client Authentication

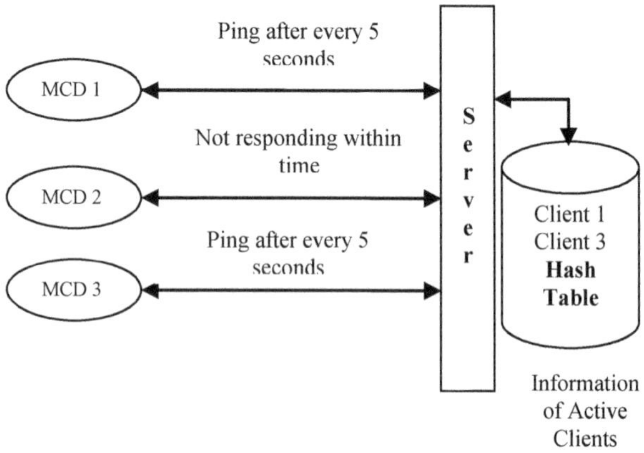

Fig. 7. Pinging of Clients at Server using hash table.

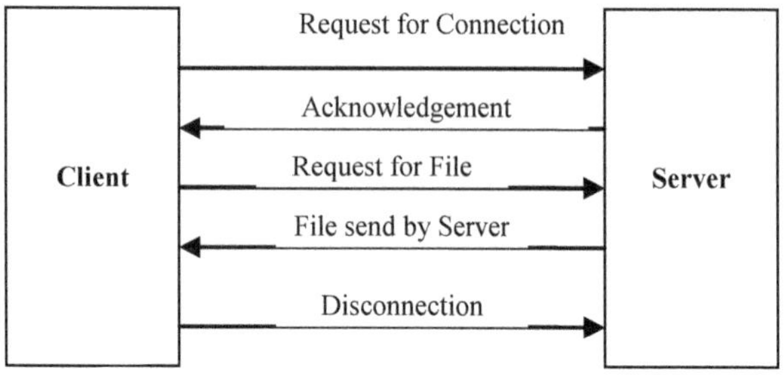

Fig. 8: Receiving file from server.

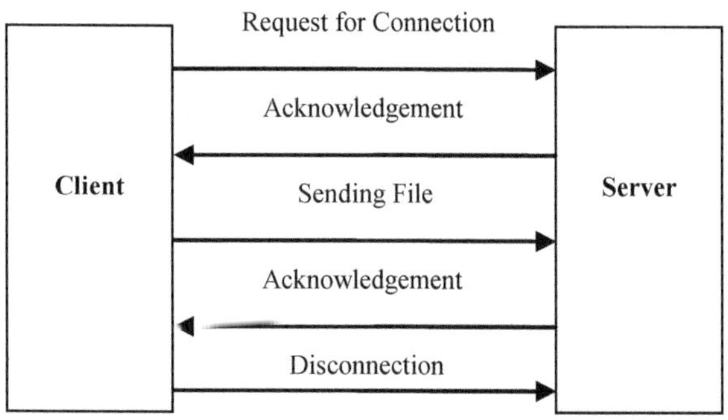

Fig. 9: Sending file to server.

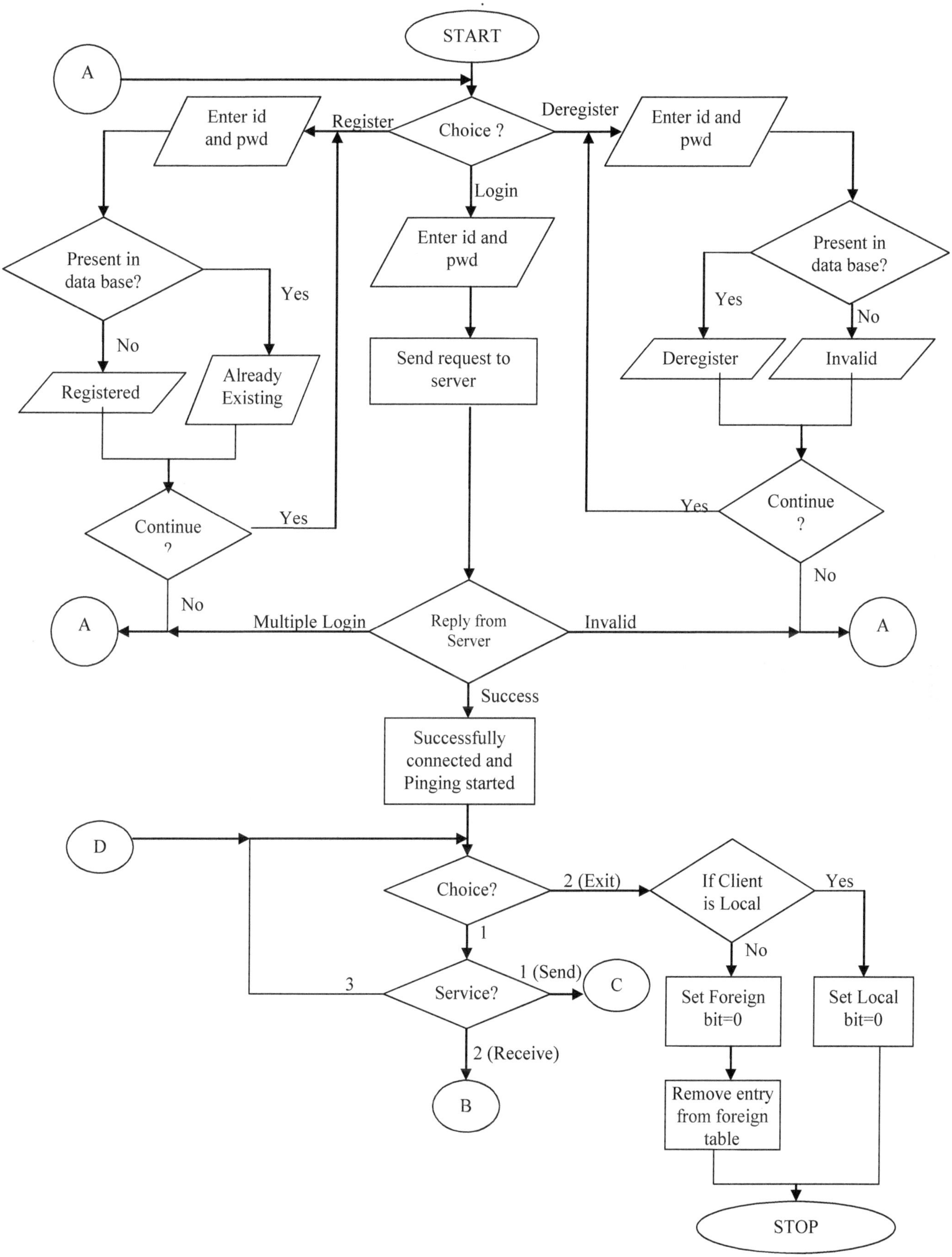
START
A
Register
Choice ?
Deregister
Enter id and pwd
Enter id and pwd
Login
Enter id and pwd
Present in data base?
Yes
No
Registered
Already Existing
Send request to server
Present in data base?
Yes
No
Deregister
Invalid
Continue ?
Yes
No
Continue ?
Yes
No
A
Multiple Login
Reply from Server
Invalid
A
Success
Successfully connected and Pinging started
D
Choice?
2 (Exit)
If Client is Local
Yes
No
1
3
Service?
1 (Send)
C
2 (Receive)
B
Set Foreign bit=0
Set Local bit=0
Remove entry from foreign table
STOP

Fig. 10: Client Side Flow Chart

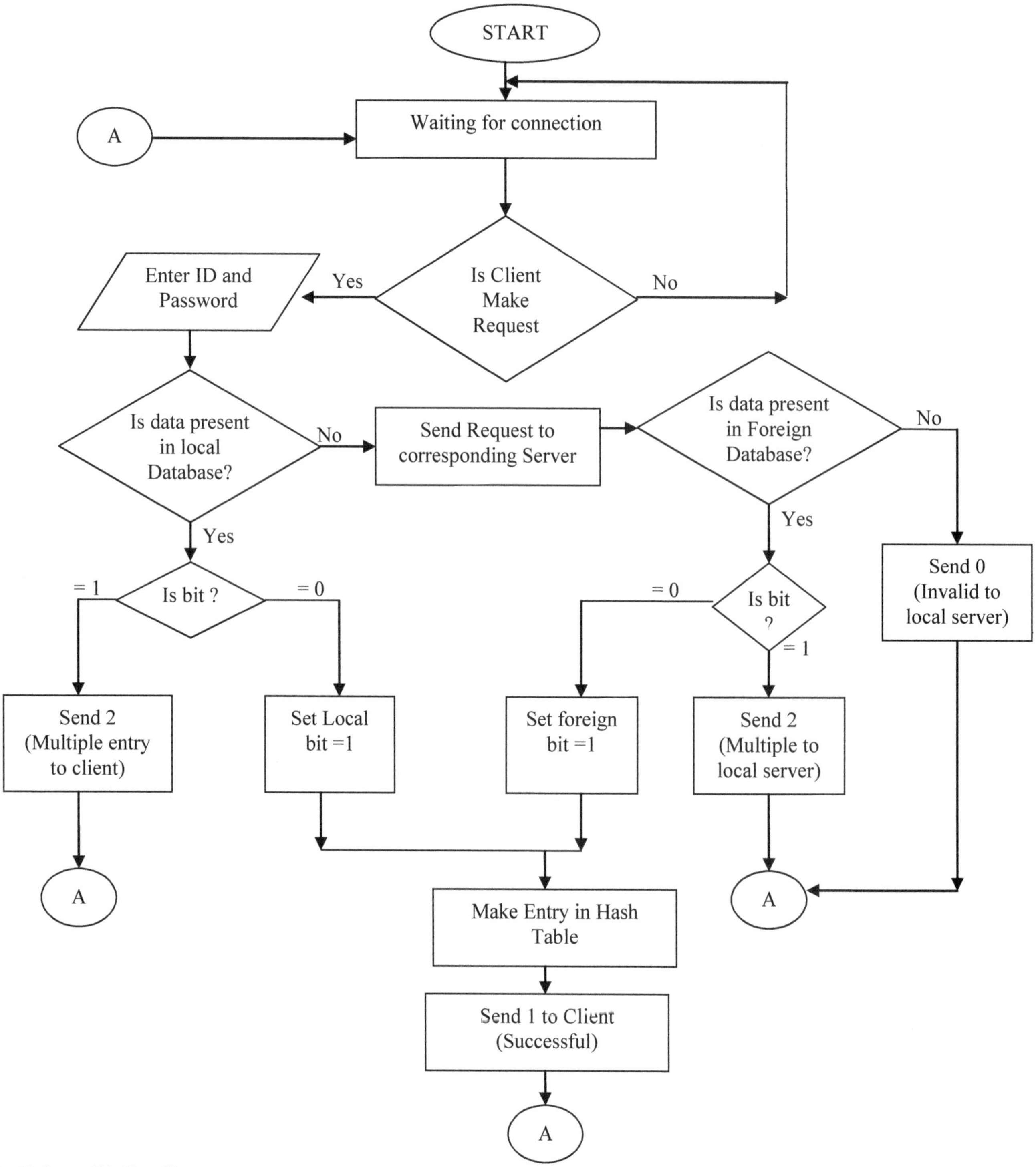

Fig. 11: Server Side Flow Chart

Figure 10 and Figure 11 describe the steps involved in the implementation of client and server side processes respectively with the help of scalable replica concepts while client made a request to the server.

VI. RESULTS

Database

We have used Microsoft- Access as Database. In this we have used following tables.

TABLE I
SERVER

Server ID	Server Name	Server Password
S1	Ser1	Pwd1
S2	Ser2	Pwd2

TABLE 2
SERVER1

Server ID	Server Name
M11	Pwd11
M12	Pwd12

TABLE 3
SERVER2

MCD_ID	MCD_PWD
M21	Pwd21
M22	Pwd22

Experimental Results

Figure 12 to 21 shows the output over command prompt while testing our proposed approach for different scenario. Figure 12 shows the registration of a new client. Client will enter the desired ID and password; if ID is available it will be assigned otherwise prompt for different ID. Figure 13 depicts the scenario of deletion of a registered ID. To delete the ID, password will be required.

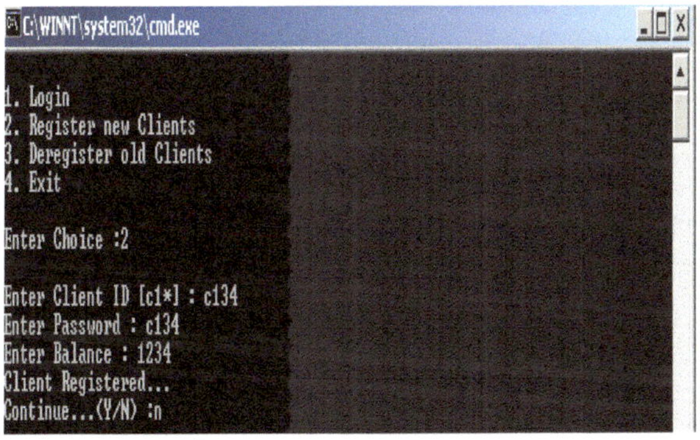

Fig. 12: Register new Clients

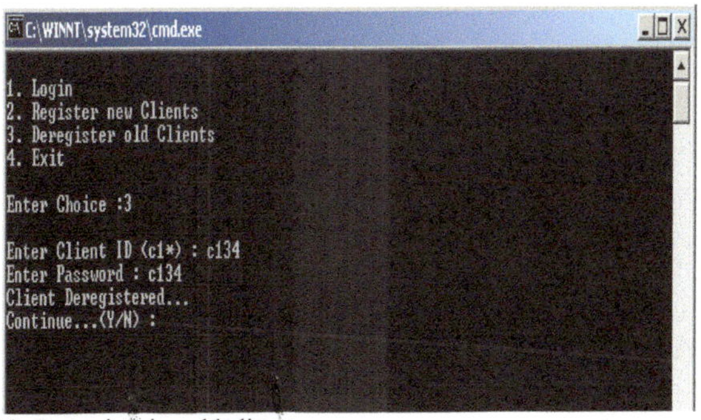

Fig. 13: De-registering old clients

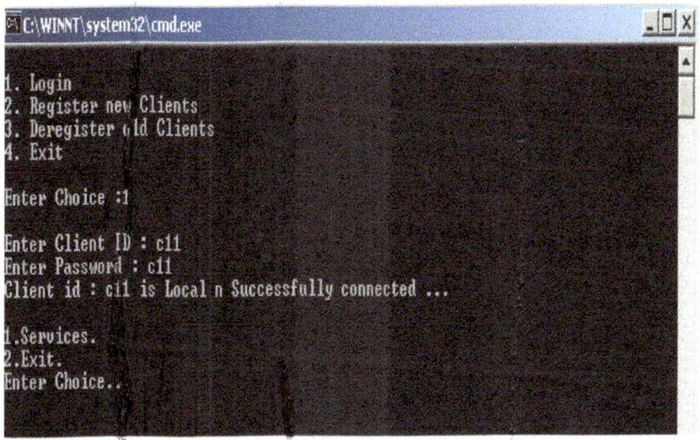

Fig. 14: Login

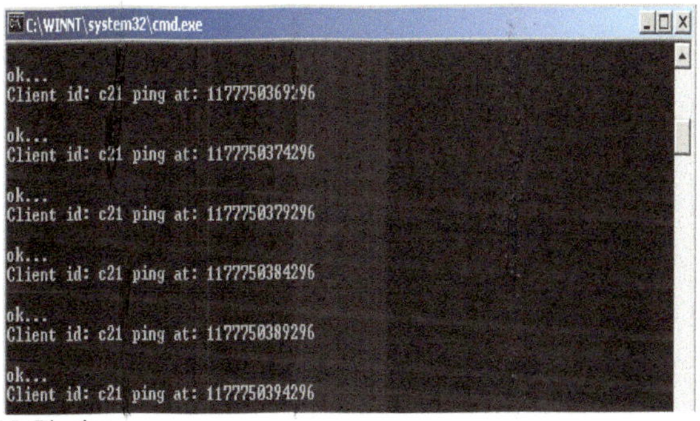

Fig. 15: Pinging

Fig. 16: Server Window

Fig. 17: Services used by clients

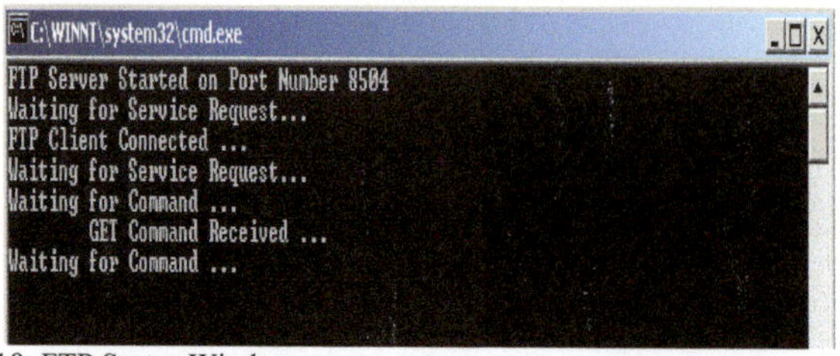

Fig. 18: FTP Server Window

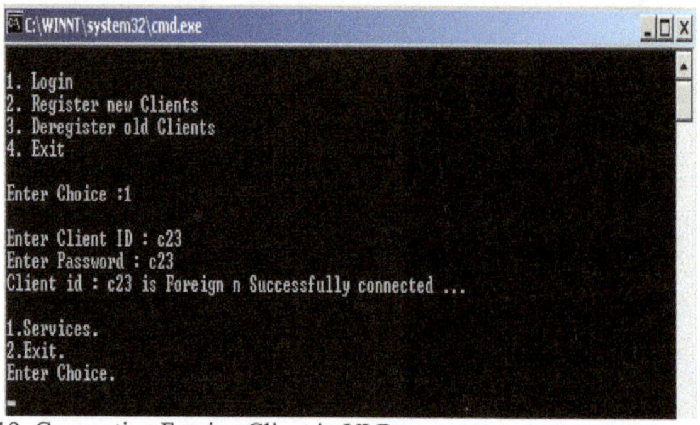

Fig. 19: Connecting Foreign Client in VLR

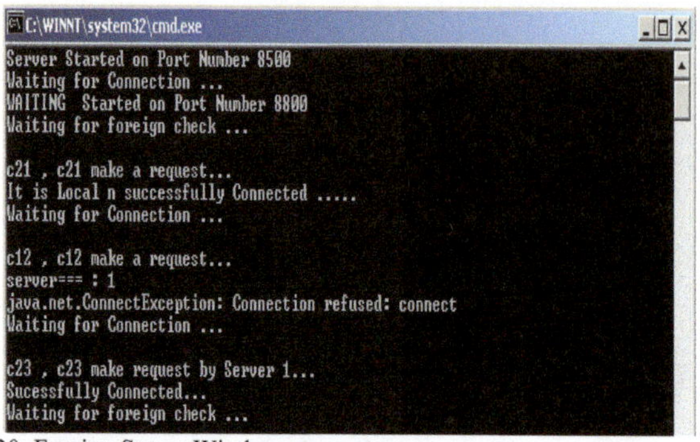

Fig. 20: Foreign Server Window

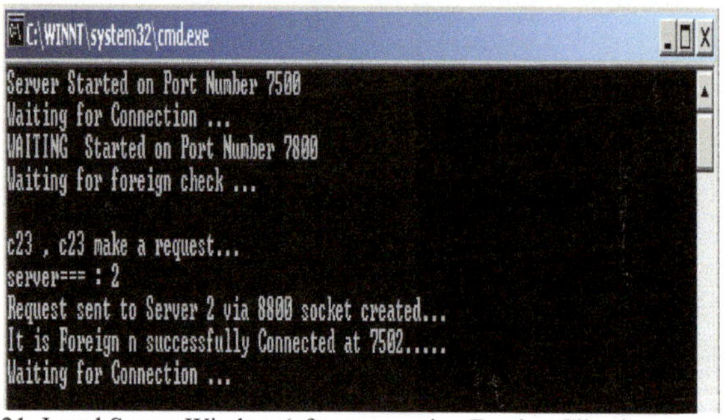

Fig. 21: Local Server Window (after connecting Foreign Client)

Figure 14 shows the steps of login; ID and password is required to login and after successful login some choices will be generated. Figure 15 is the case to test that any client is active or not. If it is pinging means it is active. Figure 16 shows the actions at the server side. At port number 8500, server is waiting for connection and if any request is made then it verifies its credential and after successful verification, client will be connected. Figure 17 shows the commands available for performing services to a client such as sending or receiving a file. Figure 18 shows the activities involved at the server side for file transfer to the clients. The example of connecting a foreign client is depicted in the Figure 19. Figure 20 shows the connection with a foreign server from different clients. Figure 21 shows the status of a local server window after connecting a foreign client. Foreign clients are connected at the port 7502 of local server.

From the experiments, it is deduced that scalable replica facilitates efficient client and server operations in the mobile computing which is purely based on the client and server architecture.

VII. Conclusion

In this paper we proposed a number of significant problems in large scale replication, and produced a scalable replication system for mobile environments. With the arrival of machines capable of supporting truly mobile computing came users wanting to access and update their data while mobile. If the user has more than one machine, for instance a laptop and a desktop, or if the data must be shared between multiple users, then the data must be replicated. Unfortunately, the existing replication systems are not mobile-compliant. Designed for stationary environments, they do not provide users with the abilities they require when mobile. Furthermore, we build into Roam's algorithms the ability to automatically handle and update outdated meta-data as part of the normal synchronization of user data. Roam makes extensive use of the optimistic approach in its management of the system. Roam is a successful piece of research, based not just on the performance analysis but also on real-world experience and use. It is our hope that Roam paves the way for real mobile use and future mobile computing research; simultaneously, we would like the underlying ideas and concepts to bear fruit and diseases and become used in other areas of computer science.

Refrences

[1] D. Ratner, P. Reiher, and G. J. Popek, "Roam: a scalable replication system for mobile computing," in: Proceedings of the 10th International Workshop on Database and Expert Systems, 1999, pp. 96 – 104.

[2] M. Satyanarayanan, J. J. Kistler, P. Kumar, M. E. Okasaki, E. H. Siegel, and D. C. Steere, "Coda: A highly available file system for a distributed work station environment," IEEE Transactions on Computers, vol. 39, no. 4, pp. 447-459, 1990.

[3] J. J. Kistler and M. Satyanarayanan, "Disconnected operation in the Coda file system," ACM Transactions on Computer Systems, vol. 10, no. 1, pp. 3-25, 1992.

[4] P. Honeyman, L. Huston, J. Rees, and D. Bachmann, "The Little Work project," in: Proceedings of the 3rd Workshop on Workstation Operating Systems, 1992, pp. 11-14.

[5] J. Howard, M. Kazar, S. Menees, D. Nichols, M. Satyanarayanan, R. Sidebotham, and M. West, "Scale and performance in a distributed file system," ACM Transactions on Computer Systems, vol. 6, no. 1, pp. 51- 81, 1988.

[6] D. B. Terry, M. M. Theimer, K. Petersen, A. J. Demers, M. J. Spreitzer, and C. H. Hauser, "Managing update conicts in Bayou, a weakly connected replicated storage system," in: Proceedings of the 15th Symposium on Operating Systems Principles, 1995, pp. 172-183.

[7] Jatav, Vinod Kumar, et al. "Wireless Sensor Networks: Attack Models and Detection." 2012 IACSIT Hong Kong Conferences, IPCSIT vol. 30 (2012)©(2012) IACSIT Press, Singapore. 2012.

[8] P. Reiher, J. Popek, M. Gunter, J. Salomone, and D. Ratner, "Peer-to-peer reconciliation based replication for mobile computers," in: Proceedings of the ECOOP Workshop on Mobility and Replication, 1996.

[9] Semwal, Vijay Bhaskar, K. Susheel Kumar, Vinay S. Bhaskar, and Meenakshi Sati. "Accurate location estimation of moving object with energy constraint & adaptive update algorithms to save data." arXiv preprint arXiv:1108.1321 (2011).

[10] M. Ferreira and M. Casquilho, "An Example in Remote Computing Over the Internet applied to Geometry," International Journal of Interactive Multimedia and Artificial Intelligence, vol. 2, no. 3, pp. 38-43, 2013.

[11] J. Solis-Martinez, N. Garcia-Menendez, C. Pelayo G-Bustelo, and J. M. Cueva-Lovelle, "BPLOM: BPM Level-Oriented Methodology for Incremental Business Process Modeling and Code Generation on Mobile Platforms," International Journal of Interactive Multimedia and Artificial Intelligence, vol. 2, no. 2, pp. 13-27, 2013.

[12] Neha Gandotra, Vishwanath Bijalwan "coexistence model of zigbee & ieee 802.11b (wlan) in ubiquitousnetwork environment" IJARCET, volume 1, Issue 4,June 2012.

[13] Vishwanath Bijalwan, Dr. Sanjay Singh "ANALYSIS &DESIGN OF JOINT PHY-MAC MODEL OF IEEE 802.15.4" IJSETR, volume 2, Issue 9, September 2013..

[14] S. Isaacman, R. Becker, R. Caceres, S. Kobourov,M. Martonosi, J. Rowland, and A. Varshavsky. Identifying important places in people's lives from cellular network data. In Proc. Int. Conf. on Pervasive Computing, San Francisco, Jun. 2011.

[15] R. G. Guy, "Ficus: a very large scale reliable distributed file system," Ph.D. dissertation, University of California, Los Angeles, June 1991.

[16] J. P. Gupta, N. Singh, P. Dixit, V. B. Semwal, and S. R. Dubey, "Human Activity Recognition using Gait Pattern," International Journal of Computer Vision and Image Processing, vol. 3, no. 3, pp. 31 – 53, 2013.

[17] K. K. Susheel, V. B. Semwal and R. C. Tripathi, "Real time face recognition using adaboost improved fast PCA algorithm," arXiv preprint arXiv:1108.1353 (2011).

[18] K. S. Kumar, V. B. Semwal, S. Prasad and R. C. Tripathi, "Generating 3D Model Using 2D Images of an Object," International Journal of Engineering Science, 2011.

[19] N. Singh, S. R. Dubey, P. Dixit and J. P. Gupta, "Semantic Image Retrieval by Combining Color, Texture and Shape Features," In the Proceedings of the International Conference on Computing Sciences, pp. 116-120, 2012.

[20] V. B. Semwal, V. B. Semwal, M. Sati and S. Verma, "Accurate location estimation of moving object in Wireless Sensor network," International Journal of Interactive Multimedia and Artificial Intelligence, vol. 1, no. 4 , pp. 71-75, 2011.

[21] E. Felembane, C. Lee and E. Ekici, "MMSPEED: Multipath Multi-SPEED Protocol for QoS Guarantee of Reliability and Timeliness in Wireless Sensor Networks," IEEE Transactions on Mobile Computing, vol. 5, no. 6, pp. 738-754, 2006.

[22] S. R. Dubey and A. S. Jalal, "Adapted Approach for Fruit Disease Identification using Images," International Journal of Computer Vision and Image Processing, vol. 2, no. 3, pp. 44-58, 2012.

[23] Sanjay Singh, Vishwanath Bijalwan, "Design Of Wireless Sensor Network Node On Zigbee For Water Level Detection" Volume 3,Issue8, August 2013.

[24] Anchit Bijalwan, Vishwanath Bijalwan, "Examining the Crimninology using Network Forensic" 8th National Conference USCSTC ,Dec 2013.

[25] Vishwanath Bijalwan, Vinay Kumar, Pinki Kumari,Jordan Pascual," KNN based Machine Learning Approach for Text and Document Mining" International Journal of Database Theory and Application Vol.7, No.1 (2014), pp.61-70.

[26] Pinki Kumari and Vikas Pareek, RAKSHITA- A Novel web based Approach for Protecting Digital Copyrights Using Public Key Digital Watermarking and Human Fingerprints" International conference on methods and models in computer science (ICM2CS-2010).

[27] Pinki Kumari and Abhishek Vaish, "Instant Face detection and attributes recognition" in International Journal of Advanced Computer Science and Applications (IJACSA - ISSN 2156 5570):2011

[28] Pinki Kumari and Abhishek Vaish,"A Comparative study of Machine Learning algorithms for Emotion State Recognition through Physiological signal", Advances in Intelligent Systems and Computing, Vol.236-Springer; ISBN 978-81-322-1601-8

[29] Pinki Kumari and Abhishek Vaish "Brainwave's Energy feature Extraction using wavelet Transform" proceeding of IEEE SCEECS, 2014, MANIT, Bhopal ISBN: 978-1-4799-2526-1.

[30] Semwal, Vijay Bhaskar, K. Susheel Kumar, Vinay S. Bhaskar, and Meenakshi Sati. "Accurate location estimation of moving object with energy constraint & adaptive update algorithms to save data." arXiv preprint arXiv:1108.1321 (2011).

Meenakshi Sati presently working as Assistant Professor, India. She is B.Tech from Graphic Era University Dehradun and pursing M.Tech from NITTR Chandigrah. Her major research work Interest in Image Processing, Sensor Network, Design and Analysis of Algorithm, Data Structure.

Vivek Vikash is presently working as Software Engineer, Honeybell, India. He completed his M.Tech form Indian Institute of Information Technology, Allahabad, his major research work Interest in Image Processing, computer sensor network & Pattern Recognition.

Vishwanath Bijalwan is working as assistance professor at Uttarakhand Technical University Dehradun. He is person with lot of potential. His research interest is wireless sensor network, WiMax, Wi-Fi, machine learning, Informtion Retrieval.So far he has published 4 high quality researches paper and work on many MHRD funded project.

Pinki Kumari received her M.Tech in Computer Science and Engineering from Banasthali University, Rajasthan, and her Ph.D in Information Security. Her research interests include Pattern recognition, Cloud Computing, Big Data and Distributed Database.

Manish Raj is currently pursuing his Ph.D fm IIITA.. He received his M.Tech from IIITA India. He has more than 5 years of research and teaching experience. He supervised 5 PG and 21 UG students. He has 2 research papers in reputed international journal and conference.

Meenu Balodhi is currently pursuing her fm Uttaranchal technical university. His research area biometric identification and wireless sensor network

Priya Gairola is currently pursuing her fm Uttaranchal technical university. His research area biometric identification and wireless sensor network, Wi-Max

Vijay Bhaskar Semwal has been severed as a PCC member in several International conferences and Journals. He was also head for 5th IEEE conference on big data at COER Roorkee. He is carrying more than 5 year of academic and industry experience. He served for various top organizations like Siemens, Newgen etc. Earlier he received his M.tech in wireless sensor domain and was awarded by gold medal. His major research work interest in Wireless Sensor Network, Artificial Intelligence, Image Processing, Computer Network &Security, and Design & Analysis of Algorithm, Machine Learning, and Information Retrieval Soft Computing.

24

A Useful Metaheuristic for Dynamic Channel Assignment in Mobile Cellular Systems

Deepak Kumar Singh, K. Srinivas and D. Bhagwan Das
Department of Electrical Engineering, Faculty of Engineering,
Dayalbagh Educational Institute

Abstract —**The prime objective of a Channel Assignment Problem (CAP) is to assign appropriate number of required channels to each cell in a way to achieve both efficient frequency spectrum utilization and minimization of interference effects (by satisfying a number of channel reuse constraints). Dynamic Channel Assignment (DCA) assigns the channels to the cells dynamically according to traffic demand, and hence, can provide higher capacity (or lower call blocking probability), fidelity and quality of service than the fixed assignment schemes. Channel assignment algorithms are formulated as combinatorial optimization problems and are NP-hard. Devising a DCA, that is practical, efficient, and which can generate high quality assignments, is challenging. Though Metaheuristic Search techniques like Evolutionary Algorithms, Differential Evolution, Particle Swarm Optimization prove effective in the solution of Fixed Channel Assignment (FCA) problems but they still require high computational time and therefore may be inefficient for DCA. A number of approaches have been proposed for the solution of DCA problem but the high complexity of these proposed approaches makes them unsuitable/less efficient for practical use. Therefore, this paper presents an effective and efficient Hybrid Discrete Binary Differential Evolution Algorithm (HDB-DE) for the solution of DCA Problem.**

Keywords —**Dynamic Channel Assignment, Mobile Networks, Optimization and Soft Computing.**

I. Introduction

It is important to efficiently utilize the scarce radio spectrum using a proper channel assignment scheme [1]. The aim of a channel assignment algorithm is to determine a spectrum efficient assignment of channels to the cells such that the traffic demand can be met as far as possible while satisfying the channel reuse constraints: co-channel constraints, channels separation constraints, and co-site constraints. A channel can be reused is that the same channel can be assigned to multiple cells simultaneously due to the radio propagation path loss. Channel assignment algorithms can be classified as static or dynamic. In a static approach, which is commonly called Fixed Channel Assignment (FCA) the channels are allocated and prefixed to each cell during the setup according to the traffic intensity estimated by the designer in the cell. FCA is still in use because it requires a moderate amount of BS radio installer and equipment and a simple monitoring procedure/algorithm. However, high efficiency of total channel usage over the whole service area if the traffic varies dynamically from cell to cell cannot be attained by FCA. To solve this problem, DCA [2] has been in trend since last twenty years. In DCA, channels are assigned dynamically over the cells in accordance with traffic load. FCA being a static technique can afford to spend more time to generate a better solution and is also easier to implement FCA in practice. However, from a resource utilization point of view, DCA is more preferable over FCA as DCA is designed to adjust resource assignment according to traffic demand, and hence, can support a higher capacity (or lower call blocking probability). The advantages of DCA are that it has a lower blocking rate than FCA at low traffic intensity and a lower forced call termination rate than FCA when blocking rates are equal [3].

DCA Problem is NP-hard and its effectiveness depends on its algorithm's efficiency in determining fast solution, good quality solution and its ease of implementation. Thus, devising a DCA, that is practical, efficient, and which can generate high quality assignments, is a challenging problem [4]. Metaheuristic Search techniques prove effective in the solution of DCA problem. A number of approaches [5-13,17,19,20,21,22] have been proposed for the solution of Channel Assignment Problem. The initial efforts for the solution of FCA were based on deterministic methods but as the problem is NP-hard these methods proved ineffective and inefficient for practical implementation for the next generation of mobile systems in which higher traffic demand was expected. To overcome the difficulties associated with the deterministic methods other heuristic methods such as Simulated Annealing [5], Tabu Search [6], Neural Networks [7,8] and Genetic Algorithms [9,11,12] were used for the solution of FCA problem. Later, Feedforward Neural Networks [19], Hopfield Neural Networks [20,21], Genetic Algorithms [10,13], Combinatorial Evolution Strategy [22], and Particle Swarm Optimization [17] have been used for the solution of DCA problem. However, the ever increasing number of mobile cellular users and the increasing demand for bandwidth call for more and more efficient Dynamic Channel Assignment strategies. Therefore, this paper presents an effective and efficient Hybrid Discrete Binary Differential Evolution (HDB-DE) for the solution of DCA Problem.

HDB-DE is a discrete binary version of Differential Evolution [23] which is an effective stochastic parallel search evolutionary algorithm for global optimization. The problem formulation and the implementation of HDB-DE take care of the soft constraints as well as hard constraints and hence focuses search only in the feasible regions of the search space resulting in fast convergence. The features of HDB-DE also help in the reduction of the population size which consequently reduces the computation time, and also results in better global/ near global solution.

Section II briefly describes the constraints in CAP, assumptions pertaining to cellular model and call arrival and then the problem formulation of DCA. Section III briefly describes Hybrid Discrete Binary Differential Evolution Algorithm (HDB-DE) for the solution of DCA Problem. The implementation details of HDB-DE algorithm are given in Section IV. The performance of HDB-DE is evaluated on standard benchmark problems. The details of these benchmark problems, simulations and the obtained results are presented in Section V. Conclusions are presented in Section VI.

II. Problem Formulation For DCA

A. Constraints in Channel Assignment Problem

In any cellular network, whenever two cells use the same channel or when two cells use channels adjacent to each other on the spectrum or when two channels are assigned to the same cell, interference occurs; these types of interference are called Co-channel Interference, Adjacent Channel Interference and Co-site Interference respectively [14].

They lower the signal-to-noise ratio at the receiving end, leading to the deterioration of system performance. Though the computation of the actual level of interference is tough, primarily owing to its dependence on the topology of the real environment, experiments show that the effect of interference is reasonably low if the following three constraints are satisfied:

Co-Channel Constraint (CCC): The same channel cannot be simultaneously allocated to a pair of cells unless there is a minimum geographical separation between them.

Adjacent Channel Constraint (ACC): Adjacent channels cannot be assigned to a pair of cells unless there is a minimum distance between them.

Co-Site Constraint (CSC): A pair of channels can be employed in the same cell only if there is a minimum separation in frequency between them. These constraints are called Electromagnetic Compatibility Constraints, which together with the traffic demand constraint, are known as *hard constraints*.

Apart from the *hard constraints*, another set of constraints called *soft constraints* is also considered, which may be described as follows:

The *packing condition* requires that a channel, in use in one cell, should be reused in another cell as close as possible (but obviously not interfering with the former) so that the number of channels used by the network is minimal, thereby lowering the probability of future call blocking in other cells.

The *resonance condition* tries to ensure that same channels are assigned to cells belonging to the same *reuse scheme* [4], as far as possible.

Another soft constraint is that, when a call arrives, minimum number of channel reassignment operations should be performed because excessive reassignment in a cell may lead to increase in blocking probability.

A solution to the CAP must satisfy the hard constraints whereas a soft constraint may be violated; the latter only helps maximize the utilization of resources and/or improve the quality of service.

Apart from the traffic demand constraint, the only other hard constraint that we have taken into account is Co-channel Constraint; other sources of interference are assumed to be absent, as reported in literature Vidyarthi et al [15], Battiti et al [16], and Chakraborty et al [17].

B. Assumption Pertaining to Cellular Model and Call Arrival

1) The topological model is a group of hexagonal cells that form a parallelogram shape (equal number of cells along x-axis and y-axis) as shown in figure 1(adapted from [15], figure 1]).

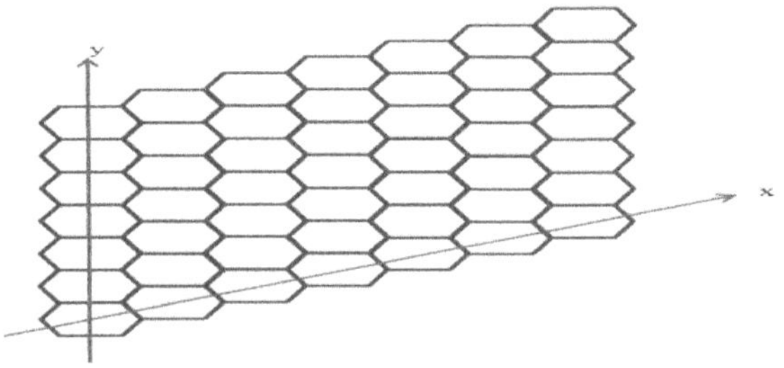

Fig.1. Cellular network model.

2) Cells are grouped in cluster of size 7 cells.
3) Each channel may serve only one call (i.e., multiplexing techniques are ignored). In DCA, all channels are put in central pool. A channel is assigned to an incoming call by a central controller that supervises the whole cellular network.
4) The selection of a channel is only subject to co-channel interference shown in figure 2. Other sources of interference are ignored.

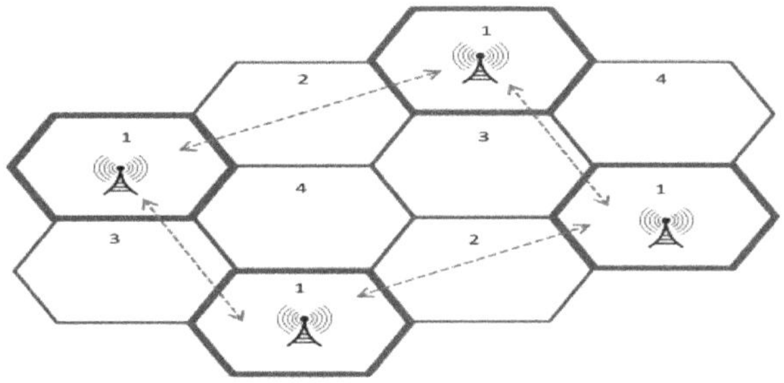

Fig.2. Co-channel interference

5) We consider not the start-up situation but the situation at a certain intermediate time-instant t when a certain number of calls is already being served by the network.
6) At time t, only one new call arrives at only one cell, called the host cell, all other conditions in the entire network remaining unaltered.
7) We set a minimum "reuse distance", which represents the minimum allowable normalized distance between two cells which may use the same channel at the same time as shown in figure 3.

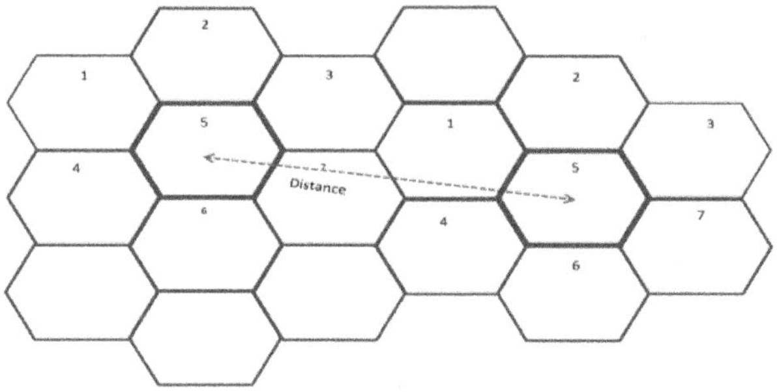

Fig.3. Frequency reuse distance.

8) A call is blocked if the entire set of channels in the network is in use in the cell involved in call arrival and its neighborhood that is there is no channel that satisfies the co-channel interference.
9) Existing calls in a cell involved in a new call arrival may be rearranged.

C. Formulation of Fitness Function

The equations 1, 2, 3 & 4 given below correspond to the different conditions i.e. co-channel interference, packing conditions, resonance conditions, and discouraging excessive rearrangement condition. These equations are combined to constitute a quadratic energy function (equation 5) whose minimization leads to an optimal solution for DCA [13].

$$f_1(X) = \sum_{j=1}^{CH} \sum_{\substack{i=1 \\ i \neq k}}^{CE} V_{k,j} * A_{i,j} * interf(i,k) \quad (1)$$

Equation (1) expresses the hard condition. Where V_k is output vector for cell k, with dimension channel (CH). $V_{k,j}=1$ if channel j is assigned to cell k, otherwise $V_{k,j}=0$. Here k signifies in which cell, call arrives. The energy function increases in case a channel j which is assigned in cell i is selected by cell k that interferes with i. It thus ensures that solutions with no interference give *better* (smaller) fitness values. $A_{i,j}$ is the ij[th] element of the assignment table A, which is 1 if channel j is assigned to cell i, and 0 otherwise.

$$f_2(X) = \sum_{j=1}^{CH} \sum_{\substack{i=1 \\ i \neq k}}^{CE} V_{k,j} * A_{i,j} * \frac{(1 - interf(i,k))}{dist(i,k)} \quad (2)$$

This term expresses the *packing condition.* The energy decreases if channel j assigned to cell k is also selected by cell and interf (i, k) =0. Energy reduction depends on the distance between i and k. The packing condition requires that a channel, in use in one cell, should be reused in another cell as close as possible (but obviously not interfering with the former) so that the number of channels used by the network is minimal, thereby lowering the probability of future call blocking in other cells.

$$f_3(X) = \sum_{j=1}^{CH} \sum_{\substack{i=1 \\ i \neq k}}^{CE} V_{k,j} * A_{i,j} * \left(1 - res(i,k)\right) \quad (3)$$

Where res(i, k) is function whose value is 1 if cells i and k belongs to the same reuse scheme, otherwise 0. Equation (3) symbolize the *resonance condition.* The resonance condition tries to ensure that same channels are assigned to cells belonging to the same reuse scheme, as far as possible.

$$f_4(X) = \sum_{j=1}^{CH} V_{k,j} * A_{k,j} \quad (4)$$

Which subtracts 1 whenever a channel already being used by cell *k*, before the arrival of the new call, is considered in the candidate solution (i.e. in the new configuration) so that a mobile terminal being served need not change its channel too often.

Finally, the fitness function *F*(*X*) is given by,

$$F(X) = W_1 * f_1(X) - W_2 * f_2(X) + W_3 * f_3(X) - W_4 * f_4(X) \quad (5)$$

Where W_1, W_2, W_3, W_4 are *weights* that determine the importance of various terms. Clearly, $f_1(X)$ accounts for the *hard constraint* which should be given a high weightage over the other terms those are associated with the *soft constraints*. The coefficient values as used in Battiti et al [16] have been used. Where W_1=7000, W_2=1.2625, W_3=0.01, W_4= 4.17625.

III. HDB-DE

HDB-DE is a discrete binary version of Differential Evolution which is an effective stochastic parallel search evolutionary algorithm for global optimization. Unlike DE here the individuals are initialized as binary strings. HDB-DE algorithm consists of three major operations – mutation, crossover and selection, which are carried out for each member of the population (called as target vector). Mutation on each target vector of the population generates a new mutant vector uniquely associated with it. The resultant mutant vector is no longer binary because of the difference operator and the control parameter. Therefore, the discretization process from a real continuous space to a binary space is done. Then the crossover operation generates a new trial vector using the mutant vector and the target vector itself. In selection phase the fitness of the trial vector is compared with the target vector and the vector with higher fitness replaces the target vector in the population for the next iteration.

A. Pseudo-Code of HDB-DE

1) Initialize parameters *t*=0, *NP* (NP is number of individuals in population), CR and F. Where $CR \in (0,1)$ is crossover constant, $F \in (0,2)$ is mutant constant.
2) Initialize target population X^t.
3) Evaluate each individual *i* in the population using the objective function.
4) Obtain the mutant population (a mutant individual, $V_i^t = [v_{i1}^t, v_{i2}^t, \ldots, v_{in}^t]$ is determined such that
$$V_i^t = X_{ai}^t + F * (rand * X_{bi}^t - rand * X_{ci}^t)$$
Where X_{ai}^t, X_{bi}^t *and* X_{ci}^t are three randomly chosen individuals from the population such that $ai \neq bi \neq ci$.
5) Discretization process from a real continuous space to binary space is done according to the following equation

$$V_i^t = \begin{cases} 1 \; if \; rand(0,1) \le sigmoid(V_i^t) \\ 0 \; if \; rand(0,1) > sigmoid(V_i^t) \end{cases}$$

Where *sigmoid()* is a sigmoid limiting transformation function.

6) Obtain the trial population (For each mutant individual, $V_i^t = [v_{i1}^t, v_{i2}^t, ..., v_{in}^t]$ an integer random number between 1 and *n*, i.e., $D_i \in (1,2,...,n)$, is chosen, and a trial individual, $U_i^t = [U_1^t, U_2^t, ..., U_{NP}^t]$ is generated such that:

$$U_i^t = \begin{cases} V_{ji}^t, if (rand(0,1) \le CR) \; or \; (j = I_{rand}) \\ X_{ji}^t, if (rand(0,1) > CR) \; and \; (j \ne I_{rand}) \end{cases}$$

Where I_{rand} refers to a randomly chosen dimension ($j=1,2,..,n$))

7) Evaluate trial population
8) Selection (The selection is based on the survival of the fittest among the trial population and target population such that:

$$X_i^{t+1} = \begin{cases} U_i^{t+1}, & if \; f(U_i^{t+1}) \ge f(X_i^t) \\ X_i^t, & otherwise \end{cases}$$

9) Repeat steps 2 to 8 While Termination condition not reached.
10) Output best solution.

IV. Implementation Details of HDB-DE For DCA

The Dynamic Channel Assignment problem is specified in the literature in terms of the number of cells in the network (N_{ce}), the number of channels in the pool (N_{ch}) and a *demand vector* ***D*** which is a vector whose i^{th} element denotes the traffic demand in cell *i*, *i*= 1,2,…, N_{ce}.

We assume that the new call demand is placed at cell *k* which is already serving demand (*k*) calls where demand (*k*) denotes the total traffic load (ongoing) in cell *k* at time *t*; and no ongoing call is terminated in the entire network. Our problem is to assign an available channel to the incoming call with possible reassignment of channels to the calls in progress in cell *k*.

A. Solution Representation

The candidate solution to the problem is represented as a binary string X which is the representation of V_k mentioned earlier where k signifies the cell in which call arrives. The size of the vector X is equal to the number of available channels (N_{ch}) and $X_{k,j}=1$ if channel j is assigned to cell k, otherwise $X_{k,j}=0$. The number of 1s in each solution vector is equal to demand(*k*)+1 i.e. the total ongoing calls plus the call that arrives at the concerned time instant.

B. Mutation

A new mutation operator has been designed for HDB-DE which is more effective than the one which has been used earlier [17] for mutation of solutions with binary representation in DE. The new mutation operator is as given below:

$$V_i^t = X_{ai}^t + F * (rand * X_{bi}^t - rand * X_{ci}^t)$$

The effectiveness of the operator can be seen from the fact that if both X_{bi}^t and X_{ci}^t assume a value 1 then in the earlier mutation operation the difference becomes 0 and therefore does not result in any change in the value of X_{ai}^t whereas in the newly designed mutation operation it does not happen so and leads to the generation of better mutants and thereby faster convergence.

V. ComputationalResults

HDB-DE algorithm was implemented in Matlab and the following benchmark problems were used for its evaluation:

1) The first benchmark problem CSys 49 [13] consists of 49 hexagonal cells that are arranged to form a parallelogram structure with 70 channels available to the system.
2) The second and third benchmark problems i.e. HEX 1 and HEX 3 [18] are based on a 21-cell system.
3) The last four benchmark problems i.e. EX 1, EX 2, KUNZ 1 and KUNZ 2 [18] are based on a 4, 5, 10 and 15-cell system respectively.

The details of the benchmark problems and the used demand vectors are summarized in Table I.

TABLE I.
Details of Benchmark Problems

Problem	N_{ce}	N_{ch}	Demand Vector (demand)
CSys 49	49	70	4,2,6,4,4,6,2,4,2,4,2,2,4,4,4,4,2,2,6,2,6,6,2,4,8,2,2,2,4, 2,2,4,2,4,6,2,4,6,4,2,6,4,4,2,2,6,4,4,4
HEX 1	21	37	2,6,2,2,2,4,4,13,19,7,4,4,7,4,9,14,7,2,2,4,2
HEX 3	21	21	1,1,1,2,3,6,7,6,10,10,11,5,7,6,4,4,7,5,5,5,6
EX 1	4	11	1,1,1,2
EX 2	5	17	2,2,2,3,3
KUNZ 1	10	30	10,11,9,5,9,4,4,7,4,8
KUNZ 2	15	44	10,11,9,5,9,4,4,7,4,8,8,9,10,7,7

For each of the considered problems it has been assumed that all cells are arranged in the form of a parallelogram, the given N_{ce} of each problem is expressed in the form $r \times c$, where *r*, *c* are integers, and hence determine the configuration of the cellular network by setting the number of rows to *r* and the number of all columns to *c* [17]. A cell is arbitrarily selected and then it is assumed that, just before a call demand arrives in this cell at time *t*, *demand(i)* calls were already in progress in the ith cell, *i* =1 ,2,…, N_{ce}, and *demand(k)* calls are ongoing in the kth cell. Accordingly, N_{ce} X N_{ch} assignment matrix, avoiding co-channel interference has been manually determined which describes the status of ongoing calls in each cell before the new call arrival and thus represents the initial condition. The assignment table used for HEX3 problem is given in Table II.

Whenever a simulation results in a solution, which violates *Co-Channel Constraint* (CCC) the call is rejected. *Call Rejection Probability* (CRP) as given in [17] is used as a parameter for determining the effectiveness of the proposed method.

$$CRP= N_{rejected}/ N_{total}$$

Where $N_{rejected}$ = number of simulations in which the incoming call is rejected in the host cell considered; N_{total} = total number of simulations.

TABLE II.
INITIAL ASSIGNMENT MATRIX FOR HEX3

Problem	Initial Assignment Matrix
	1 0 0 0 0 0 0 0 0 0 0 0 0 0 0 0 0 0 0 0 0
	0 0 0 0 0 0 0 0 0 0 0 0 0 0 0 0 0 0 0 0 1
	0 0 0 0 0 0 0 0 0 0 1 0 0 0 0 0 0 0 0 0 0
	0 0 1 0 0 0 0 0 0 0 0 0 0 0 0 0 0 0 0 0 0
	0 0 0 1 0 0 0 0 0 0 0 0 0 0 0 1 0 0 0 0 0
	1 0 1 0 0 0 1 0 1 0 0 0 1 0 0 0 0 0 0 0 0
	0 1 0 0 0 1 0 1 0 0 0 1 0 0 1 0 0 0 0 0 0
	0 0 0 1 0 0 0 0 0 1 0 0 0 1 0 1 0 1 0 0 0
	1 0 1 0 1 0 1 0 1 0 1 0 1 0 1 0 0 0 0 0 0
	0 1 0 1 0 1 0 1 0 1 0 1 0 1 0 1 0 0 0 0 0
HEX3	1 0 1 0 1 0 1 0 1 0 1 0 1 0 1 0 0 0 0 0 0
	0 0 0 0 0 0 0 0 0 0 0 1 0 1 0 1 0 1 0 1 0
	0 1 0 1 0 1 0 1 0 1 0 0 1 0 1 0 0 0 0 0 0
	1 0 0 0 1 0 1 0 0 0 0 0 0 0 0 0 0 0 0 0 0
	0 0 1 0 0 0 0 0 1 0 1 0 0 0 0 0 0 0 0 0 0
	0 0 0 0 0 0 0 0 0 0 0 0 0 1 0 1 0 1 0 1 0
	0 1 0 1 0 0 0 0 0 1 0 0 1 0 1 0 0 0 0 0 0
	0 0 1 0 1 0 0 1 0 0 0 1 0 0 0 0 1 0 0 0 0
	1 0 0 0 0 1 0 0 1 0 1 0 0 0 0 0 0 0 0 0 0
	0 0 0 1 0 0 1 0 0 0 0 0 0 1 0 0 0 1 0 0 0
	0 0 1 0 0 1 0 0 0 0 0 0 0 1 0 1 0 0 0 0 0

Thus, CRP is the cumulative proportion of simulations, for which the call is rejected, in the long run. This parameter CRP is based on but different from the *call blocking probability* used in the [13,14]. The former characterizes a particular cell under a given initial condition while the latter characterizes the cellular network as a whole.

TABLE III.
SIMULATION RESULTS

Problem	Host Cell	Number of simulations	Number of times, call is rejected				CRP			
			Method							
			PSO [17]	SGA	MGA	HDBDE	PSO [17]	SGA	MGA	HDBDE
CSys 49	6	250	-	14	0	**0**	-	0.06	0	**0**
HEX1	6	104	21	8	0	**0**	0.20	0.08	0	**0**
HEX3	5	105	32	12	0	**0**	0.31	0.11	0	**0**
EX1	4	112	32	6	0	**0**	0.28	0.05	0	**0**
EX2	4	100	28	9	0	**0**	0.28	0.09	0	**0**
KUNZ1	7	110	39	11	0	**0**	0.35	0.10	0	**0**
KUNZ2	7	100	20	11	0	**0**	0.20	0.11	0	**0**

The parameters used in HDB-DE are as follows:
Population size=30,
Mutation factor F=1.2,
Crossover constant CR= 0.6 and
Maximum number of generations=50.

The simulation results obtained by HDB-DE and those obtained by Simple Genetic Algorithm (SGA), Modified Genetic Algorithm (MGA) [24] and PSO [17] for the different benchmark problems are shown in Table III.

The results shown in table III indicate that the performance of HDB-DE and MGA is good compared to PSO. The comparison of the convergence curves and the average number of Evaluation expended to yield the best solution over a good number of runs of the algorithms will further throw light on the efficiency, efficacy and consistency of the algorithms.

Figure 4, 5, 6, 7, 8, 9, and 10 shows the convergence curve of HDB-DE, MGA and SGA for the seven benchmark problems CSys 49, HEX 1, HEX 3, EX 1, EX 2, KUNZ 1 and KUNZ 2 respectively.

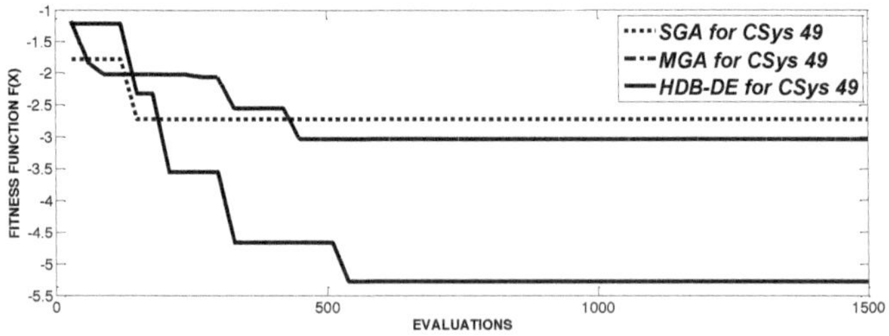

Fig.4. Convergence curve for CSys 49 consists of 49 hexagonal cells with 70 channels.

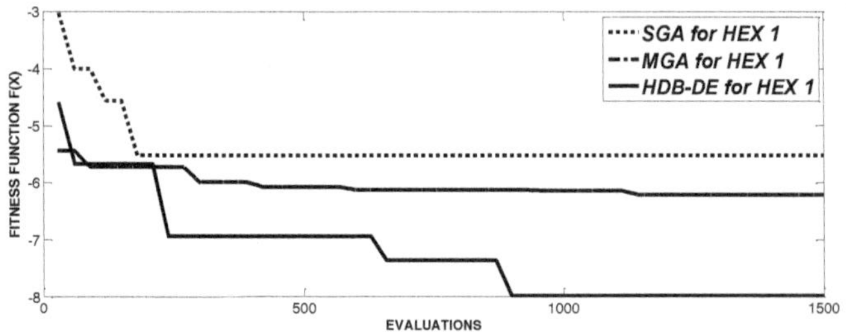

Fig.5. Convergence curve for HEX 1 consists of 21 hexagonal cells with 37 channels

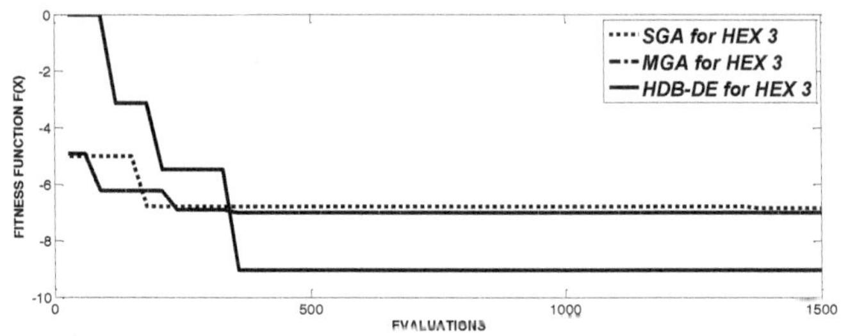

Fig.6. Convergence curve for HEX 3 consists of 21 hexagonal cells with 21 channels.

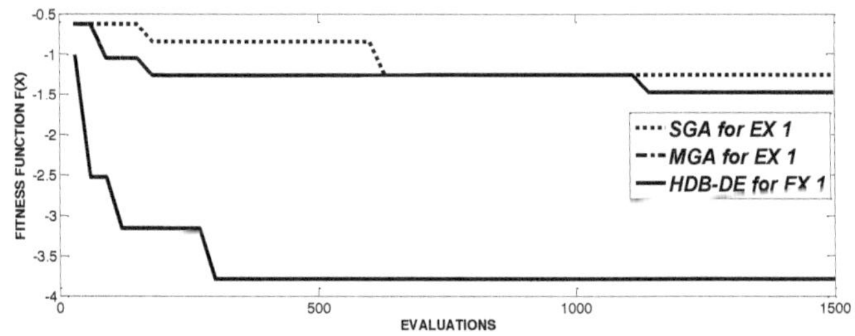

Fig.7. Convergence curve for EX 1 consists of 4 hexagonal cells with 11 channels.

TABLE IV.
SIMULATION RESULTS OBTAINED BY HDB-DE AND GA FOR ALL BENCHMARK PROBLEMS

Benchmark	**CSys 49**			**HEX 1**			**HEX 3**			**EX 1**			**EX 2**			**KUNZ 1**			**KUNZ 2**		
Algorithm	*HDBDE*	*MGA [24]*	*SGA*	*HDBDE*	*MGA [24]*	*SGA*	*HDBDE*	*MGA [24]*	*SGA*	*HDBDE*	*MGA [24]*	*SGA*	*HDBDE*	*MGA [24]*	*SGA*	*HDBDE*	*MGA [24]*	*SGA*	*HDBDE*	*MGA [24]*	*SGA*
Best # of Evaluations	**90**	200	450	**60**	150	380	**60**	100	250	**28**	65	210	**60**	85	250	**60**	160	230	**90**	150	230
Worst # of Evaluations	**300**	410	900	**170**	360	750	**130**	270	600	**65**	130	410	**95**	240	550	**160**	410	550	**180**	350	650
Average # of Fitness Evaluations	**163**	301	681	**98**	249	512	**93**	180	416	**47**	97	312	**79**	156	394	**99**	255	368	**123**	236	387

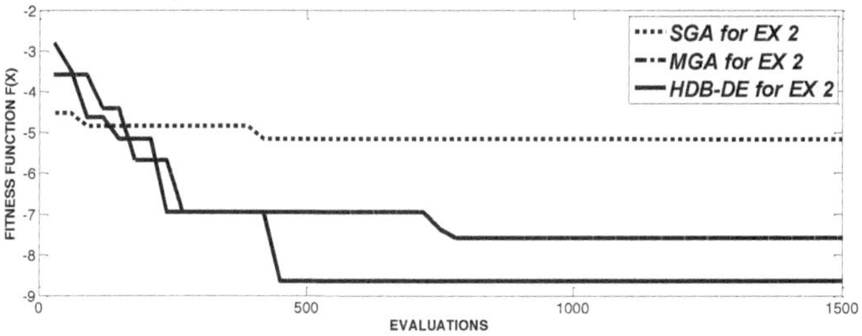

Fig.8. Convergence curve for EX 2 consists of 5 hexagonal cells with 17 channels.

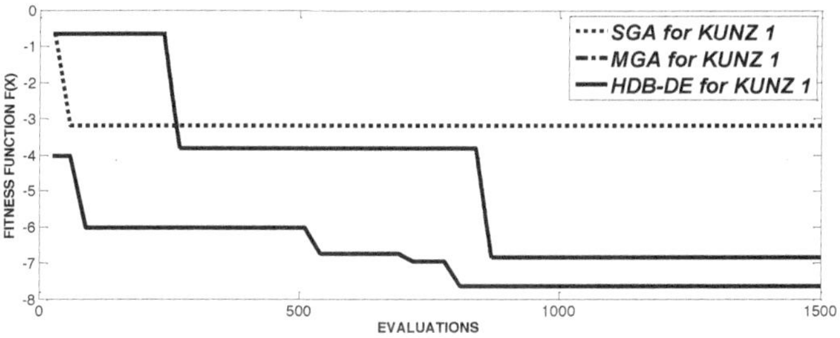

Fig.9. Convergence curve for KUNZ 1 consists of 10 hexagonal cells with 30 channels.

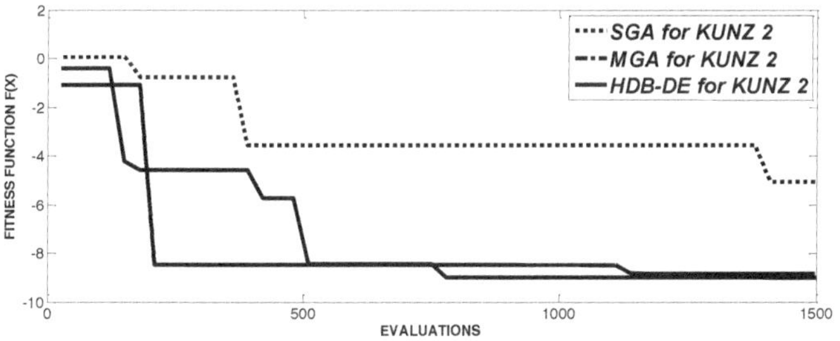

Fig.10. Convergence curve for KUNZ 2 consists of 15 hexagonal cells with 44 channels.

Table IV presents the results in terms of Best number of evaluations, Worst Number of evaluations and average number of evaluations expended by HDB-DE, MGA and SGA to give the final best solution in 30 independent runs on each benchmark problem.

The convergence curves shown in figures 4 to 10, and the results presented in Tables III and IV clearly indicate the superior and consistent performance of HDB-DE over the other methods reported in the literature in terms of quality of solutions and speed of convergence. The results are promising and indicate the suitability of HDB-DE for the solution of even bigger instances of real-world DCA problem.

VI. CONCLUSION

This paper presents an effective and efficient Hybrid Discrete Binary Differential Evolution (HDB-DE) Algorithm for the solution of Dynamic Channel Assignment Problem in cellular radio networks. The features of HDB-DE help in achieving fast convergence and good solution. The problem formulation and the implementation of HDB-DE for DCA takes care of the soft constraints as well as hard constraints and hence focuses search only in the feasible regions of the search space resulting in fast convergence. HDB-DE algorithm has been evaluated on benchmark problems for DCA. The obtained results and their comparison with those obtained by the other methods establish the superiority of HDB-DE over other methods and indicates its suitability for the solution of even bigger instances of dynamic channel assignment in cellular networks resulting in lower call rejection probability, higher capacity utilization and a good quality of service.

REFERENCES

[1] A. Mehrotra, Cellular Radio. Analog and Digital Systems. Norwood, MA: Artech House, 1994.

[2] W. C. Jakes, Microwave Mobile Communication. New York: Wiley, 1974.

[3] W. C. Y. Lee, Mobile Cellular Telecommunications Systems. New York: McGraw-Hill, 1989.

[4] R. C. V Macario, Cellular Radio: Principles and Design, 2nd ed. New York: Macmillan, 1997.

[5] M. Duque-Anton, D. Kunz, and B. Ruber, "Channel assignment for cellular radio using simulated annealing". IEEE Trans. Vehicular Technology., vol. 42, no. 1, pp. 14–21, Feb. 1993

[6] D. J. Castelino, S. Hurley, and N. M. Stephens, "A tabu search algorithm for frequency assignment". Ann. Oper. Research., vol. 63, pp. 301–319. Sept. 19, 2005.

[7] D. Kunz, "Channel assignment for cellular radio using neural networks". IEEE Trans. Vehicular Technology., vol. 40, no. 1 ,pp. 188–193. Feb. 1991

[8] N. Funabiki and Y. Takefuji, "A neural network parallel algorithm for channel assignment problems in cellular radio networks".IEEE Trans. Vehicular Technology., vol. 41,no. 4,pp. 430–437. Nov. 1992

[9] M. A. C. Lima , A. F. R. Araujo and A. C. Cesar, "Dynamic channel assignment in mobile communications based on genetic algorithms", In 13th IEEE International Symposium on Personal, Indoor and Mobile Radio Communications., no. 5, pp. 2204-2208. Sep. 2002.

[10] M. Cuppini , "A genetic algorithm for channel assignment problems", European Trans. Telecom. Rel. Technology., vol. 5, pp. 285–294, Mar 1994.

[11] K. Lai and G. G. Coghill , "Channel assignment through evolutionary optimization", IEEE Trans. Vehicular Technology., vol. 45, no. 1,pp. 91–96. Feb. 1996.

[12] R. Dorne, and J. K. Hao, "Constraint handling in evolutionary search: A case study of the frequency assignment, problem solving from nature—PPSN IV.Voigt, H. M., Ebeling, Rechenberg, W., I., and Schwefel, H.-P., Springer, vol. 1141, pp. 801–810, 1996.

[13] H. G. Sandalidis, P. P. Stavroulakis, and J. Rodriguez-Tellez, "Implementation of genetic algorithms to a channel assignment problem in cellular communications", In Proceedings of the 6th Int. Conf. Advances in Communications and Control (COMCON 6), pp. 453–460. Jun. 1997.

[14] K. Okada and F. Kubota, "On dynamic channel assignment in cellular mobile radio systems", IEEE International Symposium on Circuits and Systems. vol. 2, pp. 938-941. June 11-14, 1991.

[15] G. Vidyarthi, A. Ngom, and I. Stojmenovic´, "A hybrid channel assignment approach using an efficient evolutionary strategy in wireless mobile networks", IEEE Trans. Vehicular Technology, vol. 54, no. 5 , pp. 1887–1895. Sept. 2005.

[16] R. Battiti, A. A. Bertossi, and M. Brunato, "Cellular channel assignment: A new localized and distributed strategy, mobile networks and applications", Kluwer., no. 6, pp. 493–500. Jan 10, 1998.

[17] M. Chakraborty, R. Chowdhury, J. Basu, R. Janarthanan and A. Konar, "A particle swarm optimization-based approach towards the Solution of the dynamic channel assignment problem in mobile cellular networks", IEEE Region 10 Conference (Hyderabad, India,), pp. 1-6, Nov. 19-21, 2008.

[18] K. Smith, and M. Palaniswami, "Static and dynamic channel assignment using neural network".IEEE Journal on Selected Area in Communications, vol. 15, no. 2, pp. 238-249, Feb. 1997

[19] P. T. H. Chan, M. Palaniswami, and D. Everitt, "Neural network-based dynamic channel assignment for cellular mobile communication systems," IEEE Trans. Vehicular Technology, vol. 43, pp. 279–288. May 1994

[20] K. Nakano, M. Yokono, M. Sengoku, Y. Yamaguchi, S. Shinoda, S. Motooka and T. Abe, "An application of dynamic channel assignment to a part of a service area of a cellular mobile communication system," IEICE Trans. Fundamentals, no. 75, pp. 369–379. Mar. 1992

[21] E. Del Re, R. Fantacci, and L. Ronga, "A dynamic channel allocation technique based on hopfield neural networks," IEEE Trans. Vehicular Technology, vol. 45, pp. 26–32. Feb. 1996

[22] H. G. Sandalidis, P. P. Stavroulakis, and J. Rodriguez-Tellez, "An efficient evolutionary algorithm for channel resource management in cellular mobile systems". IEEE Trans Evolutionary Computation, vol. 2, no. 4, pp. 125–137, Nov. 1998.

[23] C. Peng, L. Jian and L. Zhiming. "Solving 0-1 knapsack problems by a discrete binary version of differential evolution", in Proc. Second International Symposium on Intelligent Information Technology Application, vol. 2, pp. 513-516, Dec. 20-22, 2008.

[24] D. K. Singh, K. Srinivas and D. B. Das, "A dynamic channel Assignment in GSM telecommunication network using modified genetic algorithm", in proc. 6th Euro-American Conference on Telematics and Informat5ion System, IEEE, pp. no. 1-5, May 23-25, 2012.

25

An Effective Approach for Mobile ad hoc Network via I-Watchdog Protocol

Nidhi Lal

Indian Institute of Information Technology

Abstract — **Mobile ad hoc network (MANET) is now days become very famous due to their fixed infrastructure-less quality and dynamic nature. They contain a large number of nodes which are connected and communicated to each other in wireless nature. Mobile ad hoc network is a wireless technology that contains high mobility of nodes and does not depend on the background administrator for central authority, because they do not contain any infrastructure. Nodes of the MANET use radio wave for communication and having limited resources and limited computational power. The Topology of this network is changing very frequently because they are distributed in nature and self-configurable. Due to its wireless nature and lack of any central authority in the background, Mobile ad hoc networks are always vulnerable to some security issues and performance issues. The security imposes a huge impact on the performance of any network. Some of the security issues are black hole attack, flooding, wormhole attack etc. In this paper, we will discuss issues regarding low performance of Watchdog protocol used in the MANET and proposed an improved Watchdog mechanism, which is called by I-Watchdog protocol that overcomes the limitations of Watchdog protocol and gives high performance in terms of throughput, delay.**

Keywords - MANET; Watchdog; AODV; Black hole; RREP; RREQ; RRER; Malicious node; PDR; I-Watchdog; Sequence number;

I. Introduction

IN the mid of 1990's, Mobile ad hoc network became very famous topic in the research area of networking. Mobile ad hoc network is a wireless technology and it does not hold any infrastructure; nodes in the MANET environment are dynamic in character and do not relay on any topology. They are scattered in nature and do not rely on any central authority. In a MANET, [11] each node can take responsibility of a router as well as take a role as a host. The nodes in the mobile ad hoc network are linking and Communicate with each other all the way through the usage of radio waves. [1]MANET supports fast establishing of networks so they encompass very high degree of flexibility, the only necessity is to provide a new set of nodes with some degree of wireless communication range. [17]If the nodes are within the same radio waves wireless communication rang than they can communicate directly otherwise they can communicate with their respective destination node with the help of intermediate nodes.

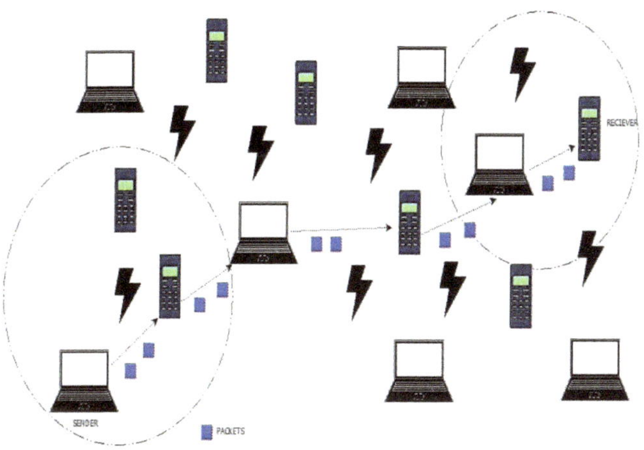

Fig 1. The sketch out of MANET

In the above figure, many numbers of nodes are in the network, in which, one is acting as a sender node and another one is the receiving node. Sender wishes to propel data packet to his subsequent receiving node. For initiating this communication, the sender can send these data packets via the help of intermediate nodes, which are within the communication range of the sender node. By following this strategy, the sender node sends all data packets to the respective receiver node. [2]There are various types of the mobile ad hoc networks, such like, vehicular ad hoc network (VANET) [12] that is used for making the communication between the vehicles, (IMANET) internet based mobile ad hoc networks which is used to link mobile nodes to network gateways and Tactical MANET which is used in the application of military. [3]Meant for routing of the packets among the mobile device nodes a routing protocol is obligatory. [37]The routing protocol should design and chooses in such a way that it provides high reliability, security, power efficient, avoid overhead and provide best quality of service as well as should consider the unidirectional links also. So by taking these points into account, there are various methodologies are proposed like, AODV, DSDV, DSR, CBRP. [4]AODV maintains route on the demand such that traffic of the network remain minimum and it uses distance sequence number for surety of the loop free route. [5] Security threats and packet loss due to transmission error are major challenges of the MANET. [18]Destination-Sequenced Distance-Vector Routing (DSDV) maintains a table for routing and follows the concept of bellman-ford algorithm, it is basically used to prevent the network from the formation of a loop. DSDV make sure that the network does not restrain any cycle or loop. DSDV has certain disadvantages like it is not power saving and does not worth for networks which are highly dynamic. [19]Dynamic Source Routing (DSR) is a source routing protocol which is similar to AODV but it does not contain latest updated information regarding the network therefore it leads to inconsistency in the routing tables. [20]A cluster-based routing protocol (CBRP) is a routing protocol in which, nodes of a network make a group and that group is called a cluster, after forming this cluster they uses a clustering algorithm to determine the cluster head among the nodes in that group. [30-32]A Mobile Location Aware Information System is also proposed for control of the presence of a non-intrusive by use of technologies which are based on the global positional system (GPS) and light weight indoor location system. [33-35]This technology can be used universally and applies very minimum cost. [5]Due to the decentralized environment of the MANET, they are constantly susceptible to black hole attack and the recital of the AODV routing protocol decrease. To triumph over this dilemma, Watchdog protocol with AODV is commenced which builds recognition of malicious nodes.

Watchdog protocol uses local information of the next hop node and overhears it. If it gets that it spending time of the packet is exceeded above the predefined threshold then it marks that node as malicious, this way Watchdog protocol detects malicious node in the network. [6, 10]Watchdog protocol has some disadvantages that it does not find link transmission error due to congestion in the network as well as it does not support high mobility of large number of nodes in the network and give a wrong report about the malicious node which eventually decreases the system throughput and performance. [7]Mobile ad hoc networks are more vulnerable to transmission errors than fixed wired network because of their wireless nature, environmental conditions, network congestion etc. In this paper, an improved Watchdog mechanism is presented which identifies the malicious node in the network as well as spots the network congestion. We give the name of this improved Watchdog protocol as I-Watchdog protocol. The proposed work is implemented in ns2-simulator and gives very high performance in terms of throughput, packet delivery ratio and end-to-end delay. Section-2 will describe the AODV protocol, Section-3 will introduce about some variety of security attacks, Section-4 describes the activity of black hole attack in the network, Section-5 will describe Watchdog protocol, Section-6 will introduce about proposed improved Watchdog protocol (I-Watchdog protocol) and section-7 is containing simulation parameters and Section-8 will contain comparison, results of Watchdog protocol and proposed improved Watchdog protocol.

II. AODV PROTOCOL

MANET applies Ad hoc On Demand Distance Vector (AODV) routing protocol for transmitting the packet from the source towards their particular target node. [8]AODV each point in time determine route when network wishes it. AODV bring into play route request message (RREQ) for creation the route request from source to target which enclose the distance sequence number. This communication is basically does neighbor discovery and it is broadcasting in nature. If the intermediary neighboring node have a path to the resultant destination, then it propels the route reply message (RREP) reverse to the source. If the intermediary adjacent nodes have no path to the destination then it generates reverse route entry towards the source and broadcast RREQ message to its neighbors. This course of action will be on recurrence until the destination route is not set up. Just the once the route is found to the target node then the RREP message is unicast from the current node to the source node and this RREP message include destination sequence number and hope count. Later than receiving the RREP message, source node brings up to date its routing table only in the provision when coming destination sequence number is bigger than the prior. Upon receiving on the several RREP, source opt for greater sequence number with smallest hop count and renew its routing table information. [22]In the AODV routing protocol, each adjacent node in the network background maintains track information about the status of the link by keeping an eye on the link and when there is found the splintering in link of the route then the RERR message is promoted by this node which detects the link breakage, to all the nodes in the network to broadcast this link status information. Figure 2 demonstrates the appearance of AODV routing protocol. Here, node marked by S represents a source node and node marked by D work as destination node.

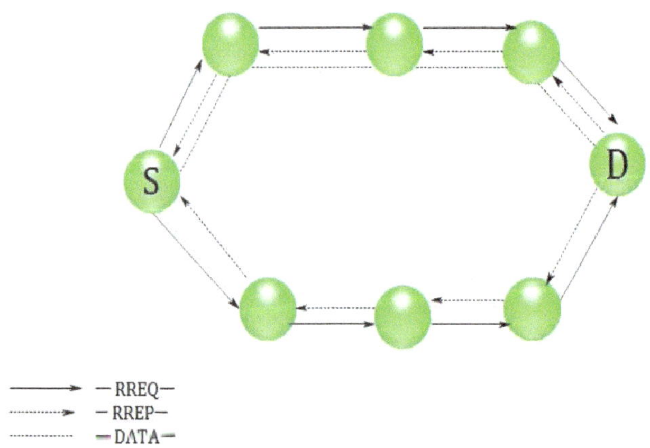

Fig: 2. AODV routing protocol method

III. VARIETY OF SECURITY ATTACKS

There is a range of attacks probable in MANET. [31] Because of the absence of any central authority and dynamic and distributed nature, mobile- ad hoc network environment can have a lot of security breaches and problems. Security does direct impact on the performance of the network because when the network is more secure then always there is a high possibility of successful transmission of packet from the sender to respective receiver. This will directly lead to high throughput and minimizes the end-to-end delay. But due to MANET is wireless in nature, there is always the possibility of attacks on the network. [21]A few of the security attacks are characterized below:

- BLACK HOLE ATTACK: In this attack, every packet which is sent out from the source towards the relevant target node, is dropping by black hole node. This attack is illustrated in detail in the section- IV
- PARTITION OF THE NETWORK: In this attack, a path is always present from the sender node to the respective receiver node, but nodes cannot communicate with each other.
- DENIAL OF SERVICE: In this attack, there is constraint on the nodes of a network for conveyance the packet as well as in receipt of the packet.
- SLEEP DEPRIVATION: [28, 36] In this attack, the nodes are forced to be sleeping, means to force for use its battery power.
- INFORMATION THEFT: [36] In this attack, the whole information which is reside inside the packet is read by unauthorized entity.
- INTRUSION: [36] In this attack, an unapproved person can have right to use services and those services are constrained to that entity.
- TAMPERING: [36] In this attack, Data is modified by an unauthorized entity.
- WORM HOLE ATTACK: [37] In this attack, basically the attacker node compromised with any host in the network and record the packet at one point of location and after tunneling to another location it again sends back to the network.

IV. BLACK HOLE ATTACK

MANET does not encompass any central authority and there is a lack of infrastructure so that is defenseless to black hole attack. The black hole attack is an attack on a network who hurriedly dwindle network performance by dropping the packets. [16]When a black hole node (malicious node) present in the network, it always advertises itself with the highest sequence number and minimum hop count. [23]Black hole node always tries to attract and capture the attention of the source node by ensuring them that it has the shortest path towards the destination node. The black hole attack is very dangerous in the network environment and it leads to the system to a denial of service (DOS) attack. When it obtains RREQ from the source node, it instantaneously propels RREP respond to the source enclose very large sequence number and lowest hop count. This is the nature of the black hole node that it tries to get attention in the network from source point of view, [17] that's why it advertises itself with very high sequence numbers. Upon receiving such RREP from this black hole node, source node thinks that it has valid fresh route to the destination because it have a high sequence number with minimum hope count and starts forwarding of data packets towards this black hole node. Upon receiving data packets, the black hole node drops all data packets and system performance degrades rapidly. Figure 3 shows the activity of black hole attack, in this figure red circles is used for denoting malicious node and green circle to represent valid nodes.

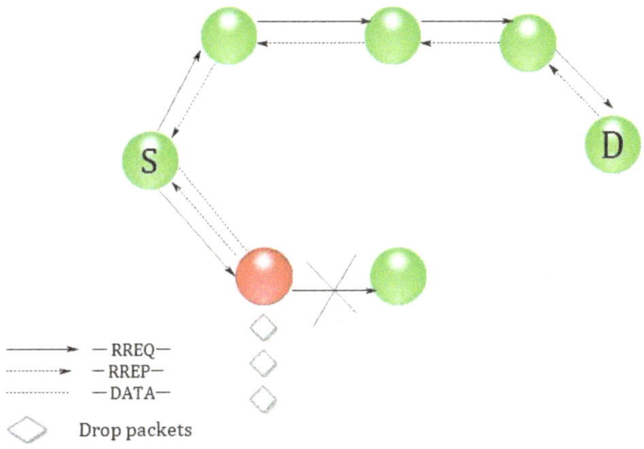

Fig: 3. The course of action of black hole attack

On top of figure, source node S launch RREQ to its neighbor node. The malicious node does not further send RREQ to its neighborhood node and propel RREP to source node S with high sequence number and least hope count. After getting such RREP, the source node sends data to this malicious node and malicious node drop all packets and system is in beneath the black hole attack.

V. WATCHDOG PROTOCOL

To avoid the problem and detection of this black hole attack, [9] Watchdog protocol is introduce. In this protocol, every node is work as an observer to watch the working of its next hope neighborhood node. It collects transmission information of this node and observes that node correctly forward to its next hope neighborhood node along with the correct destination route. [13]This protocol measures the sending time of the next hope node. If the sending time of the next hop neighbor is greater than the packet storing time and exceeds above some defined threshold of the network, then Watchdog knows that system is under black hole attack and it immediately mark this node as a malicious node. The Watchdog protocol announces the existence of the malicious node in the network by generating the alerts. The benefit of the Watchdog protocol is that, they make use of only local information and are proficient to spot the malicious node.

They can resolve the predicament of black hole attack which demonstrate the way to denial of service attack (DOS) in MANET network. [14]Watchdog protocol act as a very good intrusion detection system mechanism in the network. However, [15] there are certain disadvantages regarding to this protocol such that it decreases the network performance in terms of throughput, it does not support mobility with high number of nodes, and it doesn't detect the actual reason of the packet loss. To overcome these disadvantages of this Watchdog protocol, the improved Watchdog mechanism is proposed which perfectly distinguishes the packet loss due to congestion or due to the presence of a malicious node in the network. Our improved Watchdog protocol also supports a high degree of the mobility and enhances the performance.

VI. PROPOSED WORK

In this paper, the improved Watchdog protocol is proposed with some modifications to overcome the problem related to the early Watchdog protocol. This improved Watchdog protocol is very efficient to detect the actual reason for the packet loss. Because MANET is a wireless technology, mobile node devices of the MANET are independent to move anywhere so the mobility is very high and Watchdog protocol does not support a high degree of mobility but our I-Watchdog protocol supports a very large number of nodes with a high degree of mobility. Also, due to the wireless nature of the mobile ad hoc networks, they are more vulnerable to congestion and Watchdog protocol does not detect the network congestion and link error of transmission. It just observes that whenever the sending time of packet greater than the packet storing time, it sends alert in the system and marks the node as malicious. Watchdog protocol does not find the actual cause of packet loss and this leads to low throughput and system performance degrades. But proposed I-Watchdog protocol does not take a decision about the node very easily, because the packet loss also happens due to network congestion, it implements some modifications to the existing protocol. I-Watchdog protocol gives good results in throughput, packet delivery ratio and end-to-end delay as compared to Watchdog protocol. In the below figure shows the algorithm that we will use in the implementation of our I-Watchdog protocol.

ALGORITHM OF I-WATCHDOG PROTOCOL:

1. If (sending time of packet > Packet storing time) else go to step 8
2. Calculate d = sequence no of suspected node – sequence no of current node
3. If d is very large and within the range of suspected node's sequence number then go to step 4 else go to stop 7.
4. Then calculate % packet loss of suspected node to be malicious
5. If (% of packet loss > threshold of % packet loss)
6. Then Mark the suspected node as malicious
7. Else call local repair of link function
8. stop

In the above algorithm, Watchdog observes the next hope node activity. Whenever it detects that node's sending time exceeds the packet storing time then it does not directly mark node as malicious, it further checks for the sequence number. It calculates the difference which is denoted by d in the above algorithm, between the sequence number of suspected nodes and the sequence number of itself. If this difference d is very close to the suspected sequence number and it is very far from the sequence number by itself then it checks for the packet loss of percentage. For example, consider the below 2 cases:

- CASE 1:

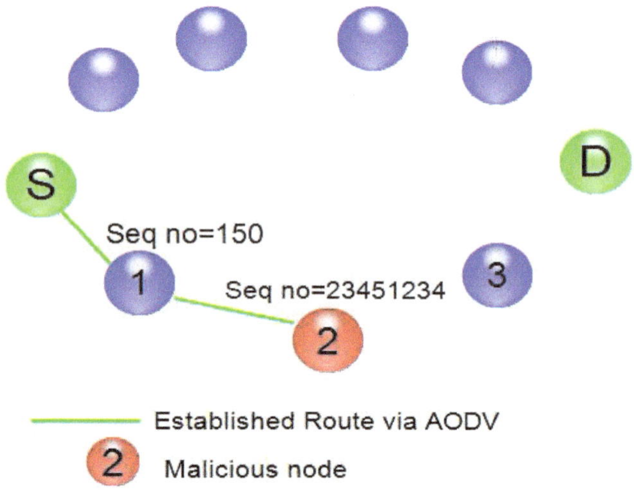

Fig 4. CASE 1

From the above figure 4, suppose node 1 act as a Watchdog and observes the next hop node that is node 2 and node 2 is suspected to be malicious because its sending time is exceeded over packet storing time. Then node 1 calculates difference d. Suppose node 2 is a malicious node so it contain a very large sequence number for example the sequence number of node 2 is 23451234 and sequence number of node 1 is 150 then difference d is evaluated to be d = (23451234-150) =>23451084, which is very close to the sequence number of node2 (suspected malicious nodes) and very far from node 1. After calculating this value of d, to ensure the suspected node is malicious it checks for the percentage of packet loss. If this percentage of packet loss has exceeded the predefined threshold value of percentage of packet loss then I-Watchdog protocol mark this node as malicious and send alert in the network about the malicious node. If suspected malicious node's percentage of packet loss is less than the threshold value than it does local repair link because it indicates that packet loss is due to the network congestion, transmission errors and the suspected malicious node is not malicious.

- CASE 2:

In case 2, when difference d is not close to the suspected malicious node as well as node which act as Watchdog, then there is a confirmation that node is not malicious and sending time exceeds the threshold due to transmission errors and congestion so local repair of link function is called. For example node 1 sequence number is 150 and sequence numbers of suspected malicious nodes is 170 then the differences will be d = (170-150) =>20, which is not close to 170 as well as 150. In this case, local repair of link function can be directly called. This algorithm of the I-Watchdog protocol gives better performance and high throughput.

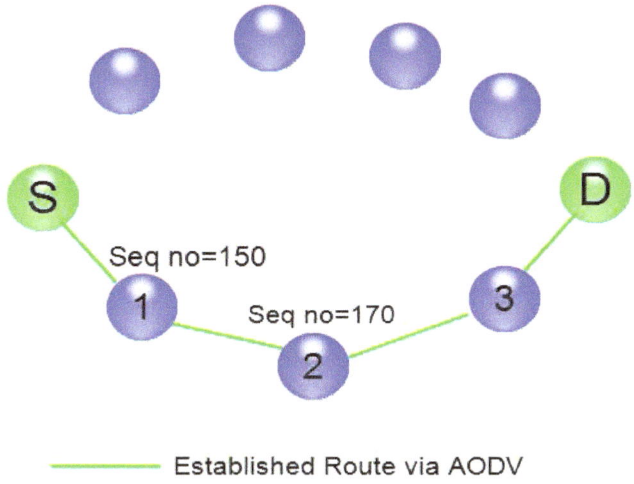

Fig 5. CASE 2

In case 2, when difference d is not close to the suspected malicious node as well as node which act as Watchdog, then there is a confirmation that node is not malicious and sending time exceeds the threshold due to transmission errors and congestion so local repair of link function is called. For example node 1 sequence number is 150 and sequence numbers of suspected malicious nodes is 170 then the differences will be d = (170-150) =>20, which is not close to 170 as well as 150. In this case, local repair of link function can be directly called. This algorithm of the I-Watchdog protocol gives better performance and high throughput.

Now for better understanding of implemented I-Watchdog protocol algorithm following figure shows a flowchart. This flowchart well describes about the mechanism of the improved watchdog protocol. It first calculates the difference d and then check this d is close to the suspected malicious sequence number or not. If it is in the range, then again check for the percentage of packet loss. If it is greater than the predefined threshold then it provides surety that the node is malicious. If d is not within the range of malicious node sequence number then it indicated that sending time exceeds the packet storing time due to the presence of congestion in the network and local repair of link function is directly called.

VII. SIMULATION

I-Watchdog protocol is implemented in network simulator (NS-2) in Ubuntu platform. In this paper, we are comparing the performance of I-Watchdog protocol with existing Watchdog protocol in terms of throughput, packet delivery ratio and end-to-end delay. The simulation parameters are shown in table 1 which we will use in the simulation of Watchdog protocol and I-Watchdog protocol.

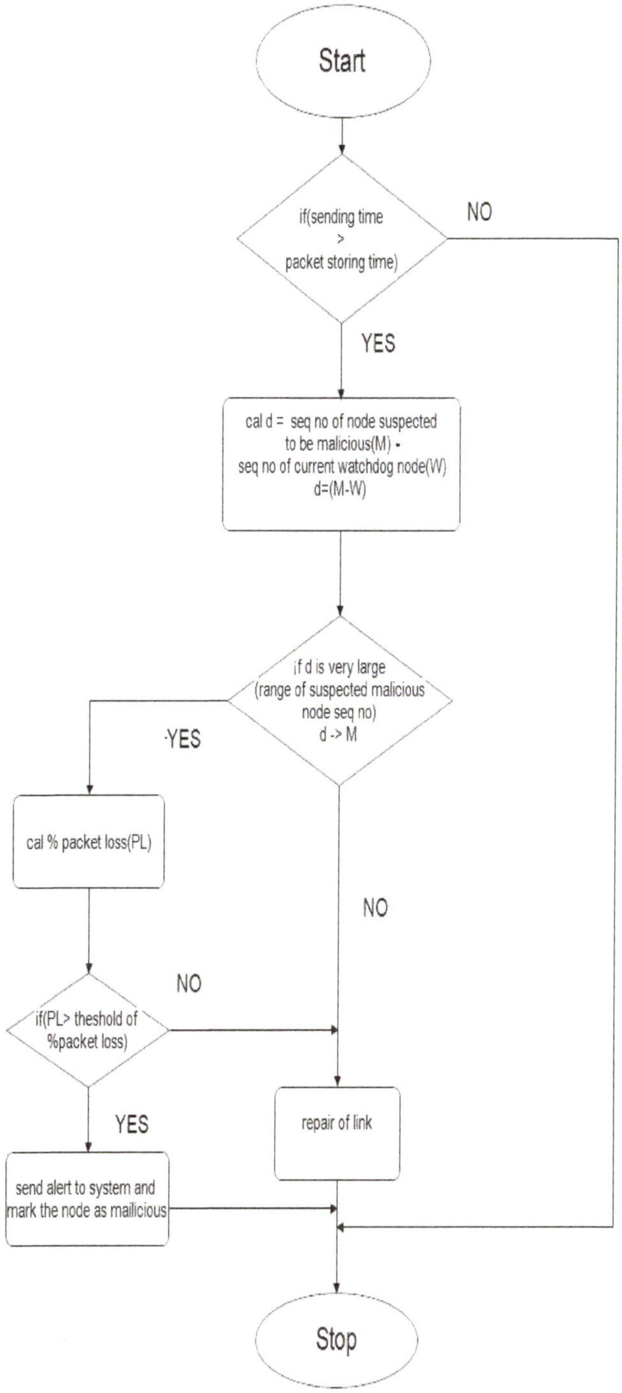

Fig 6. Flowchart of I-Watchdog protocol algorithm.

VIII. RESULTS AND COMPARISION

In this paper, I-Watchdog protocol implemented as an improved Watchdog protocol for mobile ad hoc networks. I-Watchdog protocol gives better results in terms of throughput, packet delivery ratio and end-to-end delay. Further, we will equate the performance of both the protocols by in terms of these attribute by plotting the X-graph in NS-2.

TABLE I
SIMULATION PARAMETERS FOR WATCHDOG

Sr no.	Parameter	Value
1	Simulator	NS-2
2	Channel type	Channel/Wireless channel
3	Radio Propagation Model	Propagation/ Two ray ground wave
4	Network interface type	Phy/WirelessPhy
6	MAC Type	Mac /802.11
7	Interface queue Type	Queue/Drop Tail
8	Routing procedure(protocol)	AODV
9	Antenna	Antenna/Omni Antenna
10	Type of traffic	CBR
11	Area (M*M)	500 * 500
12	Simulation Time	250 sec
13	No of Nodes	50

A. *THROUGHPUT*

The principal performance is measured in the relations of the throughput. Throughput is represented in bits per second (bps) and it is the number of packets which is received in per unit of time. Figure 7 represents the throughput of the Watchdog protocol.

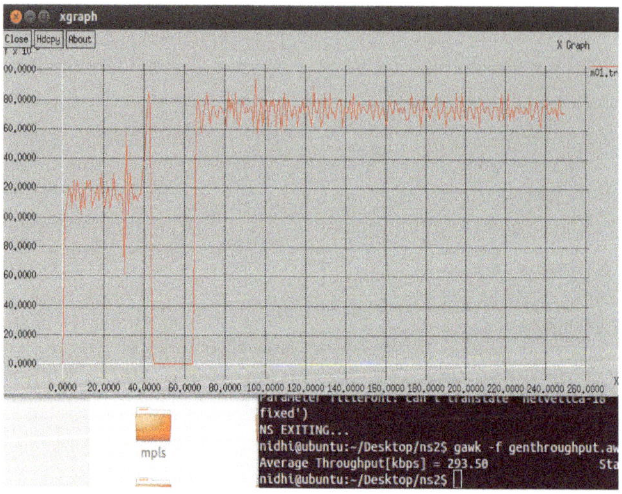

Fig 7. Throughput of the Watchdog protocol

In the above figure, shows graph of the throughput corresponding Watchdog with AODV protocol The figure represents the average throughput of the system using Watchdog protocol is 293.50 Kbps.

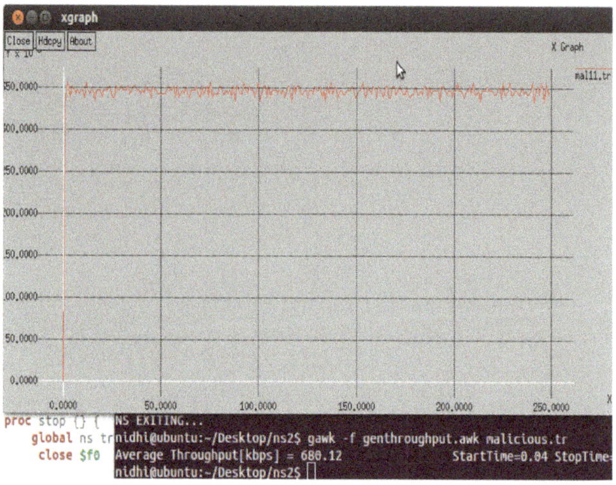

Fig 8. Throughput of I-Watchdog protocol with AODV

In the above figure, the graph for the throughput of I-Watchdog protocol is shown. It is clear from the graph is, the average throughput calculated for me-Watchdog protocol is very high and comes out to be 680.12 kbps, which is very high as compared to the existing Watchdog protocol.

B. *PACKET DELIVERY RATIO/FRACTION*

Packet delivery ratio (PDR) is the portion with reference to the data Packets received by the target node to folks propel by the source node. This evaluates the ability of the protocol performance and its efficiency.

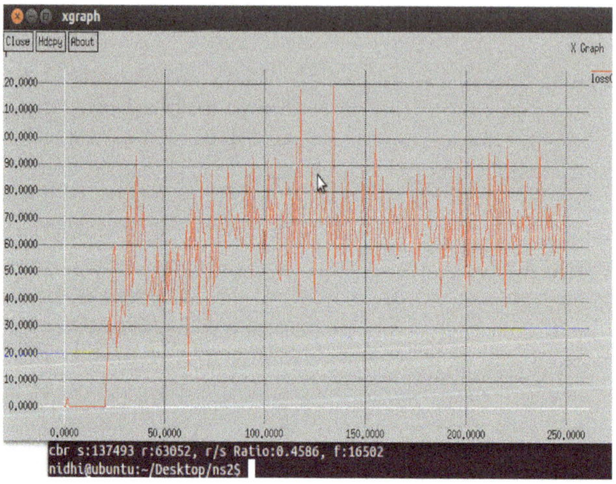

Fig 9. Packet delivery ratio of Watchdog protocol

It is shown from the above figure that the PDR of the system which uses the Watchdog protocol is less. The difference between the sending and receiving packet is 74441 means that 74441 packets are not received by the destination so that we can say the 74441 packets are dropped.

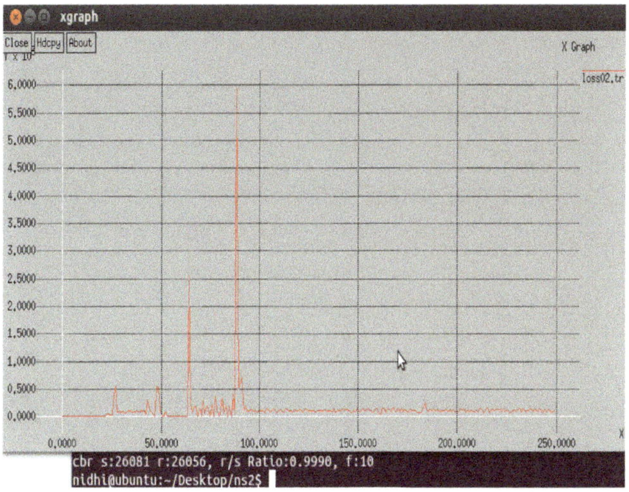

Fig 10. Packet delivery ratio of the system within Watchdog protocol

The figure shows clearly that the numbers of dropped packets are very few. The difference between the number of sending packets and the number of receiving packets is about 25, which is very less as compared to the Watchdog protocol. So that it is clear that the I-Watchdog protocol gives better results and performance than the existing Watchdog protocol.

C: *END-TO-END DELAY*

End to end delay is the quantity of the time which is occupied to sending packets from source to their respective destination to receive those packets. More delay can lead to low performance of the MANET and low delay is the indication of high efficiency and speed of the network.

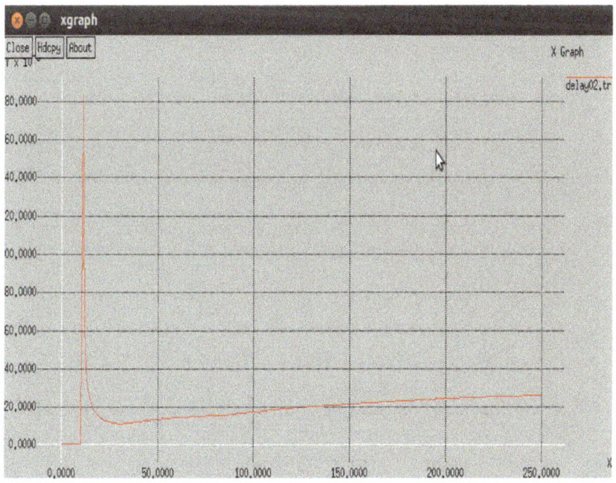

Fig 11. End-to-end delay of the network with Watchdog protocol

The above figure shows end-to-end delay of the network, and it is clear from the graph that it comes out to be about 80 ms which is very high and can highly degrade the system performance. It is fundamentally the total time, which is occupied by the network to send the all packets from source to destination. Here the delay is 80 ms represent the total time is 80ms to send packets from the source node to the destination mobile node in the MANET.

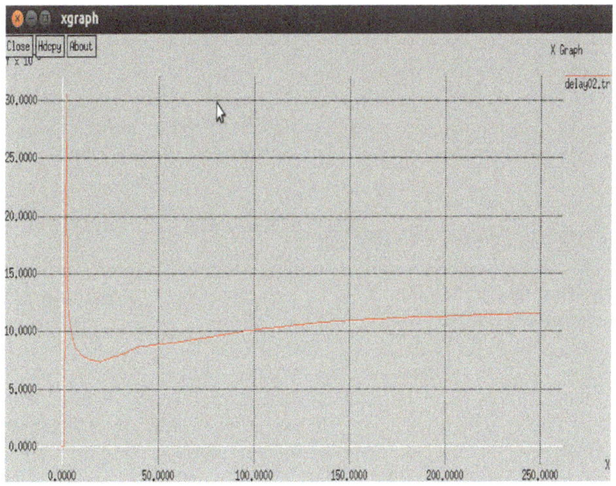

Fig 12. End-to end delay of the system using I-Watchdog protocol

The above figure shows graph of the end–to-end delay of the network which is time to sending the packets from source mobile node to the destination mobile node. The end-to-end delay is about 30ms which is represented by the above X-graph which is very low as compare to the existing Watchdog protocol. So that we can come into the conclusion that proposed I-Watchdog protocol [24-27] requires very less time to send packets from source to destination, it also drops very few packets because of PDR is high for I-Watchdog protocol as compared to existing Watchdog protocol as well as it gives very high throughput which enhances the system performance.

IX. CONCLUSION

MANET is a wireless ad hoc network which is infrastructure less, dynamic and distributed in nature. The attacks are the key sanctuary encounter of MANETs. However I-Watchdog protocol provides a way that can enhance the system performance and detects the congestion, transmission error, link error in the network. It gives high performance and supports a very high number of nodes and provides minimum delay with enhanced throughput. I-Watchdog protocol overcomes the limitation of the previous Watchdog protocol, results less number of dropped packets, high throughput and minimum delay. It can easily detect that the delay and packet drop event is occurring due to transmission error or any attack in the network. If it is due to attack, then an alert is generated by Watchdog node and broadcast information regarding malicious node in the entire network. If it is due to any transmission error then local repair of link function is called. I-Watchdog protocol supports very high degree of the mobility and also supports the dynamic and distributed nature of the MANET. The future work will be on the prevention of the packets from being alternation by the malicious node in the network, in such a way the main focus will be on the integrity and confidentiality of the contents inside the packet. For further enhancing security and efficiency, we provide some authentication techniques and repairing of link methods such that reliable delivery of packets from source to destination will take place.

ACKNOWLEDGMENT

The author would like to be grateful for Mr. Vijay Bhasker Semwal working as a PH.D Scholar in IIIT-Allahabad for the great support and all the contemporaries and fellows, for precious involvement in this paper.

REFERENCES

[1] Mobile Ad-hoc Network (MANET) Properties and Spectrum needs at thenodesG. © Oxford University Press 2007.

[2] Al-Roubaiey, A.; Sheltami, T.; Mahmoud, A.; Shakshuki, E.; Mouftah, H., "AACK: Adaptive Acknowledgment Intrusion Detection for MANET with Node Detection Enhancement," Advanced Information Networking and Applications (AINA), 2010 24th IEEE International Conference on , vol., no., pp.634,640, 20-23 April 2010.

[3] Surana, K. A. "Securing Black Hole Attack in Routing Protocol Aodv in Manet with Watchdog Mechanisms." World Re-search Journal of Computer Architecture, ISSN (2012): 2278-8514.

[4] Aarti, Dr SS. "Tyagi,"Study Of Manet: Characteristics, Challenges, Application And Security Attacks"." International Journal of Advanced Research in Computer Science and Software Engineering 3.5 (2013): 252-257.

[5] Gandhewar, Nisarg, and Rahila Patel. "Performance Evaluation of AODV protocol in MANET using NS2 Simulator." 2nd National Conference on Information and Communication Technology (NCICT), Proceedings published in International Journal of Computer Applications® (IJCA). 2011.

[6] Wage, Pratibha, and Channveer Patil. "INTRUSION-DETECTION SYSTEM FOR MANETS: A SECURE EAACK."

[7] Bhosle, Amol A., Tushar P. Thosar, and Snehal Mehatre. "Black-Hole and Wormhole Attack in Routing Protocol AODV in MANET." International Journal of Computer Science, Engineering and Applications (IJCSEA) Vol 2 (2012).

[8] Ramesh, V., et al. "Performance Comparison and Analysis of DSDV and AODV for MANET." International Journal on Computer Science and Engineering 2.02 (2010): 183-188.

[9] Ochola, Elisha O., Mariki M. Eloff, and John A. van der Poll. "The failure of watchdog schemes in MANET security: a case of an intelligent black-hole." Proceedings of the South African Institute for Computer Scientists and Information Technologists Conference. ACM, 2013.

[10] Hortelano, Jorge, et al. "Watchdog intrusion detection systems: Are they feasible in manets." XXI Jornadas de Paralelismo (CEDI'2010) (2010).

[11] Ismail, Zahian, and Rosilah Hassan. "A performance study of various mobility speed on AODV routing protocol in homogeneous and heterogeneous MANET." Communications (APCC), 2011 17th Asia-Pacific Conference on. IEEE, 2011.

[12] Makwana, Nishant P., Sunil K. Vithalani, and Jayesh D. Dhanesha. "Intrusion Detection-Watchdog: For Secure AODV Routing Protocol in VANET."

[13] Huang, Lei, and Lixiang Liu. "Extended watchdog mechanism for wireless sensor networks." Journal of Information and Computing Science 3.1 (2008): 39-48.

[14] Karygiannis, A., K. Robotis, and E. Antonakakis. "Creating Offline MANET IDS Network Traces." World of Wireless, Mobile and Multimedia Networks, 2007. WoWMoM 2007. IEEE International Symposium on a. IEEE, 2007.

[15] Hernandez-Orallo, Enrique, et al. "Improving selfish node detection in MANETs using a collaborative watchdog." Communications Letters, IEEE 16.5 (2012): 642-645.

[16] Serrat Olmos, Manuel D., et al. "Collaborative Watchdog to Improve the Detection Speed of Black Holes in MANETs." (2012).

[17] Santhosh Krishna, B. V., and A. L. Vallikannu. "Detecting malicious nodes for secure routing in MANETS using reputation based mechanism." Int. J. Sci. Eng. Res 1.3.

[18] Singh, Smita, Shradha Singh, and S. R. Biradar. "Comparison and Study of AOMDV and DSDV Routing Protocols in MANET Using NS-2." International Journal on Computer Science & Engineering 4.3 (2012).

[19] Raja, L., and S. Santhosh Baboo. "Performance Analysis and Simulation of Reactive Routing Protocols (AODV, DSR and TORA) in MANET using NS-2." (2014).

[20] Saeed, Nagham H., Maysam F. Abbod, and Hamed Saffa Al-Raweshidy. "MANET routing protocols taxonomy." Future Communication Networks (ICFCN), 2012 International Conference on. IEEE, 2012.

[21] Li, Yi, and June Wei. "Guidelines on selecting intrusion detection methods in MANET." The 21st annual conference for information systems educators (ISECON), Rhode Island, USA. 2004.

[22] Ul Islam Khan, Burhan, Rashidah Funke Olanrewaju, and Mohamed Hadi Habaebi. "Malicious Behaviour of Node and its Significant Security Techniques in MANET-A Review." Australian Journal of Basic & Applied Sciences 7.12 (2013).

[23] Tamilselvan, Latha, and V. Sankaranarayanan. "Prevention of blackhole attack in MANET." Wireless Broadband and Ultra Wideband Communications, 2007. AusWireless 2007. The 2nd International Conference on. IEEE, 2007.

[24] Gupta, Jay Prakash, et al. "Human activity recognition using gait pattern."International Journal of Computer Vision and Image Processing (IJCVIP) 3.3 (2013): 31-53.

[25] Kumar, K. Susheel, Vijay Bhaskar Semwal, and R. C. Tripathi. "Real time face recognition using adaboost improved fast PCA algorithm." arXiv preprint arXiv:1108.1353 (2011).

[26] Semwal, Vijay Bhaskar, et al. "Accurate location estimation of moving object with energy constraint & adaptive update algorithms to save data." arXiv preprint arXiv:1108.1321 (2011).

[27] Bhaskar-Semwal, V., et al. "Accurate location estimation of moving object In Wireless Sensor network." International Journal of Interactive Multimedia and Artificial Intelligence 1.4 (2011).

[28] Gomes, Maria Cecília, et al. "Accesing wireless sensor networks via dinamically reconfigurally interaction models." IJIMAI 1.7 (2012): 52-61.

[29] García, Óscar, et al. "Wireless Sensor Networks and Real-Time Locating Systems to Fight against Maritime Piracy." International Journal of Interactive Multimedia and Artificial Intelligence 1.5 (2012).

[30] Rios-Aguilar, Sergio. "Intelligent Position Aware Mobile Services for Seamless and Non-Intrusive Clocking-in." International Journal of Interactive Multimedia & Artificial Intelligence 2.5 (2014).

[31] Montañés, José Antonio Puértolas, Adriana Mendoza Rodríguez, and Iván Sanz Prieto. "Smart Indoor Positioning/Location and Navigation: A Lightweight Approach."

[32] Gómez Bey, Bernardo, and Roberto García Sanz. "Fuzzy Logic Indoor Positioning System." EDITOR'S NOTE (2008): 49.

[33] Vijay Bhaskar Semwal, Pavan Chakraborty, G.C. Nandi, Less computationally intensive fuzzy logic (type-1)-based controller for humanoid push recovery, Robotics and Autonomous Systems, Available online 16 September 2014.

[34] Gupta, Jay Prakash, et al. "Human activity recognition using gait pattern."International Journal of Computer Vision and Image Processing (IJCVIP) 3.3 (2013): 31-53.

[35] Semwal, Vijay Bhaskar, Aparajita Bhushan, and G. C. Nandi. "Robotics and AI Lab, IIIT Allahabad, UP-211012 (India)." Control, Automation, Robotics and Embedded Systems (CARE), 2013 International Conference on. IEEE, 2013.

[36] S.Dhanalakshmi, Dr.M.Rajaram, " A Reliable and Secure Framework for Detection and Isolation of Malicious Nodes in MANET", IJCSNS International Journal of Computer Science and Network Security, VOL.8 No.10, October 2008

[37] Gowda, Sumati Ramakrishna, and P. S. Hiremath. "Review of Security Approaches in Routing Protocol in Mobile Adhoc Network." International Journal of Computer Science Issues (IJCSI) 10.1 (2013).

Nidhi Lal, is a M.tech student in the Area of Wireless Communication and Computing at Indian Institute of Information Technology, Allahabad. She received her graduate degree from Galgotias college of Engineering and Technology.

Linked Data Methodologies for Managing Information about Television Content

José Luis Redondo-García[1], Vicente Botón-Fernández[2] and Adolfo Lozano-Tello[2]
[1]*Multimedia Modeling and Interaction Department,* EURECOM, Sophia-Antipolis, France
[2]*Quercus Software Engineering Group,* Universidad de Extremadura. Cáceres, Spain

***Abstract* — OntoTV is a television information management system designed for improving the quality and quantity of the information available in the current television platforms. In order to achieve this objective, OntoTV (1) collects the information offered by the broadcasters, (2) integrates it into a ontology-based data structure, (3) extracts extra data from alternative television sources, and (4) makes possible for the user to perform queries over the stored information.**

This document shows the way Linked Data methodologies have been applied in OntoTV system, and the improvements in the data consumption and publication processes that have been obtained as result. On the one hand, the possibility of accessing to information available in the Web of Data has made possible to offer more complete descriptions about the programs, as well as more detailed guides than those obtained by using classic collection methods. On the other hand, as the information of the television programs and channels is published according to the Linked Data philosophy, it becomes available not only for OntoTV clients, but also for other agents able to access Linked Data resources, who could offer the viewer more fresh and innovative features.

***Keywords* — Linked Data, OntoTV, ontology, consuming, publishing.**

I. Introduction

NOWADAYS, the number of television platforms and channels number is growing significantly, so it is not easy for the viewer to decide what he want to watch in a certain moment of the day. Even when providers offer some descriptions about the programmes they broadcast, this information is not detailed enough and does not permit to perform advanced operations like content recommendations.

The creation of a management information system that gives solution to these problems could be very beneficial for the viewers as well as their television experience. This system aims to become a universal and easy-to-use television solution, able to offer more advanced features than those implemented in classic set-top-boxes. When information is scarce, it should access to external sources in order to complete the missing data, in a transparent and flexible fashion. This way the clients have access to a common television information service, no matter the particular device that is being used: a mobile phone, a decoder, or a personal computer. In previous researches in this same direction [1], OntoTV system was created in order to fulfil these requirements.

OntoTV collects information about television contents from various sources and represents all the data using knowledge engineering and ontologies. However, there are still some problems related to the way OntoTV manages the television information. First, the system uses a kind of software components called "Crawlers", which retrieve information from non-structured sources like HTML Web pages. These components consume many computational resources, and have to be customized to fit the particularities of every of the considered data sources. Secondly, only the clients who are compatible with the OntoTV's specifications can access the information stored in his knowledge base.

In this situation, the Linked Data consuming and publishing methodology [2] is gaining presence and importance in the Web. It consists of a set of principles for structuring and interlinking data that make information more useful and easy to reuse by others. As this methodology is built on the top of widely used standards in the Web, such as URI and HTTP, the information shared in this way becomes accessible to both humans and machines. At the end, this interlinked and easy accessible information obtained from different sources is what we commonly known as "Web of Data".

The main objective in this research is to apply the Linked Data methodology in OntoTV, in order to improve the data collection processes and the viewers' television experience. To to achieve that, some components for consuming television information from Linked Data sources have been designed. More specifically, OntoTV will retrieve extra information about movies, obtaining more complete electronic programming guides than before. Also, the data stored in the knowledge base will be now published according to Linked Data principles, so it will be available in the Web of Data for all these agents who are able to access to it.

II. OntoTV System

The OntoTV system (ONTOlogy-based management system for digital TeleVision) is a television content information management system that allows the viewers to access data about programs that have been or will be broadcasted in the various digital platforms. Due to the fact that this system incorporates appropriate mechanisms for data acquisition, it can provide the user detailed content descriptions and allows him to perform advanced search and recommendation operations. The system OntoTV was previously presented in [4], where the most important features were shown:

- To **integrate** all the possible information about television content by using different collection mechanisms for accessing the different existing sources.
- To **represent** the collected data by using ontologies, making possible to perform complex reasoning processes and inferences that generate new knowledge [5].
- To **execute** operations over the knowledge base that are interesting for the user. For example searches and recommendations, with a high degree of personalization.
- To allow the user to **interact** with the system in an easy and intuitive way. The client device sends requests for the execution of certain operations, receives the results from the server, and displays them to the user. The viewers can access the system, no matter which kind of implementation is running on their devices: MHP, Google TV, Media Centers, etc.

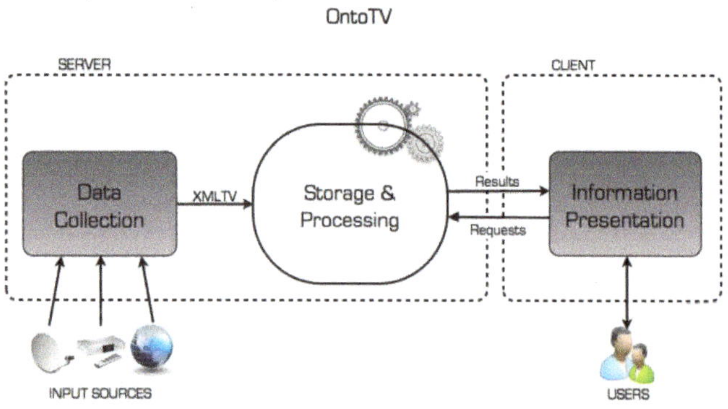

Fig. 1. Main modules inside OntoTV system.

Figure 1 shows a schema of the OntoTV system. The "Storage and Processing" module includes the television content ontology and the different search and recommendation algorithms that are executed over the knowledge base. "Data Collection" and "Information Presentation" modules will be described in more detail below, since they are the ones that will be modified for being compliant to Linked Data principles. This is done to improve the way OntoTV system consumes and publishes the data.

A. Data Collection module

This module directly reads the data from the sources supported by the system. The process consists of being able to interpret the format of a certain input source, and transform the extracted information to the XMLTV format, which is the one used in the system for representing the input files.

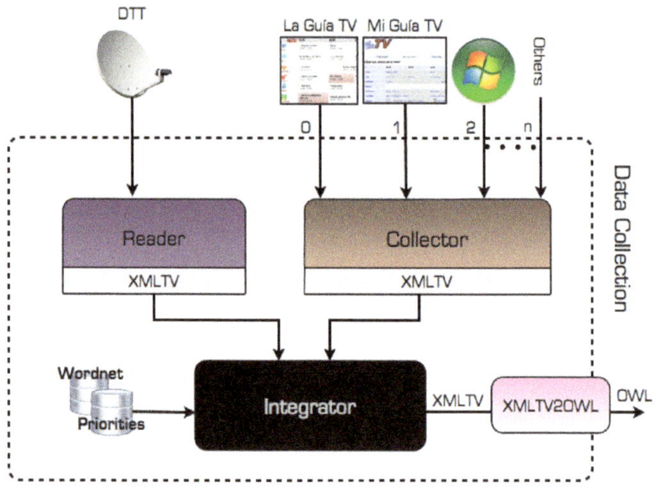

Fig. 2. "Data Collection" module in OntoTV

In Figure 2 the different components inside the "Data Collection" module can be seen. "Reader" components extract the data that television broadcasters offer in their platforms and channels. "Collector" components connect to external server, normally using the TCP/IP protocol, in order to obtain alternative programming guides. Finally, the component called "Integrator" includes in a single file all the information in XMLTV format that has been previously retrieved by the other two components.

Data Sources supported in OntoTV

According to the schema shown in Figure 2, "Reader" and "Collector" components access the following television data sources to feed OntoTV's knowledge base:

- Information included in the DTT data stream, accessed by performing the processes described in [1].
- Accessing to "La Guía TV", a Web page that contains information regarding to television contents broadcasted by the major television channels in Spain. It is necessary to perform translation processes from HTML to XMLTV.
- Accessing to "Mi Guía TV". It is a Web page with similar characteristics than the previous case.
- Accessing to "Windows Media Center" guide. Microsoft offers very complete programming guides for the main television channels in Spain. OntoTV extracts information from these guides and converts them into XMLTV format.

Figure 3 shows three fragments of XMLTV files related to the film "Blade Runner", broadcasted in Spain on the channel "Antena 3" on December 14, 2010. There are some differences in the level of detail provided by each source: for example the fragment corresponding to "Mi GuíaTV" is completely empty, while the "Windows Media Center" one contains precise information about the categories associated with that particular content.

```
<!-- DTT READER -->
<programme channel="15" start="20101214210940" stop="2
    <title>Blade Runner</title>
    <sub-title></sub-title>
    <desc></desc>
</programme>
<!-- LAGUIATV.COM -->
<programme start="20101214220000 +0100" channel="CLa
    <title lang="es">El cine de La 2: Blade Runner</
    <category lang="es">pelicula</category>
</programme>
<!-- MIGUIATV.COM -->
  <!--NO INFORMATION RETRIEVED -->

<!--WINDOWS MEDIA CENTER -->
<programme start="20101214220000 +0100" stop="2010121
    <title lang="es">El cine de La 2</title>
    <desc lang="es">Espacio que incluye la emisión de
    <date>20070427</date>
    <category lang="es">Otro</category>
    <category lang="es">Película</category>
    <length units="minutes">120</length>
</programme>
```

Fig. 3. XMLTV fragments collected from the various considered data sources.

Merging Duplicate Instances of Television Programs. OntoTV is able to detect if descriptions from different sources refer to the same content. Duplicate descriptions about the same program are identified and resolved according to mechanisms described in [4]. Various criteria are taken into account in this process: *spatio-temporal similarity* of content (if two descriptions refer to the same channel, beginning and ending almost at the same time, then it is highly possible that both belong to the same program), *similarity in the titles*, (applying relative comparison string functions as the Levenshtein one [3]), or *global similarity* (given two descriptions, we look for words that appear in both description, regardless of the exact position in the text).

```
<!-- FUSION XMLTV -->
<programme start="20101214220000 +0100" stop="20101214235000 +0
    <title lang="es">El cine de La 2: Blade Runner</title>
    <desc lang="es">Espacio que incluye la emisión de una pelíc
    <date>20070427</date>
    <category lang="es">Otro</category>
    <category lang="es">Película</category>
       <category lang="es">pelicula</category>
    <length units="minutes">120</length>
</programme>
```

Fig. 4. XMLTV description obtained after merging the information from the considered data sources.

Once all the descriptions that belong to the same content have been identificated, it is neccesary to merge them into a single instance, as shown in Figure 4. If a description provides one attribute that is missing in the rest of sources, this field is taken immediately. However, if there is some overlapped parameters in the descriptions, the involved fields are concatenated if possible. If not, those who come from less important sources are discarded. At the end of this step for each content we obtain a unique description that is more complete and detailed than those extracted individually from each source.

Disadvantages of this Approach

As can be seen, all the considered sources provide information about television content. The problem is that the consuming data strategies used in each case are different: the access to DTT is done by interpreting DVB-SI tables, information from Web pages is extracted from certain HTML tags, etc. So each time a new data source needs to be incorporated to the system, is necessary to implement a new access method, as well as integrate it into the global data collection workflow. This process usually requires considerable engineering efforts, which makes more difficult for OntoTV to access new data stores where new television information can be found.

In addition to this lack of uniformity in the collection methods, the processes involved in them are usually very resource intensive because the information is not sufficiently structured.

B. Presentation of the Information

The client-server architecture that has been implemented in OntoTV makes possible that a great variety of television devices can access to the functionalities offered by this system regardless of their particular characteristics. This fact is especially important today, given the different options that are available on the market: MHP set-top-boxes, Google TV televisions, mobile devices with Android operating system, etc. For all these platforms it is possible to develop a client application, called "OntoTV-Client", which performs all the necessary functions to present the television information to the viewer. The premises are to have an Internet connection (for stablishing the client-server communication), as well as being able to use platform-specific libraries for tracking the user's actions, generate graphical interfaces, and interchange messages between client and server.

TABLE I
MESSAGE INTERCHANGING IN ONTOTV'S CLIENTS

Type of Message	Output	Input
Content Management	- Search request, taking into account various criteria.	List of contents that match the selected criteria.
	- Request for a detailed description of a particular content.	Description of a particular content.
	- Ask for a personalized electronic programming guide.	List of contents that match the user preferences.
User Data	- Sending of local events (like button presses, menu navigation, etc.) - Sending of information available on the explicit preferences menu.	User profile that is stored on the server.
Server Connection	- Open connection request.	Confirmation of successful connection.
	- Closing conncetion request.	Confirmation of successful disconnection.

However this information exchange is done by using certain

types of messages and a communication sequence that have been defined beforehand and are exclusive for OntoTV system. Then, for establishing a valid communication with the server, a client must implement this particular set of requests and responses.

Table 1 lists the most important messages the client sends and receives when communicating with OntoTV server. The HTTP protocol and the interchange of XMLTV files over TCP/IP are the basis for implementing those messages.

Disadvantages of this Approach

The problem with this approach is that, despite being independent of the platform used by the consumer, it is always necessary to implement this specific set of messages, even when the agent is not exactly a OntoTV client but another entity that eventually needs television information.

For example, a website that offers the user some miscellaneous information can access OntoTV for retrieving the broadcast times of different television programs, but it needs to incorporate all the communication logic that an OntoTV's client's is supposed to use.

III. Applying Linked Data Methodology

After analizing the way OntoTV operates when providing different features to the viewers, various problems have arised. On the one hand, traditional mechanisms for extracting information from television sources have been probed to be inefficient, due to the heterogeneity in the access methods and the accessing to non-structured information. On the other hand, only clients that are compliant with OntoTV specification can access its television information. This section aims to solve these problems by applying the Linked Data consuming and publishing principles [6], continuing the research line initiated by other television systems that also have used semantic technologies, as Notube [7], [8].

A. Linked Data Consumption

This section shows how to incorporate new Linked Data consumption strategies in the module "Data Collection", in order to increase the amount of television content information available in its knowledge base.

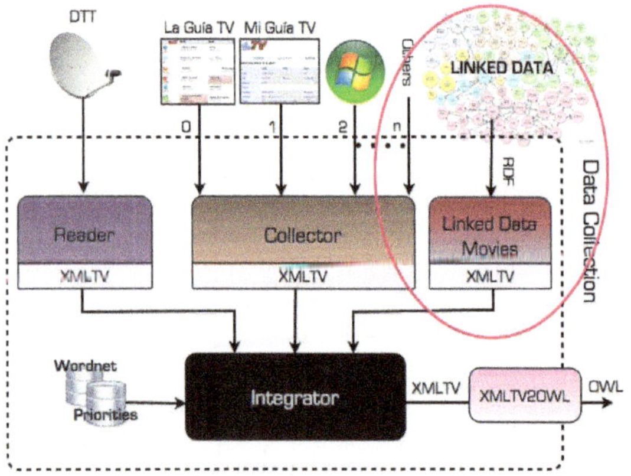

Fig. 5. Extended "Collection Data" module that accesses the Web of Data.

Specifically, the objective is to describe the way the new component called "LinkedData Movies" operates (see Figure 5). As summary, this component accesses Linked Data resources, identifies certain information about movies that is available on the Web of Data, and complete the missing parts of the XMLTV program guides that OntoTV has previously retrieved.

Alternatives for accessing the Web of Data

Several alternatives for accessing information about television content in the Web of Data have been studied. The most significant ones are shown below:

- Accessing to LinkedMDB dataset. It is possible to execute SPARQL queries over the entry point that this dataset provides, in order to obtain information about movies. However, although LinkedMDB is intended to be in the Web of Data the same than IMBD is today in Web of Documents, there are still a lot of films entries missing.
- To implement the method described in [6], which applies the "Crawling" consumption pattern. It consists of using Jena TDB[2] to create a local storage structure where the information collected by the Linked Data crawler DSpider[3] is continuously added. The disadvantage of this approach is that it has a high computational cost. In addition, the collection process is slow and must be repeated periodically to ensure that the information inside the local storage is not out-dated.
- Access the semantic mashup SIG.MA. The advantage of this alternative is that it is possible to access to relevant information from a great variety of semantic sources, without executing very intensive and slow collection processes. In addition, SIG.MA performs frequent updates in their data indexes so the obtained information about movies is updated enough.

The "LinkedData Movies" component

The component "LinkedData Movies" has been coded in Java and performs the following actions in order to extract information about movies from the Web of Data:

a) Getting the movie descriptions in RDF format. The basic mechanism to access Linked Data on the Web is to resolve HTTP URIs for retrieving a certain RDF data fragment. In the case of the SIG.MA mashup, it is necessary to perform an HTTP request to the following URL "http://sig.ma/search?q=moviename", where "moviename" is a string indicating the name of the movie we are looking for. Code 1 shows how to obtain RDF data from SIG.MA using the library "org.apache.commons.httpclient":

Code 1. Requesting RDF Data to the Server

2 http://openjena.org/TDB/
3 http://code.google.com/p/ldspider/

```
import org.apache.commons.httpclient.*;
   //Get Method
   HttpClient client = new HttpClient();
   HttpMethod method = new GetMethod(urlsigma + filmName);
   method.addRequestHeader("Accept",
   "application/rdf+xml");
   int responseCode = client.executeMethod(method);
   //Write RDF to FILE
   InputStream is = method.getResponseBodyAsStream();
   OutputStream os = new FileOutputStream(rdfFile);
   byte[] buffer = new byte[4096];
   for (int n; (n = is.read(buffer)) != -1;)
       os.write(buffer, 0, n);
```

b) Use SPARQL queries to extract the desired information from the previously obtained RDF file. The RDF file, which contains information about a particular film, is already available in the consumer side. So it is possible to extract the desired fragment of information by executing SPARQL queries over it. The "Jena ARQ" library has been used for this purpose, as shown below.

CODE 2. EXECUTING SPARQL QUERIES OVER THE RDF FILE

```
import com.hp.hpl.jena.query.*;
Model m;
m = ModelFactory.createMemModelMaker().createModel("");
model.read(in,null);
//Execute the Query
Query query = QueryFactory.create(stquery);
QueryExecution qe;
qe = QueryExecutionFactory.create(query, m);
ResultSet results = qe.execSelect();
```

Code 2 is able to execute the SPARQL query stored inside the variable "stquery". Figure 6 shows an example that extracts the name of the film's director by accessing the property "director", which is included on the SIG.MA vocabulary (http://sig.ma/property/).

```
PREFIX sigma: http://sig.ma/property/
PREFIX rdfs: http://www.w3.org/2000/01/rdf-schema#
SELECT ?director ?name
WHERE {
    ?film  sigma:director ?director.
    ?director rdfs:label ?name.
    }
-------------------------------------------------------------------
| director                                   | name               |
===================================================================
| <http://dbpedia.org/resource/Ridley_Scott> | "Sir Ridley Scott" |
-------------------------------------------------------------------
```

Fig. 6. SPARQL query for retrieving the name of the film's director.

In a similar way, it is possible to obtain also more film's attributes like the language, the country, the length, and others shown in Table 2. This way descriptions about movies that are available in OntoTV system become more detailed and complete than those obtained before accessing the Web of Data.

TABLE II
OTHER ITEMS ABOUT FILMS THAT CAN BE RETRIEVED FROM SIG.MA

Item	XMLTV Element	SIG.MA Property
Language	tv.programme.language	<sigma:language>
Length	tv.programme.length	<sigma:runtime>
Country	tv.programme.country	<sigma:country
Rating	tv.programme.rating	<sigma:ratings>
Director	tv.programme.credits.director	<sigma:director>
Actor	tv.programme.credits.actor	<sigma:starring>
Writer	tv.programme.credits.writer	<sigma:writer>
Producer	tv.programme.credits.producer	<sigma:producer>
Composer	tv.programme.credits.composer	<sigma:music_composer>
Image	tv.programme.icon	<sigma:picture>

c) Accessing to Other Datasets. The Linked Data philosophy is based on the idea of navigating through the global knowledge. For this reason, if the information that SIG.MA offers is insufficient, it is possible to retrieve alternative data by following the links available in the RDF triples. For example, in Figure 6, the URI for the director Ridley Scott refers to a document in the DBpedia dataset. Additional data can be obtained when URI is resolved with the same process described avobe, as seen in Figure 7:

```
PREFIX dbpedia-owl: <http://dbpedia.org/ontology/>
PREFIX rdfs: <http://www.w3.org/2000/01/rdf-schema#>
SELECT ?date
WHERE {
    ?director  dbpedia-owl:birthDate ?date.
    }
Finding for more information in http://dbpedia.org/resource/Ridley_Scott
-----------------------------------------------------------
| date                                                    |
===========================================================
| "1937-11-30"^^<http://www.w3.org/2001/XMLSchema#date>   |
-----------------------------------------------------------
```

Fig. 7. SPARQL query for retrieving the director's birth date.

In the end, OntoTV stores much more information about the film on which the collection process has been applied. For example Figure 8 shows how the description of "Blade Runner" is much more detailed than before the access to the Web of Data (check again Figure 4 for a better comparison). All this extra information allows the system to offer the viewers more accurate results when executing operations, such as search and recommendations.

```
<!-- FINAL XMLTV -->
<programme start="20101214220000 +0100" stop="20101214235000 +0100"
    <title lang="es">El cine de La 2: Blade Runner</title>
    <desc lang="es">Espacio que incluye la emisión de una película.<
    <category lang="es">Otro</category> <category lang="es">Película
    <category lang="es">pelicula</category>
    <date>20070427</date>
    <language>English</language>
    <country>United States</country>
    <credits>
      <director>Ridley Scott, 1927-11-30, South Shields.</director>
      <actor>Harrison Ford</actor> <actor>Rutger Hauer</actor>
      <actor>Sean Young</actor> <actor>Edward James Olmos</actor>
      <actor>Daryl Hannah</actor> <actor>M. Emmet Walsh</actor>
      <writer>Philip K. Dick</writer> <producer>Michael Deeley</prod
      <composer>Vangelis</composer>
    </credits>
    <icon src="http://getmovielink.com/images/covers/BladeRunner.jpg
    <length units="minutes">120</length>
</programme>
```

Fig. 8. Description available in OntoTV system about the movie "Blade Runner", after accessing information in the Web of Data.

Analyzing the entire collection workflow, it is clear the benefits obtained when consuming information available on the Web of Data over traditional accesses to unstructured data sources. The use of URIs and the HTTP protocol provides a more uniform access to different datasets and makes easier to incorporate new sources in OntoTV system. Likewise, the fact that the data is represented in RDF format and structured according to certain vocabularies (such as SIG.MA), greatly facilitates the way the information is interpreted and processed.

B. Publishing Data according to Linked Data principles

As noted in paragraph 2.b, the only way to access the information stored in OntoTV's knowledge base is to implement a predefined and specific communication logic for the interchange of information between the client and the server. This section explains the changes made in OntoTV in order to publish television content descriptions by following Linked Data principles. This way any agent that is able to access the Web of Data can also take profit of them.

Television Domain Ontology

The first step in order to publish data using Linked Data principles is to choose a valid domain vocabulary that allows representation of television information. In previous works, OntoTV used the ontology proposed in AVATAR [9]. However, for the current research this ontology has been replaced by the one used by the BBC (British Broadcasting Corporation), called BBC Programmes. This organization has created this vocabulary by using its wide experience in the use of semantic technologies. This background knowledge has led to consider this alternative as the most suitable one for representing television programs and channels in a standard way, that is one of the main principles in the Linked Data philosophy.

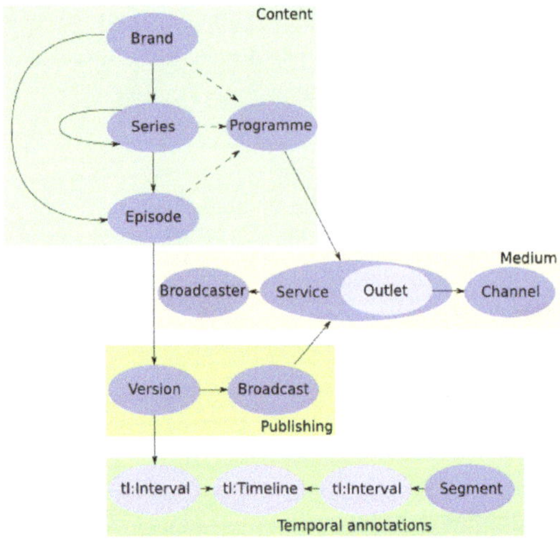

Fig. 9. BBC Programmes.

This ontology provides a simple vocabulary that includes multiple classes related to the television content and broadcasters domain. In the Figure 9, the box "Content" (Programme, Brand, Series, Episode) contains classes for representing different types of television content. Inside the "Medium" box we can find the class "Channel" for representing the different kinds of transmission mediums, as well as the class "Broadcaster " for modelling the television organization. The "Publishing" box includes the class "Version", which is very important for representing the different occurrences of a particular program in a certain channel, date and time. Classes inside the box "Temporal Annotations" have not been considered in this research.

Generating the RDF Data

This section describes the different steps to be performed in order to transform the XMLTV information about television channels and programs (previously extracted by the "Data Collection" module) into instances of the BBC Programes ontology.

The software component that performs this translation process is XMLTV2OWL. As seen in Figure 2, this component is included inside the "Data Collection" module. However, it plays an important role in the process of making this information available in the Web of Data, because it is the one who generates the instances that will conform the RDF code. The stages of this process are described in more detail below:

a) Step 1. Each element of type "<channel>" in the XMLTV file is transformed into an instance of the class "Service" in the BBC ontology. Before including this new individual in the knowledge base, it is necessary to check that there are no collisions with those instances that are already stored in OntoTV (because the channel has been previously inserted in the system). Also, an instance of the class "Broadcast" is created in order to relate the current instance of the program with the particular channel that broadcasts it. Figure 10 shows the RDF example code for the film "Blade Runner" in Turtle notation.

```
<http://data.ontotv.es/service/0a566f0d-27f4-9648-adf5-03a0cabf365a>
      a      ' po:Service
      po:broadcaster "RTVE"
      po:parent_service <http://data.ontotv.es/service/RTVE>
      po:channel <TDT>.

<http://data.ontotv.es/broadcast/f3dafbb0-8407-47c0-8267-8d87381b95ba>
      a        po:Broadcast ;
      po:broadcast_of <http://data.ontotv.es/version/2dfe2b07-1df3-4111-8
      po:broadcast_on <http://data.ontotv.es/service/0a566f0d-27f4-9648-a
```

Fig. 10. Instances of the classes "Service" and "Broadcast" in the BBC Programmes" ontology.

b) *Step 2.* For each "<programme>" XMLTV element:

An instance of the class "Programme" is created in the OntoTV's knowledge database, by transforming certain XMLTV fields into their corresponding properties in the BBC ontology. Before including that instance in the knowledge base, it is necessary to check if this particular program has not been previously added to the system. The Figure 11 shows the RDF example code for the movie "Blade Runner":

```
<http://data.ontotv.es/episode/7ffdb885-fcf4-44cd-80a7-7c137c8d457a>
      a        po:Episode ;
      dc:title "El cine de la 2: Blade Runner"" ;
      po:id   "7ffdb885-fcf4-44cd-80a7-7c137c8d457a" ;
      po:long_synopsis "Espacio que incluye la emision de una pelicula. Pa
      po:masterbrand "La 2" ;
      po:microsites <http://www.rtve.es/alacarta/tve/la2/> ;
      po:subject "Pelicula" , "pelicula" , "Otro" ;
      po:version <http://data.ontotv.es/version/2dfe2b07-1df3-4111-8f2d-70
      po:actor <http://data.ontotv.es/person/Harrison_Ford> <http://data.o
      po:director <http://data.ontotv.es/person/Ridley_Scott> ;
      po:duration "120"^^xsd:int ;
      po:executive_producer <http://data.ontotv.es/person/Michael_Deelay.
```

Fig.11. Instance of the class "Episode" for the movie "Blade Runner".

- An instance of class "Version" in the BBC ontology is created. This instance stores the attributes "start" and "stop" that are present in every "<programme>" XMLTV element. Also, this instance is associated with the one created in the previous step by using the property "po:version" in the class "Program". Before including it in the knowledge base, XMLTV2OWL looks again for possible collisions between individuals. If some duplicates are found, only the most recent instance is maintained. Figure 12 shows the corresponding RDF code for the "Blade Runner" example:

```
<http://data.ontotv.es/version/2dfe2b07-1df3-4111-8f2d-70adde8d2097>
      po:sound_format "urn:ard:tva:metadata:cs:ARDFormatCS:2008:3.2" ;
      po:subtitle_language "Spanish" ;
      po:aspect_ratio "urn:ard:tva:metadata:cs:ARDFormatCS:2008:1.24" ;
      po:time [ a       event:Interval ;
              event:end "2012-07-23T18:48:29.959Z"^^xsd:dateTime ;
              event:start "2011-11-15T20:45:00Z"^^xsd:dateTime
            ] .
```

Fig. 12. Instance of the class "Version" for the movie "Blade Runner".

Interlinking with other Linked Data Datasets

The Linked Data methodology put special emphasis on the need of establishing links between data fragments that are semantically related in some way [10]. This makes possible to browse the entire knowledge, jumping from one concept to another. For this reason, OntoTV executes some special processes that try to match the local instances available in the RDF base with other similar individuals from external datasets. This way it is possible to create links between OntoTV's triples and other resources in the Web of Data:

- Links to the DBpedia dataset: DPpedia is considered to be the core of the Web of Data cloud. It contains information about any domain, so it has become a reference dataset in the Linked Data research field. Here, the instance matching process has been performed by applying simple lexical similarity functions over the textual attributes in the classes "Service" and "Programme" (like for example, "producer "," director "," actor ", etc.) Figure 13 shows examples of such links:

```
<http://data.ontotv.es/person/Harrison_Ford>
      a        foaf:Person ;
       foaf:gender "m" ;
       foaf:nick "Harrison Ford" ;
       owl:sameAs <http://dbpedia.org/resource/Harrison_Ford>

<http://data.ontotv.es/person/Ridley_Scott>
      a        foaf:Person ;
       foaf:gender "m" ;
       foaf:nick "Ridley Scott" ;
       owl:sameAs <http://dbpedia.org/resource/Ridley_Scott> .
```

Fig. 13. Persons and their corresponding links to instances in DBpedia.

- Links to the "Geonames" dataset. Certain individuals in the knowledge base refer to geographical places. In these cases, OntoTV checks whether these instances are geographically equivalent to others in "Geonames" dataset, which contains over eight million names of places that are available for search.
- Links to "LinkedMDB" dataset. Although this dataset still contains only a few records of certain movies, it will become the reference dataset for information about films in the Web of Data. For this reason, OntoTV will try to identify possible alignments between the local instances and the ones stored in this dataset, especially for some attributes like "director", "actor" and "film". Again, string similarity functions on the titles will be applied.

The module "LD Publishing" (see Figure 14) is responsible of accessing to external datasets in order to execute all these the instance matching processes. As shown in Figure 13, the links found with this method are expressed in the form of <owl:sameas> triplets.

Finally, it is neccesary to mention the existence of some data publishing frameworks such as Openlink Virtuoso[4], which stores RDF triples, generates HTML pages containing the data (so they can be browsed online), and creates a SPARQL endpoint where this kind of queries can be executed. However, this possibility has not been addressed in this research.

C. OntoTV after applying Linked Data methodologies

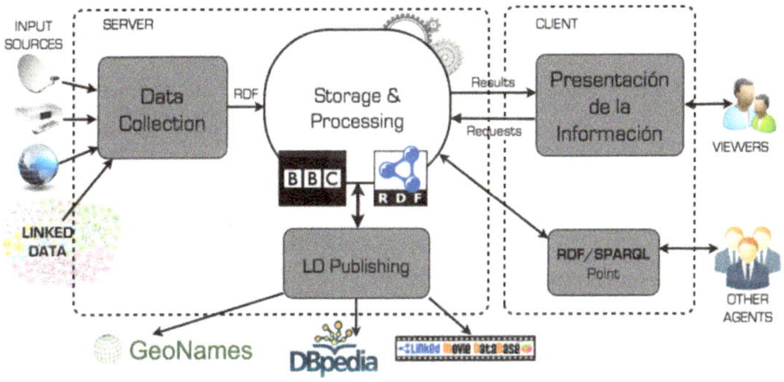

Fig. 14. OntoTV schema after applying Linked Data principles.

Figure 14 shows the changes occurred in OntoTV after the application of Linked Data methodologies. On the one hand, the "Collection Data" module adds a new source: The Web of Data cloud. Furthermore, the "Storage and Processing" module now contains the RDF information represented according to the BCC ontology and conveniently linked to other external dataset. Regarding the way the information is presented to the user, not only the OntoTV's clients have access to the data, but also all the agents who are able to access resources in the Linked Data cloud.

IV. Conclusions

Nowadays viewers have to do a considerable effort every time they want to find, access or compare television programs, due to the great variety of them available in the different platforms. OntoTV system has been designed for giving a solution to this problem. It uses advanced data collection techniques and ontology-based representation methods.

4 http://virtuoso.openlinksw.com/

However, the previous version of OntoTV accessed to non-structured data sources, so the collection mechanisms had to be fully customized for each considered resource. Furthermore, only the clients that were compatible with OntoTV's information interchange protocol could access the data stored in the system.

This paper describes how Linked Data principles have been applied in OntoTV system in order to solve these problems. On the one hand, Linked Data resources have been accessed to complete the information about movies available in the system; on the other hand, a mechanism for publishing information about television content and channels has been designed.

Regarding the *data consumption*, it has been probed that the data collected from Linked Data sources has been useful to enrich the scarce content descriptions originally sent by the providers. As the considered sources are compliant with Linked Data principles, the data extracted from them is well structured and includes semantic links between concepts that are not present in classic HTML links. In this situation it is straightforward to extract the desired information, not only in the case of film description, but also for other types of content. Furthermore, the decision of accessing a resource like SIG.MA, which automatically integrates many others Linked Data sources, has provided advantages over the crawling strategies and the execution of SPARQL queries. As the information comes from various sources, it is possible to find movie descriptions for almost any title.

Regarding the *data publishing*, information can now be accessed not only by OntoTV's clients, but also by any other agent able to consume Linked Data resources. And all of this without having to implement a specific logic for message interchange or interpret particular formats like XMLTV. Also, the decision of using the BBC's ontology, which is widely agreed in the television domain, has been very appropriate because the information collected by OntoTV system becomes available in the Web of Data in a more standard way.

Despite the improvements achieved, it is still necessary to continue enhancing the processes that transform the collected XMLTV data into instances of the BBC ontology. Other future research line is trying to incorporate better mechanisms for finding inconsistences in the data and detecting instance collisions, especially when adding instances of the programs. Finally, the algorithms for aligning information with LinkedMDB and DBpedia datasets can be also improved because until the moment they only use simple lexical comparisons.

In conclusion, the application of Linked Data methodologies has been very beneficial for improving the performance of systems that consume and publish data, like OntoTV does. With these information management strategies applied to the television domain, viewers will have access to a more accurate, complete and useful information.

V. Acknowledgment

This work has been partially funded by the Spanish Ministry of Education and Science and ERDF (the European Regional Development Fund), under contract TIN2011-27340.

References

[1] Redondo-Garcia, J.L, Valiente-Rocha, P., Lozano-Tello, A.: Ontology-based system for content management in Digital Television. CISTI, pp. 277--283, (2010).

[2] Berners-Lee, T.: Linked Data. International Journal on Semantic Web and Information Systems, vol. 4, no. 2, W3C (2006).

[3] Li Yujian and Liu Bo.: A normalized levenshtein distance metric. IEEE Trans. Pattern Analysis and Machine Intelligence, pp. 1091–1095, (2007).

[4] Redondo-Garcia, J.L., Lozano-Tello, A.: Recolección de Datos sobre Contenidos Televisivos en el Sistema OntoTV. CISTI, Accepted in press (2011).

[5] Gomez-Perez, A., Corcho, O., and Fernandez-Lopez, M.: Ontological Engineering: With Examples from the Areas of Knowledge Management, e-Commerce and the Semantic Web. Springer-Verlag, New York (2004).

[6] Heath, T., Bizer, C.: Linked Data: Evolving the Web into a Global Data Space. Synthesis Lectures on the Semantic Web: Theory and Technology, Morgan & Claypool (2011).

[7] Schopman, B., Brickley, D., Aroyo, L., van Aart, C., Buser, V., Siebes, R.: NoTube: making the Web part of personalised TV. Proceedings of the WebSci10 (2010).

[8] Buser, V.: NoTube: experimenting with Linked Data to improve user experience, Summer School on Multimedia Semantics, Amsterdam, 3 September 2010

[9] Y. Blanco-Fernández, et al, "Exploiting synergies between semantic reasoning and personalization strategies in intelligent recommender systems" Journal of Systems and Software, vol.81, pp. 2371-2385, 2008.

[10] Volz, J., Bizer, T., Gaedke, M., Kobilarov, G.: Discovering and Maintaining Links on the Web of Data. International Semantic Conference (ISWC2009), Westfields, USA, pp. 650—665, (2009).

Permissions

All chapters in this book were first published in IJIMAI, by Imai-Software Research Group; hereby published with permission under the Creative Commons Attribution License or equivalent. Every chapter published in this book has been scrutinized by our experts. Their significance has been extensively debated. The topics covered herein carry significant findings which will fuel the growth of the discipline. They may even be implemented as practical applications or may be referred to as a beginning point for another development.

The contributors of this book come from diverse backgrounds, making this book a truly international effort. This book will bring forth new frontiers with its revolutionizing research information and detailed analysis of the nascent developments around the world.

We would like to thank all the contributing authors for lending their expertise to make the book truly unique. They have played a crucial role in the development of this book. Without their invaluable contributions this book wouldn't have been possible. They have made vital efforts to compile up to date information on the varied aspects of this subject to make this book a valuable addition to the collection of many professionals and students.

This book was conceptualized with the vision of imparting up-to-date information and advanced data in this field. To ensure the same, a matchless editorial board was set up. Every individual on the board went through rigorous rounds of assessment to prove their worth. After which they invested a large part of their time researching and compiling the most relevant data for our readers.

The editorial board has been involved in producing this book since its inception. They have spent rigorous hours researching and exploring the diverse topics which have resulted in the successful publishing of this book. They have passed on their knowledge of decades through this book. To expedite this challenging task, the publisher supported the team at every step. A small team of assistant editors was also appointed to further simplify the editing procedure and attain best results for the readers.

Apart from the editorial board, the designing team has also invested a significant amount of their time in understanding the subject and creating the most relevant covers. They scrutinized every image to scout for the most suitable representation of the subject and create an appropriate cover for the book.

The publishing team has been an ardent support to the editorial, designing and production team. Their endless efforts to recruit the best for this project, has resulted in the accomplishment of this book. They are a veteran in the field of academics and their pool of knowledge is as vast as their experience in printing. Their expertise and guidance has proved useful at every step. Their uncompromising quality standards have made this book an exceptional effort. Their encouragement from time to time has been an inspiration for everyone.

The publisher and the editorial board hope that this book will prove to be a valuable piece of knowledge for researchers, students, practitioners and scholars across the globe.

List of Contributors

Renato Barahona Neri
Universidad Pontificia de Salamanca, Madrid, Spain

Gustavo Millán García
Universidad Pontificia de Salamanca, Madrid, Spain

Holman D. Bolibar Barón
Universidad Católica de Colombia, Bogotá, Colombia

Rubén González Crespo
Universidad Pontificia de Salamanca, Madrid, Spain

Alexander Pacheco
Universidad Católica de Colombia, Bogotá, Colombia

Holman Bolívar
Universidad Católica de Colombia, Bogotá, Colombia

Jordán Pascual Espada
Universidad de Oviedo, Asturias, Spain

Rubén González Crespo
Universidad Internacional de La Rioja - UNIR, Madrid, Spain

Jesús Iglesias Feijoo
Zed World Wide

Guillermo Amat Gomariz
Zed World Wide

Rharon M. Guedes
Centro de Informática, Universidade Federal da Paraíba (UFPB), João Pessoa, Brazil

Marcus R. Laurentino
Centro de Informática, Universidade Federal da Paraíba (UFPB), João Pessoa, Brazil

Carlos S. Dias
Centro de Informática, Universidade Federal da Paraíba (UFPB), João Pessoa, Brazil

Alisson V. Brito
Centro de Informática, Universidade Federal da Paraíba (UFPB), João Pessoa, Brazil

Luis de-la-Fuente-Valentín
UNIR Research, International University of La Rioja, Paseo de la Castellana 163, Madrid, Spain

Aurora Carrasco
UNIR Research, International University of La Rioja, Paseo de la Castellana 163, Madrid, Spain

Kinga Konya
UNIR Research, International University of La Rioja, Paseo de la Castellana 163, Madrid, Spain

Daniel Burgos
UNIR Research, International University of La Rioja, Paseo de la Castellana 163, Madrid, Spain

Edward Rolando Nuñez-Valdez
University of Oviedo, Computer Science Department, Sciences Building, Oviedo, Spain

Oscar Sanjuan
University Carlos III of Madrid, Computer Science and Engineering Department, Madrid, Spain

Begoña Cristina Pelayo Garcia-Bustelo
University of Oviedo, Computer Science Department, Sciences Building, Oviedo, Spain

Juan Manuel Cueva-Lovelle
University of Oviedo, Computer Science Department, Sciences Building, Oviedo, Spain

Guillermo Infante Hernandez
University of Oviedo, Computer Science Department, Sciences Building, Oviedo, Spain

Jaime Solís-Martínez
Department of Computer Science, University of Oviedo, Spain

Natalia García-Menéndez
Department of Computer Science, University of Oviedo, Spain

B. Cristina Pelayo G-Bustelo
Department of Computer Science, University of Oviedo, Spain

Juan Manuel Cueva Lovelle
Department of Computer Science, University of Oviedo, Spain

Luis F. Chiroque
IMDEA Networks Institute, Madrid, Spain

Héctor Cordobés
IMDEA Networks Institute, Madrid, Spain

Antonio Fernández Anta
IMDEA Networks Institute, Madrid, Spain

Nidhi Lal
Indian Institute of Information Technology

José Luis Redondo-García
Multimedia Modeling and Interaction Department, EURECOM, Sophia-Antipolis, France

Vicente Botón-Fernández
Quercus Software Engineering Group, Universidad de Extremadura. Cáceres, Spain

Adolfo Lozano-Tello
Quercus Software Engineering Group, Universidad de Extremadura. Cáceres, Spain